Brassinosteroids: A Class of Plant Hormone

Shamsul Hayat • Aqil Ahmad
Editors

Brassinosteroids: A Class of Plant Hormone

Springer

Editors
Shamsul Hayat
Associate Professor
Department of Botany
Aligarh Muslim University
ALIGARH 202 002
India
hayat_68@yahoo.co.in

Aqil Ahmad
Professor
Department of Botany
Aligarh Muslim University
ALIGARH 202 002
India
aqil_ahmad@lycos.com

ISBN 978-94-007-0188-5 e-ISBN 978-94-007-0189-2
DOI 10.1007/978-94-007-0189-2
Springer Dordrecht Heidelberg London New York

Cover design: deblik

Printed on acid-free paper

Springer is part of Springer Science+Business Media (www.springer.com)

Dedicated To The
Aligarh Muslim University
Aligarh

Contents

Foreword

Brassinosteroids (BRs) are endogenous plant growth-promoting hormones found throughout the plant kingdom that influence cellular expansion and proliferation and the phenotype of mutants affected in BR biosynthesis or signaling clearly show that these plant steroids are essential regulators of a variety of physiological processes including organ elongation, vascular differentiation, male fertility, timing of senescence, and leaf development. BRs are structurally similar to animal steroid hormones, and like their animal counterparts, BRs regulate the expression of hundreds of genes, impact the activity of numerous metabolic pathways, and help control overall developmental programs leading to morphogenesis. Several books covering various aspects of BR biology and chemistry appeared in 1991, 1999 and 2003. However, in the past seven years a great deal of progress has been made in understanding specific components of BR signal transduction and in clarifying mechanisms by which BR perception ultimately results in changes in the expression of specific genes associated with different developmental programs. The number of physiological processes known to involve BR action has also expanded and significant experiments quantifying the utility of BR application in practical agricultural have been documented. Therefore, it is very timely that the editors of the current volume have collected and integrated a series of informative chapters on BR molecular biology, physiology, metabolism and practical applications.

BR signaling has been intensively studied over the past ten years, becoming one of the best-characterized plant hormone pathways. The role of membrane-bound receptor kinases and their cytoplasmic targets in perceiving and propagating the BR signal have been elucidated and novel BR-dependent transcription factors that regulate hundreds of different genes have been characterized. Two chapters in the current volume expound on these advances in BR signaling and a third chapter is devoted to the use of global proteomic and genomic technologies that have further advanced our understanding of BR mechanisms. Besides signal transduction, regulation of endogenous BR levels in the plant is critical to BR function and two chapters in the current edition thoroughly address these issues. The known and proposed physiological processes involving BR action are then documented in detail in a total of six chapters, covering general physiology, interactions with light and the role of BRs in abiotic and biotic stress responses. In view of the multitude of demonstrated effects BR application has on crop plants, it is highly appropriate that the remaining six chapters of this book are devoted to a variety of topics involving practical BR applications including horticulture, phytoremediation, herbicide protection and medicine.

The extensive coverage of a range of BR topics presented in this volume, written by active researchers in the field, will provide a useful resource for plant hormone biologists and graduate students in plant physiology, biochemistry, horticulture and agronomy. The editors should be commended for adding another useful edition to the ever-growing body of BR literature.

Steven D. Clouse
Professor,
Department of Horticultural Science,
North Carolina State University,
Raleigh, NC 27695-7609
USA

Preface

The entire range of the developmental processes in plants is regulated by a shift in the hormonal concentration, tissue sensitivity and their interaction with the factors operating around the plants. Out of the recognized hormones, attention has largely been focused on five (Auxins, Gibberellins, Cytokinin, Abscisic acid and Ethylene). However, in this book, the information about the most recent group of phytohormone (Brassinosteroids) has been incorporated by us. It is a class of over 40 polyhydroxylated sterol derivatives, ubiquitously distributed throughout the plant kingdom. A large portion of these steroids is restricted to the reproductive organs (pollens and immature seeds). Moreover, their strong growth-inducing capacity, recognized as early as prior to their identification in 1979, tempted the scientists to visualize the practical importance of this group of phytohormones.

Chapter 1 of this book (which embodies a total of 17 chapters), gives a comprehensive survey of the occurrence and chemical structure of brassinosteroids. Chapter 2 deals with currently available data on brassinosteroid signaling pathway and the latest findings on brassinosteroid signaling regulatory mechanisms in plants. The recent progress in brassinosteroid research in relation to the regulation of brassinosteroid metabolism is discussed in Chapter 3. Chapter 4 summarized the recent advances in brassinosteroids signaling research in rice. The Chapter 5 views the physiological action of brassinosteroids which depends on their concentration in a plant, the medium, the method of exogenous treatment, light of different spectral composition and the mechanism of regulation of the photomorphogenesis. Chapter 6 summarizes the current knowledge of the impact of brassinosteroids on the chloroplasts, and on various components of photosynthetic apparatus. Chapter 7 discusses the physiological modifications that occur in cells, tissue or whole plants when exposed to brassinosteroids and their practical use in agriculture, describing the analogues and the dosages used in field and laboratory experiments. An insight into the genomic and non-genomic events involved in the brassinosteroid promoted plant cell growth is covered in chapter 8. Practical application of BRs to horticultural crops for enhancing crop production and their protection is discussed in Chapter 9. Chapter 10 specifically discusses potential roles for heat shock proteins, antioxidant metabolites and enzymes in brassinosteroid-induced thermal tolerance. In addition, as stress mechanisms are not exclusive of plants, therefore this chapter also discusses the possible way of involving BRs in protection during normal growth stimulation. The Chapter 11 reports data on the protective effect of brassinosteroids on plants, treated with herbicides that inhibit photosynthetic electron transport at the PSII level. It also discusses the biochemical and physiological causes of their protective effects. Role of brassinosteroids under biotic stress is discussed in Chapter 12. Chapter 13 describes how to use a variation in the controls in solving the problems of the interpretation, where it may be another factor contributing to the non reproducibility and/or ambiguity in the results, obtained in relation to brassinosteroids. The Chapter 14 summarizes the latest developments in the field of immunoassay of brassinosteroids. The use of transcriptomics and proteomics techniques to study the regulation of brassinosteroids in plants is introduced in Chapter 15. Role of brassinosteroids for phytoremediation application is discussed in Chapter 16. Finally in Chapter 17 prospects of brassinosteroids in clinical application has been described.

This book is not an encyclopedia of reviews but includes a selected collection of newly written, integrated, illustrated reviews describing our knowledge of brassinosteroids. The aim of this book is to tell all about brassinosteroids, by the present time. The various chapters incorporate both theoretical and practical aspects and may serve as baseline information for future researches through which significant developments are possible. It is intended that this book will be useful to the students, teachers and researchers, both in universities and research institutes, especially in relation to biological and agricultural sciences.

With great pleasure, we extend our sincere thanks to all the contributors for their timely response, their excellent and up-to-date contributions and consistent support and cooperation. The authors are thankful to Aligarh Muslim University, Aligarh (U.P.) India that gave us the employment and the seat to work. We are also thankful to Dr. Zaki A. Siddiqui and Dr. Qazi Fariduddin, Department of Botany, Aligarh Muslim University, Aligarh for their encouragement. Thanks are also due to Mr. Mohd Irfan, Research student in the Department of Botany for typesetting and proof reading of the manuscript. We are extremely thankful to Springer, The Netherlands for expeditious acceptance of our proposal and completion of the review process. Subsequent cooperation and understanding of their staff is also gratefully acknowledged. We express our sincere thanks to the members of our family for all the support they provided and the neglect and loss they suffered during the preparation of this book.

Finally, we are thankful to the Almighty God who provided and guided all the channels to work in cohesion and coordination right from the conception of the idea to the development of the final version of this treatise 'Brassinosteroids: A Class of Plant Hormone' until the successful completion of the job.

Shamsul Hayat
Aqil Ahmad

Contributors

Andrzej Bajguz
University of Bialystok
Institute of Biology
Swierkowa 20 B, 15-950 Bialystok
POLAND

M.-C. Codreanu
Department of Plant Systems Biology
VIB, 9052 Ghent
BELGIUM
And
Department of Plant Biotechnology and Genetics
Ghent University
Technologiepark 927, 9052 Ghent
BELGIUM

E. Russinova
Department of Plant Systems Biology
VIB, 9052 Ghent
BELGIUM
And
Department of Plant Biotechnology and Genetics
Ghent University
Technologiepark 927, 9052 Ghent
BELGIUM

L. Hategan
Institute of Plant Biology
Biological Research Center of the Hungarian Academy of Sciences
P.O.B.521. H-6701, Szeged
HUNGARY

B. Godza
Institute of Plant Biology
Biological Research Center of the Hungarian Academy of Sciences
P.O.B.521. H-6701, Szeged
HUNGARY

M. Szekeres
Institute of Plant Biology
Biological Research Center of the Hungarian Academy of Sciences
P.O.B.521. H-6701, Szeged
HUNGARY

H. Nakagawa
Division of Plant Sciences
National Institute of Agrobiological Sciences
2-1-2 Kannondai, Tsukuba, Ibaraki 305-8602
JAPAN

A. Tanaka
Division of Plant Sciences
National Institute of Agrobiological Sciences
2-1-2 Kannondai, Tsukuba, Ibaraki 305-8602
JAPAN

M. Mori
Division of Plant Sciences
National Institute of Agrobiological Sciences
2-1-2 Kannondai, Tsukuba, Ibaraki 305-8602
JAPAN

I.F. Golovatskaya
Plant Physiology and Biotechnology Department
Tomsk State University
Lenin Avenue, 36, 634050, Tomsk
RUSSIA

D. Holá
Charles University in Prague
Faculty of Science
Department of Genetics and Microbiology
Viničná 5, Prague 2 – Nové Město, CZ 128 43
CZECH REPUBLIC

M.M.A. Gomes
Instituto Superior de Tecnologia em
Ciências Agrárias/Fundação de Apoio à Escola
Técnica do Estado do Rio de Janeiro (FAETEC)
Av. Wilson Batista, s/n, Parque Aldeia
Campos dos Goytacazes-RJ
CEP28070-620
BRAZIL

A.B. Pereira-Netto
Department of Botany-SCB
Centro Politecnico-UFPR
C.P. 19031 Curitiba, PR, 81531-970
BRAZIL

Y.Y. Kang
College of Horticulture
South China Agricultural University
Guangzhou, 510642
CHINA

S.R. Guo
College of Horticulture
Nanjing Agricultural University
Nanjing, 210095
CHINA

L.M. Mazorra
Department of Plant Physiology and Biochemistry
National Institute of Agricultural Sciences
Carretera a Tapaste Km 3½
San José de las Lajas, Havana
CUBA

R. Piñol
Cátedra de Fisiología Vegetal
Facultad de Biología
Universidad de Barcelona
Diagonal 645, E-08028 Barcelona
SPAIN

E. Simón
Cátedra de Fisiología Vegetal
Facultad de Biología
Universidad de Barcelona
Diagonal 645, E-08028 Barcelona
SPAIN

Shamsul Hayat
Department of Botany
Aligarh Muslim University
Aligarh 202002
INDIA

Mohammad Irfan
Department of Botany
Aligarh Muslim University
Aligarh 202002
INDIA

Aqil Ahmad
Department of Botany
Aligarh Muslim University
Aligarh 202002
INDIA

Anna Janeczko
The Franciszek Górski Institute of Plant Physiology
Polish Academy of Sciences
Niezapominajek 21, 30-239 Krakow
POLAND

V.A. Khripach
Institute of Bioorganic Chemistry
Academy of Sciences of Belarus
Kuprevich str., 5/2, 220141 Minsk
BELARUS

V.N. Zhabinskii
Institute of Bioorganic Chemistry
Academy of Sciences of Belarus
Kuprevich str., 5/2, 220141 Minsk
BELARUS

R.P. Litvinovskaya
Institute of Bioorganic Chemistry
Academy of Sciences of Belarus
Kuprevich str., 5/2, 220141 Minsk
BELARUS

H.F. Juan
Department of Life Science
Institute of Molecular and Cellular Biology
and Graduate Institute of Biomedical Electronics and Bioinformatics
National Taiwan University
No. 1, Sec. 4, Roosevelt Rd., Taipei 106
TAIWAN

M. Barbafieri
National Research Council
Institute of Ecosystem Study
Pisa Section
ITALY

E. Tassi
National Research Council
Institute of Ecosystem Study
Pisa Section
ITALY

R. Bhardwaj
Department of Botanical & Environmental Sciences
Guru Nanak Dev University
Amritsar 143005
INDIA

N. Arora
Department of Botanical & Environmental Sciences
Guru Nanak Dev University
Amritsar 143005
INDIA

P. Uppal
Department of Botanical & Environmental Sciences
Guru Nanak Dev University
Amritsar 143005
INDIA

I. Sharma
Department of Botanical & Environmental Sciences
Guru Nanak Dev University
Amritsar 143005
INDIA

M.K. Kanwar
Department of Botanical & Environmental Sciences
Guru Nanak Dev University
Amritsar 143005
INDIA

About the Editors

About Shamsul Hayat

Dr. Shamsul Hayat is an Associate Professor, in the Department of Botany, Aligarh Muslim University, Aligarh, India. He received his Ph.D. degree in Botany from Aligarh Muslim University, Aligarh, India. Before joining the Department as faculty, he has worked as Research Associate and Young Scientist in the same Department. His major areas of research are phytohormones and heavy metal stress in plants Dr. Hayat has been awarded Prof. Hira Lal Chakravorty Award by Indian Science Congress Association, Kolkata, India, Associate of National Academy of Agricultural Sciences, New Delhi, India, BOYSCAST fellow by Department of Science & Technology, Government of India, New Delhi and young scientist by Association of the Advancement of Science, Aligarh, India. He has been the Principal Investigator of the various projects sanctioned by different agencies and guided three students for the award of Ph.D. degree and two students for the award of M.Phil degree. Dr. Hayat has published more than eighty research papers in leading journal of the world with high impact factor and also edited six books published by Kluwer Academic, Springer, Wiley-VCH, Science Publisher and Narosa Publishing House.

About Aqil Ahmad

Dr. Aqil Ahmad is a Professor, in the Department of Botany, Aligarh Muslim University, Aligarh, India. He gained his Ph.D. from Aligarh Muslim University, Aligarh in 1975. Dr. Ahmad worked as Post-Doctorate Research Fellow of Danish International Development Agency (DANIDA) in Plant Physiology at Royal Vet. & Agric. University, Denmark. He has worked at Department of Applied Sciences, Higher College of Technology, Al-Khuwair, Sultanate of Oman. He has been the Principal Investigator of the various projects sanctioned by different agencies and guided six students for the award of Ph.D. degree. His current research area include, phytohormones and heavy metal stress. Dr. Ahmad has published about 90 research papers and edited four books.

Chapter 1

BRASSINOSTEROIDS – OCCURENCE AND CHEMICAL STRUCTURES IN PLANTS

ANDRZEJ BAJGUZ
University of Bialystok, Institute of Biology, Swierkowa 20 B, 15-950 Bialystok, Poland

Abstract: Brassinosteroids (BRs) are a class of plant polyhydroxysteroids that have been recognized as a new kind of phytohormones that play an essential role in plant development. BRs occur at low concentrations throughout the plant kingdom. They have been detected in all plant organs (pollen, anthers, seeds, leaves, stems, roots, flowers, and grains) and also in the insect and crown galls. BRs are structurally related to animal and insect steroid hormones. Natural 69 BRs identified so far, have a common 5α-cholestan skeleton, and their structural variations come from the kind and orientation of oxygenated functions in rings A and B. As regards the B-ring oxidation, BRs are divided into 7-oxalactone, 6-ketone (6-oxo) and 6-deoxo (non-oxidized). These steroids can be classified as C_{27}, C_{28} or C_{29} BRs depending on the alkyl-substitution on the C-24 in the side chain. In addition to free BRs, sugar and fatty acid conjugates have been also identified in plants.

Key words: chemical structures, conjugates, occurrence

1. INTRODUCTION

Brassinosteroids (BRs) are a class of plant polyhydroxysteroids that have been recognized as a new kind of phytohormones that play an essential role in plant development. Brassinosteroids are structurally related to animal and insect steroid hormones. Intensive research conducted on BRs reveals that they elicit a broad spectrum of physiological and morphological responses in plants, including stem elongation, leaf bending and epinasty, induction of ethylene biosynthesis and proton pump activation, synthesis of nucleic acid

S. Hayat and A. Ahmad (eds.), *Brassinosteroids: A Class of Plant Hormone*,
DOI 10.1007/978-94-007-0189-2_1, © Springer Science+Business Media B.V. 2011

and proteins, regulation of carbohydrate assimilation and allocation, and activation of photosynthesis. Furthermore, BRs can protect plants from various biotic and abiotic stresses, such as those caused by salt, high temperatures, and heavy metals (Sasse, 2003; Bajguz and Hayat, 2009).

2. OCCURENCE OF BRASSINOSTEROIDS

Brassinosteroids are plant growth promoting molecules found at low concentrations throughout the plant kingdom and are widely distributed in lower and higher plants (Tables 1–6). They have been detected in all plant organs such as pollens, anthers, seeds, leaves, stems, roots, flowers, and grains. BRs are also present in the insect and crown galls, for example the galls of *Castanea crenata*, *Distylium racemosum* or *Catharanthus roseus*. These plants have higher levels of BRs than the normal tissues. Also, young growing tissues contain higher levels of BRs than mature tissues. Pollen and immature seeds are the richest sources of BRs with a range of 1–100 μg kg^{-1} fresh weight, while shoots and leaves usually have a lower amounts of 0.01–0.1 μg kg^{-1} fresh weight. In the pollen of *Cupressus arizonica* the concentration of 6-deoxotyphasterol (6-deoxoTY) can be about 6400-fold greater than brassinolide (BL). Furthermore, the highest concentration of BR, 6.4 mg 6-deoxoTY per kilogram pollen, has been detected in *Cupressus arizonica*. BRs occur endogenously at quite low levels. Compared to the pollen and immature seeds, the other plant parts contain BRs in the microgram or nanogram levels of BRs per kilogram fresh weight. Since the discovery of BL in 1979, 69 BRs have been isolated from 64 plant species including 53 angiosperms (12 monocotyledons and 41 dicotyledons) (Tables 1–4), 6 gymnosperms (Table 5), 1 pteridophyte (*Equisetum arvense*), 1 bryophyte (*Marchantia polymorpha*) and 3 algae (*Chlorella vulgaris*, *Cystoseira myrica* and *Hydrodictyon reticulatum*) (Table 6) (Fujioka, 1999; Bajguz and Tretyn, 2003).

Table 1. The occurrence of brassinosteroids in the monocotyledons

Family/Species	Plant parts	Brassinosteroid	Isolated quantity (μg/kg fresh wt.)	References
Arecaceae				
Phoenix dactylifera L.	Pollen	24-epiCS		Zaki *et al.* (1993)
Gramineae				
Lolium perenne L.	Pollen	25-MeCS	0.001	Taylor *et al.* (1993)
Oryza sativa L.	Shoots	CS	0.014	Abe *et al.* (1984b)
		DS	0.008	Abe (1991)
		BL		

	Bran	6-deoxoCS 28-homoTE 28-homoTY		Abe *et al.* (1995a)
	Seeds	CS TE 6-deoxoCS		Park *et al.* (1994b)
Phalaris canariensis L.	Seeds	CS TE	5 0.7	Shimada *et al.* (1996)
Secale cereale L.	Seeds	CS TY TE 6-deoxoCS 28-norCS SE		Schmidt *et al.* (1995b)
	Seedlings	SE 2,3-diepiSE secasterol		Antonchick *et al.* (2003, 2005)
Triticum aestivum L.	Grain	CS TY TE 6-deoxoCS 3-DT		Yokota *et al.* (1994)
Gramineae				
Zea mays L. dent corn	Pollen	CS TY TE	120 6.6 4.1	Suzuki *et al.* (1986)
sweet corn	Pollen	CS 28-norCS DS	27.2 18.3 16.9	Gamoh *et al.* (1990)
	Primary roots	BL CS 6-deoxoCS 6-deoxoCT 6-deoxoTE 6-deoxoTY 28-norCS		Kim *et al.* (2000a, 2005, 2006)
Liliaceae				
Erythronium japonicum Decne	Pollen anthers	TY	5	Yasuta *et al.* (1995)
Lilium elegans Thunb.	Pollen	BL CS TY TE	1–5 10–50 10–50 1–5	Suzuki *et al.* (1994b) Yasuta *et al.* (1995)
Lilium longiflorum Thunb.	Pollen	BL CS TY		Abe (1991)
	Anthers	3-DT TE-3-La TE-3-My TE-Glu	 720	Abe *et al.* (1994) Asakawa *et al.* (1994, 1996) Soeno *et al.* (2000)
Tulipa gesneriana L.	Pollen	TY		Abe (1991)
Typhaceae				
Typha latifolia Mey	Pollen	TY TE	68	Schneider *et al.* (1983) Abe (1991)

Table 2. The occurrence of brassinosteroids in the dicotyledons – the Apetalae

Family/Species	Plant parts	Brassinosteroid	Isolated quantity (μg/kg fresh wt.)	References
Betulaceae				
Alnus glutinosa (L.) Gaertn.	Pollen	BL CS		Plattner *et al.* (1986)
Cannabaceae				
Cannabis sativa L.	Seeds	CS TE	600 1800	Takatsuto *et al.* (1996b)
Caryophyllaceae				
Gypsophilla perfoliata L.	Seeds	24-epiBL		Schmidt *et al.* (1996)
Lychnis viscaria L.	Seeds	24-epiCS 24-epiSE		Friebe *et al.* (1999)
Chenophyllaceae				
Beta vulgaris L.	Seeds	CS 24-epiCS		Schmidt *et al.* (1994)
Fagaceae				
Castanea crenata Sieb. et Zucc.	Galls	CS BL 6-deoxoCS	1 4–12 9–26	Yokota *et al.* (1982a) Ikeda *et al.* (1983)
	Shoots leaves	CS 6-deoxoCS	2–6 15–30	Arima *et al.* (1984)
Polygonaceae				
Fagopyrum esculentum Moench	Pollen	BL CS	5 7.1	Takatsuto *et al.* (1990b)
Rheum rhabarbarum L.	Panicles	BL CS 24-epiCS		Schmidt *et al.* (1995a)

Table 3. The occurrence of brassinosteroids in the dicotyledons – the Chloripetalae

Family/Species	Plant parts	Brassinosteroid	Isolated quantity (μg/kg fresh wt.)	References
Apiaceae				
Apium graveolens L.	Seeds	2-deoxyBL		Schmidt *et al.* (1995c)
Daucus carota Ssp. *sativus* L.	Seeds	BL CS 24-epiCS		Schmidt *et al.* (1998)

Brassicaceae				
Arabidopsis thaliana (L.) Heynh.	Shoots ecotype Columbia (wild-type)	CS	0.75	Fujioka *et al.* (1996, 1997, 2000a) Nomura *et al.* (2001) Katsumata *et al.* (2008)
		6-deoxoCS	0.71	
		TY	0.11	
		6-deoxoTY	0.95	
		BL	0.04	
		TE	0.025	
		6-deoxoCT	1.96	
		6-deoxoTE	0.1	
		6-deoxo-3DT	0.13	
		3-epi-6-deoxo -28-norCT		
		28-norCS		
		28-norTY		
		7-oxTY		
		7-oxTE		
	Seeds ecotype Columbia (wild-type)	BL	0.5–1.9	Fujioka *et al.* (1998)
		24-epiBL	0.22	
		CS	0.4–5	
		6-deoxoCS	1.5–3	
		TY	1.3	
		6-deoxoTY	0.5–5.4	
		6-deoxoTE	0.5–1	
	Seeds (ecotype 24)	24-epiBL	0.22	Schmidt *et al.* (1997)
		CS	0.36	
Brassicaceae				
Arabidopsis thaliana (L.) Heynh.	Root callus	BL		Konstantinova *et al.* (2001)
		3-epiBL		
Brassica campestris var. *pekinensis* L.	Seeds	BL	940	Abe *et al.* (1982, 1983) Ikekawa *et al.* (1984)
		28-norBL	1300	
		CS	1600	
		28-norCS	780	
		28-homoCS	130	
Brassica napus L.	Pollen	BL	100	Grove *et al.* (1979)
Raphanus sativus L.	Seeds	BL	0.3	Schmidt *et al.* (1991, 1993b)
		CS	0.8	
		TE		
		28-homoTE		
Fabaceae				
Cassia tora L.	Seeds	BL	0.018	Park *et al.* (1994a)
		CS	0.16	
		TY	0.007	
		TE	0.04	
		28-norCS	0.008	
Robinia pseudo-acacia L.	Pollen	CS		Abe *et al.* (1995b)
		TY		
		6-deoxoCS		

(*continued*)

(*continued Table 3.*)

Family/Species	Plant parts	Brassinosteroid	Isolated quantity (μg/kg fresh wt.)	References
Fabaceae				
Dolichos lablab L.	Seeds	DL	160	Baba *et al.* (1983)
		DS	50	Yokota *et al.* (1982b, 1983b, 1984)
		28-homoDS	20	
		28-homoDL	12	
		BL		
		CS		
		6-deoxoCS		
		6-deoxoDS		
Vicia faba L.	Seeds	BL	190	Park *et al.* (1987)
		24-epiBL	5	Ikekawa *et al.* (1988)
		CS		
		28-norCS		
	Pollen	BL	181	Gamoh *et al.* (1989)
		CS	134	
		28-norCS	628	
		DS	537	
	Shoots	6-deoxoCT	1.27	Fukuta *et al.* (2004)
		6-deoxoTE	0.17	
		6-deoxoTY	0.40	
		6-deoxoCS	3.57	
		TY	0.02	
		CS	0.37	
Psophocarpus tetragonolobus (Stickm.) DC.	Seeds	BL		Takatsuto (1994)
		CS		
		6-deoxoCS		
		6-deoxoDS		
Fabaceae				
Ornithopus sativus Brot.	Seeds	CS	5	Schmidt *et al.* (1993a)
		24-epiCS	25	
	Shoots	CS		Spengler *et al.* (1995)
		6-deoxoCS		
		24-epiCS		
		6-deoxo-24-epiCS		
		6-deoxo-28-norCS		
Theaceae				
Thea sinensis L.	Leaves	28-norCS	0.002	Abe *et al.* (1983, 1984a)
		28-homoCS	<0.001	Morishita *et al.* (1983)
		BL	0.006	Ikekawa *et al.* (1984)
		CS	0.1	
		TY	0.06	
		TE	0.02	

Camellia sinensis (L.) Kuntze	Leaves	6-deoxoCS 24-epiBL 3-DT TY 3-deoxoTY 28-homoDL	Gupta *et al.* (2004)
	Immature seeds	6-deoxo-28-norCT 6-deoxo-28-norTE 3-dehydro-6-deoxo-28-norTE 6-deoxo-28-norTY 6-deoxo-28-norCS	Bhardwaj *et al.* (2007)
Fabaceae			
Phaseolus vulgaris L.	Seeds	BL CS 2-epiCS 3-epiCS 2,3-diepiCS 3,24-diepiCS TY TE 6-deoxoCS 3-epi-6-deoxoCS 1β-OH-CS 3-epi-1α-OH-CS DL DS 6-deoxoDS 6-deoxo-28-homoDS 25-MeDS 2-epi-25-MeDS 2,3-diepi-25-MeDS 2-deoxy-25-MeDS 2-epi-2-deoxy-25-MeDS 3-epi-2-deoxy-25-MeDS 6-deoxo-25-MeDS 25-MeDS-Glu 2-epi-25-MeDS-Glu	Yokota *et al.* (1983c, 1987c) Kim *et al.* (1987, 1988, 2000b); Kim (1991) Park *et al.* (2000)
Fabaceae			
Pisum sativum L.	Seeds	BL CS TY 6-deoxoCS 2-deoxyBL	Yokota *et al.* (1996)

(*continued*)

(*continued Table 3.*)

Family/Species	**Plant parts**	**Brassinosteroid**	**Isolated quantity (μg/kg fresh wt.)**	**References**
	Shoots	BL CS 6-deoxoCS TY 6-deoxoCT 6-deoxoTE 6-deoxo-3DT 6-deoxoTY	0.2–0.8 0.4–2.4 5.2 1 3.75 0.047 0.074 0.8	Nomura *et al.* (1997, 1999, 2001)
Myrtaceae				
Eucalyptus calophylla R. Br.	Pollen	BL		Takatsuto (1994)
Eucalyptus marginata Sn.	Pollen	DS		Takatsuto (1994)
Rosaceae				
Eriobotrya japonica (Thunb.) Lindl.	Flower buds	CS		Takatsuto (1994)
Rutaceae				
Citrus unshiu Marcov.	Pollen	BL CS TY TE		Abe (1991)
Citrus sinensis Osbeck	Pollen	BL CS	36.2 29.4	Motegi *et al.* (1994)
Aegle marmelos (L.) Correa	Leaves	24-epiBL		Sondhi *et al.* (2008)

Table 4. The occurrence of brassinosteroids in the dicotyledons – the Sympetalae

Family/Species	**Plant parts**	**Brassinosteroid**	**Isolated quantity (μg/kg fresh wt.)**	**References**
Apocynaceae				
Catharanthus roseus G. Don.	Cultured cells	BL CS 6-deoxoTY 6-deoxoTE 6-deoxoCS CT 6-deoxoCT 3-epi-6-deoxoCT 3-epi-6-deoxo-28-norCT 3-DT TY TE	0.4–8.7 0.6–4.5 0.76 0.047 5.9–18.9 2–4 30	Choi *et al.* (1993, 1996, 1997) Fujioka *et al.* (1995, 2000b) Park *et al.* (1989) Suzuki *et al.* (1993, 1994a, 1994c, 1995) Yokota *et al.* (1990)

Asteraceae				
Zinnia elegans L.	Cultured cells	CS	0.2	Yamamoto *et al.* (2001, 2007)
		TY		
		6-deoxoCS	0.8	
		6-deoxoTY	0.3	
		6-deoxoTE	0.3	
		6-deoxoCT	8.7	
Helianthus annuus L.	Pollen	BL	106	Takatsuto *et al.* (1989)
		CS	21	
		28-norCS	65	
Solidago altissima L.	Shoots	BL		Takatsuto (1994)
Boraginaceae				
Echium plantagineum L.	Pollen	BL		Takatsuto (1994)
Convolvulaceae				
Pharbitis purpurea Voigt	Seeds	CS	1.1	Suzuki *et al.* (1985)
		28-norCS	0.2	
Cucurbitaceae				
Cucurbita moschata Duch.	Seeds	BL		Jang *et al.* (2000)
		CS		
	Pollen	BL		Pachthong *et al.* (2006)
		CS		
Lamiaceae				
Perilla frutescens (L.) Britt.	Seeds	CS		Park *et al.* (1994b)
Solanaceae				
Nicotiana tabacum L.	Cultured cells	CS		Park *et al.* (1994b)
Lycopersicon esculentum Mill.	Shoots	CS	0.2	Yokota *et al.* (1997d)
		6-deoxoCS	1.7	
		28-norCS	0.03	
	Root	6-deoxo-28-norCT	0.22	Yokota *et al.* (2001)
		6-deoxo-28-norTY	0.13	
		6-deoxo-28-norCS	0.09	
-dwarf mutant	Shoots	6-deoxoCT	1.1	Bishop *et al.* (1999)
		6-deoxoTE	0.04	
		6-deoxo-3DT	0.03	
		6-deoxoTY		
		6-deoxoCS	0.5	
		6α-OH-CS	5.2	
		CS	0.2	
		BL	< 0.001	
		TY	< 0.001	
		3-DT	< 0.001	
		TE	< 0.001	
		CT	< 0.001	

Table 5. The occurrence of brassinosteroids in gymnosperms

Family/Species	Plant parts	Brassinosteroid	Isolated quantity (μg/kg fresh wt.)	References
Cupressaceae				
Cupressus arizonica Greene	Pollen	6-deoxoTY	6400	Griffiths *et al.* (1995)
		6-deoxo-3DT	2300	
		6-deoxoCS	1200	
		CS	1000	
		TY	460	
		TE	5	
		28-homoCS	4	
		3-DT	2	
		BL	<1	
Ginkgoaceae				
Ginkgo biloba L.	Seeds	TE	15	Takatsuto *et al.* (1996a)
Pinaceae				
Piceae sitchensis Trantv. ex Mey	Shoots	CS	5	Yokota *et al.* (1985)
		TY	7	
Pinus silvestris L.	Cambial region	BL		Kim *et al.* (1990)
		CS		
Pinus thunbergii Parl.	Pollen	TY	89.5	Yokota *et al.* (1983a)
Taxodiaceae				
Cryptomeria japonica D. Don.	Pollen anthers	TY		Yokota *et al.* (1998) Watanabe *et al.* (2000)
		DL		
		3-DT		
		28-homoBL		
		28-homoDL		
		23-dehydroBL (cryptolide)		
		2-epi-23-dehydroBL		
		3-epi-23-dehydroBL		
		2,3-diepi-23-dehydroBL		

Table 6. The occurrence of brassinosteroids in lower plants

Family/Species	Plant parts	Brassinosteroid	Isolated quantity (μg/kg fresh wt.)	References
Hydrodictyaceae (green alga)				
Hydrodictyon reticulatum (L.) Lager.	Whole plant	24-epiCS	0.3	Yokota *et al.* (1987b)
		28-homoCS	4.0	

Trebouxiophyceae (green alga)				
Chlorella vulgaris Beijerinck	Cultured cells	TY	0.39	Bajguz (2009)
		TE	0.26	
		6-deoxoTE	0.22	
		6-deoxoTY	0.18	
		6-deoxoCS	0.32	
		CS	0.47	
		BL	0.07	
Crystoseiraceae (brown alga)				
Cystoseira myrica (Gmelin) Agardh	Whole plant	3-keto-22-epi-28-norCT		Hamdy *et al.* (2009)
Marchantiaceae (liverwort)				
Marchantia polymorpha L.	Cultured cells	TE		Park *et al.* (1999)
		3-DT		
		TY		
Equisetaceae (horsetail)				
Equisetum arvense L.	Whole plant	CS	0.17	Takatsuto *et al.* (1990a)
		DS	0.75	
		28-norBL	0.15	
		28-norCS	0.35	

Table 7. Division of free brassinosteroids according to number of carbon in structure and different types of B-ring and substituents in the A-ring

No. of carbon	Type of B-ring	Substituent in A-ring	Representative(s)
C_{27}	7-Oxalactone	C(2α,3α)-OH	28-norBL
	6-Oxo	C(2α,3α)-OH	28-norCS
		C3α-OH	28-norTY
	6-Deoxo	C(2α,3α)-OH	6-deoxo-28-norCS
		C3α-OH	6-deoxo-28-norTY, 3-epi-6-deoxo-28-norCT
		C3β-OH	6-deoxo-28-norTE, 6-deoxo-28-norCT
		C3-oxo group	3-dehydro-6-deoxo-28-norTE, 3-keto-22-epi-28-norCT
C_{28}	7-Oxalactone	C(2α,3α)-OH	BL, 24-epiBL, 23-dehydroBL, DL
		C(2α,3β)-OH	3-epi-23-dehydroBL, 3-epiBL
		C(2β,3α)-OH	2-epi-23-dehydroBL
		C(2β,3β)-OH	2,3-diepi-23-dehydroBL
		C3α-OH	2-deoxyBL, 7-oxTY
		C3β-OH	7-oxTE
	6-Oxo	C(2α,3α)-OH	CS, 24-epiCS, DS
		C(2α,3β)-OH	3-epiCS, 3,24-diepiCS
		C(2β,3α)-OH	2-epiCS
		C(2β,3β)-OH	2,3-diepiCS
		C(1β,2α,3α)-OH	1β-OH-CS
		C(1α,2α,3β)-OH	3-epi-1α-OH-CS
		C3α-OH	TY

(*continued*)

(*continued Table 7.*)

No. of carbon	Type of B-ring	Substituent in A-ring	Representative(s)
		C3β-OH	TE, CT
		C3-oxo group	3-DT (3-dehydroTE)
		C(2β,3β)-epoxide	SE, 24-epiSE
		C(2α,3α)-epoxide	2,3-diepiSE
		$\Delta^{2,3}$	Secasterol
	6-Deoxo	C(2α,3α)-OH	6-deoxoCS, 6-deoxo-24-epiCS, 6-deoxoDS
		C(2α,3β)-OH	3-epi-6-deoxoCS
		C3α-OH	6-deoxoTY, 3-epi-6-deoxoCT
		C3β-OH	6-deoxoTE, 6-deoxoCT
		C3-oxo group	6-deoxo-3DT (3-dehydro-6-deoxoTE)
	6-Hydroxy	C(2α,3α)-OH	6α-OH-CS
C_{29}	7-Oxalactone	C(2α,3α)-OH	28-homoBL, 28-homoDL
	6-Oxo	C(2α,3α)-OH	28-homoCS, 28-homoDS, 25-MeDS, 25-MeCS
		C(2β,3α)-OH	2-epi-25-MeDS, 2-epi-25-MeCS
		C(2β,3β)-OH	2,3-diepi-25-MeDS, 2,3-diepi-25-MeCS
		C3α-OH	28-homoTY, 2-deoxy-25-MeDS
		C3β-OH	28-homoTE, 3-epi-2-deoxy-25-MeDS
	6-Deoxo	C(2α,3α)-OH	6-deoxo-28-homoDS, 6-deoxo-25-MeDS

Table 8. Division of free brassinosteroids according to different substituents in the side chain

Type	Representative(s)
23-Oxo	23-Dehydrobl, 2-epi-23-dehydroBL, 3-epi-23-dehydroBL, 2,3-diepi-23-dehydroBL
24*S*-Methyl	BL, 3-epiBL, CS, 2-epiCS, 3-epiCS, 2,3-diepiCS, TY, TE, 6-deoxoCS, 3-epi-6-deoxoCS, 3-DT, SE, 2,3-diepiSE, 6-deoxoTY, 2-deoxyBL, 3-epi-1α-OH-CS, 1β-OH-CS, 6α-OH-CS, 6-deoxoTE, 6-deoxo-3-DT, secasterol, 7-xoTY, 7-oxTE
24*R*-Methyl	24-epiBL, 24-epiCS, 3,24-diepiCS, 6-deoxo-24-epiCS, 24-epiSE
24-Methylene	DL, DS, 6-deoxoDS
24*S*-Ethyl	28-Homobl, 28-homoCS, 28-homoTE, 28-homoTY
24-Ethylidene	28-Homodl, 28-homoDS, 6-deoxo-28-homoDS
24-Methylene-25-Methyl	25-MeDS, 2-epi-25-MeDS, 2,3-diepi-25-MeDS, 2-deoxy-25-MeDS, 3-epi-2-deoxy-25-MeDS, 6-deoxo-25-MeDS
24*S*-Methyl-25-Methyl	25-MeCS, 2-epi-25-MeCS, 2,3-diepi-25-MeCS
Without substituent at C-23	CT, 6-deoxoCT, 3-epi-6-deoxoCT
Without substituent at C-24	28-norBL, 28-norCS, 28-norTY, 6-deoxo-28-norCS, 6-deoxo-28-norTY, 3-epi-6-deoxo-28-norCT, 6-deoxo-28-norTE, 3-dehydro-6-deoxo-28-norTE
Without substituents at C-23 and C-24	6-Deoxo-28-norCT, 3-keto-22-epi-28-norCT

Among the BRs, CS is the most widely distributed (53 plant species), followed by BL (37), TY (28), 6-deoxoCS (23), TE (20), 28-norCS (13), and 6-deoxoTY (10). Furthermore from 2 to 9 BRs are distributed in a limited number of plant species, it means that 6-deoxoTE – 9 compounds, 3-DT – 8, 24-epiCS, DS, 6-deoxoCT – 7, 24-epiBL – 6, 6-deoxo-3-DT, 28-homoCS – 4, DL, 6-deoxoDS, 28-homoDL – 3, SE, 2-deoxyBL, 28-homoTE, CT, 28-norBL, 28-norTY, 6-deoxo-28-norTY, 6-deoxo-28-norCT, 3-epi-6-deoxo-28-norCT – 2. To the present day 36 other BRs and 5 BR conjugates have been found in only one plant species. Among all naturally occurring BRs, CS and BL are the most important BRs because of their wide distribution as well as their potent biological activity (Bajguz and Tretyn, 2003).

Among the plant sources investigated, immature seeds of *Phaseolus vulgaris* contain a wide array of BRs, this is 23 free BRs and 2 conjugates. The wide occurrence of BRs was also found in the shoots of *Arabidopsis thaliana* (13 compounds), cultured cells of *Catharanthus roseus* (13 compounds), dwarf mutant of *Lycopersicon esculentum* (12 compounds), pollen and anthers of *Cryptomeria japonica* (9 compounds), pollen of *Cupressus arizonica* (9 compounds), shoots of *Pisum sativum* (8 compounds), seeds of *Dolichos lablab* (8 compounds), cultured cells of *Chlorella vulgaris* (7 compounds), primary roots of *Zea mays* (7 compounds), seeds of *Secale cereale* (6 compounds), and leaves of *Thea sinensis* and *Camellia sinensis* (6 compounds) (Bajguz and Tretyn, 2003).

3. CHEMICAL STRUCTURE OF BRASSINOSTEROIDS

Natural BRs identified so far, have a common 5α-cholestane skeleton, and their structural variations come from the kind and orientation of oxygenated functions in rings A and B (Figure 1). These modifications are produced by oxidation and reduction reactions during biosynthesis. Generally, BRs are divided into free (64) (Table 7, Figures 2–4) and conjugated (5) (Figure 5) compounds. These steroids can be classified as C_{27}, C_{28} or C_{29} BRs (10, 38 and 16 compounds, respectively) depending on the alkyl-substitutions in the side chain (Figures 2–4). These side chain structures are all common in plants sterols. According to the cholestane side chain (Table 8), BRs are divided into 11 types with different substituents at C-23, C-24 and C-25: 23-oxo (4 compounds), 24*S*-methyl (23 compounds), 24*R*-methyl (5 compounds), 24-methylene (3 compounds), 24*S*-ethyl (4 compounds), 24-ethylidene (3 compounds), 24-methylene-25-methyl (6 compounds), 24-methyl-25-methyl (3 compounds), without substituent at C-23 (3 compounds), without substituent at C-24 (8 compounds) and without substituents at C-23, C-24

(2 compounds) (Yokota *et al.*, 1987a; Yokota, 1995, 1997, 1999; Abe *et al.*, 2001; Schneider 2002; Bajguz and Tretyn, 2003).

Brassinolide [(22*R*,23*R*,24*S*)-2α,3α,22,23-tetrahydroxy-24-methyl-homo-7-oxa-5α-cholestan-6-one] is the most active compound. It is a C_{28} BR bearing an *S*-methyl group at C-24 in the side chain of its 5α-ergostane structure. It has a lactone function at C-6/C-7 in ring B, 2α,3α vicinal hydroxyl at A-ring and in the lateral chain exhibits *R* configuration of the diol at C-22/C-23 and 24*S* methyl substitution. Although C_{28} BRs are the most ubiquitous in nature, other BRs with a different steroidal side chain are known. Among them, 28-homoBL and 28-homocastasterone, which have a 5α-stigmastane structure, are the most active C_{29} BRs (Yokota, 1999; Bajguz and Tretyn, 2003).

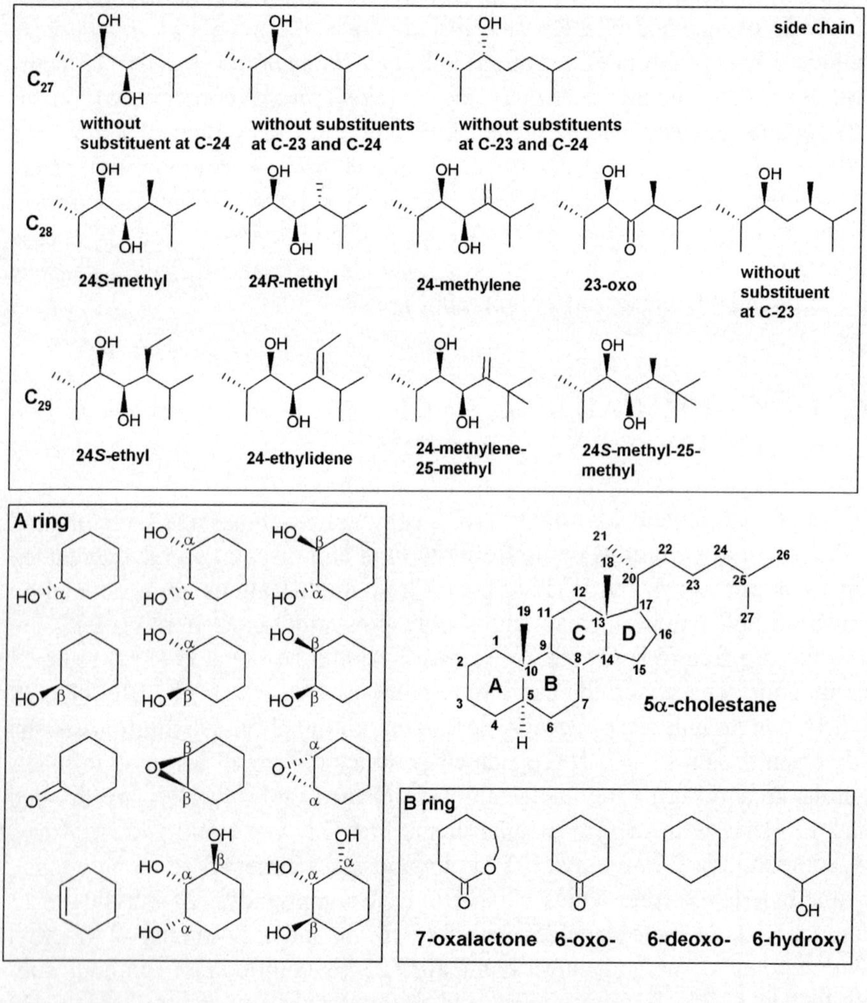

Figure 1. Different substituents in the A- and B-rings and side chain of naturally occurring brassinosteroids.

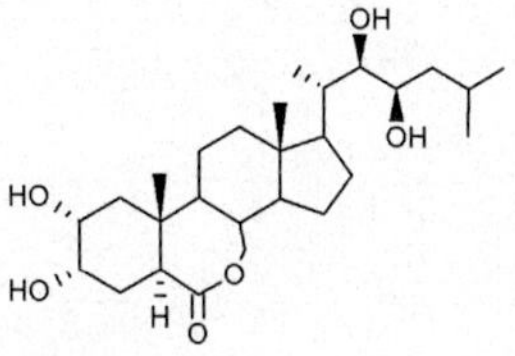

28-norbrassinolide (28-norBL)

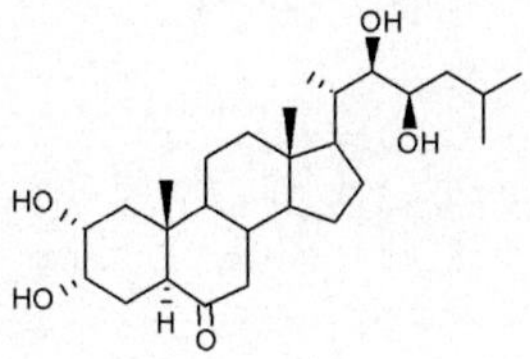

28-norcastasterone (28-norCS)

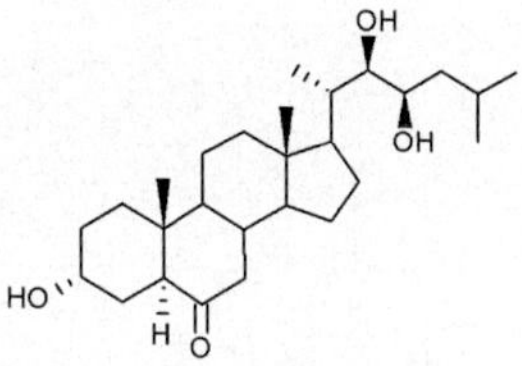

28-nortyphasterol (28-norTY)

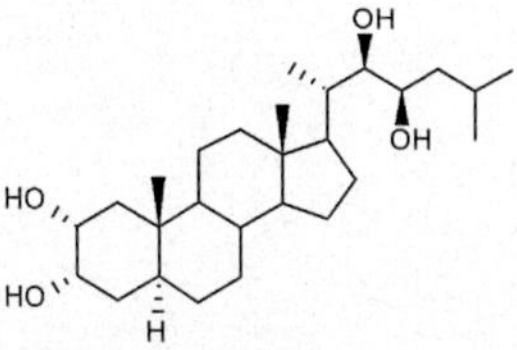

6-deoxo-28-norcastasterone (6-deoxo-28-norCS)

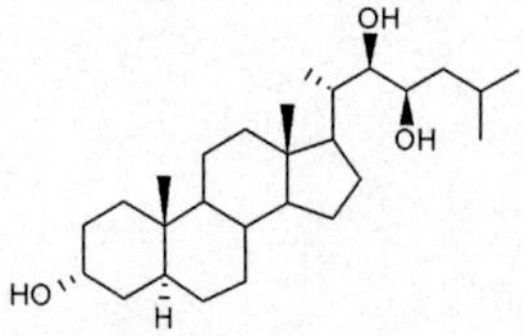

6-deoxo-28-nortyphasterol (6-deoxo-28-norTY)

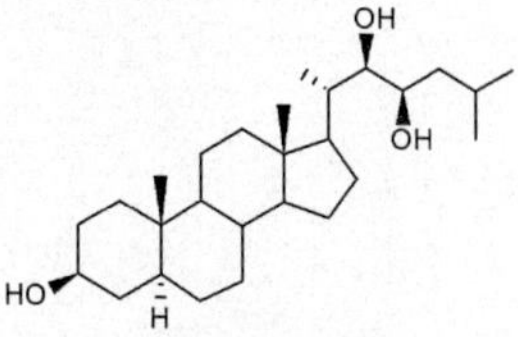

6-deoxo-28-norteasterone (6-deoxo-28-norTE)

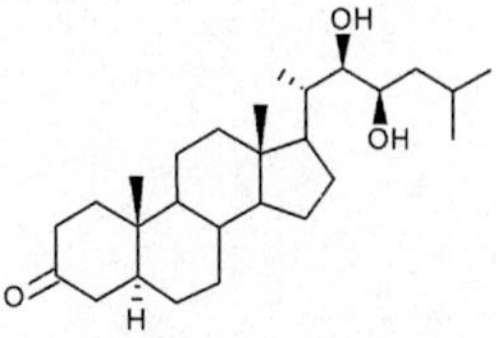

3-dehydro-6-deoxo-28-norteasterone (3-dehydro-6-deoxo-28-norTE)

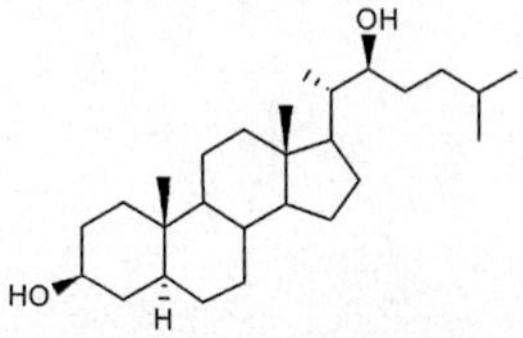

6-deoxo-28-norcathasterone (6-deoxo-28-norCT)

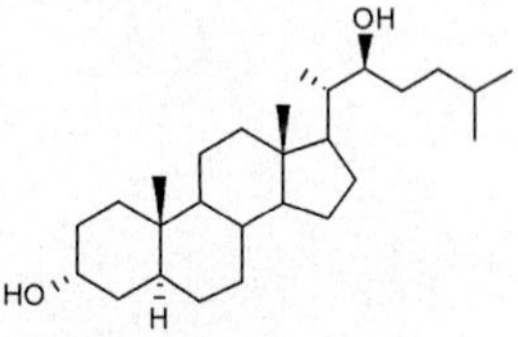

3-epi-6-deoxo-28-norcathasterone (3-epi-6-deoxo-28-norCT)

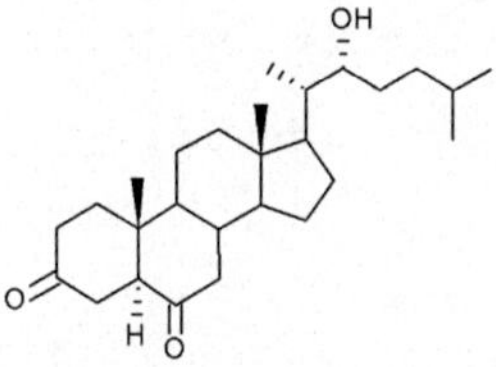

3-keto-22-epi-28-norcathasterone (3-keto-22-epi28-norCT)

Figure 2. Chemical structures of C_{27} brassinosteroids.

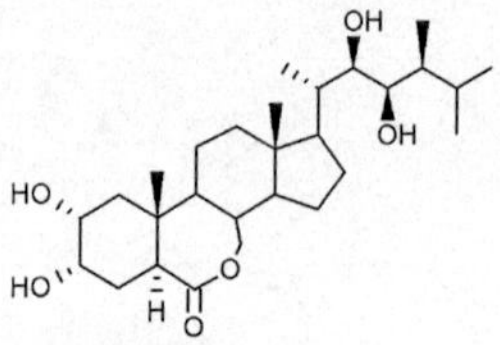

brassinolide
(BL)

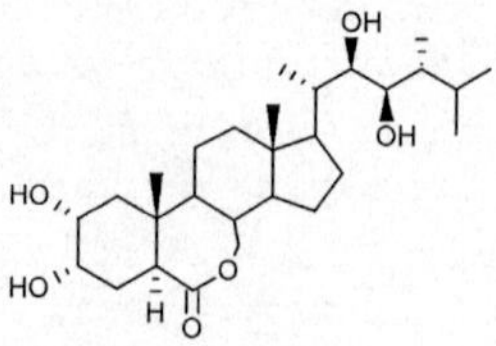

24-epibrassinolide
(24-epiBL)

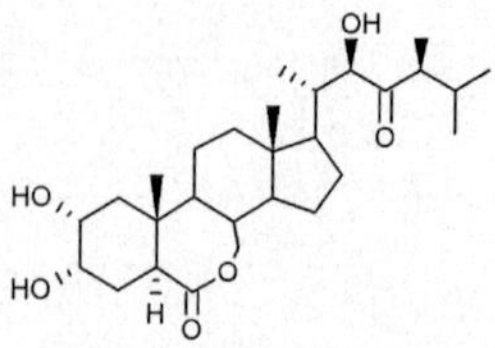

23-dehydrobrassinolide
(23-dehydroBL) (cryptolide)

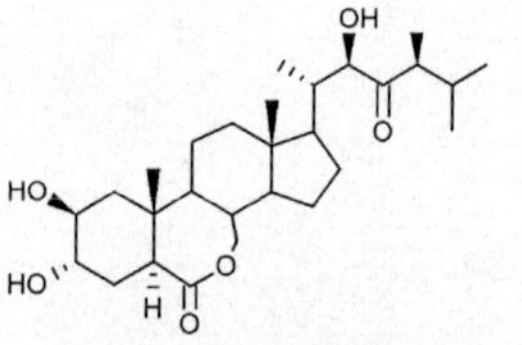

2-epi-23-dehydrobrassinolide
(2-epi-23-dehydroBL)

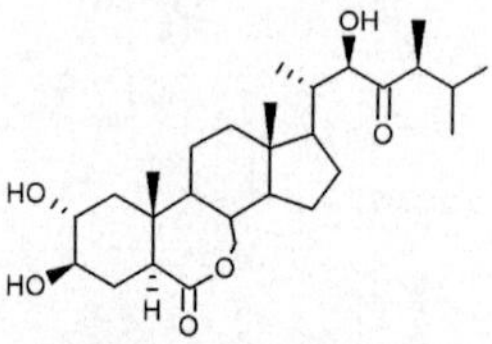

3-epi-23-dehydrobrassinolide
(3-epi-23-dehydroBL)

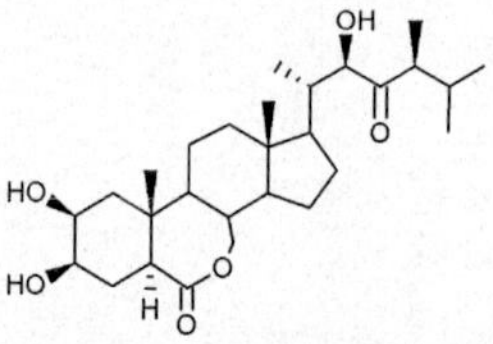

2,3-diepi-23-dehydrobrassinolide
(2,3-diepi-23-dehydroBL)

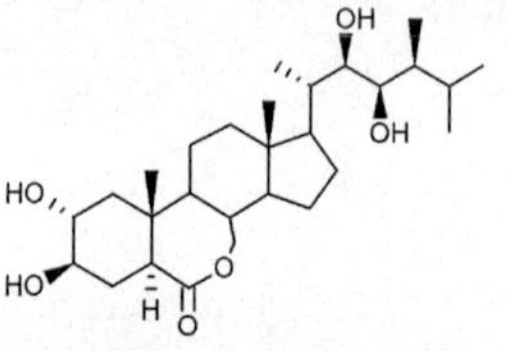

3-epibrassinolide
(3-epiBL)

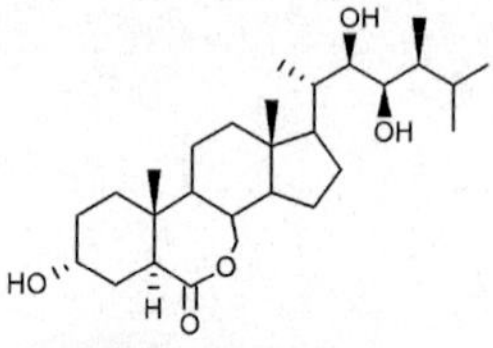

2-deoxybrassinolide
(2-deoxyBL)

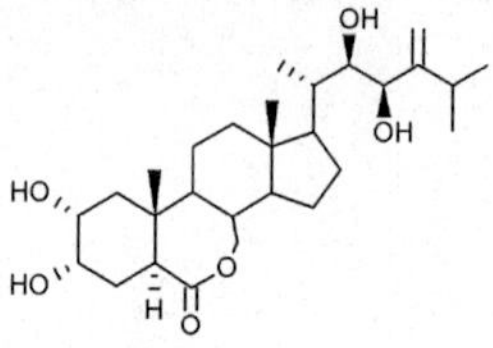

dolicholide
(DL)

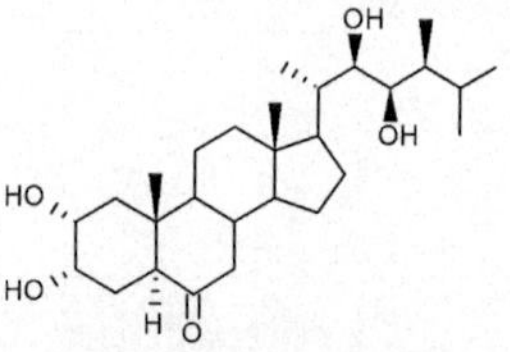

castasterone
(CS)

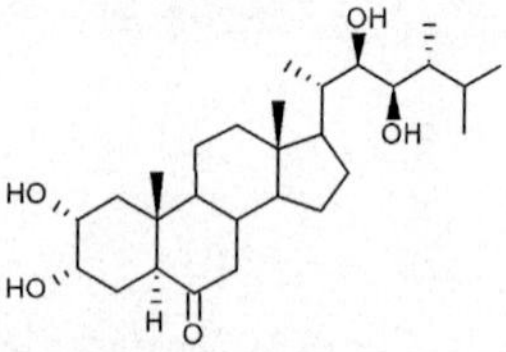

24-epicastasterone
(24-epiCS)

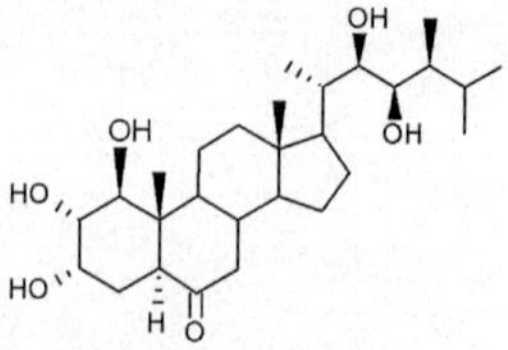

1β-hydroxycastasterone
(1β-OH-CS)

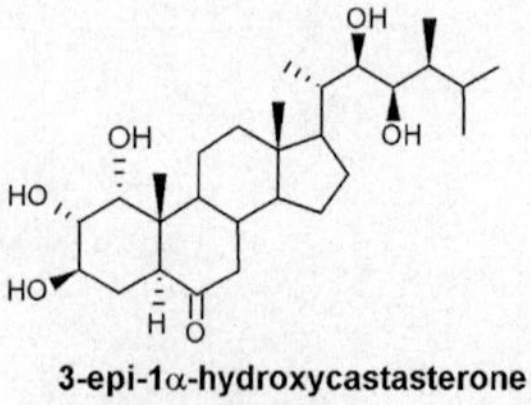

3-epi-1α-hydroxycastasterone
(3-epi-1α-OH-CS)

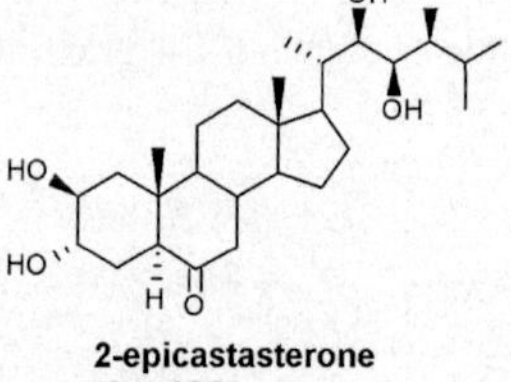

2-epicastasterone
(2-epiCS)

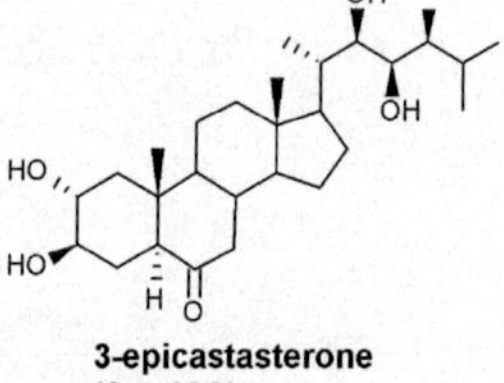

3-epicastasterone
(3-epiCS)

Figure 3. Chemical structures of C_{28} brassinosteroids.

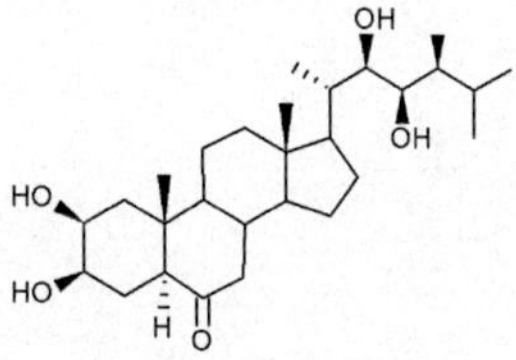

2,3-diepicastasterone
(2,3-diepiCS)

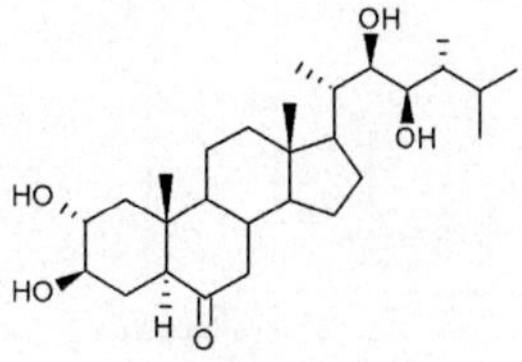

3,24-diepicastasterone
(3,24-diepiCS)

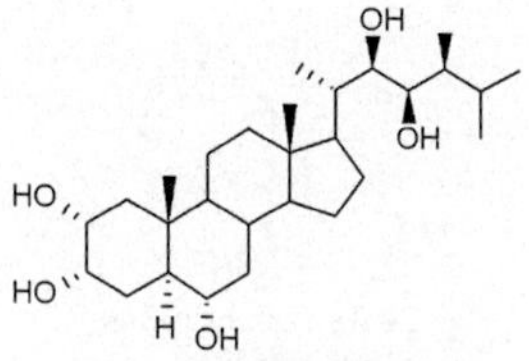

6-α-hydroxycastasterone
(6-α-OH-CS)

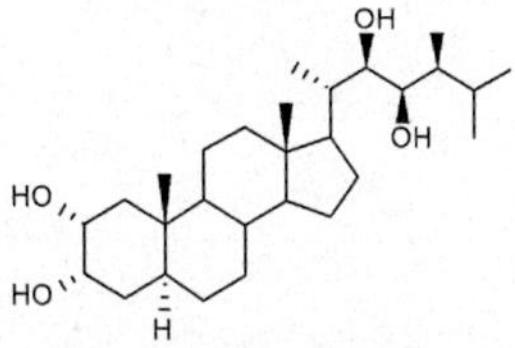

6-deoxocastasterone
(6-deoxoCS)

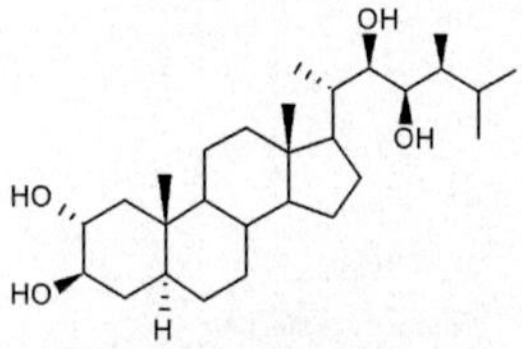

3-epi-6-deoxocastasterone
(3-epi-6-deoxoCS)

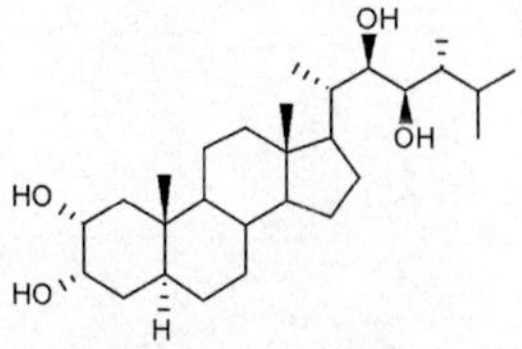

6-deoxo-24-epicastasterone
(6-deoxo-24-epiCS)

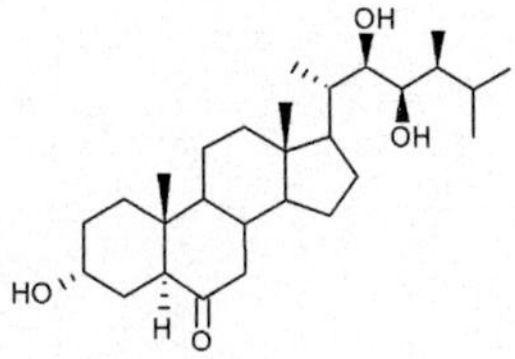

typhasterol
(TY)

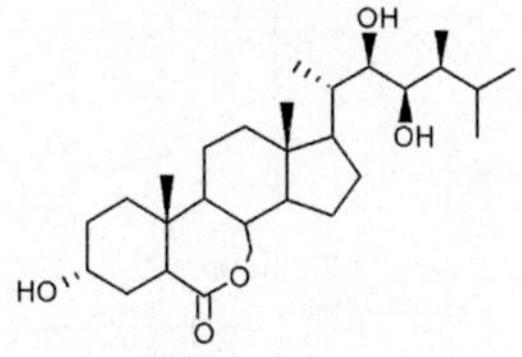

7-oxotyphasterol
(7-oxTY)

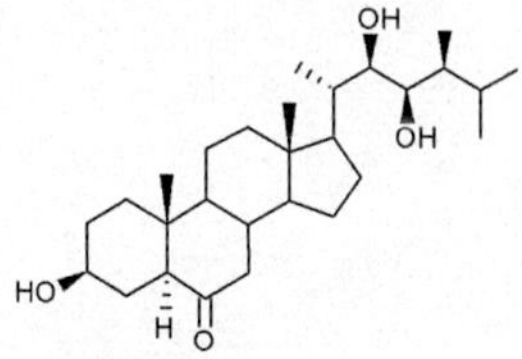

teasterone
(TE)

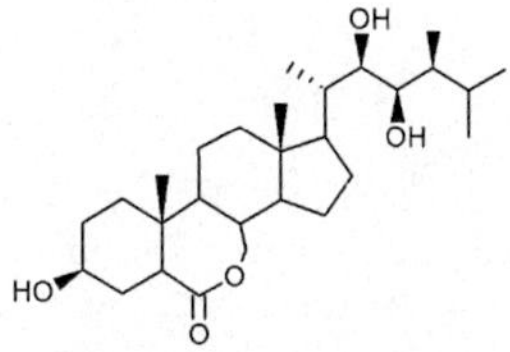

7-oxoteasterone
(7-oxTE)

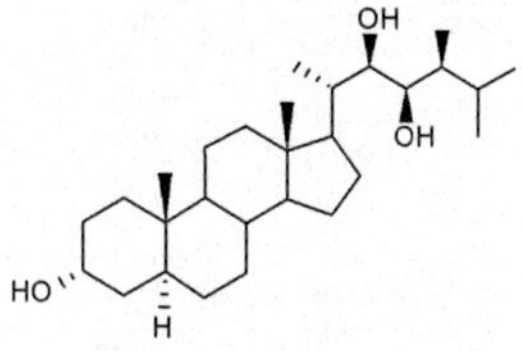

6-deoxotyphasterol
(6-deoxoTY)

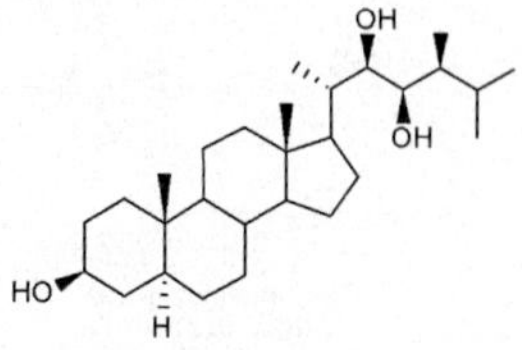

6-deoxoteasterone
(6-deoxoTE)

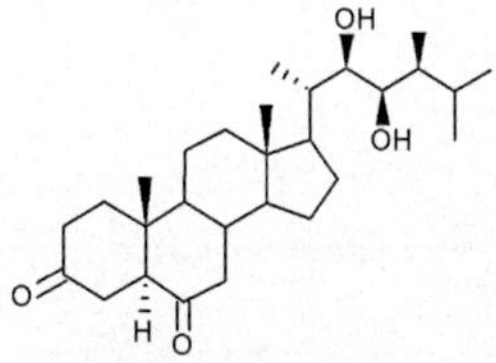

3-dehydroteasterone
(3-DT)

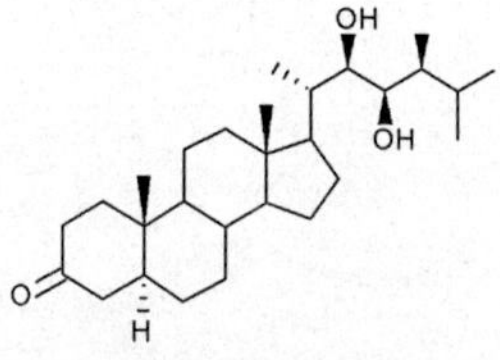

6-deoxo-3-dehydroteasterone
(6-deoxo-3-DT)

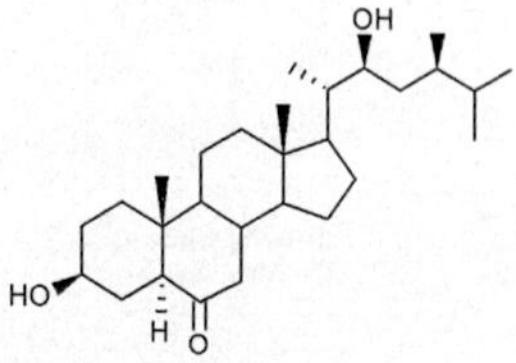

cathasterone
(CT)

Figure 3. Chemical structures of C_{28} brassinosteroids – *continued*.

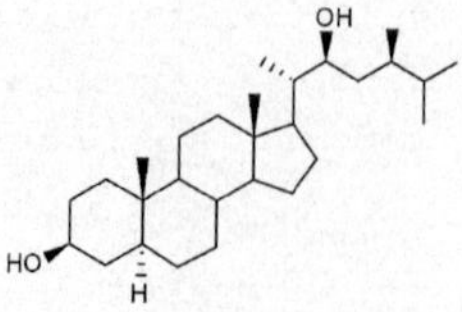

6-deoxocathasterone
(6-deoxoCT)

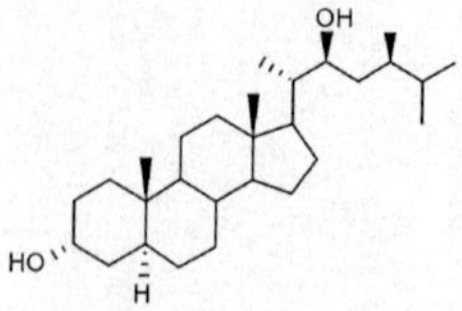

3-epi-6-deoxocathasterone
(3-epi-6-deoxoCT)

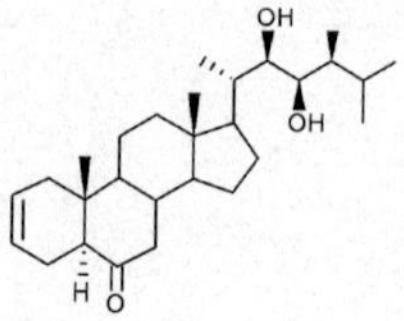

secasterol

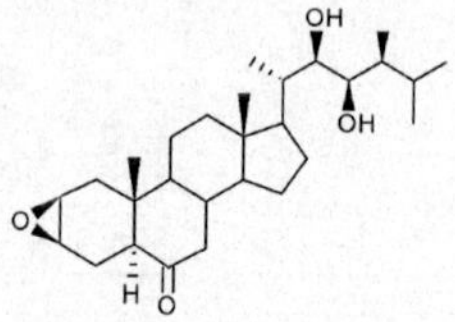

secasterone
(SE)

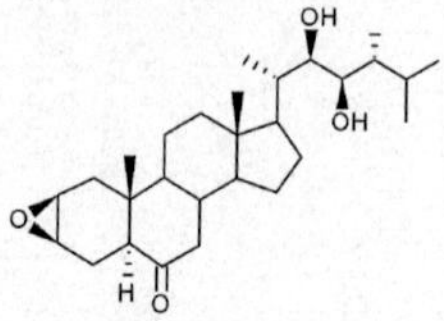

24-episecasterone
(24-epiSE)

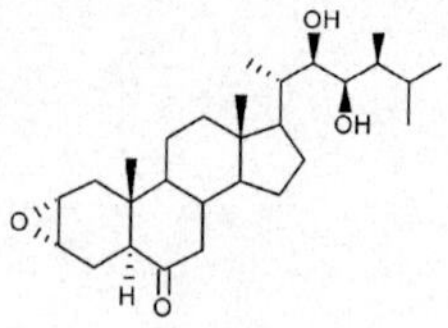

2,3-diepisecasterone
(2,3-diepi-SE)

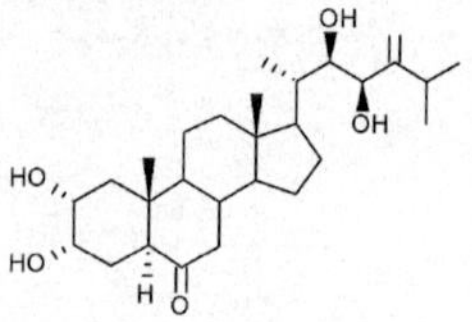

dolichosterone
(DS)

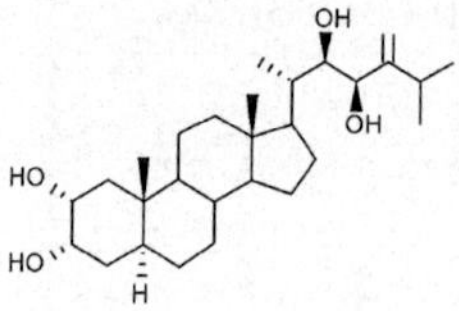

6-deoxodolichosterone
(6-deoxoDS)

Figure 3. Chemical structures of C_{28} brassinosteroids – *continued.*

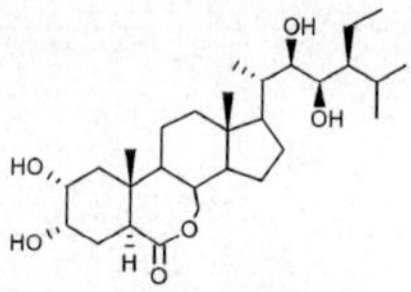

28-homobrassinolide
(28-homoBL)

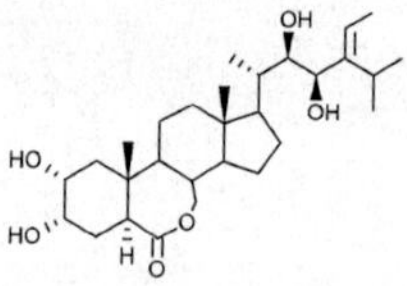

28-homodolicholide
(28-homoDL)

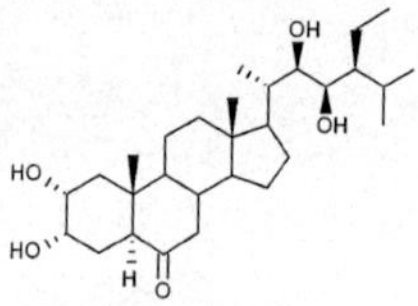

28-homocastasterone
(28-homoCS)

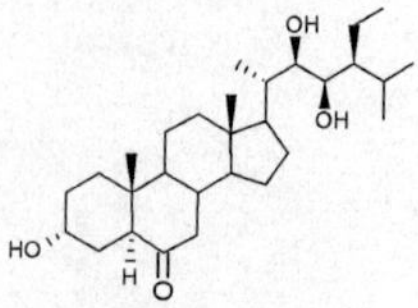

28-homotyphasterol
(28-homoTY)

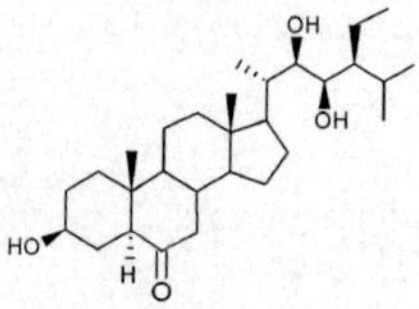

28-homoteasterone
(28-homoTE)

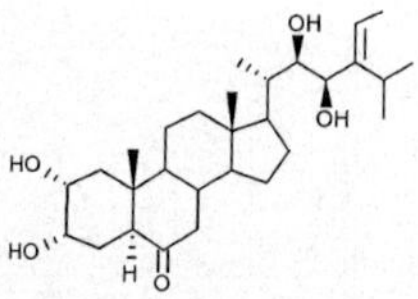

28-homodolichosterone
(28-homoDS)

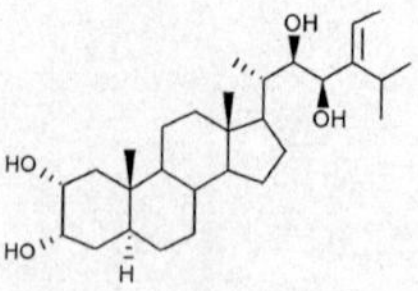

6-deoxo-28-homodolichosterone
(6-deoxo-28-homoDS)

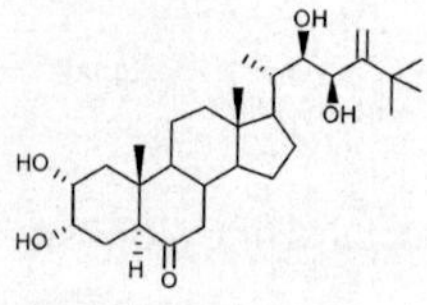

25-methyldolichosterone
(25-MeDS)

Figure 4. Chemical structures of C_{29} brassinosteroids.

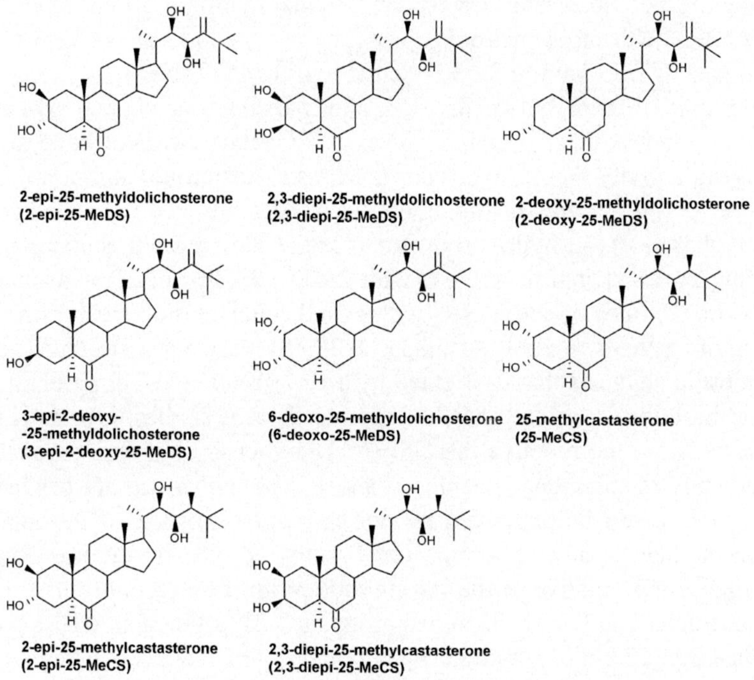

Figure 4. Chemical structures of C_{29} brassinosteroids – *continued.*

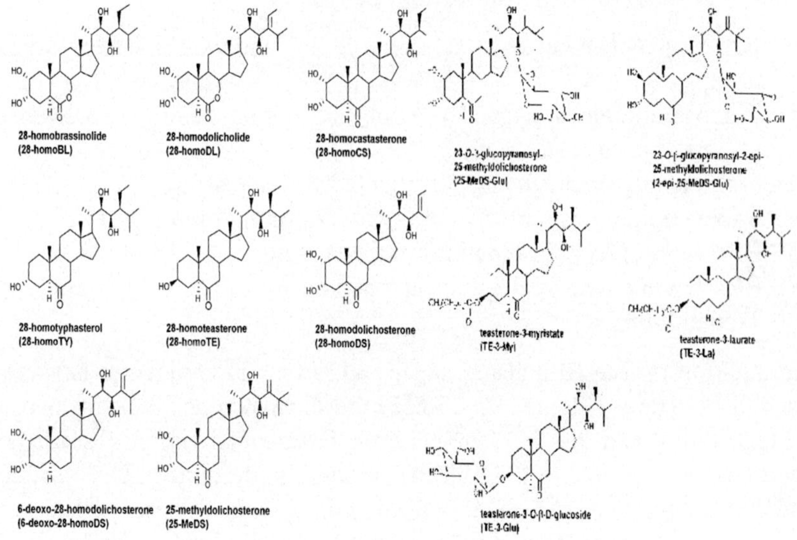

Figure 5. Chemical structures of brassinosteroid conjugates.

With respect to A-ring, BRs having vicinal hydroxyl groups at C-2α and C-3α. BRs with an α-hydroxyl, β-hydroxyl or ketone at position C-3 are precursors of BRs having 2α,3α-vicinal hydroxyls (Table 7). As regards the B-ring oxidation, BRs are divided into 7-oxalactone, 6-oxo (6-ketone), 6-deoxo (non-oxidized), and 6-hydroxy. 7-Oxalactone BRs have stronger biological activity than 6-oxo types (such as castasterone) and non-oxidized BRs reveal no activity in biological tests. BRs with either a 3α-hydroxyl (e.g. typhasterol), 3β-hydroxyl or 3-oxo in the A-ring are precursors of BRs carrying 2α,3α-vicinal diol; those with 2α,3β-, 2β,3α-, or 2β,3β-vicinal diol can be metabolites of active BRs with a 2α,3α-vicinal diol. Decreasing order of activity (2α,3α > 2α,3β > 2β,3α > 2β,3β) shown by structure-activity relationship suggests that α-oriented hydroxyl group at C-2 is essential for a greater biological activity of BRs in plants. Recent studies have shown that 3α,4α-diols are more active than 2α,3α. This fact is in strong contrast with the structure requirements, mentioned above. The higher activity of unnatural 3α,4α-diols could be explained by twisting and distortion of the molecule due to the seven- or eight-membered B-ring and also by the position of a carbonyl group, relative to the A-ring diol. With the exception of some not fully characterized BRs with an oxo group at C-23, all bioactive BRs possess a vicinal 22*R*,23*R* diol structural functionality, which appears essential for a high biological activity (Yokota, 1999; Bajguz and Tretyn, 2003).

Furthermore, with respect to A/B ring functionalities the hitherto clarified members can be divided into following groups:

- BRs with 7-membered 7-oxalactone-B-ring and vicinal 2α,3α-hydroxyl groups;
- 6-oxo compounds with a 6-membered B-ring having two hydroxyl groups at position C-2 and C-3;
- 6-oxo compounds with 2β,3β-oriented epoxide group;
- 6-oxo compounds with an additional hydroxyl group at position C-1;
- BRs without oxygen functions in the B-ring;
- BRs having hydroxyl group at position C-6 (Table 7) (Bajguz and Tretyn 2003).

In addition to free BRs also 5 sugar and fatty acid conjugates have been identified in plants (Figure 5). 25-Methyldolichosterone-23-β-*D*-glucoside (25-MeDS-Glu) and its 2β isomer from *Phaseolus vulgaris* seeds and teasterone-3β-*D*-glucoside (TE-3-Glu), teasterone-3-laurate (TE-3-La) and teasterone-3-myristate (TE-3-My) from *Lilium longiflorum* pollen were isolated as endogenous BRs (Yokota *et al*., 1987a; Abe *et al*., 2001).

4. REFERENCES

Abe, H. 1991. Rice-lamina inclination, endogenous levels in plant tissues and accumulation during pollen development of brassinosteroids. *In* HG Cutler, T Yokota, G Adam (Eds.) Brassinosteroids: Chemistry, Bioactivity and Applications. *American Chemical Society, Washington*, pp. 200–207.

Abe, H., Honjo, C., Kyokawa, Y., Asakawa, S., Narsume, M., and Narushima, M. 1994. 3-Oxoteasterone and the epimerization of teasterone: identification in lily anthers and *Distylium racemosum* leaves and its biotransformation into typhasterol. *Biosci. Biotech. Biochem.*, **58**: 986–989.

Abe, H., Morishita, T., Uchiyama, M., Marumo, S., Munakata, K., Takatsuto, S., and Ikekawa, N. 1982. Identification of brassinolide-like substances in Chinese cabbage. *Agric. Biol. Chem.*, **46**: 2609–2611.

Abe, H., Morishita, T., Uchiyama, M., Takatsuto, S., and Ikekawa, N. 1984a. A new brassinolide-related steroid in the leaves of *Thea sinensis*. *Agric. Biol. Chem.*, **48**: 2171–2172.

Abe, H., Morishita, T., Uchiyama, M., Takatsuto, S., Ikekawa, N., Ikeda, M., Sassa, T., Kitsuwa, T., and Marumo, S. 1983. Occurrence of three new brassinosteroids: brassinone, 24(*S*)-24-ethylbrassinone and 28-norbrassinolide, in higher plants. *Experientia*, **39**: 351–353.

Abe, H., Nakamura, K., Morishita, T., Uchiyama, M., Takatsuto, S., and Ikekawa, N. 1984b. Endogenous brassinosteroids of the rice plant: castasterone and dolichosterone. *Agric. Biol. Chem.*, **48**: 1103–1104.

Abe, H., Soeno, K., Koseki, N.-N., and Natsume, M. 2001. Conjugated and unconjugated brassinosteroids. *In* DR Baker, NK Umetsu (Eds.) Agrochemical Discovery. Insect, Weed, and Fungal Control. *American Chemical Society, Washington*, pp. 91–101.

Abe, H., Takatsuto, S., Nakayama, M., and Yokota, T. 1995a. 28-Homotyphasterol, a new natural brassinosteroid from rice (*Oryza sativa* L.) bran. *Biosci. Biotech. Biochem.*, **59**: 176–178.

Abe, H., Takatsuto, S., Okuda, R., and Yokota, T. 1995b. Identification of castasterone, 6-deoxocastasterone, and typhasterol in the pollen of *Robinia pseudo-acacia* L. *Biosci. Biotech. Biochem.*, **59**: 309–310.

Antonchick, A., Svatoš, A., Schneider, B., Konstantinova, O.V., Zhabinskii, V.N., and Khripach, V.A. 2005. 2,3-Epoxybrassinosteroids are intermediates in the biosynthesis of castasterone in seedlings of *Secale cereale*. *Phytochemistry*, **66**: 65–72.

Antonchick, A.P., Schneider, B., Zhabinskii, V.N., Konstantinova, O.V., and Khripach, V.A. 2003. Biosynthesis of 2,3-epoxybrassinosteroids in seedlings of *Secale cereale*. *Phytochemistry*, **63**: 771–776.

Arima, M., Yokota, T., and Takahashi, N. 1984. Identification and quantification of brassinolide-related steroids in the insect gall and healthy tissues of the chestnut plant. *Phytochemistry*, **23**: 1587–1591.

Asakawa, S., Abe, H., Kyokawa, Y., Nakamura, S., and Natsume, M. 1994. Teasterone 3-myristate: a new type of brassinosteroid derivative in *Lilium longiflorum* anthers. *Biosci. Biotech. Biochem.*, **58**: 219–220.

Asakawa, S., Abe, H., Nishikawa, N., Natsume, M., and Koshioka, M. 1996. Purification and identification of new acyl-conjugated teasterones in lilly pollen. *Biosci. Biotech. Biochem.*, **60**: 1416–1420.

Baba, J., Yokota, T., and Takahashi, N. 1983. Brassinolide-related new bioactive steroids from *Dolichos lablab* seed. *Agric. Biol. Chem.*, **47**: 659–661.

Bajguz, A., and Hayat, S. 2009. Effects of brassinosteroids on the plant responses to environmental stresses. *Plant Physiol. Biochem.*, **47**: 1–8.

Bajguz, A., and Tretyn, A. 2003. The chemical characteristic and distribution of brassinosteroids in plants. *Phytochemistry*, **62**: 1027–1046.

Bajguz, A. 2009. Isolation and characterization of brassinosteroids from algal cultures of *Chlorella vulgaris* Beijerinck (Trebouxiophycaea). *J. Plant Physiol.*, **166**: 1946–1949.

Bhardwaj, R., Kaur, S., Nagar, P.K., and Arora, H.K. 2007. Isolation and characterization of brassinosteroids from immature seeds of *Camellia sinensis* (O) Kuntze. *Plant Growth Regul.*, **53**: 1–5.

Bishop, G.J., Nomura, T., Yokota, T., Harrison, K., Noguchi, T., Fujioka, S., Takatsuto, S., Jones, J.D.G., and Kamiya, Y. 1999. The tomato DWARF enzyme catalyzes C-6 oxidation in brassinosteroid biosynthesis. *Proc. Natl. Acad. Sci. USA*, **96**: 1761–1766.

Choi, Y.-H., Fujioka, S., Harada, A., Yokota, T., Takatsuto, S., and Sakurai, A. 1996. A brassinolide biosynthetic pathway via 6-deoxocastasterone. *Phytochemistry*, **43**: 593–596.

Choi, Y.-H., Fujioka, S., Nomura, T., Harada, A., Yokota, T., Takatsuto, S., and Sakurai, A. 1997. An alternative brassinolide biosynthetic pathway via late C-6 oxidation. *Phytochemistry*, **44**: 609–613.

Choi, Y.-H., Inoue, T., Fujioka, S., Saimoto, H., and Sakurai, A. 1993. Identification of brassinosteroid-like active substances in plant-cell cultures. *Biosci. Biotech. Biochem.*, **57**: 860–861.

Friebe, A., Volz, A., Schmidt, J., Voigt, B., Adam, G., and Schnabl, H. 1999. 24-*Epi*-secasterone and 24-*epi*-castasterone from *Lychnis viscaria* seeds. *Phytochemistry*, **52**: 1607–1610.

Fujioka, S. 1999. Natural occurrence of brassinosteroids in the plant kingdom. *In* A Sakurai, T Yokota, SD Clouse (Eds.) Brassinosteroids: Steroidal Plant Hormones. *Springer-Verlag, Tokyo*, pp. 21–45.

Fujioka, S., Choi, Y.-H., Takatsuto, S., Yokota, T., Li, J., Chory, J., and Sakurai, A. 1996. Identification of castasterone, typhasterol, and 6-deoxotyphasterol from the shoots of *Arabidopsis thaliana*. *Plant Cell Physiol.*, **37**: 1201–1203.

Fujioka, S., Inoue, T., Takatsuto, S., Yanagisawa, T., Sakurai, A., and Yokota, T. 1995. Identification of a new brassinosteroid, cathasterone, in cultured cells of *Catharanthus roseus* as a biosynthetic precursor of teasterone. *Biosci. Biotech. Biochem.*, **59**: 1543–1547.

Fujioka, S., Li, J., Choi, Y.-H., Seto, H., Takatsuto, S., Noguchi, T., Watanabe, T., Kuriyama, H., Yokota, T., Chory, J., and Sakurai, A. 1997. The *Arabidopsis deetiolated2* mutant is blocked early in brassinosteroid biosynthesis. *Plant Cell*, **9**: 1951–1962.

Fujioka, S., Noguchi, T., Sekimoto, M., Takatsuto, S., and Yoshida, S. 2000a. 28-Norcastasterone is biosynthesized from castasterone. *Phytochemistry*, **55**: 97–101.

Fujioka, S., Noguchi, T., Watanabe, T., Takatsuto, S., and Yoshida, S. 2000b. Biosynthesis of brassinosteroids in cultured cells of *Catharanthus roseus*. *Phytochemistry*, **53**: 549–553.

Fujioka, S., Noguchi, T., Yokota, T., Takatsuto, S., and Yoshida, S. 1998. Brassinosteroids in *Arabidopsis thaliana*. *Phytochemistry*, **48**: 595–599.

Fukuta, N., Fujioka, S., Takatsuto, S., Yoshida, S., Fukuta, Y., and Nakayama, M. 2004. 'Rinrei', a brassinosteroid-deficient dwarf mutant of faba bean (*Vicia faba*). *Physiol. Plant.*, **121**: 506–512.

Gamoh, K., Okamoto, N., Takatsuto, S., and Tejima, I. 1990. Determination of traces of natural brassinosteroids as dansylaminophenylboronates by liquid chromatography with fluorimetric detection. *Anal. Chim. Acta*, **228**: 101–105.

Gamoh, K., Omote, K., Okamoto, N., and Takatsuto, S. 1989. High-performance liquid chromatography of brassinosteroids in plants with derivatization using 9-phenanthreneboronic acid. *J. Chromatogr.*, **469**: 424–428.

Griffiths, P.G., Sasse, J.M., Yokota, T., and Cameron, D.W. 1995. 6-Deoxotyphasterol and 3-dehydro-6-deoxoteasterone, possible precursors to brassinosteroids in pollen of *Cupressus arizonica*. *Biosci. Biotech. Biochem.*, **59**: 956–959.

Grove, M.D., Spencer, G.F., Rohwedder, W.K., Mandava, N., Worley, J.F., Warthen Jr, J.D., Steffens, G.L., Flippen-Anderson, J.L., and Cook Jr, J.C. 1979. Brassinolide, a plant growth-promoting steroid isolated from *Brassica napus* pollen. *Nature*, **281**: 216–217.

Gupta, D., Bhardwaj, R., Nagar, P.K., and Kaur, S. 2004. Isolation and characterization of brassinosteroids from leaves of *Camellia sinensis* (L.) O. Kuntze. *Plant Growth Regul.*, **43**: 97–100.

Hamdy, A.-H.A., Aboutabl, E.A., Sameer, S., Hussein, A.A., Díaz-Marrero, A.R., Darias, J., and Cueto, M. 2009. 3-Keto-22-epi-28-nor-cathasterone, a brassinosteroid-related metabolite from *Cystoseira myrica*. *Steroids*, **74**: 927–930.

Ikeda, M., Takatsuto, S., Sassa, T., Ikekawa, N., and Nukina, M. 1983. Identification of brassinolide and its analogues in chestnut gall tissue. *Agric. Biol. Chem.*, **47**: 655–657.

Ikekawa, N., Nishiyama, F., and Fujimoto, Y. 1988. Identification of 24-epibrassinolide in bee pollen of the broad bean, *Vicia faba* L. *Chem. Pharm. Bull.*, **36**: 405–407.

Ikekawa, N., Takatsuto, S., Kitsuwa, T., Saito, H., Morishita, T., and Abe, H. 1984. Analysis of natural brassinosteroids by gas chromatography and gas chromatography-mass spectrometry. *J. Chromatogr.*, **290**: 289–302.

Jang, M.-S., Han, K.-S., and Kim, S.-K. 2000. Identification of brassinosteroids and their biosynthetic precursors from seeds of pumpkin. *Bull. Korean Chem. Soc.* **21**: 161–164.

Katsumata, T., Hasegawa, A., Fujiwara, T., Komatsu, T., Notomi, N., Abe, H., Natsume, M., and Kawaide, H. 2008. Arabidopsis CYP85A2 catalyzes lactonization reactions in the biosynthesis of 2-deoxy-7-oxalactone brassinosteroids. *Biosci. Biotechnol. Biochem.*, **72**: 2110–2117.

Kim, S.-K. 1991. Natural occurrences of brassinosteroids. *In* HG Cutler, T Yokota, G Adam (Eds.) Brassinosteroids: Chemistry, Bioactivity and Applications. *American Chemical Society, Washington*, pp. 26–35.

Kim, S.-K., Abe, H., Little, C.H.A., and Pharis, R.P. 1990. Identification of two brassinosteroids from the cambial region of scots pine (*Pinus silvestris*) by gas chromatography-mass spectrometry, after detection using a dwarf rice lamina inclination bioassay. *Plant Physiol.*, **94**: 1709–1713.

Kim, S.-K., Akihisa, T., Tamura, T., Matsumoto, T., Yokota, T., and Takahashi, N. 1988. 24-Methylene-25-methylcholesterol in *Phaseolus vulgaris* seed: structural relation to brassinosteroids. *Phytochemistry*, **27**: 629–631.

Kim, S.-K., Chang, S.C., Lee, E.J., Chung, W.-S., Kim, Y.-S., Hwang, S., and Lee, J.S. 2000a. Involvement of brassinosteroids in the gravitropic response of primary root of maize. *Plant Physiol.*, **123**: 997–1004.

Kim, S.-K., Yokota, T., and Takahashi, N. 1987. 25-Methyldolichosterone, a new brassinosteroid with tertiary butyl group from immature seed of *Phaseolus vulgaris*. *Agric. Biol. Chem.*, **51**: 2303–2305.

Kim, T.-W., Park, S.-H., Han, K.-S., Choo, J., Lee, J.S., Hwang, S., and Kim, S.-K. 2000b. Occurrence of teasterone and typhasterol, and their enzymatic conversion in *Phaseolus vulgaris*. *Bull. Korean Chem. Soc.*, **21**: 373–374.

Kim, Y.-S., Kim, T.-W., and Kim, S.-K. 2005. Brassinosteroids are inherently biosynthesized in the primary roots of maize, *Zea mays* L. *Phytochemistry*, **66**: 1000–1006.

Kim, Y.-S., Kim, T.-W., Chang, S.C., Pharis, R.P., Lee, J.S., Han, T.-J., Takatsuto, S., Cheong, H., and Kim, S.-K. 2006. Regulation of castasterone level in primary roots of maize, *Zea mays*. *Physiol. Plant.*, **127**: 28–37.

Konstantinova, O.V., Antonchick, A.P., Oldham, N.J., Zhabinskii, V.N., Khripach, V.A., and Schneider, B. 2001. Analysis of underivatized brassinosteroids by HPLC/APCI-MS. Occurrence of 3-epibrassinolide in *Arabidopsis thaliana*. *Collect. Czech. Chem. Commun.*, **66**: 1729–1734.

Morishita, T., Abe, H., Uchiyama, M., Marumo, S., Takatsuto, S., and Ikekawa, N. 1983. Evidence for plant growth promoting brassinosteroids in leaves of *Thea sinensis*. *Phytochemistry*, **22**: 1051–1053.

Motegi, C., Takatsuto, S., and Gamoh, K. 1994. Identification of brassinolide and castasterone in the pollen of orange (*Citrus sinensis* Osbeck) by high-performance liquid chromatography. *J. Chromatogr. A*, **658**: 27–30.

Nomura, T., Kitasaka, Y., Takatsuto, S., Reid, J.B., Fukami, M., and Yokota, T. 1999. Brassinosteroid/sterol synthesis and plant growth as affected by *lka* and *lkb* mutations of pea. *Plant Physiol.*, **119**: 1517–1526.

Nomura, T., Nakayama, M., Reid, J.B., Takeuchi, Y., and Yokota, T. 1997. Blockage of brassinosteroid biosynthesis and sensitivity causes dwarfism in garden pea. *Plant Physiol.*, **113**: 31–37.

Nomura, T., Sato, T., Bishop, G.J., Kamiya, Y., Takatsuto, S., and Yokota, T. 2001. Accumulation of 6-deoxocathasterone and 6-deoxocastasterone in *Arabidopsis*, pea and tomato is suggestive of common rate-limiting steps in brassinosteroid biosynthesis. *Phytochemistry*, **57**: 171–178.

Pachthong, C., Supyen, D., Buddhasukh, D., and Jatisatienr, A. 2006. Isolation and characterization of brassinolide and castasterone in the pollen of pumpkin. *Chiang Mai J. Sci.*, **33**: 95–101.

Park, K.-H., Park, J.-D., Hyun, K.-H., Nakayama, M., and Yokota, T. 1994a. Brassinosteroids and monoglycerides in immature seeds of *Cassia tora* as the active principles in the rice lamina inclination bioassay. *Biosci. Biotech. Biochem.*, **58**: 1343–1344.

Park, K.-H., Park, J.-D., Hyun, K.-H., Nakayama, M., and Yokota, T. 1994b. Brassinosteroids and monoglycerides with brassinosteroid-like activity in immature seeds of *Oryza sativa* and *Perilla frutescens* and in cultured cells of *Nicotiana tabacum*. *Biosci. Biotech. Biochem.*, **58**: 2241–2243.

Park, K.-H., Saimoto, H., Nakagawa, S., Sakurai, A., Yokota, T., Takahashi, N., and Syono, K. 1989. Occurrence of brassinolide and castasterone in crown gall cells of *Catharanthus roseus*. *Agric. Biol. Chem.*, **53**: 805–811.

Park, K.-H., Yokota, T., Sakurai, A., and Takahashi, N. 1987. Occurrence of castasterone, brassinolide and methyl 4-chloroindole 3-acetate in immature *Vicia faba* seeds. *Agric. Biol. Chem.*, **54**: 3081–3086.

Park, S.C., Kim, T.-W., and Kim, S.-K. 2000. Identification of brassinosteroids with 24*R*-methyl in immature seeds of *Phaseolus vulgaris*. *Bull. Korean Chem. Soc.*, **21**: 1274–1276.

Park, S.-H., Han, K.-S., Kim, T.-W., Shim, J.-K., Takatsuto, S., Yokota, T., and Kim, S.-K. 1999. In vivo and in vitro conversion of teasterone to typhasterol in cultured cells of *Marchantia polymorpha*. *Plant Cell Physiol.*, **40**: 955–960.

Plattner, R.D., Taylor, S.L., and Grove, M.D. 1986. Detection of brassinolide and castasterone in *Alnus glutinosa* (European alder) by mass spectrometry/mass spectrometry. *J. Nat. Prod.*, **49**: 540–545.

Sasse, J.M. 2003. Physiological actions of brassinosteroids: an update. *J. Plant Growth Regul.*, **22**: 276–288.

Schmidt, J., Altmann, T., and Adam, G. 1997. Brassinosteroids from seeds of *Arabidopsis thaliana*. *Phytochemistry*, **45**: 1325–1327.

Schmidt, J., Böhme, F., and Adam, G. 1996. 24-Epibrassinolide from *Gypsophila perfoliata*. *Z. Naturforsch.*, **51c**: 897–899.

Schmidt, J., Himmelreich, U., and Adam, G. 1995a. Brassinosteroids, sterols and lup-20(29)-en-2α,3β,28-triol from *Rheum rhabarbarum*. *Phytochemistry*, **40**: 527–531.

Schmidt, J., Kuhnt, C., and Adam, G. 1994. Brassinosteroids and sterols from seeds of *Beta vulgaris*. *Phytochemistry*, **36**: 175–177.

Schmidt, J., Porzel, A., and Adam, G. 1998. Brassinosteroids and a pregnane glucoside from *Daucus carota*. *Phytochem. Anal.*, **9**: 14–20.

Schmidt, J., Spengler, B., Yokota, T., and Adam, G. 1993a. The co-occurrence of 24-epi-castasterone and castasterone in seeds of *Ornithopus sativus*. *Phytochemistry*, **32**: 1614–1615.

Schmidt, J., Spengler, B., Yokota, T., Nakayama, M., Takatsuto, S., Voigt, B., and Adam, G. 1995b. Secasterone, the first naturally occurring 2,3-epoxybrassinosteroid from *Secale cereale*. *Phytochemistry*, **38**: 1095–1097.

Schmidt, J., Voigt, B., and Adam, G. 1995c. 2-Deoxybrassinolide – a naturally occurring brassinosteroid from *Apium graveolens*. *Phytochemistry*, **40**: 1041–1043.

Schmidt, J., Yokota, T., Adam, G., and Takahashi, N. 1991. Castasterone and brassinolide in *Raphanus sativus* seeds. *Phytochemistry*, **30**: 364–365.

Schmidt, J., Yokota, T., Spengler, B., and Adam, G. 1993b. 28-Homoteasterone, a naturally occurring brassinosteroid from seeds of *Raphanus sativus*. *Phytochemistry*, **34**: 391–392.

Schneider, B. 2002. Pathways and enzymes of brassinosteroid biosynthesis. *Progress Bot.*, **63**: 286–306.

Schneider, J.A. Yoshihara, K., Nakanishi, K., and Kato, N., 1983. Typhasterol (2-deoxycastasterone): a new plant growth regulator from cat-tail pollen. *Tetrahedron Lett.*, **24**: 3859–3860.

Shimada, K., Abe, H., Takatsuto, S., Nakayama, M., and Yokota, T. 1996. Identification of castasterone and teasterone from seeds of canary grass (*Phalaris canariensis*). *Rec. Res. Dev. Chem. Pharm. Sci.*, **1**: 1–5.

Soeno, K., Kyokawa, Y., Natsume, M., and Abe, H. 2000. Teasterone-3-*O*-β-D-glucopyranoside, a new conjugated brassinosteroid metabolite from lily cell suspension cultures and its identification in lily anthers. *Biosci. Biotechnol. Biochem.*, **64**: 702–709.

Sondhi, N., Bhardwaj, R., Kaur, S., Kumar, N., and Singh, B. 2008. Isolation of 24-epibrassinolide from leaves of *Aegle marmelos* and evaluation of its antigenotoxicity employing *Allium cepa* chromosomal aberration assay. *Plant Growth Regul.*, **54**: 217–224.

Spengler, B., Schmidt, J., Voigt, B., and Adam, G. 1995. 6-Deoxo-28-norcastasterone and 6-deoxo-24-epicastasterone – two new brassinosteroids from *Ornithopus sativus*. *Phytochemistry*, **40**: 907–910.

Suzuki, H., Fujioka, S., Takatsuto, S., Yokota, T., Murofushi, N., and Sakurai, A. 1993. Biosynthesis of brassinolide from castasterone in cultured cells of *Catharanthus roseus*. *J. Plant Growth Regul.*, **12**: 101–106.

Suzuki, H., Fujioka, S., Takatsuto, S., Yokota, T., Murofushi, N., and Sakurai, A. 1994a. Biosynthesis of brassinolide from teasterone via typhasterol and castasterone in cultured cells of *Catharanthus roseus*. *J. Plant Growth Regul.*, **13**: 21–26.

Suzuki, H., Fujioka, S., Takatsuto, S., Yokota, T., Murofushi, N., and Sakurai, A. 1995. Biosynthesis of brassinosteroids in seedlings of *Catharanthus roseus, Nicotiana tabacum* and *Oryza sativa*. *Biosci. Biotech. Biochem.* **59**: 168–172.

Suzuki, H., Fujioka, S., Yokota, T., Murofushi, N., and Sakurai, A. 1994b. Identification of brassinolide, castasterone, typhasterol, and teasterone from the pollen of *Lilium elegans*. *Biosci. Biotech. Biochem.*, **58**: 2075–2076.

Suzuki, H., Inoue, T., Fujioka, S., Takatsuto, S., Yanagisawa, T., Yokota, T., Murofushi, N., and Sakurai, A. 1994c. Possible involvement of 3-dehydroteasterone in the conversion of teasterone to typhasterol in cultured cells of *Catharanthus roseus*. *Biosci. Biotech. Biochem.*, **58**: 1186–1188.

Suzuki, Y., Yamaguchi, I., and Takahashi, N. 1985. Identification of castasterone and brassinone from immature seeds of *Pharbitis purpurea*. *Agric. Biol. Chem.*, **49**: 49–54.

Suzuki, Y., Yamaguchi, I., Yokota, T., and Takahashi, N. 1986. Identification of castasterone, typhasterol and teasterone from the pollen of *Zea mays*. *Agric. Biol. Chem.*, **50**: 3133–3138.

Takatsuto, S. 1994. Brassinosteroids: distribution in plants, bioassays and microanalysis by gas chromatography – mass spectrometry. *J. Chromatogr. A*, **658**: 3–15.

Takatsuto, S., Abe, H., and Gamoh, K. 1990a. Evidence for brassinosteroids in strobilus of *Equisetum arvense* L. *Agric. Biol. Chem.*, **54**: 1057–1059.

Takatsuto, S., Abe, H., Shimada, K., Nakayama, M., and Yokota, T. 1996a. Identification of teasterone and 4-desmethylsterols in the seeds of *Ginkgo biloba* L. *J. Jpn. Oil Chem. Soc.*, **45**: 1349–1351.

Takatsuto, S., Abe, H., Yokota, T., Shimada, K., and Gamoh, K. 1996b. Identification of castasterone and teasterone in seeds of *Cannabis sativa* L. *J. Jpn. Oil Chem. Soc.*, **45**: 871–873.

Takatsuto, S., Omote, K., Gamoh, K., and Ishibashi, M. 1990b. Identification of brassinolide and castasterone in buckwheat (*Fagopyrum esculentum* Moench) pollen. *Agric. Biol. Chem.*, **54**: 757–762.

Takatsuto, S., Yokota, T., Omote, K., Gamoh, K., and Takahashi, N. 1989. Identification of brassinolide, castasterone and norcastasterone (brassinone) in sunflower (*Helianthus annuus* L.) pollen. *Agric. Biol. Chem.*, **53**: 2177–2180.

Taylor, P.E., Spuck, K., Smith, P.M., Sasse, J.M., Yokota, T., Griffiths, P.G., and Cameron, D.W. 1993. Detection of brassinosteroids in pollen of *Lolium perenne* L. by immunocytochemistry. *Planta*, **189**: 91–100.

Watanabe, T., Yokota, T., Shibata, K., Nomura, T., Seto, H., and Takatsuto, S. 2000. Cryptolide, a new brassinolide catabolite with a 23-oxo group from Japanese cedar pollen/anther and its synthesis. *J. Chem. Res.*, **2000(1)**: 18–19.

Yamamoto, R., Fujioka, S., Demura, T., Takatsuto, S., Yoshida, S., and Fukuda, H. 2001. Brassinosteroid levels increase drastically prior to morphogenesis of tracheary elements. *Plant Physiol.*, **125**: 556–563.

Yamamoto, R., Fujioka, S., Iwamoto, K., Demura, T., Takatsuto, S., Yoshida, S., and Fukuda, H. 2007. Co-regulation of brassinosteroid biosynthesis-related genes during xylem cell differentiation. *Plant Cell Physiol.*, **48**: 74–83.

Yasuta, E., Terahata, T., Nakayama, M., Abe, H., Takatsuto, S., and Yokota, T. 1995. Free and conjugated brassinosteroids in the pollen and anthers of *Erythronium japonicum* Decne. *Biosci. Biotech. Biochem.*, **59**: 2156–2158.

Yokota, T. 1995. Chemical structures of plant hormones. *Chem. Regul. Plants*, **30**: 185–196.

Yokota, T. 1997. The structure, biosynthesis and function of brassinosteroids. *Trends Plant Sci.*, **2**: 137–143.

Yokota, T. 1999. Brassinosteroids. *In* PJJ Hooykaas, MA Hall, KR Libbenga (Eds.) Biochemistry and Molecular Biology of Plant Hormones. *Elsevier Science, London*, pp. 277–293.

Yokota, T., Arima, M., and Takahashi, N. 1982a. Castasterone, a new phytosterol with plant-hormone potency, from chestnut insect gall. *Tetrahedron Lett.*, **23**: 1275–1278.

Yokota, T., Arima, M., Takahashi, N., and Crozier, A. 1985. Steroidal plant growth regulators, castasterone and typhasterol (2-deoxycastasterone) from the shoots of Sitka spruce (*Picea sitchensis*). *Phytochemistry*, **24**: 1333–1335.

Yokota, T., Arima, M., Takahashi, N., Takatsuto, S., Ikekawa, N., and Takematsu, T. 1983a. 2-Deoxycastasterone, a new brassinolide-related bioactive steroid from *Pinus* pollen. *Agric. Biol. Chem.*, **47**: 2419–2420.

Yokota, T., Baba, J., and Takahashi, N. 1982b. A new steroidal lactone with plant growth-regulatory activity from *Dolichos lablab* seed. *Tetrahedron Lett.*, **23**: 4965–4966.

Yokota, T., Baba, J., and Takahashi, N. 1983b. Brassinolide-related bioactive sterols in *Dolichos lablab*: brassinolide, castasterone and a new analog, homodolicholide. *Agric. Biol. Chem.*, **47**: 1409–1411.

Yokota, T., Baba, J., Koba, S., and Takahashi, N. 1984. Purification and separation of eight steroidal plant-growth regulators from *Dolichos lablab* seed. *Agric. Biol. Chem.*, **48**: 2529–2534.

Yokota, T., Higuchi, K., Takahashi, N., Kamuro, Y., Watanabe T., and Takatsuto, S. 1998. Identification of brassinosteroids with epimerized substituents and / or the 23-oxo group in pollen and anthers of Japanese cedar. *Biosci. Biotechnol. Biochem.*, **62**: 526–531.

Yokota, T., Kim, S.K., Fukui, Y., Takahashi, N., Takeuchi, Y., and Takematsu, T. 1987a. Conjugation of brassinosteroids. *In* K Schreiber, HR Schuette, G Sembdner (Eds.) Conjugated Plant Hormones. Structure, Metabolism and Function. *VED Deutscher Verlag der Wissenschaften, Berlin*, pp. 288–296.

Yokota, T., Kim, S.K., Fukui, Y., Takahashi, N., Takeuchi, Y., and Takematsu, T. 1987b. Brassinosteroids and sterols from a green alga, *Hydrodictyon reticulatum*: configuration at C-24. *Phytochemistry*, **26**: 503–506.

Yokota, T., Koba, S., Kim, S.K., Takatsuto, S., Ikekawa, N., Sakakibara, M., Okada, K., Mori, K., and Takahashi, N. 1987c. Diverse structural variations of the brassinosteroids in *Phaseolus vulgaris*. *Agric. Biol. Chem.*, **51**: 1625–1631.

Yokota, T., Matsuoka, T., Koarai, T., and Nakayama, M. 1996. 2-Deoxybrassinolide, a brassinosteroid from *Pisum sativum* seed. *Phytochemistry*, **42**: 509–511.

Yokota, T., Morita, M., and Takahashi, N. 1983c. 6-Deoxocastasterone and 6-deoxodolichosterone: putative precursors for brassinolide-related steroids from *Phaseolus vulgaris*. *Agric. Biol. Chem.*, **47**: 2149–2151.

Yokota, T., Nakayama, M., Wakisaka, T., Schmidt, J., and Adam, G. 1994. 3-Dehydroteasterone, a 3,6-diketobrassinosteroid as a possible biosynthetic intermediate of brassinolide from wheat grain. *Biosci. Biotech. Biochem.*, **58**: 1183–1185.

Yokota, T., Nomura, T., and Nakayama, M. 1997d. Identification of brassinosteroids that appear to be derived from campesterol and cholesterol in tomato shoots. *Plant Cell Physiol.*, **38**: 1291–1294.

Yokota, T., Ogino, Y., Takahashi, N., Saimoto, H., Fujioka, S., and Sakurai, A. 1990. Brassinolide is biosynthesized from castasterone in *Catharanthus roseus* crown gall cells. *Agric. Biol. Chem.*, **54**: 1107–1108.

Yokota, T., Sato, T., Takeuchi, Y., Nomura, T., Uno, K., Watanabe, T., and Takatsuto, S. 2001. Roots and shoots of tomato produce 6-deoxo-28-cathasterone, 6-deoxo-28-nortyphasterol and 6-deoxo-28-norcastasterone, possible precursors of 28-norcastasterone. *Phytochemistry*, **58**: 233–238.

Zaki, A.K., Schmidt, J., Hammouda, F.M., and Adam, G. 1993. Steroidal constituents from pollen grains of *Phoenix dactylifera*. *Planta Med. (Suppl.)*, **59**: A 613.

Chapter 2

REGULATORY MECHANISMS OF BRASSINOSTEROID SIGNALING IN PLANTS

M.-C. CODREANU[1,2] AND E. RUSSINOVA[1,2]
[1]*Department of Plant Systems Biology, VIB, 9052 Ghent, Belgium;* [2]*Department of Plant Biotechnology and Genetics, Ghent University, Technologiepark 927, 9052 Ghent, Belgium*

Abstract: Brassinosteroids (BRs) are steroidal hormones essential for plant growth and development. Over the past decade, genetic, molecular, and proteomic studies have established the complete BR signaling pathway and revealed important regulatory mechanisms. In *Arabidopsis thaliana*, BRs are perceived by receptor kinases that transduce the signal from the cell surface to the nucleus by an intracellular cascade of protein-protein interactions, involving kinases, phosphatases, 14-3-3 proteins, and nuclear transcription factors. In addition, the BR signaling is regulated by the plant endocytic machinery because the increased endosomal localization of the BR receptor enhances the signaling. As several counterparts of the BR signaling proteins of *Arabidopsis* are found in rice (*Oryza sativa*), the BR signaling is apparently conserved between monocotyledonous and dicotyledonous plants. In this chapter, we discuss the currently available data on BR signaling pathway and the latest findings on BR signaling regulatory mechanisms in plants.

Key words: *Arabidopsis*, brassinosteroids, receptors, BRI1, signal transduction, brassinosteroid signaling

1. INTRODUCTION

Brassinosteroids (BRs) are plant-specific steroidal hormones with strong growth-promoting properties. Since the first discovery of the most active BR, called brassinolide (BL) (Grove *et al.*, 1979), more than 50 BL analogs have been identified throughout the plant kingdom (Fujioka and Yokota, 2003). BRs

S. Hayat and A. Ahmad (eds.), *Brassinosteroids: A Class of Plant Hormone*,
DOI 10.1007/978-94-007-0189-2_2, © Springer Science+Business Media B.V. 2011

act in multiple processes throughout the plant life cycle, including germination, root and stem elongation, seedling photomorphogenesis, vascular different-tiation, male fertility, timing of senescence and flowering, and resistance to biotic and abiotic stresses (Clouse and Sasse, 1998). The discovery of BR-deficient *Arabidopsis thaliana* mutants as severe dwarfs displaying a light-grown morphology in the dark that could be reverted to a wild type by the exogenous application of BL (Chory *et al.*, 1991; Szekeres *et al.*, 1996), established BRs as essential plant hormones. Concurrently, the BR biosynthetic pathway was elucidated through chemical analysis and the isolation of additional BR-biosynthetic mutants, defective in genes encoding proteins that catalyze the conversion of plant steroids to BR precursors (Asami *et al.*, 2005). Forward genetics screens for BR mutants showing dwarf phenotypes similar to those of the BR-deficient mutants, but BR insensitive, led to the identification of the BR perception mutants in *Arabidopsis* (Clouse *et al.*, 1996; Kauschmann *et al.*, 1996) and the isolation of the BR receptor, the leucine rich-repeat (LRR) receptor-like kinase (RLK), BRASSINOSTEROID-IN-SENSITIVE 1 (BRI1) (Li and Chory, 1997). Another powerful tool for the elucidation of the BR signaling pathway was the discovery of the first BR biosynthesis inhibitor, brassinazole that induced a BR-deficient phenotype in *Arabidopsis* (Asami *et al.*, 2000). In genetic screens for mutants conferring resistance to brassinazole, new BR signal transduction components were isolated (Wang *et al.*, 2001, 2002; Yin *et al.*, 2002). Over the past decade, a combination of genetic, genomic, and proteomic approaches have established the complete BR signaling pathway, providing important insights into the mechanisms of receptor activation and regulation of downstream components by phosphorylation (Tang *et al.*, 2010). In addition to the completion of the major gap in the BR signaling by understanding how BR signals are transmitted from the plasma membrane to the nucleus in *Arabidopsis*, the subcellular com-partmentalization and trafficking of BR receptor complexes was introduced as an integral part of the BR signal transduction (Irani and Russinova, 2009).

In this chapter, we overview the most recent advances in the study of the BR signaling pathway in *Arabidopsis* and its regulatory mechanisms. The interest in this pathway as tool to modify growth of important crops has led to the discovery of several downstream BR signaling components in rice (*Oryza sativa*), suggesting that the BR signaling pathway is conserved between monocotyledonous and dicotyledonous plants.

2. THE BR SIGNALING PATHWAY

Early extensive genetic screens for loss-of-function BR signaling mutants in *Arabidopsis* identified only one locus, *BRI1* that encodes an LRR RLK (Clouse *et al.*, 1996; Kauschmann *et al.*, 1996; Li and Chory, 1997; Noguchi *et al.*, 1999; Friedrichsen *et al.*, 2000). The phenotypes of the *bri1* mutants are identical to those of the BR-deficient mutants, but they cannot be rescued by the addition of BRs. Several other components of the BR signal transduction pathway have been characterized in additional suppressor and gain-of-function screens, including a second LRR RLK, the BRI1-ASSOCIATED RECEPTOR KINASE-1 (BAK1) (Li *et al.*, 2002); the serine/-carboxypeptidase BRI1 SUPPRESSOR-1 (BRS1) (Li *et al.*, 2001a), the GLYCOGEN SYNTHASE KINASE-3 (GSK3)-like kinase, BR INSENSITIVE-2 (BIN2) (Li *et al.*, 2001b; Li and Nam, 2002), the phosphatase BRI1 SUPPRESSOR-1 (BSU1) (Mora-García *et al.*, 2004), and two transcription factors, brassinazole-resistant 1 (BZR1) (Wang *et al.*, 2002) and BZR2/BRI1-EMS suppressor 1 (BES1) (Yin *et al.*, 2002). Later, in yeast two-hybrid screens, the transthyretin-like (TTL) protein (Nam and Li, 2004), the BRI1 kinase inhibitor-1 (BKI1) (Wang and Chory, 2006), and the 14-3-3 proteins (Gampala *et al.*, 2007) were isolated

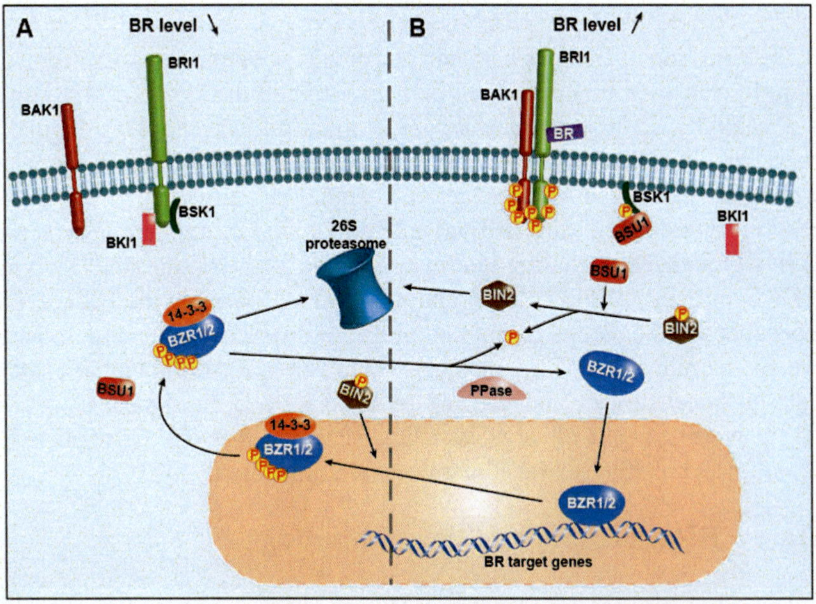

Figure 1. The BR signal transduction pathway in *Arabidopsis*. (A) Inactive BR pathway. In the absence of BRs, BRI1 is inactive and associates with BKI1. The BRI1-bound BSK1 and BSU1 are inactive, and consequently BIN2 is active. BIN2 phosphorylates BZR1 and BZR2/BES1, which cannot bind DNA and are retained in the cytoplasm by the 14-3-3 proteins to finally be degraded by the proteasome. (B) Active BR pathway. In the presence of BRs, BRI1 is activated through dissociation of BKI1 and oligomerization/transphosphorylation with BAK1. Activated BRI1 phosphorylates BSK1, which activates BSU1. The activated BSU1 inhibits BIN2 through dephosphorylation and, hence, BZR1 and BZR2/BES1 are dephosphorylated, possibly with the help of an unknown phosphatase (PPase). The unphosphorylated BZR1 and BZR2 accumulate in the nucleus and regulate the BR responses (Picture adapted from Tang *et al.*, 2010).

as part of the BR signaling cascade. Thanks to molecular and biochemical studies of these components, a model for the BR signaling pathway has been established that differs from that of mammals. Whereas animal steroids are perceived by nuclear receptors, influencing directly transcriptional activity (Thummel and Chory, 2002; Lösel and Wehling, 2003), in plants BRs bind plasma membrane-localized receptors and initiate a phosphorylation-mediated signaling cascade that affects novel transcription factors (Vert *et al.*, 2005; Belkhadir *et al.*, 2006). More recently, proteomics studies have refined the BR signaling model by identifying components not found in previous screens, such as the BR SIGNALING KINASEs (BSKs) that bridged the last gap in the genetically defined BR signaling pathway, generating a complete pathway from a RLK to transcription factors in plants (Tang *et al.*, 2008) (Figure 1).

3. UPSTREAM MECHANISMS OF THE BR SIGNALING REGULATION

Identification of BRI1 as a protein containing an extracellular LRR domain, a single transmembrane domain and an intracellular domain that consist of a catalytic serine/threonine kinase domain and the flanking regulatory sequences – the juxta-membrane (JM) and the carboxyl-terminal (CT) domains – suggested that BR hormones activate signaling similarly to the classic receptor tyrosine kinases (RTKs) and the transforming growth factor-β (TGF-β)-mediated signaling pathways in animals (Feng and Derynck, 1997; Schlessinger, 2000, 2002) and that BRI1 is possibly the BR receptor. Subsequently, the complete annotation of the *Arabidopsis* genome sequence revealed more than 600 RLK members (Shiu *et al.*, 2004), hinting at common signaling mechanisms, and that the plant RLKs and the signaling pathways they activate are part of large signaling networks that can be regulated by hormones and multiple extracellular cues (De Smet *et al.*, 2009).

3.1 BR receptors and ligand binding

The extracellular domain of BRI1 was initially predicted to contain a leucine-zipper motif, 25 tandem LRRs, two cysteine pairs before and after the LRRs, and a 70-amino-acid island inserted between LRR21 and LRR22 (Li and Chory, 1997). However, a recent reannotation reported that BRI1 had no leucine zipper and only 24 LRRs, with LRR21 being an unusual methionine-rich repeat (Vert *et al.*, 2005). Several mutations in the extra-cellular part of BRI1, more exactly on the island domain, determine a severe dwarf and male-sterile phenotype, revealing the importance of this domain for the BR signaling pathway (Li and Chory, 1997; Friedrichsen *et al.*, 2000).

A chimeric receptor, consisting on the extracellular LRR and transmembrane domains of BRI1, fused to the serine/threonine kinase domain of Xa21 (a rice LRR RLK for disease resistance), elicited defense responses in rice cells upon treatment with BL (He *et al.*, 2000) and the BR-binding activity co-immunoprecipitated with BRI1 (Wang *et al.*, 2001). Because LRR is a well-known protein-protein interaction domain, BRI1 might need a protein co-ligand to interact with BRs. Such a theory had previously been supported by the discovery of BRS1, a secreted serine/carboxypeptidase that, when overproduced, suppresses an extracellular *bri1* mutation without effect on an intracellular *bri1* allele (Li *et al.*, 2001a). Furthermore, at least one putative membrane steroid-binding protein (MSBP) is present in the *Arabidopsis* genome, raising the possibility that BRs might bind MSBPs that interact directly with BRI1 (Yang *et al.*, 2005). Recently, a biotin-tagged photoaffinity castasterone, a partially active biosynthetic precursor of BL, has been used to show that BRs directly bind to BRI1 in *Arabidopsis* and that BRI1 is the *bona fide* BR receptor (Kinoshita *et al.*, 2005). BR-binding assays with recombinant BRI1 fragments indicated that the minimal BR-binding region in BRI1 consists of the 70-amino-acid island domain and the CT-flanking LRR22. This domain is conserved in the two close BRI1 homologs that function as BR receptors in the vascular differentiation, BRI1-like 1 (BRL1) and BRI1-like 3 (BRL3) (Caño-Delgado *et al.*, 2004).

3.2 Activation of BRI1 by phosphorylation

Ligand binding to animal RTKs induces their dimerization, resulting in autophosphorylation of their cytoplasmic domains. The tyrosine autophosphorylation is not only essential for the intracellular kinase activation, but also crucial for recruitment and activation of various signaling proteins because most of the autophosphorylation sites are located in noncatalytic regions (Schlessinger, 2000, 2002). Although, unlike animal RTKs, plant RLKs belong to the group of serine/threonine kinases, they also undergo oligomerization, auto- and transphosphorylation upon ligand binding (Wang *et al.*, 2008). BRI1 is present in the plant membranes as a ligand-independent homodimer that is stabilized and activated by exogenous BR application, as demonstrated by fluorescent spectroscopy approaches and by co-immunoprecipitation experiments (Russinova *et al.*, 2004; Wang *et al.*, 2005a; Hink *et al.*, 2008). Both mobility shift assays of BRI1 in sodium dodecyl sulfate-polyacrylamide gel electrophoresis and increased signal detection by anti-phospho-serine/threonine antibodies in BR-treated samples showed that BRI1 kinase is activated by autophosphorylation (Wang *et al.*, 2001). Several phosphorylation sites of BRI1 have been identified by mass spectrometry analysis of recombinant BRI1 proteins produced in *Escherichia coli* or immunoprecipitated from plant extracts. The autophosphorylation sites of BRI1 kinase domain characterized *in vitro* (Oh *et al.*, 2000) were highly predictive for *in vivo* phosphorylation (Wang *et al.*, 2005b). In addition to the phosphorylated serine/threonine residues, the occurrence of *in vivo* tyrosine phosphorylation was also confirmed (Oh *et al.*, 2009), suggesting that BRI1 is a dual specificity kinase. At least three BR-stimulated phosphorylation sites in the activation loop of BRI1 (T1049, S1044, and T1045) are crucial for BRI1 activation because mutations in these residues completely inhibit the *in vivo* BRI1 function (Wang *et al.*, 2005b). Interestingly, whereas the deletion of the JM region inactivated BRI1, the deletion of the CT domain or phosphorylation mimicry mutations in the CT serine/threonine residues significantly increased the *in vivo* receptor activity of BRI1, suggesting an autoinhibitory role for the CT region (Wang *et al.*, 2005a). In contrast, loss-of-function mutations of several phosphorylated serine/threonine residues in the JM and CT domains had little effect on the BRI1 activity, but dramatically affected the phosphorylation of a peptide substrate *in vitro* (Wang *et al.*, 2005b), hinting at the importance of JM or CT domains for the BRI1 downstream signaling.

3.3 The function of BRI1 co-receptors

A second LRR RLK required for BR signaling and designated as BAK1 was identified simultaneously in a *bri1* suppressor screen (Li *et al.*, 2002) and in a yeast two-hybrid screen for BRI1 interactors (Nam and Li, 2002). Genetic analysis demonstrated a role for BAK1 in BR signaling because the overexpression of a dominant negative mutant allele of *BAK1* in a weak *bri1-5* caused a severe dwarf phenotype similar to that of the strong *bri1*. However, a loss-of-function mutation of *BAK1* resulted in a weak dwarf phenotype probably due to a gene redundancy (Nam and Li, 2002; Li *et al.*, 2002). The direct interaction of BAK1 with BRI1 was confirmed by several independent approaches (Nam and Li, 2002; Li *et al.*, 2002; Russinova *et al.*, 2004). BAK1 belongs to the small LRR RLK family of five SOMATIC EMBRYOGENESIS RECEPTOR-LIKE KINASEs (SERKs) in *Arabidopsis* that possesses a small extracellular domain consisting of only five LRRs (Hecht *et al.*, 2001). BAK1 itself is not required for the BR binding of BRI1 (Kinoshita *et al.*, 2005), but requires a ligand-dependent activation of its partner BRI1 for the formation of a heterotetramer (Wang *et al.*, 2005a; Karlova *et al.*, 2006). BAK1 is *trans*-phosphorylated on multiple residues *in vitro* and *in vivo* by BRI1 in the activation loop of the kinase domain (Wang *et al.*, 2008; Karlova *et al.*, 2009; Yun *et al.*, 2009) and, by contrast, activated BAK1 *trans*-phosphorylates BRI1 in the CT and JM domains, which enhances the BRI1 kinase activity and promotes it toward downstream substrates (Wang *et al.*, 2008). A recent proteomics approach revealed that another member of the SERK family, SERK1, could also interact with BRI1 and modulate BRI1 signaling in a manner similar to that observed for BAK1 (also known as SERK3) (Karlova *et al.*, 2006, 2009). A complementary genetic study uncovered that combinations of different SERK family members have multiple functions in distinct pathways, including flagellin/flagellin-sensitive2 (FLS2) signaling (Chinchilla *et al.*, 2007; Heese *et al.*, 2007), cell death (He *et al.*, 2007; Kemmerling *et al.*, 2007), and male fertility (Albrecht *et al.*, 2005; Colcombet *et al.*, 2005). Only SERK1, BAK1/SERK3, and SERK4/ BKK1 have a co-receptor function in BR signaling because they all interact with BRI1 and their loss-of-function mutations contribute additively to the BR-insensitive phenotype (Karlova *et al.*, 2006; He *et al.*, 2007; Albrecht *et al.*, 2008).

3.4 Other plasma membrane-associated BR signaling components

Another protein that regulates the signaling output of BRI1 was identified in a yeast two-hybrid screen and designated BRI1 KINASE INHIBITOR-1 (BKI1) (Wang and Chory, 2006). The sequence of BKI1 is similar to that of a rice protein previously known to be phosphorylated by the rice BRI1 homolog (Hirabayashi *et al.*, 2004). Overexpression of the *BKI1* gene provoked dwarfism and reduced BR sensitivity, but RNA interference-mediated BKI1 silencing resulted in a dose-dependent long-hypocotyl phenotype, suggesting that BKI1 functions as a negative regulator in the BR signaling. BKI1 localized both at the plasma membrane and the cytoplasm and BR treatment rapidly dissociated BKI1 from the plasma membrane. No BR-induced dissociation of BKI1 from the plasma membrane was observed in a *bri1* kinase–inactive mutant background, suggesting that the kinase activity of BRI1 is required for BKI1 re-localization. It is hypothesized that BKI1 interacts with BRI1 to prevent the association of the BR receptor with BAK1 and other BRI1 substrates, thus blocking the BR signaling output from the plasma membrane. Actually, BKI1 has been shown to interact directly with the kinase domain of BRI1, through the C-terminal domain of BKI1 (Wang and Chory, 2006).

In addition to BKI1, two other *Arabidopsis* proteins, TTL and a homolog of the TGF-β-receptor-interacting protein-1 (TRIP-1), were found to interact and to be phosphorylated by BRI1 *in vitro* and, therefore, also to play a role in regulating the signaling output from BRI1 (Jiang and Clouse, 2001; Nam and Li, 2004; Ehsan *et al.*, 2005; Wang and Chory, 2006). TTL is a plasma membrane-associated protein and a negative regulator of BR responses, but its exact function in BR signal transduction remains unclear.

TRIP-1 is an *Arabidopsis* homolog of the mammalian TGF-β receptor-interacting protein that in mammals also functions as a subunit of the eIF3 translation initiation factor (Jiang and Clouse, 2001). It contains two WD40 motifs and also binds to the UV-DAMAGED DNA BINDING PROTEIN 1a in yeast two-hybrid and in co-immunoprecipitation assays (Zhang *et al.*, 2008). Although BRI1, but not BAK1, can phosphorylate and interact with TRIP-1 *in vitro* (Ehsan *et al.*, 2005), a clear function of TRIP-1 in BR signaling has not been demonstrated and it remains to be verified whether it is involved in the ubiquitin proteasome-mediated degradation of the BR receptor complex as a part of the CULLIN4-based E3 ubiquitin ligase.

Despite the identification of several upstream BR signaling components, the substrates of the BRI1 kinase that transduce the signal to downstream components remained unknown until recently. Of late, proteomic studies of plasma membrane proteins facilitated the identification of the homologous BSK1, BSK2, and BSK3-proteins (Tang *et al.*, 2008) that are members of the receptor-like cytoplasmic kinase subfamily RLCK-XII of 12 proteins. Each BSK contains one kinase domain at the N terminus and tetratricopeptide repeat domains at the C terminus that are known to mediate protein-protein interactions and to be present in components of steroid receptor complexes in animals. Consistent with their identification in the plasma membrane fractions, BSK1 proteins were localized at the cell surface without impact by the BL treatment. The bimolecular fluorescence complementation and co-immunoprecipitation assays revealed that BSK1 interacts with BRI1. BSK1, BSK2 and BSK3 function immediately downstream of BRI1 because when overexpressed they suppress the dwarf phenotypes of the biosynthetic (*det2-1*) and signaling (*bri1-5* and *bri1-116*) mutants, but had no effect on another downstream signaling mutant *bin2-1*. BSK1 is also a BRI1 kinase substrate that is phosphorylated upon BR activation of BRI1. BSKs might play roles in BR signaling distinct from those of BAK1 because BRs induce BRI1-BAK1 interactions, but reduce those of BRI1-BSK3 (Tang *et al.*, 2008). Identification of the direct downstream targets of all BSKs will be essential to fully understand how they transduce the BR signal.

4. REGULATION OF BR SIGNALING BY DOWNSTREAM COMPONENTS

4.1 The function of the GSK3/shaggy-like kinases in BR signaling

Three isoforms of highly conserved serine/threonine GSK3/shaggy-like kinases are implicated in several signaling pathways in mammals, including

Wnt, Hedgehog and insulin signaling, mitosis, and apoptosis (Kockeritz *et al.*, 2006; Patel *et al.*, 2008; Jin *et al.*, 2009b; Kim *et al.*, 2009b). In contrast, in plants, the set of divergent GSK3 proteins is much larger; for example, analysis of the *Arabidopsis* genome revealed the existence of 10 GSK3 members divided in four groups and annotated as *Arabidopsis* SHAGGY-like kinases (ASKs) (Jonak and Hirt, 2002). Genetic screens in *Arabidopsis* for BR-insensitive and new leaf mutants led to the isolation of eight gain-of-function *BR-insensitive2* (*bin2*)/*dwarf12*/*ultracurvata1* (*ucu1*) alleles (Li *et al.*, 2001b; Choe *et al.*, 2002; Pérez-Pérez *et al.*, 2002). Seven *bin2* alleles carried gain-of-function mutations in the highly conserved TREE motif within the catalytic domain of the BIN2/ASK21 (Choe *et al.*, 2002; Li and Nam, 2002). In addition, overexpression of the *BIN2* gene inhibited BR signaling, resulting in plants that resembled *bri1*, whereas reduced *BIN2* expression partially rescued a weak *bri1* mutation, suggesting that BIN2 functions as negative regulator of the BR signaling (Li and Nam, 2002). No single *bin2* loss-of-function mutation was discovered in any forward genetic screen in *Arabidopsis*, implying that the *Arabidopsis* GSK3-like kinases function redundantly. Only in a recent suppressor screen of the weak *ucu1* phenotype, eight new loss-of-function *bin2* alleles were identified (Yan *et al.*, 2009). Genetic analysis of T-DNA insertional mutants of BIN2/ASK21 and its closed homologs from group II, ASK22 and ASK23, revealed that they indeed function redundantly to negatively regulate BR signaling (Vert and Chory, 2006), but that BIN2 plays a dominant role (Yan *et al.*, 2009). Recently, chemical genomics has demonstrated its power in overcoming gene redundancy in plants (Hicks and Raikhel, 2009). In a phenotype-based chemical screen for compounds that induce BR responses, a small molecule, designated bikinin had been characterized that activated the BR signaling downstream of the BR receptor (De Rybel *et al.*, 2009). Bikinin directly binds BIN2 by acting as an ATP competitor and also inhibits the activity of six other GSK3 proteins of *Arabidopsis*, all ASKs of groups I and II, and one from group III. The genome-wide transcript analysis of the early effect of BL and bikinin demonstrated that they have largely overlapping responses, suggesting that the inhibition of seven ASKs is the sole activation mode of BR signaling and argues against GSK3-independent BR responses in *Arabidopsis* (De Rybel *et al.*, 2009). These findings were also confirmed by genetic and molecular studies, revealing that indeed, besides group-II ASKs, members of group-I ASKs, for example ASK12, also negatively regulate the BR signaling (Kim *et al.*, 2009a). However, the significance of multiple ASKs functioning in the BR signaling pathways remains to be determined. The substrate specificity of the different plant GSK3 homologs has not been addressed because until now only a limited number of substrates have been identified. A strong evidence that BIN2 regulates the auxin response factor 2

(ARF2) by phosphorylation was provided (Vert *et al.*, 2008), implying a direct link between auxin and BR signaling at the level of the GSK3 proteins in plants.

4.2 Mechanisms of BIN2 regulation

Because BIN2 is a major BR signaling component, assignment of regulation mechanisms is essential for understanding the downstream events in this pathway. The animal GSK3 is a constitutively active kinase that phosphorylates a variety of protein substrates, including transcription factors and cytoskeleton proteins, but becomes inactive in response to various stimuli (Jope and Johnson, 2004). The activity of the animal GSK3 is regulated by phosphorylation, protein complex formation, ubiquitin/proteasome-mediated protein degradation, and subcellular distribution (Failor *et al.*, 2007; Forde and Dale, 2007; Meares and Jope, 2007). Many substrates of the animal GSK3 must be first phosphorylated by another protein kinase at priming phosphorylation sites. The kinase activity of GSK3 is inhibited through phosphorylation of a serine residue, located at the N terminus of the protein that acts as a pseudosubstrate preventing the binding of the primed substrates and, thus, keeping the GSK3 inactive (Jope and Johnson, 2004). Genetic data suggested that BIN2 is a constitutively active kinase that is inactivated in response to BR treatment (Li and Nam, 2002). The ASK sequences differ substantially from the animal GSK3 sequences in the length of the N-terminal domains that is the shortest in BIN2 (Choe *et al.*, 2002). As all ASKs also lack the conserved inhibitory serine residue, the regulation of the plant GSK3 proteins might differ from that of their animal counterparts. The N-terminal inhibitory domain and the primed phosphorylation are not essential for the activity of BIN2 (Zhao *et al.*, 2002), but instead the phosphorylation of the highly conserved TREE domain could have an inhibitory effect, or alternatively, define an interaction with another protein (Choe *et al.*, 2002; Li and Nam, 2002).

How upstream BR signaling regulates BIN2 has been an outstanding question until recently although previous studies have shown that a different subcellular localization (Vert and Chory, 2006) and proteasome-mediated degradation of BIN2 (Peng *et al.*, 2008) might be important mechanisms to regulate its activity. Revision of the previously proposed function of another downstream BR signaling component, a Kelch-repeat-containing phosphatase, BSU1 (Mora-García *et al.*, 2004), elucidated the mechanism of the BIN2 inactivation in response to BRs (Kim *et al.*, 2009a). BSU1 was identified as an activation-tagged suppressor of the weak *bri1-5* allele and originally proposed to act downstream of BIN2 (Mora-García *et al.*, 2004). Recently its function was revised and BSU1 was shown to dephosphorylate all ASKs functioning in the BR signaling pathway at a conserved phosphotyrosine residue (pY200) (Kim *et al.*, 2009a). The *bin2* mutant protein was not dephosphorylated by BSU1, indicating that the mutation in the TREE domain of BIN2 blocks the BSU1 dephosphorylation of pY200. Furthermore, BSU1 directly interacted with BSK1 and this binding was enhanced by the BSK1 phosphorylation (Kim *et al.*, 2009a)

4.3 Regulation of downstream BR signaling by BZR1 and BES1/BZR2 transcription factors

BIN2 controls the phosphorylation status of two transcription factors, BZR1 and BES1/BZR2 (He *et al.*, 2002, 2005; Wang *et al.*, 2002; Yin *et al.*, 2002; Zhao *et al.*, 2002), also designated as BR response factors (BRFs) (Belkhadir *et al.*, 2006). BZR1 and BES1/BZR2 are closely related nuclear proteins that have been identified as positive regulators of the BR signaling through genetic screens for mutants insensitive to the BR biosynthesis inhibitor brassinazole and for *bri1* suppressors, respectively (Wang *et al.*, 2002; Yin *et al.*, 2002). BZR1 and BES1/BZR2 are members of a small gene family with at least four other proteins in *Arabidopsis*, designated as BES1/BZR1 homolog 1 to 4 (BEH1 to BEH4) (He *et al.*, 2005; Yin *et al.*, 2005). In the absence of BR signaling, BZR1 and BES1/BZR2 exist mostly in phosphorylated forms and accumulate as dephosphorylated proteins in response to BR treatment (He *et al.*, 2002; Wang *et al.*, 2002). BIN2 interacted with BZR1 and BES1/BZR2 in yeast two-hybrid assays and directly phosphorylated both transcription factors *in vitro* (He *et al.*, 2002; Zhao *et al.*, 2002). In the gain-of function *bin2* mutant, the accumulation of BZR1 and BES1/BZR2 is reduced and their BR-induced dephosphorylation is attenuated. The large number of putative GSK3 phosphorylation sites in the BZR1 and BES1/BZR2 sequences and the mobility shift caused by the BIN2 phosphorylation suggest that BIN2 phosphorylates the substrates at multiple amino acid residues (He *et al.*, 2002). The positions of these phosphorylation sites and their influence on the BZR1 and BES1/BZR2 accumulation and activity are still to be determined experimentally. The phosphorylated BZR1 is seemingly degraded by the 26S proteasome-mediated proteolysis, because treatment of seedlings with the proteasome inhibitor MG132 preferentially increased the accumulation of the phosphorylated BZR1 protein (He *et al.*, 2002). Hence, phosphorylation might promote degradation and inhibit the nuclear localization of BZR1 and BES1/BZR2 (He *et al.*, 2002; Yin *et al.*, 2002). This model was refined to take into account that BES1 phosphorylation inhibits its DNA binding and transcriptional activity (Vert and Chory, 2006), but does not affect its cellular distribution (Zhao *et al.*, 2002; Vert and Chory, 2006). However, recent studies in *Arabidopsis* and rice revised this model again and confirmed the importance of the BR-regulated nuclear localization of BZR1 by identifying the 14-3-3 proteins as new components of the BR pathway (Bai *et al.*, 2007; Gampala *et al.*, 2007; Ryu *et al.*, 2007). The 14-3-3 proteins are highly conserved phospho-binding domain proteins that play an important role in many signaling pathways in all eukaryotes (Chevalier *et al.*, 2009). A conserved, BIN2-phosphorylated 14-3-3 binding site was detected in BZR1. Mutations in the 14-3-3 binding site of BZR1 abolished its interaction with 14-3-3, increased the nuclear accumulation, and induced BR responses, indicating that the role for 14-3-3 proteins is essential for inhibiting the phosphorylated BZR1 by increasing its cytoplasmic retention (Bai *et al.*, 2007; Gampala *et al.*, 2007).

Despite their high sequence similarity and functional redundancy, BZR1 and BES1/BZR2 have different modes of action because the BR biosynthetic genes and other BR target genes are apparently regulated differentially by BZR1 and BES1/BZR2. Whereas BZR1 binds to the CGTG(T/C)G motif (also known as BR-Response Element or BRRE) to suppress the expression of the BR-biosynthetic genes, such as the cytochrome P450-encoding *CPD* and *DWARF4* (He *et al.*, 2005), BES1/BZR2 binds the CANNTG sequence (also known as E box) of the SAUR-AC1 promoter for gene expression activation (Yin *et al.*, 2005). In yeast two-hybrid assays, BES1/BZR2, which also contains an atypical basic-helix-loop-helix (bHLH)-like domain, interacts with three homologous bHLH transcription factors, designated BES1-interacting *Myc*-like-1 (BIM1), BIM2, and BIM3. The BIM1 protein bound the same E box and synergism between BES1/BZR2 and BIM1 was observed for DNA binding *in vitro* and activation of the SAUR-AC1 promoter in transient assays. Overexpression of *BIM1* partially suppressed a weak *bri1* allele, and the *bim1bim2bim3* triple mutant had a reduced cell elongation, suggesting a role for the BIM1 to BIM3 proteins in mediating BR responses. BES1/BZR2 and the BIM1 to BIM3 proteins have been proposed to bind synergistically to E-box elements to activate the BR-induced gene expression (Yin *et al.*, 2005). Although different DNA-binding motifs have been identified for BZR1 and BES1/BZR2, their binding specificities might not be as distinct as it appears. Both *bes1-D* and *bzr1-1D* mutants exhibit the brassinazole-resistant phenotype in the dark, most probably, through the regulation of an overlapping set of target genes. Indeed, several of the BES1/BZR2 target genes were up-regulated in the *bzr1-1D* mutant (He *et al.*, 2005) and several feedback-regulated BR-biosynthetic genes were down-regulated in the *bes1-D* mutant (Vert *et al.*, 2005). Several examples demonstrated that besides BIM proteins, BES1/BZR2 can heterodimerize with other transcriptional regulators to modulate their DNA binding and transcriptional activities and, thus, to coordinate BR responses in different developmental pathways. In a yeast genetics screen of *Arabidopsis*, two flowering time transcriptional regulators, the Jumonji N/C (JmjN/C) domain-containing proteins, EARLY FLOWERING 6 (ELF6) and its close homolog, RELATIVE OF EARLY FLOWERING 6 (REF6) (Noh *et al.*, 2004; Yu *et al.*, 2008) were identified as BES1-interacting partners. Further genetic data suggested that BES1 recruits ELF6 and REF6 to positively regulate BR-induced gene expression and coordinate both the BR and flowering pathways (Yu *et al.*, 2008).

Other microarray and chromatin immunoprecipitation experiments revealed a MYB family transcription factor, *AtMYB30*, as a direct target gene of BES1/BZR2 (Li *et al.*, 2009b). Besides its function in BR-regulated gene responses, *AtMYB30* is implicated in pathogen responses and cell death in adult plants (Daniel *et al.*, 1999; Vailleau *et al.*, 2002) and interacts directly with BES1/BZR2 to regulate synergistically a subset of genes with the two binding sites (Li *et al.*, 2009b).

4.4 Regulation of downstream BR signaling by atypical bHLH proteins

The bHLH proteins are important transcriptional regulators that play key roles in developmental pathways, including light response, cell fate, and stomatal development in plants (Toledo-Ortiz *et al.*, 2003; Pillitteri and Torii, 2007). Recent reports stressed the importance of a subclass of atypical bHLH proteins homologous to the human Inhibitor of DNA binding-1 (Id-1) as transcriptional regulators of the BR signaling (Wang *et al.*, 2009; Zhang *et al.*, 2009). These transcription factors lack the basic DNA-binding motif and negatively regulate transcription by heterodimerization with other bHLH transcription factors. By means of an activation-tagging screen for suppressors of the weak *bri1-301* allele and a genetics screen for BR-response mutants in rice, two atypical bHLH transcription factors were identified that positively regulate the BR signaling, the ACTIVATION-TAGGED BRI1 SUPPRESSOR 1 (ATBS1) (Zhang *et al.*, 2009) and the *Arabidopsis* PACLOBUTRAZOL-RESISTANT1 (PRE1), homologous to the rice transcription factor, Increased Leaf Inclination-1 (ILI1) (Wang *et al.*, 2009). ATBS1/PRE3 and ILI1/PRE1 are members of a previously described PRE family of six atypical bHLH proteins involved in gibberellin and light signaling pathways in *Arabidopsis* (Hyun and Lee, 2006; Lee *et al.*, 2006). All PRE1 to PRE6 proteins function redundantly in the BR signaling and suppress the *bri1-301* phenotype when overexpressed (Wang *et al.*, 2009; Zhang *et al.*, 2009). The PRE family might enhance the BR signaling by heterodimerization with other bHLH proteins that negatively regulate it. This hypothesis was supported by the identification of four additional atypical bHLH proteins, designated ATBS1 INTERACTING FACTOR-1 (AIF1), AIF2, AIF3, and AIF4, in a yeast two-hybrid screen (Wang *et al.*, 2009) and another bHLH transcription factor, designated ILI1

BINDING BHLH-1 (IBH1). Indeed, AIF1 and IBH1 negatively regulated the BR signaling because their overexpression led to BR-deficient dwarf phenotypes (Wang *et al.*, 2009; Zhang *et al.*, 2009). Preliminary experiments revealed that AIF1 did not interact with BES, BZR1, and BIM proteins, but with other previously identified BR-regulated bHLH proteins, designated BR-ENHANCED EXPRESSION1 (BEE1), BEE2, and BEE3 (Friedrichsen *et al.*, 2002). In addition, IBH1 is a BR-repressed direct target of BZR1 and PRE1 is also induced directly by BRs through the BZR1 (Zhang *et al.*, 2009). It remains to be established whether any of the newly identified HLH transcription factors and IBH1 are controlled by BIN2 because AIF1 has previously been identified as a BIN2-interacting protein that is phosphorylated by BIN2 *in vitro* (Wang *et al.*, 2009).

5. ENDOMEMBRANE TRAFFICKING REGULATES THE BR SIGNALING

5.1 The endocytosis of BRI1 is ligand independent and BFA sensitive

Over the last decade, studies have revealed how important endocytosis is for cell signaling. In mammals, endocytic membrane trafficking and cell signaling are intimately connected because endocytosis controls the number of plasma membrane receptors available for activation and, conversely, receptor activation or downstream signaling effectors stimulate endocytosis. Endosomes can provide a platform for the assembly of specific signaling complexes that are required for signaling (Sorkin and Von Zastrow, 2009). Growing evidence suggests that plants adapted similar mechanisms to regulate signal transduction through endocytosis (Irani and Russinova, 2009).

In *Arabidopsis* roots, a functional BRI1-Green Fluorescent Protein (GFP) fusion was detected on both the plasma membrane and mobile vesicle compartments (Friedrichsen *et al.*, 2000; Russinova *et al.*, 2004; Geldner *et al.*, 2007). Some of the BRI1-GFP-containing vesicles colocalized with the endocytic tracer FM4-64 after 5–10 min of uptake and with the *trans*-Golgi network (TGN)/early endosome (EE) marker VHAa1, suggesting that EEs are the first recipients of the internalized BRI1 receptor (Russinova *et al.*, 2004; Dettmer *et al.*, 2006; Geldner *et al.*, 2007). The localization of BRI1-GFP was sensitive to the fungal toxin Brefeldin A (BFA) (Russinova *et al.*, 2004; Geldner *et al.*, 2007) that inhibits secretion and endosomal recycling of plasma membrane proteins in plants (Geldner *et al.*, 2001), hinting at a continuous recycling of BRI1. The BRI1 recycling hypothesis was supported by the long half-life time (approximately 5 hours) of the BRI1 protein (Geldner *et al.*, 2007) that corresponded to the earliest observed accumulation of BRI1-GFP in the vacuole in the dark or after treatment with the specific inhibitor of the vacuolar H^+-ATPases, concanamycin A (Kleine-Vehn *et al.*, 2008). How- ver, the mechanism of BRI1 recycling and the identity of the intermediate recycling endosomes are still unknown. Recently, BRI1 has been found o colocalize with the putative retromer component SORTIN NEXIN1 in *Arabidopsis* roots (Jaillais *et al.*, 2008), for which a function in retrieving proteins from prevacuolar compartments to the recycling pathway has been proposed (Kleine-Vehn *et al.*, 2008). Although BRI1 accumulated in the prevacuolar compartments and lytic vacuole (Jaillais *et al.*, 2008; Kleine-Vehn *et al.*, 2008), the BRI1 vacuolar trafficking, degradation, recycling, and its significance for the BR signaling are still not understood. When the endogenous BR levels were depleted by using brassinazole or increased by addition of exogenous BRs, the endosomal localization and the turnover of BRI1 remained unchanged (Russinova *et al.*, 2004; Geldner *et al.*, 2007), suggesting that the BRI1 endocytosis is ligand-independent.

Similar to BRI1, its co-receptor BAK1 also localized to the plasma membrane and endosomal compartments (Li *et al.*, 2002; Nam and Li, 2002; Russinova *et al.*, 2004). While BAK1 is required for the endocytosis of another plant LRR-RLK, FLS2 (Chinchilla *et al.*, 2009), its function in BRI1 endocytosis remains unclear. The coexpression of *BRI1* and *BAK1* in plant protoplasts accelerates endocytosis, but the rate of BRI1 endocytosis in the *serk3/bak1* null mutant is still unchanged. Thus, the presence or absence of the BAK1 protein does not affect BRI1 endocytosis, probably because of functional redundancy (Russinova *et al.*, 2004). It has to be demonstrated whether BRI1 endocytosis is affected in multiple *SERK* loss-of-function mutants.

5.2 BRI1 signals from endosomal compartments

The concept of endosomal signaling originated from biochemical fractionation of animal signaling components that placed active receptors and various downstream effectors in endosomes. Since then, numerous studies supported the hypothesis that endosomal signaling can occur either as a continuation of the signaling from the plasma membrane or as signaling that requires receptor endocytosis and occurs exclusively on endosomal membranes (Sorkin and von Zastrow, 2009). The functional significance of endocytosis for signaling has recently been demonstrated in plants in which an increased endosomal pool of BRI1 by application of BFA activated the BR signaling pathway, as revealed by the enhanced dephosphorylation of BES1 and repressed *DWF4* gene. Furthermore, BKI1 did not accumulate in the BRI1-containing endosomal compartments, suggesting that the endosomal BRI1 pool is active (Geldner *et al.*, 2007). Although such results are exciting and corroborate the role of endosomes in signaling , direct evidence for the presence of an active BRI1 complex in endosomes is still lacking.

Recently, in a high-throughput screen for chemical inhibitors of pollen germination, a bioactive limonoid, designated ENDOSIDIN1 (ES1), has been identified that induces the rapid agglomeration of BRI1 into distinct endomembrane compartments (Robert *et al.*, 2008). In addition to BRI1, ES1 also affected the trafficking of the auxin translocators PINFORMED2 (PIN2) and AUX1, but not PIN1, implying differences in the early endocytic routes of these proteins. Consistently, BRI1 and AUX1 had previously been shown to share recycling routes (Kleine-Vehn *et al.*, 2006). ES1-inhibited trafficking at the EE and the ES1 agglomerates contained the TGN/EE proteins and excluded the Golgi apparatus and recycling endosomes. In contrast to BFA, ES1 blocked the BR signaling because ES1 treatment phenocopied the loss-of-function *bri1* and repressed the expression of the BR-specific transcription factor (Robert *et al.*, 2008). Although the direct targets of ES1 remain to be identified, the effect of ES1 on the BR signaling and the localization of BRI1 to the TGN/EEs strengthens the prospect that endocytosis regulates the BR signaling.

5.3 The function of endoplasmic reticulum (ER) quality control (ERQC) in BR signaling

Newly synthesized secretory and plasma membrane proteins move in a vectorial manner from the ER to the Golgi complex and then to the cell surface. The ER assists in folding, assembly, and posttranslational modifications of these proteins, such as glycosylation. To ensure that the ER produces functional proteins, the secreted proteins are subjected to an ER quality control (ERQC) and only properly folded proteins are allowed to reach their final destination. The ERQC mechanisms can either retain misfolded proteins into the ER to initiate their refolding or retrotranslocate them into the cytosol for ER-associated degradation (ERAD) (Vembar and Brodsky, 2008). Several recent reports demonstrated the importance of the ERQC in BR signaling. The BR-insensitive phenotype of two mutants, *bri1-5* and *bri1-9*, both carrying mutations in the extracellular BRI1 domain C69Y and S662F, respectively, is caused by the retention of structurally defective BR receptors in the ER by the ERQC system (Jin *et al.*, 2007, 2009a; Hong *et al.*, 2008, 2009). Different mechanisms are responsible for the ER retention and degradation of the *bri1-5* and *bri1-9* mutants, including the calnexin/calreticulin (CNX/CRT) cycle, the luminal-binding protein system, and the thiol-mediated mechanism (Jin *et al.*, 2007, 2009a; Hong *et al.*, 2008). In contrast to *bri1-5*, which is degraded by a proteasome-independent ERAD (Hong et al., 2008), *bri1-9* is degraded by a proteasome-mediated ERAD (Jin *et al.*, 2007; Hong *et al.*,

2009). These observations are supported by the identification of several components of the *Arabidopsis* ERQC system in a genetics screen for extragenic suppressors of the *bri1-9* allele. Loss-of-function mutations in the *EMS-mutagenized bri1 suppressor* (*ebs1*), *ebs2*, and *ebs4* that code for the *Arabidopsis* homolog of the ER luminal enzyme, UDP-glucose:glycoprotein glucosyltransferase (Jin *et al.*, 2007), the *Arabidopsis* calreticulin 3 (Jin *et al.*, 2009a), and the *Arabidopsis* homolog of the yeast asparagine-linked glycosylation 12 enzyme (Hong *et al.*, 2009), respectively, significantly compromised the ERQC of *bri1-9* to allow mutated, but functional, BRI1 receptors to be correctly targeted to the plasma membrane. In summary, all these data reveal a previously unsuspected role of a specific subset of the ERQC machinery in the BR signaling pathway.

6. CONSERVATION OF BR SIGNALING MECHANISMS IN PLANTS

6.1 Conservation in upstream BR signaling components

BR signaling via LRR RLKs appears to be a common mechanism in both dicotyledonous and monocotyledonous plants because, besides *Arabidopsis*, BRI1 orthologs were found to mediate sensitivity to BRs in tomato (*Solanum lycopersicum*), pea (*Pisum sativum*), rice, and barley (*Hordeum vulgare*) (Yamamuro *et al.*, 2000; Montoya *et al.*, 2002; Chono *et al.*, 2003; Nomura *et al.*, 2003). All identified BR-insensitive mutants *ika* in pea (Nomura *et al.*, 2003), *curl3* (*cu3*) in tomato (Montoya *et al.*, 2002), *d61* in rice (Yamamuro *et al.*, 2000), and *uzu* in barley (Chono *et al.*, 2003) carried mutations in the orthologous *BRI1* genes and showed dwarfism, abnormal skotomorphogenesis, had erected leaves, and lacked BR responses, including the BR-induced internode elongation and bending of the lamina joint for monocotyledonous mutants. The sequence analysis of available *BRI1* orthologs revealed that the domains present in BRI1 are conserved across species (Nomura *et al.*, 2003). All predicted BRI1 proteins have a signal peptide, two cysteine pairs, an LRR domain, a transmembrane domain, and a kinase domain. However, the extracellular domain of BRI1 is poorly conserved between species when compared with the kinase domain that has a 44% and 93% identity to that of *Arabidopsis* BRI1 in rice and pea, respectively (Nomura *et al.*, 2003). The *BRI1* genes from tomato, tobacco (*Nicotiana benthamiana)*, and potato (*Solanum tuberosum)* fully complemented the *cu3* mutant, but unlike the *BRI1* ortholog of cotton (*Gossypium hirsutum)* (Sun *et al.*, 2004), only partially rescued the *Arabidopsis bri1*-5 mutant allele when expressed under the control of the *Arabidopsis BRI1* promoter (Holton *et al.*, 2007). A full *bri1*-5 complementation was achieved when the extracellular domain of the *Arabidopsis BRI1* gene was fused to the tomato kinase domain, suggesting that the extracellular region of *Arabidopsis* BRI1 is required for full function (Holton *et al*., 2007). The responsible ligand-binding extracellular domain (Kinoshita *et al*., 2005) is present in all BRI1 proteins and the tomato BRI1 could bind BRs (Holton *et al*., 2007), hinting at a functional conservation. However, the rice and barley BRI1 proteins lack three LRR domains (Yamamuro *et al*., 2000; Chono *et al*., 2003) and their function as BR receptors still remains to be determined. Similarly to *Arabidopsis* of which two of the three identified *BRI1* homologous genes *BRL1* and *BRL3*, could rescue the *bri1* phenotype when expressed under the *BRI1* promoter (Caño-Delgado *et al*., 2004), three homologous *BRI1* genes exist in rice, *OsBRL1, OsBRL2*, and *OsBRL3* (Nakamura *et al*., 2006). Furthermore, the unique sequence in the island domain (NGSM), which is important for the BR binding (Kinoshita *et al*., 2005), is conserved among BRI1, BRL1, and BRL3, but not BRL2, and their orthologs in *Arabidopsis,* rice, and other species (Nakamura *et al*., 2006; Holton *et al*., 2007).

Another conserved key component in the BR signal transduction pathway is the BRI1 co-receptor, BAK1. A BLAST search provided four closely related BAK1 homologs in rice, OsSERK1, OsSERK2, OsSERK3, and OsSERK4 that share from 77% to 51% sequence identity with *Arabidopsis* BAK1 and OsSERK1/OsBAK1 being the closest (Li *et al*., 2009a). The protein sequences of the OsSERK family has a structure similar to that of the *Arabidopsis* BAK1: five extracellular LRRs, a serine-proline-proline motif, a hydrophobic transmembrane domain, and a cytoplasmic serine/threonine kinase domain (Li *et al*., 2009a). Several lines of evidence support a conserved function for OsBAK1 with that of *Arabidopsis* in the BR signaling. OsBAK1 is localized on the plasma membrane and interacts with OsBRI1 *in vivo* (Li *et al*., 2009a). Similarly to the *Arabidopsis* BAK1, overexpression of the *OsSERK1* to *OsSERK4* genes partially rescued the weak *bri1-5* phenotype and OsBAK1 suppressed the weak BR-insensitive *d61-1* rice mutant. The BR signaling was enhanced in transgenic rice overexpressing both the rice and *Arabidopsis BAK1* genes and was inhibited either by the overexpression of the truncated intracellular domain of *OsBAK1* or by silencing of *OsBAK1* (Wang *et al*., 2007; Li *et al*., 2009a). Thus, rice and *Arabidopsis* might share a similar BR perception mechanism via the BRI1/BAK1 complexes.

A second BZR1-independent BR signaling pathway mediated by the heterotrimeric G protein α1 (RGA1), previously involved in gibberellic acid signaling and disease resistance, was proposed to control BR responses in rice (Wang *et al*., 2006; Oki *et al*., 2009). Although the loss-of-function *RGA1* mutant *d1* displayed less sensitivity to BRs, the expression patterns of the *BZR1*-regulated genes in *d1* mutants were not altered by BRs, suggesting a distinct RGA1-mediated BR signaling pathway (Oki *et al*., 2009). Therefore, elucidating the molecular mechanism of the G protein signaling could provide evidence for a functional link with the BR signaling.

6.2 Conservation in downstream BR signaling components

Although sequence analyses have identified orthologs for some of the *Arabidopsis* BR downstream signaling components in other plant species (Sun and Allen, 2005), the actual proof that these proteins are functional is still missing. The interest in the BR signaling pathway as tool to modify growth of important crops (Morinaka *et al.*, 2006) has stimulated BR research in monocotyledonous plants and the characterization of several downstream BR-signaling components mainly in rice.

Nine *GSK3/SHAGGY*-like genes in rice have been identified (Yoo *et al.*, 2006) and an orthologous gene of the *Arabidopsis BIN2*, designated *OsGSK1*, was isolated and structurally characterized (Koh *et al.*, 2007). Overexpression of the *OsGSK1* gene in rice resulted in a dwarfed phenotype suggesting that *OsGSK1* might be a functional rice ortholog that serves as a negative regulator of the BR signaling. Furthermore, *OsGSK1* knockout rice mutants were more sensitive to BR treatment, as indicated by a BR-responsive gene expression (Koh *et al.*, 2007).

RNA interference suppression of the expression of a rice ortholog of the *Arabidopsis BZR1*, *OsBZR1*, has demonstrated its essential role in BR responses (Bai *et al.*, 2007). Furthermore, in a yeast two-hybrid assay, 14-3-3 proteins interacted with OsBZR1 and bound to OsBZR1 through a consensus 14-3-3-binding site. Mutations of the putative 14-3-3-binding site in *OsBZR1* increased its nuclear localization and activity similarly to *Arabidopsis*, implying a conserved function for OsBZR1 and an important role of 14-3-3-protiens in the BR signaling (Bai *et al.*, 2007). OsBZR1 directly binds to the promoter of the GRAS family gene *DWARF AND LOW TILLERING* (*DTL*) to repress its expression. *DTL* also promotes the *OsBZR1* expression and is required for the full BR responses in rice (Tong *et al.*, 2009).

The importance of the bHLH-type transcriptional regulators in the BR signaling of rice has recently been shown. In a screen for a gain-of-function BR phenotype in rice, the *increased lamina inclination1-D* (*ili1-D*) mutant was identified (Zhang *et al.*, 2009). The *ili1-D* mutation was caused by the overexpression of the rice HLH homolog of the *Arabidopsis* PRE1 (Hyun and Lee, 2006; Lee *et al.*, 2006). Antisense suppression of the *ILI1* expression caused an erect leaf phenotype, confirming an essential role of atypical bHLH-transcriptional regulators in the BR signaling of rice (Zhang *et al.*, 2009). Another *ILI1* rice homolog, designated BR UNREGULATED 1 (BU1) has recently been identified by transcriptomics analysis of the *brassinosteroid deficient1* (*brd1*) mutant after BR treatment (Tanaka *et al.*, 2009). BU1, similarly to ILI1 and all *Arabidopsis* PRE family proteins, positively regulated the BR responses in rice, hence indicating a conservation of the BR signaling responses in *Arabidopsis* and rice.

7. CONCLUSION

More than a decade after the identification of the *Arabidopsis* BRI1 as a putative BR receptor (Li and Chory, 1997), intensive research by many laboratories established the complete BR signaling pathway from the cell surface-localized LRR-RLKs to nuclear transcription factors (Kim *et al.*, 2009a). As plants possess numerous RLKs (more than 600 members in *Arabidopsis* and 1100 in rice) that play key roles in plant development and defense responses, it is essential to understand their multiple signaling pathways and regulation mechanisms. The BR signal transduction pathway appears to be conserved between dicotyledonous and monocotyledonous plant species and became a new paradigm for other RLK-mediated signaling pathways in plants. Despite its plant-specific properties, the BR signaling pathway shares similar regulation mechanisms with signaling pathways activated by the mammalian EGF and TGF-β-type receptors. The BR signaling is initiated by a ligand-induced kinase activation followed by receptor oligomerization. The signal transduction in the cell is mediated through phosphorylation and dephosphorylation by kinases, phosphatases, 14-3-3 proteins, and multiple transcription factors (Tang *et al.*, 2010). Additionally, evidence exists that, as in mammals, the BR signaling can regulate endomembrane trafficking and endocytosis (Irani and Russinova, 2009). Furthermore, different signaling pathways activated by plant RLKs are interconnected and key signaling components are shared by multiple signaling cascades, leading to redundancy (Chinchilla *et al.*, 2009). Nevertheless, future research is required to figure out the role of endocytosis, protein degradation, and cellular compartmentalization in signaling regulation and to identify specificity-determining factors helping us to understand how the BR signaling regulates multiple processes in the plant.

8. ACKNOWLEDGEMENTS

The authors thank Gert Van Isterdael and Martine De Cock for help in preparing the manuscript. This work was supported by a grant from the Research Foundation-Flanders (no. G.0065.08).

9. REFERENCES

Albrecht, C., Russinova, E., Hecht, V., Baaijens, E., and de Vries, S., 2005. The *Arabidopsis thaliana* SOMATIC EMBRYOGENESIS RECEPTOR-LIKE KINASES1 and 2 control male sporogenesis. *Plant Cell*, **17**: 3337–3349.

Albrecht, C., Russinova, E., Kemmerling, B., Kwaaitaal, M., and de Vries, S.C., 2008. *Arabidopsis* SOMATIC EMBRYOGENESIS RECEPTOR KINASE proteins serve brassinosteroid-dependent and –independent signaling pathways. *Plant Physiol.*, **148**: 611–619.

Asami, T., Min, Y.K., Nagata, N., Yamagishi, K., Takatsuto, S., Fujioka, S., Murofushi, N., Yamaguchi, I., and Yoshida, S., 2000. Characterization of brassinazole, a triazole-type brassinosteroid biosynthesis inhibitor. *Plant Physiol.*, **123**: 93–99.

Asami, T., Nakano, T., and Fujioka, S., 2005. Plant brassinosteroid hormones. *Vitam. Horm.*, **72**: 479–504.

Bai, M.-Y., Zhang, L.-Y., Gampala, S.S., Zhu, S.-W., Song, W.-Y., Chong, K., and Wang, Z.-Y., 2007. Functions of OsBZR1 and 14-3-3 proteins in brassinosteroid signaling in rice. *Proc. Natl. Acad. Sci. USA*, **104**: 13839–13844.

Belkhadir, Y., Wang, X., and Chory, J., 2006. *Arabidopsis* brassinosteroid signaling pathway. *Science STKE*, **2006**: cm5 (doi:10.1126/stke.3642006cm5).

Caño-Delgado, A., Yin, Y., Yu, C., Vafeados, D., Mora-García, S., Cheng, J.-C., Nam, K.H., Li, J., and Chory, J., 2004. BRL1 and BRL3 are novel brassinosteroid receptors that function in vascular differentiation in *Arabidopsis*. *Development* **131**: 5341–5351.

Chevalier, D., Morris, E., and Walker, J.C., 2009. 14-3-3 and FHA domains mediate phosphoprotein interactions. *Annu. Rev. Plant Biol.*, **60**: 67–91.

Chinchilla, D., Shan, L., He, P., de Vries, S., and Kemmerling, B., 2009. One for all: the receptor-associated kinase BAK1. *Trends Plant Sci.*, **14**: 535–541.

Chinchilla, D., Zipfel, C., Robatzek, S., Kemmerling, B., Nürnberger, T., Jones, J.D.G., Felix, G., and Boller, T., 2007. A flagellin-induced complex of the receptor FLS2 and BAK1 initiates plant defence. *Nature*, **448**: 497–500.

Choe, S., Schmitz, R.J., Fujioka, S., Takatsuto, S., Lee, M.-O., Yoshida, S., Feldmann, K.A., and Tax, F.E., 2002. *Arabidopsis* brassinosteroid-insensitive *dwarf12* mutants are semidominant and defective in a glycogen synthase kinase 3β-like kinase. *Plant Physiol.*, **130**: 1506–1515.

Chono, M., Honda, I., Zeniya, H., Yoneyama, K., Saisho, D., Takeda, K., Takatsuto, S., Hoshino, T., and Watanabe, Y., 2003. A semidwarf phenotype of barley uzu results from a nucleotide substitution in the gene encoding a putative brassinosteroid receptor. *Plant Physiol.*, **133**: 1209–1219.

Chory, J., Nagpal, P., and Peto, C.A., 1991. Phenotypic and genetic analysis of *det2*, a new mutant that affects light-regulated seedling development in *Arabidopsis*. *Plant Cell*, **3**: 445–459.

Clouse, S.D., and Sasse, J.M., 1998. Brassinosteroids: Essential regulators of plant growth and development. *Annu. Rev. Plant Physiol. Plant Mol. Biol.*, **49**: 427–451.

Clouse, S.D., Langford, M., and McMorris, T.C., 1996. A brassinosteroid-insensitive mutant in *Arabidopsis thaliana* exhibits multiple defects in growth and development. *Plant Physiol.*, **111**: 671–678.

Colcombet, J., Boisson-Dernier, A., Ros-Palau, R., Vera, C.E., and Schroeder, J.I., 2005. *Arabidopsis* SOMATIC EMBRYOGENESIS RECEPTOR KINASES1 and 2 are essential for tapetum development and microspore maturation. *Plant Cell*, **17**: 3350–3361.

Daniel, X., Lacomme, C., Morel, J.-B., and Roby, D., 1999. A novel *myb* oncogene homologue in *Arabidopsis thaliana* related to the hypersensitive cell death. *Plant J.*, **20**: 57–66.

De Rybel, B., Audenaert, D., Vert, G., Rozhon, W., Mayerhofer, J., Peelman, F., Coutuer, S., Denayer, T., Jansen, L., Nguyen, L., Vanhoutte, I., Beemster, G.T.S., Vleminckx, K., Jonak, C., Chory, J., Inzé, D., Russinova, E., and Beeckman, T., 2009. Chemical inhibition of a subset of *Arabidopsis thaliana* GSK3-like kinases activates brassinosteroid signaling. *Chem. Biol.*, **16**: 594–604.

De Smet, I., Voβ, U., Jürgens, G., and Beeckman, T., 2009. Receptor-like kinases shape the plant. *Nat. Cell Biol.*, **11**: 1166–1173.

Dettmer, J., Hong-Hermesdorf, A., Stierhof, Y.-D., and Schumacher, K., 2006. Vacuolar H^+-ATPase activity is required for endocytic and secretory trafficking in *Arabidopsis*. *Plant Cell*, **18**: 715–730.

Ehsan, H., Ray, W.K., Phinney, B., Wang, X., Huber, S.C., and Clouse, S.D., 2005. Interaction of *Arabidopsis* BRASSINOSTEROID-INSENSITIVE 1 receptor kinase with a homolog of mammalian TGF-β receptor interacting protein. *Plant J.*, **43**: 251–261.

Failor, K.L., Desyatnikov, Y., Finger, L.A., and Firestone, G.L., 2007. Glucocorticoid-induced degradation of glycogen synthase kinase-3 protein is triggered by serum- and glucocorticoid-induced protein kinase and Akt signaling and controls β-catenin dynamics and tight junction formation in mammary epithelial tumor cells. *Mol. Endocrinol.*, **21**: 2403–2415.

Feng, X.-H., and Derynck, R., 1997. A kinase subdomain of transforming growth factor-β (TGF-β) type I receptor determines the TGF-β intracellular signaling specificity. *EMBO J.*, **16**: 3912–3923.

Forde, J.E., and Dale, T.C., 2007. Glycogen synthase kinase 3: a key regulator of cellular fate. *Cell. Mol. Life Sci.*, **64**: 1930–1944.

Friedrichsen, D.M., Joazeiro, C.A.P., Li, J., Hunter, T., and Chory, J., 2000. Brassinosteroid-insensitive-1 is a ubiquitously expressed leucine-rich repeat receptor serine/threonine kinase. *Plant Physiol.*, **123**: 1247–1255.

Friedrichsen, D.M., Nemhauser, J., Muramitsu, T., Maloof, J.N., Alonso, J., Ecker, J.R., Furuya, M., and Chory, J., 2002. Three redundant brassinosteroid early response genes encode putative bHLH transcription factors required for normal growth. *Genetics*, **162**: 1445–1456.

Fujioka, S., and Yokota, T., 2003. Biosynthesis and metabolism of brassinosteroids. *Annu. Rev. Plant Biol.*, **54**: 137–164.

Gampala, S.S., Kim, T.-W., He, J.-X., Tang, W., Deng, Z., Bai, M.-Y., Guan, S., Lalonde, S., Sun, Y., Gendron, J.M., Chen, H., Shibagaki, N., Ferl, R.J., Ehrhardt, D., Chong, K., Burlingame, A.L., and Wang, Z.-Y., 2007. An essential role for 14-3-3 proteins in brassinosteroid signal transduction in *Arabidopsis*. *Dev. Cell*, **13**: 177–189.

Geldner, N., Friml, J., Stierhof, Y.-D., Jürgens, G., and Palme, K., 2001. Auxin transport inhibitors block PIN1 cycling and vesicle trafficking. *Nature*, **413**: 425–428.

Geldner, N., Hyman, D.L., Wang, X., Schumacher, K., and Chory, J., 2007. Endosomal signaling of plant steroid receptor kinase BRI1. *Genes Dev.*, **21**: 1598–1602.

Grove, M.D., Spencer, G.F., Rohwedder, W.K., Mandava, N., Worley, J.F., Warthen, J.D. Jr, Steffens, G.L., Flippen-Anderson, J.L., and Cook, J.C. Jr, 1979. Brassinolide, a plant growth-promoting steroid isolated from *Brassica napus* pollen. *Nature*, **281**: 216–217.

He, J.-X., Gendron, J.M., Sun, Y., Gampala, S.S., Gendron, N., Sun, C.Q., and Wang, Z.-Y., 2005. BZR1 is a transcriptional repressor with dual roles in brassinosteroid homeostasis and growth responses. *Science*, **307**: 1634–1638 [Err. **308**: 1743].

He, J.-X., Gendron, J.M., Yang, Y., Li, J., and Wang, Z.-Y., 2002. The GSK3-like kinase BIN2 phosphorylates and destabilizes BZR1, a positive regulator of the brassinosteroid signaling pathway in *Arabidopsis*. *Proc. Natl. Acad. Sci. USA*, **99**: 10185–10190.

He, K., Gou, X., Yuan, T., Lin, H., Asami, T., Yoshida, S., Russell, S.D., and Li, J., 2007. BAK1 and BKK1 regulate brassinosteroid-dependent growth and brassinosteroid-independent cell-death pathways. *Curr. Biol.*, **17**: 1109–1115.

He, Z., Wang, Z.-Y., Li, J., Zhu, Q., Lamb, C., Ronald, P., and Chory, J., 2000. Perception of brassinosteroids by the extracellular domain of the receptor kinase BRI1. *Science*, **288**: 2360–2363.

Hecht, V., Vielle-Calzada, J.-P., Hartog, M.V., Schmidt, E.D.L., Boutilier, K., Grossniklaus, U., and de Vries, S.C., 2001. The *Arabidopsis SOMATIC EMBRYOGENESIS RECEPTOR KINASE 1* gene is expressed in developing ovules and embryos and enhances embryogenic competence in culture. *Plant Physiol.*, **127**: 803–816.

Heese, A., Hann, D.R., Gimenez-Ibanez, S., Jones, A.M.E., He, K., Li, J., Schroeder, J.I., Peck, S.C., and Rathjen, J.P., 2007. The receptor-like kinase SERK3/BAK1 is a central regulator of innate immunity in plants. *Proc. Natl. Acad. Sci. USA*, **104**: 12217–12222.

Hicks, G.R., and Raikhel, N.V., 2009. Opportunities and challenges in plant chemical biology. *Nat. Chem. Biol.*, **5**: 268–272.

Hink, M.A., Shah, K., Russinova, E., de Vries, S.C., and Visser, A.J.W.G., 2008. Fluorescence fluctuation analysis of *Arabidopsis thaliana* somatic embryogenesis receptor-like kinase and brassinosteroid insensitive 1 receptor oligomerization. *Biophys. J.*, **94**: 1052–1062.

Hirabayashi, S., Matsushita, Y., Sato, M., Oh-I, R., Kasahara, M., Abe, H., and Nyunoya, H., 2004. Two proton pump interactors identified from a direct phosphorylation screening of a rice cDNA library by using a recombinant BRI1 receptor kinase. *Plant Biotechnol.*, **21**: 35–45.

Holton, N., Caño-Delgado, A., Harrison, K., Montoya, T., Chory, J., and Bishop, G.J., 2007. Tomato *BRASSINOSTEROID INSENSITIVE1* is required for systemin-induced root elongation in *Solanum pimpinellifolium* but is not essential for wound signaling. *Plant Cell*, **19**: 1709–1717.

Hong, Z., Jin, H., Fitchette, A.-C., Xia, Y., Monk, A.M., Faye, L., and Li, J., 2009. Mutations of an α1,6 mannosyltransferase inhibit endoplasmic reticulum–associated degradation of defective brassinosteroid receptors in *Arabidopsis*. *Plant Cell*, **21**: 3792–3802.

Hong, Z., Jin, H., Tzfira, T., and Li, J., 2008. Multiple mechanism-mediated retention of a defective brassinosteroid receptor in the endoplasmic reticulum of *Arabidopsis*. *Plant Cell*, **20**: 3418–3429.

Hyun, Y., and Lee, I., 2006. *KIDARI*, encoding a non-DNA binding bHLH protein, represses light signal transduction in *Arabidopsis thaliana*. *Plant Mol. Biol.*, **61**: 283–296.

Irani, N.G., and Russinova, E., 2009. Receptor endocytosis and signaling in plants. *Curr. Opin. Plant Biol.*, **12**: 653–659.

Jaillais, Y., Fobis-Loisy, I., Miège, C., and Gaude, T., 2008. Evidence for a sorting endosome in *Arabidopsis* root cells. *Plant J.*, **53**: 237–247.

Jiang, J., and Clouse, S.D., 2001. Expression of a plant gene with sequence similarity to animal TGF-β receptor interacting protein is regulated by brassinosteroids and required for normal plant development. *Plant J.*, **26**: 35–45.

Jin, H., Hong, Z., Su, W., and Li, J., 2009a. A plant-specific calreticulin is a key retention factor for a defective brassinosteroid receptor in the endoplasmic reticulum. *Proc. Natl. Acad. Sci. USA*, **106**: 13612–13617.

Jin, Y.H., Kim, H., Oh, M., Ki, H., and Kim, K., 2009b. Regulation of Notch1/NICD and Hes1 expressions by GSK-3α/β. *Mol. Cells*, **27**: 15–19.

Jin, H., Yan, Z., Nam, K.H., and Li, J., 2007. Allele-specific suppression of a defective brassinosteroid receptor reveals a physiological role of UGGT in ER quality control. *Mol. Cell*, **26**: 821–830.

Jonak, C., and Hirt, H., 2002. Glycogen synthase kinase 3/SHAGGY-like kinases in plants: an emerging family with novel functions. *Trends Plant Sci.*, **7**: 457–461.

Jope, R.S., and Johnson, G.V.W., 2004. The glamour and gloom of glycogen synthase kinase-3. *Trends Biochem. Sci.*, **29**: 95–102.

Karlova, R., Boeren, S., Russinova, E., Aker, J., Vervoort, J., and de Vries, S., 2006. The *Arabidopsis* SOMATIC EMBRYOGENESIS RECEPTOR-LIKE KINASE1 protein complex includes BRASSINOSTEROID INSENSITIVE1. *Plant Cell*, **18**: 626–638.

Karlova, R., Boeren, S., van Dongen, W., Kwaaitaal, M., Aker, J., Vervoort, J., and de Vries, S., 2009. Identification of *in vitro* phosphorylation sites in the *Arabidopsis thaliana* somatic embryogenesis receptor-like kinases. *Proteomics,* **9**: 368–379.

Kauschmann, A., Jessop, A., Koncz, C., Szekeres, M., Willmitzer, L., and Altmann, T., 1996. Genetic evidence for an essential role of brassinosteroids in plant development. *Plant J.*, **9**: 701–713.

Kemmerling, B., Schwedt, A., Rodriguez, P., Mazzotta, S., Frank, M., Abu Qamar, S., Mengiste, T., Betsuyaku, S., Parker, J.E., Müssig , C., Thomma, B.P.H.J., Albrecht, C., de Vries, S.C., Hirt, H., and Nürnberger, T., 2007. The BRI1-associated kinase 1, BAK1, has a brassinolide-independent role in plant cell-death control. *Curr. Biol.*, **17**: 1116–1122.

Kim, T.-W., Guan, S., Sun, Y., Deng, Z., Tang, W., Shang, J.-X., Sun, Y., Burlingame, A.L., and Wang, Z.-Y., 2009a. Brassinosteroid signal transduction from cell-surface receptor kinases to nuclear transcription factors. *Nat. Cell Biol.*, **11**: 1254–1260.

Kim, W.-Y., Wang, X., Wu, Y., Doble, B.W., Patel, S., Woodgett, J.R., and Snider, W.D., 2009b. GSK-3 is a master regulator of neural progenitor homeostasis. *Nat. Neurosci.*, **12**: 1386–1393.

Kinoshita, T., Caño-Delgado, A., Seto, H., Hiranuma, S., Fujioka, S., Yoshida, S., and Chory, J., 2005. Binding of brassinosteroids to the extracellular domain of plant receptor kinase BRI1. *Nature*, **433**: 167–171.

Kleine-Vehn, J., Dhonukshe, P., Swarup, R., Bennett, M., and Friml, J., 2006. Subcellular trafficking of the *Arabidopsis* auxin influx carrier AUX1 uses a novel pathway distinct from PIN1. *Plant Cell*, **18**: 3171–3181.

Kleine-Vehn, J., Leitner, J., Zwiewka, M., Sauer, M., Abas, L., Luschnig, C., and Friml, J., 2008. Differential degradation of PIN2 auxin efflux carrier by retromer-dependent vacuolar targeting. *Proc. Natl. Acad. Sci. USA*, **105**: 17812–17817.

Kockeritz, L., Doble, B., Patel, S., and Woodgett, J.R., 2006. Glycogen synthase kinase-3-- an overview of an over-achieving protein kinase. *Curr. Drug Targets*, **7**: 1377–1388.

Koh, S., Lee, S.-C., Kim, M.-K., Koh, J.H., Lee, S., An, G., Choe, S., and Kim, S.-R., 2007. T-DNA tagged knockout mutation of rice *OsGSK1*, an orthologue of *Arabidopsis BIN2*, with enhanced tolerance to various abiotic stresses. *Plant Mol. Biol.*, **65**: 453–466.

Lee, S., Lee, S., Yang, K.-Y., Kim, Y.-M., Park, S.-Y., Kim, S.Y., and Soh, M.-S., 2006. Overexpression of *PRE1* and its homologous genes activates gibberellin-dependent responses in *Arabidopsis thaliana*. *Plant Cell Physiol.*, **47**: 591–600.

Li, D., Wang, L., Wang, M., Xu, Y.-Y., Luo, W., Liu, Y.-J., Xu, Z.-H., Li, J., and Chong, K., 2009a. Engineering *OsBAK1* gene as a molecular tool to improve rice architecture for high yield. *Plant Biotechnol. J.*, **7**: 791–806.

Li, J., and Chory, J., 1997. A putative leucine-rich repeat receptor kinase involved in brassinosteroid signal transduction. *Cell*, **90**: 929–938.

Li, J., and Nam, K.H., 2002. Regulation of brassinosteroid signaling by a GSK3/SHAGGY-like kinase. *Science*, **295**: 1299–1301.

Li, J., Lease, K.A., Tax, F.E., and Walker, J.C., 2001a. BRS1, a serine carboxypeptidase, regulates BRI1 signaling in *Arabidopsis thaliana*. *Proc. Natl. Acad. Sci. USA*, **98**: 5916–5921.

Li, J., Nam, K.H., Vafeados, D., and Chory, J., 2001b. *BIN2*, a new brassinosteroid-insensitive locus in *Arabidopsis*. *Plant Physiol.*, **127**: 14–22.

Li, J., Wen, J., Lease, K.A., Doke, J.T., Tax, F.E., and Walker, J.C., 2002. BAK1, an *Arabidopsis* LRR receptor-like protein kinase, interacts with BRI1 and modulates brassinosteroid signaling. *Cell*, **110**: 213–222.

Li, L., Yu, X., Thompson, A., Guo, M., Yoshida, S., Asami, T., Chory, J., and Yin, Y., 2009b. *Arabidopsis* MYB30 is a direct target of BES1 and cooperates with BES1 to regulate brassinosteroid-induced gene expression. *Plant J.*, **58**: 275–286.

Lösel, R., and Wehling, M., 2003. Nongenomic actions of steroid hormones. *Nat. Rev. Mol. Cell Biol.*, **4**: 46–56.

Meares, G.P., and Jope, R.S., 2007. Resolution of the nuclear localization mechanism of glycogen synthase kinase-3: functional effects in apoptosis. *J. Biol. Chem.*, **282**: 16989–17001.

Montoya, T., Nomura, T., Farrar, K., Kaneta, T., Yokota, T., and Bishop, G.J., 2002. Cloning the tomato *curl3* gene highlights the putative dual role of the leucine-rich repeat receptor kinase tBRI1/ SR160 in plant steroid hormone and peptide hormone signaling. *Plant Cell*, **14**: 3163–3176.

Mora-García, S., Vert, G., Yin, Y., Caño-Delgado, A., Cheong, H., and Chory, J., 2004. Nuclear protein phosphatases with Kelch-repeat domains modulate the response to brassinosteroids in *Arabidopsis*. *Genes Dev.*, **18**: 448–460.

Morinaka, Y., Sakamoto, T., Inukai, Y., Agetsuma, M., Kitano, H., Ashikari, M., and Matsuoka, M., 2006. Morphological alteration caused by brassinosteroid insensitivity increases the biomass and grain production of rice. *Plant Physiol.*, **141**: 924–931.

Nakamura, A., Fujioka, S., Sunohara, H., Kamiya, N., Hong, Z., Inukai, Y., Miura, K., Takatsuto, S., Yoshida, S., Ueguchi-Tanaka, M., Hasegawa, Y., Kitano, H., and Matsuoka, M., 2006. The role of *OsBRI1* and its homologous genes, *OsBRL1* and *OsBRL3*, in rice. *Plant Physiol*, **140**: 580–590.

Nam, K.H., and Li, J., 2002. BRI1/BAK1, a receptor kinase pair mediating brassinosteroid signaling. *Cell*, **110**: 203–212.

Nam, K.H., and Li, J., 2004. The Arabidopsis transthyretin-like protein is a potential substrate of BRASSINOSTEROID-INSENSITIVE 1. *Plant Cell*, **16**: 2406–2417.

Noguchi, T., Fujioka, S., Choe, S., Takatsuto, S., Yoshida, S., Yuan, H., Feldmann, K.A., and Tax, F.E., 1999. Brassinosteroid-insensitive dwarf mutants of *Arabidopsis* accumulate brassinosteroids. *Plant Physiol.*, **121**: 743–752.

Noh, B., Lee, S.-H., Kim, H.-J., Yi, G., Shin, E.-A., Lee, M., Jung, K.-J., Doyle, M.R., Amasino, R.M., and Noh, Y.-S., 2004. Divergent roles of a pair of homologous jumonji/zinc-finger-class transcription factor proteins in the regulation of *Arabidopsis* flowering time. *Plant Cell*, **16**: 2601–2613.

Nomura, T., Bishop, G.J., Kaneta, T., Reid, J.B., Chory, J., and Yokota, T., 2003. The *LKA* gene is a *BRASSINOSTEROID INSENSITIVE 1* homolog of pea. *Plant J.*, **36**: 291–300.

Oh, M.-H., Ray, W.K., Huber, S.C., Asara, J.M., Gage, D.A., and Clouse, S.D., 2000. Recombinant brassinosteroid insensitive 1 receptor-like kinase autophosphorylates on serine and threonine residues and phosphorylates a conserved peptide motif in vitro. *Plant Physiol.*, **124**: 751–765.

Oh, M.-H., Wang, X., Kota, U., Goshe, M.B., Clouse, S.D., and Huber, S.C., 2009. Tyrosine phosphorylation of the BRI1 receptor kinase emerges as a component of brassinosteroid signaling in *Arabidopsis*. *Proc. Natl. Acad. Sci. USA*, **106**: 658–663.

Oki, K., Inaba, N., Kitano, H., Takahashi, S., Fujisawa, Y., Kato, H., and Iwasaki, Y., 2009. Study of novel *d1* alleles, defective mutants of the α subunit of heterotrimeric G-protein in rice. *Genes Genet. Syst.*, **84**: 35–42.

Patel, S., Doble, B.W., MacAulay, K., Sinclair, E.M., Drucker, D.J., and Woodgett, J.R., 2008. Tissue-specific role of glycogen synthase kinase 3β in glucose homeostasis and insulin action. *Mol. Cell. Biol.*, **28**: 6314–6328.

Peng, P., Yan, Z., Zhu, Y., and Li, J., 2008. Regulation of the *Arabidopsis* GSK3-like kinase BRASSINOSTEROID-INSENSITIVE 2 through proteasome-mediated protein degradation. *Mol. Plant*, **1**: 338–346.

Pérez-Pérez, J.M., Ponce, M.R., and Micol, J.L., 2002. The *UCU1 Arabidopsis* gene encodes a SHAGGY/GSK3-like kinase required for cell expansion along the proximodistal axis. *Dev. Biol.*, **242**: 161–173.

Pillitteri, L.J., and Torii, K.U., 2007. Breaking the silence: three bHLH proteins direct cell-fate decisions during stomatal development. *BioEssays*, **29**: 861–870.

Robert, S., Chary, S.N., Drakakaki, G., Li, S., Yang, Z., Raikhel, N.V., and Hicks, G.R., 2008. Endosidin1 defines a compartment involved in endocytosis of the brassinosteroid receptor BRI1 and the auxin transporters PIN2 and AUX1. *Proc. Natl. Acad. Sci. USA*, **105**: 8464–8469.

Russinova, E., Borst, J.W., Kwaaitaal, M., Caño-Delgado, A., Yin, Y., Chory, J., and de Vries, S.C., 2004. Heterodimerization and endocytosis of *Arabidopsis* brassinosteroid receptors BRI1 and AtSERK3 (BAK1). *Plant Cell*, **16**: 3216–3229.

Ryu, H., Kim, K., Cho, H., Park, J., Choe, S., and Hwang, I., 2007. Nucleocytoplasmic shuttling of BZR1 mediated by phosphorylation is essential in *Arabidopsis* brassinosteroid signaling. *Plant Cell*, **19**: 2749–2762.

Schlessinger, J., 2000. Cell signaling by receptor tyrosine kinases. *Cell*, **103**: 211–225.

Schlessinger, J., 2002. Ligand-induced, receptor-mediated dimerization and activation of EGF receptor. *Cell*, **110**: 669–672

Shiu, S.-H., Karlowski, W.M., Pan, R., Tzeng, Y.-H., Mayer, K.F.X., and Li, W.-H., 2004. Comparative analysis of the receptor-like kinase family in *Arabidopsis* and rice. *Plant Cell*, **16**: 1220–1234.

Sorkin, A., and von Zastrow, M., 2009. Endocytosis and signalling: intertwining molecular networks. *Nat. Rev. Mol. Cell Biol.*, **10**: 609–622.

Sun, Y., and Allen, R.D., 2005. Functional analysis of the *BIN2* genes of cotton. *Mol. Genet. Genomics*, **274**: 51–59.

Sun, Y., Fokar, M., Asami, T., Yoshida, S., and Allen, R.D., 2004. Characterization of the *Brassinosteroid insensitive 1* genes of cotton. *Plant Mol. Biol.*, **54**: 221–232.

Szekeres, M., Németh, K., Koncz-Kálmán, Z., Mathur, J., Kauschmann, A., Altmann, T., Rédei, G.P., Nagy, F., Schell, J., and Koncz, C., 1996. Brassinosteroids rescue the deficiency of CYP90, a cytochrome P450, controlling cell elongation and de-etiolation in *Arabidopsis*. *Cell*, **85**: 171–182.

Tanaka, A., Nakagawa, H., Tomita, C., Shimatani, Z., Ohtake, M., Nomura, T., Jiang, C.-J., Dubouzet, J.G., Kikuchi, S., Sekimoto, H., Yokota, T., Asami, T., Kamakura, T., and Mori, M., 2009. *BRASSINOSTEROID UPREGULATED1*, encoding a helix-loop-helix protein, is a novel gene involved in brassinosteroid signaling and controls bending of the lamina joint in rice. *Plant Physiol.*, **151**: 669–680.

Tang, W., Deng, Z., and Wang, Z.-Y., 2010. Proteomics shed light on the brassinosteroid signaling mechanisms. *Curr. Opin. Plant Biol.*, **13**: 27–33.

Tang, W., Kim, T.-W., Oses-Prieto, J.A., Sun, Y., Deng, Z., Zhu, S., Wang, R., Burlingame, A.L., and Wang, Z.-Y., 2008. BSKs mediate signal transduction from the receptor kinase BRI1 in *Arabidopsis*. *Science*, **321**: 557–560.

Thummel, C.S., and Chory, J., 2002. Steroid signaling in plants and insects – common themes, different pathways. *Genes Dev.*, **16**: 3113–3129.

Toledo-Ortiz, G., Huq, E., and Quail, P.H., 2003. The *Arabidopsis* basic/helix-loop-helix transcription factor family. *Plant Cell*, **15**: 1749–1770.

Tong, H., Jin, Y., Liu, W., Li, F., Fang, J., Yin, Y., Qian, Q., Zhu, L., and Chu, C., 2009. DWARF AND LOW-TILLERING, a new member of the GRAS family, plays positive roles in brassinosteroid signaling in rice. *Plant J.*, **58**: 803–816.

Vailleau, F., Daniel, X., Tronchet, M., Montillet, J.-L., Triantaphylidès, C., and Roby, D., 2002. A R2R3-MYB gene, *AtMYB30*, acts as a positive regulator of the hypersensitive cell death program in plants in response to pathogen attack. *Proc. Natl. Acad. Sci. USA*, **99**: 10179–10184.

Vembar, S.S., and Brodsky, J.L., 2008. One step at a time: endoplasmic reticulum-associated degradation. *Nat. Rev. Mol. Cell Biol.*, **9**: 944–957.

Vert, G., and Chory, J., 2006. Downstream nuclear events in brassinosteroid signalling. *Nature*, **441**: 96–100.

Vert, G., Nemhauser, J.L., Geldner, N., Hong, F., and Chory, J., 2005. Molecular mechanisms of steroid hormone signaling in plants. *Annu. Rev. Cell Dev. Biol.*, **21**: 177–201.

Vert, G., Walcher, C.L., Chory, J., and Nemhauser, J.L., 2008. Integration of auxin and brassinosteroid pathways by Auxin Response Factor 2. *Proc. Natl. Acad. Sci. USA*, **105**: 9829–9834.

Wang, L., Wang, Z., Xu, Y., Joo, S.-H., Kim, S.-K., Xue, Z., Xu, Z., Wang, Z., and Chong, K., 2009. *OsGSR1* is involved in crosstalk between gibberellins and brassinosteroids in rice. *Plant J.*, **57**: 498–510.

Wang, L., Xu, Y.-Y., Li, J., Powell, R.A., Xu, Z.-H. and Chong, K., 2007. Transgenic rice plants ectopically expressing *AtBAK1* are semi-dwarfed and hypersensitive to 24-epibrassinolide. *J. Plant Physiol.*, **164**: 655–664.

Wang, L., Xu, Y.-Y., Ma, Q.-B., Li, D., Xu, Z.-H., and Chong, K., 2006. Heterotrimeric G protein α subunit is involved in rice brassinosteroid response. *Cell Res.*, **16**: 916–922.

Wang, X., and Chory, J., 2006. Brassinosteroids regulate dissociation of BKI1, a negative regulator of BRI1 signaling, from the plasma membrane. *Science*, **313**: 1118–1122.

Wang, X., Li, X., Meisenhelder, J., Hunter, T., Yoshida, S., Asami, T., and Chory, J., 2005a. Autoregulation and homodimerization are involved in the activation of the plant steroid receptor BRI1. *Dev. Cell*, **8**: 855–865.

Wang, X., Goshe, M.B., Soderblom, E.J., Phinney, B.S., Kuchar, J.A., Li, J., Asami, T., Yoshida, S., Huber, S.C., and Clouse, S.D., 2005b. Identification and functional analysis of in vivo phosphorylation sites of the *Arabidopsis* BRASSINOSTEROID-INSENSITIVE1 receptor kinase. *Plant Cell*, **17**: 1685–1703.

Wang, X., Kota, U., He, K., Blackburn, K., Li, J., Goshe, M.B., Huber, S.C., and Clouse, S.D., 2008. Sequential transphosphorylation of the BRI1/BAK1 receptor kinase complex impacts early events in brassinosteroid signaling. *Dev. Cell*, **15**: 220–235.

Wang, Z.-Y., Nakano, T., Gendron, J., He, J., Chen, M., Vafeados, D., Yang, Y., Fujioka, S., Yoshida, S., Asami, T., and Chory, J., 2002. Nuclear-localized BZR1 mediates brassinosteroid-induced growth and feedback suppression of brassinosteroid biosynthesis. *Dev. Cell*, **2**: 505–513.

Wang, Z.-Y., Seto, H., Fujioka, S., Yoshida, S., and Chory, J., 2001. BRI1 is a critical component of a plasma-membrane receptor for plant steroids. *Nature*, **410**: 380–383.

Yamamuro, C., Ihara, Y., Wu, X., Noguchi, T., Fujioka, S., Takatsuto, S., Ashikari, M., Kitano, H., and Matsuoka, M., 2000. Loss of function of a rice *brassinosteroid insensitive1* homolog prevents internode elongation and bending of the lamina joint. *Plant Cell*, **12**: 1591–1605.

Yan, Z., Zhao, J., Peng, P., Chihara, R.K., and Li, J., 2009. BIN2 functions redundantly with other *Arabidopsis* GSK3-like kinases to regulate brassinosteroid signaling. *Plant Physiol.*, **150**: 710–721.

Yang, X.-H., Xu, Z.-H., and Xue, H.-W., 2005. *Arabidopsis* membrane steroid binding protein 1 is involved in inhibition of cell elongation. *Plant Cell*, **17**: 116–131.

Yin, Y., Vafeados, D., Tao, Y., Yoshida, S., Asami, T., and Chory, J., 2005. A new class of transcription factors mediates brassinosteroid-regulated gene expression in *Arabidopsis*. *Cell*, **120**: 249–259.

Yin, Y., Wang, Z.-Y., Mora-Garcia, S., Li, J., Yoshida, S., Asami, T., and Chory, J., 2002. BES1 accumulates in the nucleus in response to brassinosteroids to regulate gene expression and promote stem elongation. *Cell*, **109**: 181–191.

Yoo, M.-J., Albert, V.A., Soltis, P.S., and Soltis, D.E., 2006. Phylogenetic diversification of *glycogen synthase kinase 3/SHAGGY-like kinase* genes in plants. *BMC Plant Biol.*, **6**: 3.

Yu, X., Li, L., Li, L., Guo, M., Chory, J., and Yin, Y., 2008. Modulation of brassinosteroid-regulated gene expression by jumonji domain-containing proteins ELF6 and REF6 in *Arabidopsis*. *Proc. Natl. Acad. Sci. USA*, **105**: 7618–7623.

Yun, H.S., Bae, Y.H., Lee, Y.J., Chang, S.C., Kim, S.-K., Li, J., and Nam, K.H., 2009. Analysis of phosphorylation of the BRI1/BAK1 complex in *Arabidopsis* reveals amino acid residues critical for receptor formation and activation of BR signaling. *Mol. Cells*, **27**: 183–190.

Zhang, L.-Y., Bai, M.-Y., Wu, J., Zhu, J.-Y., Wang, H., Zhang, Z., Wang, W., Sun, Y., Zhao, J., Sun, X., Yang, H., Xu, Y., Kim, S.-H., Fujioka, S., Lin, W.-H., Chong, K., Lu, T., and Wang, Z.-Y., 2009. Antagonistic HLH/bHLH transcription factors mediate brassinosteroid regulation of cell elongation and plant development in rice and *Arabidopsis*. *Plant Cell*, **21**: 3767–3780.

Zhang, Y., Feng, S., Chen, F., Chen, H., Wang, J., McCall, C., Xiong, Y., and Deng, X.W., 2008. *Arabidopsis* DDB1-CUL4 ASSOCIATED FACTOR1 forms a nuclear E3 ubiquitin ligase with DDB1 and CUL4 that is involved in multiple plant developmental processes. *Plant Cell*, **20**: 1437–1455.

Zhao, J., Peng, P., Schmitz, R.J., Decker, A.D., Tax, F.E., and Li, J., 2002. Two putative BIN2 substrates are nuclear components of brassinosteroid signaling. *Plant Physiol.*, **130**: 1221–1229.

Chapter 3

REGULATION OF BRASSINOSTEROID METABOLISM

L. HATEGAN, B. GODZA AND M. SZEKERES
Institute of Plant Biology, Biological Research Center of the Hungarian Academy of Sciences, P.O.B.521. H-6701, Szeged, Hungary

Abstract: Brassinosteroids (BRs) participate in the regulation of important physiological processes, such as germination, photomorphogenesis, elongation, and the development of reproductive organs. Unlike other phytohormones, BRs are not subject to active transport within the plant, therefore, their levels are determined by the balance between local biosynthetic and inactivation reactions. BR accumulation shows good correlation with the induction of biosynthetic genes, which are stringently regulated, primarily at the transcriptional level. These genes function under developmental, organ-specific and diurnal control, and respond to homeostatic feedback adjustment. Excess amounts of bioactive BRs are removed by reversible or irreversible inactivation mechanisms. The regulation of BR-inactivating genes is also fairly complex and often opposite to those of the biosynthetic functions. Though many details are yet to be clarified, the known regulatory mechanisms of BR metabolism reveal a finely orchestrated system capable of ensuring optimal BR supply from germination to seed production.

Key words: *Arabidopsis*, biosynthesis, brassinosteroids, gene expression, inactivation

1. INTRODUCTION

Plant development requires complex morphogenic and metabolic changes. These events are stringently controlled by phytohormones, which determine and coordinate cellular functions in a concentration-dependent manner. The main factors determining endogenous hormone levels are the local metabolic processes involved in their synthesis and inactivation, and the physiological

S. Hayat and A. Ahmad (eds.), *Brassinosteroids: A Class of Plant Hormone*,
DOI 10.1007/978-94-007-0189-2_3,

mechanisms ensuring their mobilization within the plant. A unique feature of brassinosteroids (BRs) is that, unlike other plant hormones, they are not subject to long distance transport (Symons and Reid, 2004; Montoya *et al.*, 2005), but exert their effect in a paracrine way at the sites of their synthesis. Consequently, the regulation of BR metabolism has an instrumental role in controlling local levels and physiological effects of the steroid hormone.

Whereas the distribution of bioactive BRs and BR intermediates in plants is still poorly known, an impressive amount of information has been published regarding the regulation of the genes involved in the biosynthesis and removal of these hormones. The data reveal developmental, organ- and tissue-specific, hormonal, light- and time-of-the-day-specific expressional control of BR metabolic functions. Local BR levels were shown to depend on the activities of key biosynthetic or inactivating genes, suggesting a tight connection between the transcriptional regulation of BR metabolism and the differential physiological effects of the hormone.

Our knowledge of the regulation of BR metabolism has substantially increased since the comprehensive review of Fujioka and Yokota (2003). Here we assess some new developments, highlighting connections between the metabolic control and physiological effects of the hormone. In order to give a physiologically consistent view of BR biosynthesis and inactivation, this overview will focus mainly on the metabolism in the laboratory model plant *Arabidopsis thaliana*, but will also include important results obtained in other plant species

2. BR BIOSYNTHESIS

The basic pathway of BR synthesis was deduced from metabolite conversion results obtained with *Catharanthus roseus* suspension cultures and *Arabidopsis* seedlings. These results were essential for the subsequent identification of several biosynthetic genes and the conversion steps catalyzed by the encoded enzymes. It was established that the physiologically most important C_{28} BRs are synthesized from campesterol (CR), an abundant phytosterol, via multiple, mostly oxidative reactions. These oxidative steps are performed by cytochrome P450-type monooxygenases belonging to the closely related CYP85 and CYP90 families. While most of these enzymes were originally identified in *Arabidopsis*, several of their orthologs were soon recognized in other species. The similarity between BR metabolite profiles and biosynthetic enzymes in a wide variety of higher plants served as clear indications for the evolutionary conservation of BR synthesis. (For reviews of BR synthesis see: Fujioka and Yokota, 2003; Choe, 2006; Szekeres and Bishop, 2006).

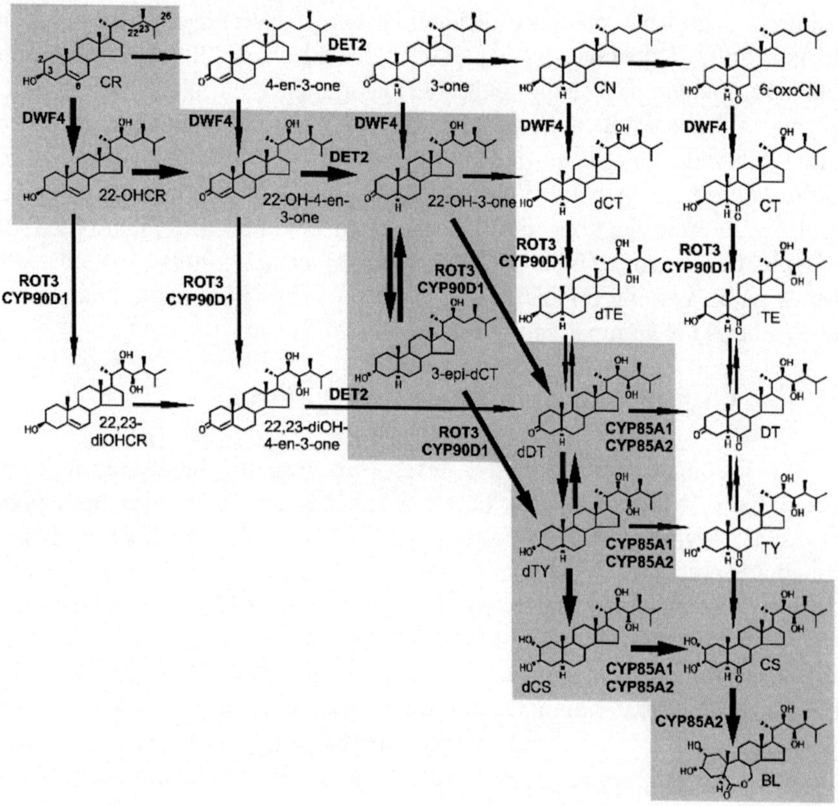

Figure 1. The pathways of BR biosynthesis. Arrows correspond to conversion steps, fat arrows denote reactions which, based on enzymological data, constitute the main synthesis routes (highlighted by gray background). The numbering of the important substituted carbon atoms is shown in the structure of campesterol (CR). Other steroid compounds are: 22-hydroxycampesterol (22-OHCR), 22,23-dihydroxycampesterol (22,23-diOHCR), (22S,24R)-hydroxyergost-4-en-3-one (4-en-3-one), (22S,24R)-22-hydroxyergost-4-en-3-one (22-OH-4-en-3-one), (22S,24R)-22,23-dihydroxyergost-4-en-3-one (22,23-diOH-4-en-3-one), (22S,24R)-hydroxyergost-3-one (3-one), (22S,24R)-22-hydroxyergost-3-one (22-OH-3-one), 3-epi-6-deoxocathasterone (3-epi-dCT), campestanol (CN), 6-oxocampestanol (6-oxoCN), 6-deoxocathasterone (dCT), cathasterone (CT), 6-deoxoteasterone (dTE), teasterone (TE), 3-dehydro-6-deoxoteasterone (dDT), 3-dehydroteasterone (DT), 6-deoxotyphasterol (dTY), typhasterol (TY), 6-deoxocastasterone (dCS), castasterone (CS), brassinolide (BL). The *Arabidopsis* enzymes with *in vitro* confirmed functions are the C-22 hydroxylase DWARF 4 (DWF4)/CYP90B1 (At3g50660), the C-23 hydroxylases ROTUNDIFOLIA 3 (ROT3)/ CYP90C1 (At4g36380) and CYP90D1 (At3g13730), the steroid 5α-reductase DE-ETIOLATED 2 (DET2; At2g38050), the C-6 oxidase CYP85A1 (At5g38970), and the C-6 oxidase, BL synthase CYP85A2 (At3g30180).

Analyses of the endogenous BR contents and biochemical assays of heterologously expressed BR-biosynthetic P450s revealed that these enzymes have broad substrate-specificities, allowing the conversion of multiple, structurally related intermediates. As a result, BR synthesis can proceed via parallel pathways that form a complex metabolic network (Nomura and Bishop, 2006; Bishop, 2007; Katsumata *et al.*, 2008; Ohnishi *et al.*, 2009). Furthermore, detailed enzymological studies indicated that, despite their relaxed specificities, BR-biosynthetic P450 activities can differ substantially depending on the actual substrates (Fujita *et al.*, 2006; Ohnishi *et al.*, 2006b; Hwang *et al.*, 2009). These findings pointed out preferential routes of the synthetic reactions, resulting in modifications of the earlier, metabolite conversion-based BR biosynthetic schemes (Bishop, 2007; Ohnishi *et al.*, 2009). The currently known pathways, intermediates and *in vitro* confirmed enzyme functions of BR synthesis are summarized in Figure 1.

2.1 Organ- and tissue-specific regulation

Gas chromatography-coupled mass spectrometric analyses revealed differential distribution of BRs between aerial and non-photosynthetic plant organs. Biologically active brassinolide (BL), castasterone (CS), and their immediate precursor 6-deoxoCS were found primarily in leaves, whereas upstream intermediates mainly in the roots. These differences, which were detected in *Arabidopsis*, pea and tomato seedlings (Bancos *et al.*, 2002), as well as in fully grown *Arabidopsis* and pea plants (Yokota *et al.*, 2001; Shimada *et al.*, 2003; Nomura *et al.*, 2007), were consistent with the unequal BR sensitivities of roots and shoots and, considering the lack of active hormone transport, suggested differential regulation of BR synthesis in these plant parts.

In agreement with the differential accumulation of BR forms, RT-PCR analyses of BR biosynthetic mRNAs also showed varying organ-specific distribution. In *Arabidopsis* seedlings, *CPD/CYP90A1* and *CYP85A2* transcripts were detected mainly in shoots, *ROT3/CYP90C1* and *CYP90D1* transcripts preferentially in roots, while *DET2* and *DWF4/CYP90B1* mRNAs were found in comparable amounts in both seedling parts (Bancos *et al.*, 2002). Similar partitioning of the orthologous *CYP90A9*, *CYP90A10*, *CYP85A1*, *CYP85A6*, *CYP90D7*, *LK* and *CYP90B8* transcripts was observed in pea seedlings (Nomura *et al.*, 2007). A quantitative comparison of mRNA levels in organs of mature *Arabidopsis* indicated that each of the BR-biosynthetic P450 genes has unique organ-specific expression pattern (Shimada *et al.*, 2003). More precise localization of these gene activities was made possible by using transgenic plants that carry promoter-reporter gene fusions.

The precise localization of *DWF4* (At3g50660) gene activity was determined in transgenic plants carrying its promoter fused to the β-glucuronidase (GUS) coding sequence. *DWF4*, which codes for the enzyme catalyzing the first committed reaction of BR synthesis (Choe *et al.*, 1998; Fujita *et al.*, 2006), shows generally weak expression. Histochemical staining could visualize (duplication) GUS activity in the root system, but remained undetectable in photosynthesizing vegetative organs. *DWF4* activity was present in flower primordia, joints of the pedicels and developing seeds. In seedlings the expression was localized mostly in the root tips, root-hypocotyl joints and the elongation zone of the hypocotyls, but in etiolated seedlings also in the cotyledons (Kim H.B. *et al.*, 2006).

The enzyme encoded by the *CPD* (At5g05690) gene was shown to be required for the synthesis of C-23-hydroxylated BRs (Szekeres *et al.*, 1996), but heterologously expressed CPD did not show C-23 hydroxylase activity *in vitro* (Ohnishi *et al.*, 2007). In transgenic plants the *CPD* promoter:*GUS* gene construct was highly active in expanding rosette leaves, particularly in the adaxial parenchimatic tissues, axillary leaves and sepals. No detectable expression was observed in any part of the root system. In young seedlings *CPD* activity was present in the cotyledons, and also in the apical hooks of etiolated seedlings (Mathur *et al.*, 1998).

Of the *ROT3* (At4g36380) and *CYP90D1* (At3g13730) genes, which encode functionally redundant C-23 hydroxylases (Ohnishi *et al.*, 2006b), only the expression of *ROT3* was studied with a *GUS* fusion construct. In young plants it was found ubiquitous and almost equal in all vegetative organs. In flowers *ROT3* was preferentially active at the basis of petals and in stamen filaments, but almost undetectable in developing tassels (Kim G.T. *et al.*, 1999). Consistent with these results, petals and stamina were found considerably shortened in *rot3* mutants, but elongated in a *ROT3* over-expressing line (Kim G.T. *et al.*, 1999; Ohnishi *et al.*, 2006b). The organ-specific distribution of the less abundant *CYP90D1* transcript is known only from RT-PCR data, which indicate its over-representation in the inflorescence stem (Shimada *et al.*, 2003). In agreement with this, decreased apical dominance was observed in the inflorescences of CYP90D1-deficient plants (Ohnishi *et al.*, 2006b).

CYP85A1 (At5g38970) encodes the C-6 oxidase, and *CYP85A2* (At3g30180) the C-6 oxidase and BL synthase that produce the bioactive BR forms CS, or CS and also BL, respectively (Shimada *et al.*, 2001, 2003; Kim T.W. *et al.*, 2005; Nomura *et al.*, 2005). In promoter:*GUS* transgenic lines *CYP85A1* activity was detected only in seedlings, mainly in their hypocotyls. By contrast, *CYP85A2* is expressed throughout the whole life cycle of the plant. In seedlings it is most active in the cotyledons, emerging leaves and root tips, while in mature plants it is also functional in flower parts, namely the sepals, anthers and ovules (Castle *et al.*, 2005). A very similar expression pattern was observed with *Dwarf*, the *CYP85A1* gene of tomato, which was also most active in meristematic regions and developing organs (Montoya *et al.*, 2005).

The organ specificities of BR biosynthetic genes, except that of the steroid 5α-reductase-encoding *DET2* (At2g38050; Li *et al.*, 1996), are quite well known. The general picture is that each of them has its own distinct spatial expression pattern, which seems to be important for the differential synthesis and proper distribution of the bioactive hormone. Despite their unique expression profiles, most biosynthetic genes are preferentially expressed in meristematic regions and differentiating organs, where BRs are key regulators of morphogenic events. In addition, differential expression of the enzymes with partially redundant functions, such as ROT3 and CYP90D1 or CYP85A1 and CYP85A2, can ensure subtle, organ-specific modulation of intermediate pools and biosynthetic capacity.

2.2 Developmental regulation

BRs are involved in complex hormonal regulatory mechanisms that are required for organ initiation and development (Vert *et al.*, 2005). Morphogenic events, such as germination, lateral shoot and root initiation, flower and fruit development, are accompanied by local, transient induction of multiple BR biosynthetic genes and, as a result, accumulation of the steroid hormone.

RT-PCR analyses revealed strong induction of all *CYP90* and *CYP85* genes during dark germination and early seedling development in *Arabidopsis*. Following imbibition, *DWF4*, *CPD* and *CYP85A2* transcripts started accumulating on the second day, and were then followed by *ROT3*, *CYP90D1* and *CYP85A1* to reach maximum abundance between the third and fifth days. After their transient induction all mRNA levels, except that of *CPD*, declined to less then 10% of the preceding maxima by the end of the first week (Bancos *et al.*, 2002). In the etiolated seedlings, the timing of elevated BR biosynthetic gene expression coincided with that of intense hypocotyl elongation. Similar strong induction of the *CYP85* and *CYP90* genes was observed in dark germinated pea seeds, where concomitant CS accumulation was also demonstrated (Nomura *et al.*, 2007).

Dry pea seeds are devoid of active BRs, but accumulate 6-deoxocathasterone that constitutes about 80% of their BR content. The rapid accumulation of CS from the first day of germination is accompanied by a substantial decrease in the 6-deoxocathasterone content, indicating that this BR is a storage form that is readily available for the hormone synthesis required during early development (Nomura *et al.*, 2007). Although recent *in vitro* enzymological data suggest that the direct, campestanol-independent route of BR synthesis bypasses 6-deoxocathasterone (Bishop, 2007; Ohnishi *et al.*, 2009), this compound can be efficiently mobilized via the campestanol-dependent pathway.

Organ initiation and development is often accompanied by elevated BR biosynthetic gene expression. GUS reporter constructs indicated the induction of *DWF4*, *ROT3* and *CYP85A2* in root (Kim H.B. *et al.*, 2006; Kim G.T. *et al.*, 1999; Castle *et al.*, 2005), and that of *CPD* and *CYP85A2* in leaf primordia (Mathur *et al.*, 1998; Castle *et al.*, 2005). Similarly, elevated *Dwarf/CYP85A1* expression was detected in tomato at the initiation of lateral roots (Montoya *et al.*, 2005), as well as in the apical meristematic region (Pien *et al.*, 2001). In *Zinnia elegans*, vascular differentiation was shown to depend on the increase of CS level. This results from the coordinated upregulation of several BR biosynthetic genes, and particularly those of the preferentially procambium-expressed ZeDWF4/CYP90B9 and ZeCPD1/ CYP90A11 (Yamamoto *et al.*, 2007).

Several reports substantiate the role of upregulated BR synthesis in the formation of reproductive organs. In *Arabidopsis*, *DWF4*, *ROT3* and *CYP85A2* (Kim G.T. *et al.*, 1999; Castle *et al.*, 2005; Kim H.B. *et al.*, 2006), whereas in tomato *Dwarf* (Montoya *et al.*, 2005) were shown to be induced in various parts of the flowers. In developing tomato fruits a transient increase of *Dwarf* activity is followed by strong, and also transient accumulation of CS and BL (Montoya *et al.*, 2005). This elevated hormone level is partly due to the concomitant expression of the fruit-specific CYP85A3, another C-6 oxidase with additional BL synthase function (Nomura *et al.*, 2005). In grapes (*Vitis vinifera*) similar temporary activation of *VvBROX1/CYP85A1*, and increase of CS content was observed in ripening berries (Symons *et al.*, 2006; Pilati *et al.*, 2007). The role of BRs in this developmental process could be confirmed by the enhanced ripening of berries exogenously treated with 24-epiBL (Symons *et al.*, 2006). A detailed study of BR synthesis in developing seeds of garden pea also demonstrated transient induction of the biosynthetic genes, and particularly those of *CYP90A10*, *LK* (the pea ortholog of *DET2*) and *CYP85A1*, during the enlargement stage. Once seeds reached their full size, the activities of these genes rapidly declined. These expressional changes were paralleled by a fivefold increase of the active BR content, and its subsequent depletion following the full expansion of the seeds (Nomura *et al.*, 2007). These developmentally controlled changes in BR biosynthetic gene activities, and others observed in cucumber in relation to fruit setting and in rice upon microspore formation (Fu *et al.*, 2008; Hirano *et al.*, 2008) clearly illustrate connections between the regulation of BR synthesis and the onset of some important differentiational processes.

2.3 Hormonal regulation

One important physiological function of BRs is the regulation of their own biosynthesis. As in the cases of other phytohormones, homeostasis that avoids harmful accumulation of the hormone is aided by feedback regulatory mechanisms, which can downregulate biosynthesis in a concentration-dependent manner. Such homeostatic control of BR synthesis was suggested by the suppression of *CPD* transcription upon BR treatment (Mathur *et al.*, 1998), the elevated mRNA level of *DWF4* in the partially BR-insensitive *bri1-5* mutant (Choe *et al.*, 2001), and also the massive accumulation of BRs in this and other *bri1* lines (Noguchi *et al.*, 1999).

In 1-week-old *Arabidopsis* seedlings the level of *CPD* mRNA rapidly decreased in response to exogenous BR treatment. After 2 hours of incubation with 100 nM BL its amount was down to 10%, whereas even with only 1 nM of the hormone to 50% of the untreated control value. This regulation was abolished in the presence of cycloheximide, suggesting its requirement of *de novo* protein synthesis. In a promoter:*GUS* fusion the 5′ regulatory regions of *CPD* were sufficient for conferring BR-responsive expression, indicating that this is controlled primarily at the level of transcription (Mathur *et al.*, 1998). A similar transcriptional feedback mechanism was shown to control gibberellin biosynthesis via the GA 20- and GA 3-oxidases, which catalyze the last two reactions of the pathway (Hedden and Phillips, 2000; Yamaguchi and Kamiya, 2000).

RT-PCR analyses revealed that in *Arabidopsis* all *CYP90* and *CYP85* genes are subject to BR feedback regulation (Bancos *et al.*, 2002; Tanaka *et al.*, 2005). And while the abundance of *DET2* mRNA was not reduced by excess BL, it increased in the BR-deficient *dwf4* and *bri1* mutants, and also in plants treated with brassinazole, an inhibitor of BR synthesis (Choe *et al.*, 2001; Tanaka *et al.*, 2005). The wide range of the BR concentration dependence of the *CYP90* and *CYP85* transcript levels, and their increase in BR-deficient lines indicated a vital role for such feedback in the regulation of BR synthesis. With each *CYP90* and *CYP85* gene the hormone response had similar extent and time course (Bancos *et al.*, 2002; Tanaka *et al.*, 2005) and required functional BR signaling (Li *et al.*, 2001; Bancos *et al.*, 2002; Tanaka *et al.*, 2005), therefore it seemed likely that the feedback control is based on the same transcriptional regulatory mechanism. Although Tanaka *et al.* (2005) found that, unlike other biosynthetic genes, *DWF4* remained partially BR-regulated even in a *bri1* null mutant, this response may also depend on BR signaling pathways. The observed weak BR response of the mainly vascular *DWF4* activity could actually be controlled via two functional BRI1 homologs, the vascular localized BRL1 and BRL3 (Caño-Delgado *et al.*, 2004).

Wang *et al.* (2002) identified BZR1, a transcription factor that is one final, effector element of BR signaling, and found that its overexpression enhanced the feedback response of *CPD*. Subsequent studies revealed that BZR1 represses *CPD* transcription by directly binding to the CGTG(T/C)G hexanucleotide motif, the so-called BR response element (BRRE) localized in the *CPD* 5′-UTR sequence. This *cis*-regulatory element is present, at variable positions, in all *CYP90* and *CYP85* promoters (He *et al.*, 2005). In rice OsBZR1 was shown to play a similar role in controlling the expression of BR biosynthetic genes, indicating the widespread conservation of the feedback regulatory mechanism in higher plants (Bai *et al.*, 2007).

The interaction between different phytohormone groups are believed to be important in the regulation of their metabolism, though many details of these crosstalks still need to be elucidated. One well known such interaction, however, is that of BRs and auxin, which have synergistic functions in many physiological processes (Mockaitis and Estelle, 2004; Nemhauser *et al.*, 2004; Vert *et al.*, 2005). In *Arabidopsis* roots auxin promotes the transport of BREVIS RADIX (BRX) to the nucleus (Scacchi *et al.*, 2009), thereby inducing *CPD* expression, and through it, BR biosynthesis. In turn, the level of active BRs determines the activity of auxin-responsive genes (Mouchel *et al.*, 2006). The regulatory role of auxin is supported by the observations that BR biosynthetic genes of *Arabidopsis* are upregulated in response to prolonged auxin exposure (Sibout *et al.*, 2006; Paponov *et al.*, 2008), and that in *Zinnia* cell cultures *ZeCPD1* and *ZeDWF4* are auxin-inducible (Yamamoto *et al.*, 2007). These findings highlight the importance of auxin-BR crosstalk and the resulting hormone biosynthetic effects in determining differentiational events (Hardtke *et al.*, 2007).

Unlike organ-specific and developmental regulation, which establishes differential expression of BR biosynthetic genes, hormonal control mechanisms modulate multiple gene activities coordinately. Intriguingly, BR feedback regulates the expression of enzymes involved in four distinct conversion steps, whereas in gibberellin synthesis only the abundance of the enzymes catalyzing the last two reactions are controlled by the bioactive end product (Hedden and Phillips, 2000; Yamaguchi and Kamiya, 2000). It is not clear why the regulation of BR synthesis requires such redundancies, but because of the highly networked pathways (Fujioka *et al.*, 2002; Shimada *et al.*, 2001) this may be necessary for efficient blocking of the flow, and also the diversion, of metabolites.

2.4 Diurnal regulation

Daily light/dark cycles provide important cues for proper timing of developmental and metabolic functions in plants. Optimal coordination of these physiological processes with light conditions is based on complex interactions between light, circadian and hormonal signaling events. A comprehensive analysis of temporal changes in the transcriptome of *Arabidopsis* seedlings showed that most of the genes involved in the metabolism and signaling of phytohormones are co-expressed in a time period that precedes maximal hypocotyl elongation. This synchronization of most growth hormone-related gene activities is the combined effect of direct diurnal light regulation and circadian-gated transcriptional responses (Michael *et al.*, 2008).

Based on the elongated stature of etiolated seedlings, it was tempting to assume that BR synthesis would be upregulated in the absence of light. In etiolated pea seedlings dark-induced DWF-like protein 1 (DDWF/CYP92A6) was proposed to function as BR C-2 hydroxylase responsible for preferential CS production in the elongation zone of the epicotyl (Kang *et al.*, 2001). Similarly, light-controlled gene expression data from *Arabidopsis* led to suggesting that light downregulates several genes assumed to be involved in BR synthesis (Ma *et al.*, 2001). These results, however, do not support the notion of negative light regulation, since subsequent studies did not verify the role of any of these genes in BR biosynthesis (Fujioka and Yokota, 2003; Szekeres and Bishop, 2006).

Paradoxically, while elongation processes are suppressed by light, BR synthesis and associated biosynthetic gene expression proved light inducible. Compared to the etiolated control, light-grown pea seedlings were shown to contain 17- and 4-fold higher amounts of BL and CS, respectively (Symons *et al.*, 2002). Similarly, in *Arabidopsis* and rice bioactive BRs were more abundant in light-, than in dark-grown plants (Choe *et al.*, 2001; Fujioka and Yokota, 2003). The transcription of *Arabidopsis CPD* and *CYP85A2* was found light induced and diurnally regulated. Red/dark photoperiods were sufficient to establish the daily pattern of *CPD* which, by contrast, was not appreciably influenced by blue/dark cycles. In line with this observation, the light induction of *CPD* was almost completely abolished in a mutant deficient in the two most important phytochrome species, suggesting a decisive role for phytochrome signaling in the light-control of this biosynthetic gene (Bancos *et al.*, 2006).

In addition to acute light effect, circadian regulation also contributes to the periodic daily expression profile of *CPD*. *In vivo* measurements using *CPD* promoter:luciferase-carrying transgenic seedlings showed that the circadian minima and maxima coincide with the light and dark phases of the days, respectively, thereby partially compensating the impact of direct light regulation. This combined effect results in a biphasic expression curve with maximal values after the onset of light, and before the start of dark periods. Very similar diurnal cycling could be observed in the activity of the *CYP85A2* promoter. Though it is not known how these transcriptional changes influence the levels of the encoded enzymes or that of BR synthesis, strong bioactive BR accumulation at the middle of the light phases suggests close correlation with the diurnal regulation of transcript levels (Bancos *et al.*, 2006).

Comprehensive microarray-based analyses of phytohormone-related and P450-encoding *Arabidopsis* transcripts revealed that most of the BR biosynthetic genes are subject to diurnal regulation (Michael *et al.*, 2008; Pan *et al.*, 2009). The observed daily expression profiles of *DWF4*, *DET2*, *CPD*, *ROT3* and *CYP85A2* showed both diurnal and circadian rhythmicity, while such regulation of *CYP90D1* and *CYP85A1* could not be ascertained. The circadian-controlled BR biosynthetic genes were found to be expressed with closely synchronized phases, which coincided with those of numerous other genes related to hormone metabolism and signaling, and which were found to have common transcriptional regulatory motifs over-represented in their promoters (Michael *et al.*, 2008).

The activities of BR biosynthetic genes are influenced by the interactions between light, circadian and feedback regulatory effects. Light periods determine the phasing of the circadian rhythm, while this latter gates light responses. Light alleviates, whereas dark enhances transcriptional feedback. Combined light and circadian cues define the diurnal profile of gene expression, which modulates the synthesis of BRs, thereby providing feedback control over the transcription. As a result, despite their distinct molecular mechanisms, the effects of light, circadian and hormonal regulations are mutually inter-dependent (Bancos *et al.*, 2006; Nemhauser, 2008).

2.5 Functional aspects of BR synthesis

Though more than 50 BR forms have been described from various plant sources (Bajguz and Tretyn, 2003), the intermediates of the C_{28} biosynthetic pathway leading to CS and BL were found most abundant and ubiquitous in all angiosperm species (Bishop and Yokota, 2001; Fujioka and Yokota, 2003). With increasing availability of genome sequence data it also became clear that in these plants evolutionarily related genes of BR synthesis encode identical sets of enzymes, namely orthologs of DWF4, DET2, CPD, ROT3 and/or CYP90D1, and CYP85A1 and/or CYP85A2 (Szekeres and Bishop, 2006; for recent listing of plant P450s: http://drnelson.utmem.edu/P450.statsfile.html). Furthermore, the similar proportioning of BR intermediate pools in *Arabidopsis*, pea and tomato strongly suggested that even the regulation of the pathway is well conserved (Nomura *et al.*, 2001). Among higher plants, the only notable phylogenetic differences in BR synthesis seems to be the dominance of the 6-deoxotyphasterol pool (Hong *et al.*, 2002) and the lack of BL (Kim B.K. *et al.*, 2008) in monocot species.

In the plant species studied so far there is little functional redundancy between BR biosynthetic enzymes. In *Arabidopsis* the C-23 hydroxylation by ROT3 and CYP90D1, as well as C-6 oxidation by CYP85A1 and CYP85A2, represent identical or largely overlapping functions (Figure 1). The genes coding for enzymes of redundant functions have distinct expression profiles, and in mature plants the roles of *CYP90D1* and *CYP85A1* are subordinate compared to their respective counterparts. Nevertheless, due to the feedback regulation of their activities, they can efficiently compensate for the inactivation of their functional equivalents in *rot3* or *cyp85a2* mutants which, therefore, show only very mild symptoms of BR deficiency (Kim G.T. *et al.*, 2005; Nomura *et al.*, 2005). The *DET2*, *DWF4* and *CPD* genes encode unique functions, and their mutational inactivation causes severe dwarfism. Intriguingly, *det2*, *dwf4* and *cpd* null mutants, just as double null *rot1 cyp90d1* and *cyp85a1 cyp85a2* lines, all contain detectable amounts of BRs produced downstream of the biosynthetic lesions (Noguchi *et al.*, 1999; Choe *et al.*, 2001; Nomura *et al.*, 2005; Ohnishi *et al.*, 2006b). This shows that yet unidentified alternative or non-specific enzyme activities also contribute to BR synthesis. The residual amounts of CS present in all biosynthetic mutants also indicate the indispensability of BR signaling for the viability of plants.

The expression of BR biosynthetic genes is controlled by multiple regulatory layers that act primarily at the transcriptional level (Fujioka and Yokota, 2003). Whereas organ-specific and developmental regulation influence BR distribution mostly in differential, gene-specific manner, hormonal and diurnal control mechanisms seem to modulate BR production coordinately. It is not clear how directly transcript levels determine the abundance of biosynthetic enzymes and hormone accumulation, but the complex regulation of gene activities, and especially the feedback-based homeostatic control of active BR levels, suggest a tight correlation. In some cases the induction of key BR biosynthetic genes and concomitant increase in the level of active BRs could be well demonstrated (Montoya *et al.*, 2005; Bancos *et al.*, 2006; Symons *et al.*, 2006; Nomura *et al.*, 2005; Yamamoto *et al.*, 2007).

In addition to the expressional control, BR synthesis is also regulated biochemically. Recent *in vitro* studies using heterologously expressed CYP90 and CYP85 monooxygenases revealed that these enzymes can utilize multiple substrates, which are hydroxylated or oxidized with reaction rates differing by orders of magnitudes (Bishop *et al.*, 1999; Shimada *et al.*, 2001; Fujita *et al.*, 2006; Ohnishi *et al.*, 2006b, 2007). This suggests that under *in vivo* conditions not only the concentration, but also the molar ratio of multiple substrates can influence the rate of each enzymatic reaction.

A challenging question of BR synthesis has been how, through which conversion steps, is the biosynthetic rate of the pathway controlled. A comparison of intermediate pool sizes revealed 6-deoxoCS as the most abundant, and 6-deoxocathasterone as the second most abundant BR forms in three dicot species (Nomura *et al.*, 2001). This implicates C-6 oxidation, producing bioactive CS, and C-23 hydroxylation as likely bottleneck reactions in the pathway, which is in good agreement with the very severe BR deficiency and dwarf phenotype in the mutants lacking these functions (Nomura *et al.*, 2005; Kwon *et al.*, 2005; Ohnishi *et al.*, 2006b). By contrast, 6-deoxotyphasterol is the most abundant intermediate in rice, a monocot species, indicating a controlling role for the C-2 hydroxylation that leads to 6-deoxoCS (Hong *et al.*, 2002). In *Arabidopsis*, DWF4-catalyzed C-22 hydroxylation was proposed to be the flux-determining reaction of BR synthesis (Choe *et al.*, 2001; Kim H.B. *et al.*, 2006; Choe, 2007), a notion that was based on the elongated phenotype of an overexpression line. While the DWF4 reaction is likely the first committed step of the BR pathway, its rate-limiting role does not seem supported by either the available intermediate pool data (Choe *et al.*, 2001; Nomura *et al.*, 2001), or the relatively mild BR-deficient phenotype of the *dwf4-1* null mutant (Azpiroz *et al.*, 1998).

In any case, identifying the rate-limiting reaction of BR biosynthesis can be an elusive task. Considering the differential expression of the enzymes, and the differential distribution and conversion rates of their potential sets of substrates, it seems quite possible that under varying conditions, or in particular organs, the efficiency of synthesis is limited by different enzymatic steps.

3. BR INACTIVATION

In addition to biosynthesis, homeostasis and physiological adjustment of BR levels also requires removal of the bioactive hormone. This can involve reversible glycosylation or acylation through hydroxyl groups, resulting in temporary inactivation, or additional hydroxylation, which is then followed by enzymatic degradation of the steroid structure or its targeting to salvage pathways (Fujioka and Yokota, 2003; Bajguz, 2007). Recent studies indicate that CS and BL levels are also influenced by specific enzymatic depletion of the pools of upstream BR intermediates (Marsolais *et al.*, 2007; Yuan *et al.*, 2007). Compared to BR synthesis, the mechanisms and enzymes of BR inactivation are poorly known, but in order to ensure proper hormonal effects and homeostasis these reactions need to be regulated with the same precision as those of the biosynthesis.

3.1 Organ- and tissue-specific regulation

Optimal adjustment of BR levels requires proper coordination of biosynthetic and inactivating functions. Accordingly, the genes of BR removal also need flexible and differential organ-specific control. Our knowledge of the underlying mechanisms is far less coherent than that related to biosynthetic genes, but all available data suggest similar importance of the transcriptional regulation.

BAS1/CYP734A1 (formerly: CYP72B1; At2g26710) is an *Arabidopsis* cytochrome P450 monooxygenase that is only distantly related to BR biosynthetic P450s. BAS1 can abolish the biological activity of CS and BL by hydroxylating them at the C-26 position of the steroid side chain (Neff *et al.*, 1999; Turk *et al.*, 2003). RT-PCR transcript analyses could detect low amounts of the *BAS1* transcript in all organs of mature plants (Shimada *et al.*, 2003; Turk *et al.*, 2005). In etiolated seedlings the GUS::BAS1 fusion protein, expressed under the control of the *BAS1* promoter, accumulated mainly in the elongation zone and apical hook of the hypocotyl, and also in the root-hypocotyl transition zone (Turk *et al.*, 2003). In tomato two *BAS1* homologs, *CYP734A7* and *CYP734A8* encode BR inactivating functions. Of these CYP734A7, which was shown to be a functional C-26 hydroxylase of CS and BL, is preferentially expressed in leaves and flowers of mature plants (Ohnishi *et al.*, 2006a).

Overexpression of another, closely related *Arabidopsis* P450, CHI2/SHK1/SOB7 (CYP72C1; At1g17060), was found to cause BR deficiency by depleting the pools of CS and its immediate precursors via an unknown molecular mechanism (Nakamura *et al.*, 2005; Takahashi *et al.*, 2005; Turk *et al.*, 2005). RT-PCR analyses detected highest *CYP72C1* transcript levels in roots and siliques. GUS histochemical assays showed elevated promoter activity in roots and hypocotyls of etiolated, whereas in roots and cotyledons of light-grown seedlings. In mature plants strong expression was seen in stamen filaments and at the basal joints of tassels (Takahashi *et al.*, 2005; Turk *et al.*, 2005).

UGT73C5 (At2g36790) is a UDP-glycosyltransferase that catalyzes 23-O-glucosylation of BL and CS. Transgenic overexpression of the *UGT73C5* gene leads to dwarfness via the severe reduction of CS and 6-deoxoCS pools. Conversely, RNA interference-mediated suppression of *UGT73C5* activity results in elongated phenotype (Poppenberger *et al.*, 2005). In transgenic seedlings carrying *UGT73C5* promoter:*GUS* fusion the gene activity was found mainly in the elongation zone of the hypocotyls (Poppenberger *et al.*, 2005).

BEN1 (At2g45400), a dihydroflavonol 4-reductase homolog, was identified in an activation-tagging screen for phenotypic modifiers of the weak BR insensitive *bri1-5* mutation. When overexpressed, BEN1 removes bioactive BRs by an unknown inactivating reaction. GUS histochemical analysis detected preferential *BEN1* promoter activity in roots, especially root tips, and in the pedicel regions of flowers and tassels. In rosette leaves *BEN1* expression was confined to vascular tissues (Yuan *et al.*, 2007).

Based on *in vitro* enzymological studies two sulfotransferases, AtST4a (At2g14920) and AtST1 (At2g03760) are also believed to be involved in BR inactivation. AtST4a is the only member of the small AtST4 subfamily that accepts BR substrates, sulfating specifically CS, BL and their C-24 epimers. By contrast, AtST1 was found to be specific for the 24-epimeric forms of BRs, among which the early intermediate 6-deoxocathasterone was its preferred substrate (Marsolais *et al.*, 2007). RT-PCR-based analyses showed that in seedlings *AtST4a* mRNA accumulated mostly in roots, whereas *AtST1* transcripts were localized primarily in the aerial parts (Marsolais *et al.*, 2007). The related BNST3 sulfotransferase from *Brassica napus*, with similar substrate specificity, showed similar organ-specific distribution when heterologously expressed in *Arabidopsis* (Rouleau *et al.*, 1999; Marsolais *et al.*, 2004).

3.2 Developmental regulation

Despite its apparent physiological importance, the developmental control of BR inactivation is very poorly understood. In this respect the best characterized expression profile is that of *CYP734A11*, the *BAS1* ortholog of pea, which was studied together with several biosynthetic genes during seed development and germination. While most of the biosynthetic genes become induced in the early, enlargement stages of seed development, *CYP734A11* activity is maintained roughly at the same level. These expressional changes bring about a fivefold increase in the active BR content. In fully grown seeds biosynthetic genes are abruptly downregulated, while the amount of the *CYP734A11* transcript declines at a much slower rate. By the final, drying stage the level of CS becomes undetectable, and the seeds lose about 75% of their total BR content. This decrease in the hormone level is certainly caused by BR inactivating enzymes, but the contribution of CYP734A11 to this effect is unclear (Nomura *et al.*, 2007). During dark germination BR biosynthetic enzymes are rapidly activated and CS starts accumulating from the first day following imbibition. During this period the amount of *CYP734A11* mRNA also increases substantially, implying that elevated BR synthesis and hormone levels may necessitate higher BR inactivating capacity (Nomura *et al.*, 2007).

The developmental pattern of *BNST3*, a BR sulfotransferase gene from *Brassica*, was studied using a promoter:*GUS* fusion expressed in transgenic *Arabidopsis*. Following germination *BNST3* activity was first observed at the root-hypocotyl transition zone, but as seedlings developed it became more prominent at the tips of the cotyledons and emerging leaves. In the flowers it was also apparent in pollen grains toward the later stages of their development (Marsolais *et al.*, 2004).

3.3 Hormonal regulation

The expression level of *BAS1* is positively regulated by active BRs, suggesting an important role for the encoded enzyme in maintaining the homeostatic balance of the hormone (Choe *et al.*, 2001; Goda *et al.*, 2002; Tanaka *et al.*, 2005). This transcriptional control can prevent unfavorable accumulation of CS and BL by increasing BAS1 levels and C-26 hydroxylation in response to elevated concentrations of these BR forms. The expression of CYP734A7 and CYP734A8, two tomato orthologs of BAS1, is similarly upregulated by BRs (Ohnishi *et al.*, 2006a). Among the known genes of BR removal only those belonging to the *CYP734A* subfamily were found BR-inducible.

Other phytohormones can also modulate the expression of BR inactivating genes, and such crosstalks can be very important for ensuring optimal hormone interactions during developmental and morphogenic events. Microarray hybridization data showed that *BAS1* and *CYP72C1* were rapidly induced by auxin, the hormone having overlapping and synergistic effects with those of BRs (Paponov *et al.*, 2008). The BR sulfotransferase-encoding *AtST4a* gene was also found hormone-responsive, showing transient induction following treatment by trans-zeatin (Marsolais *et al.*, 2007).

3.4 Diurnal regulation

As light conditions influence the expression of BR biosynthetic genes and the levels of active BRs, they also control the activities of the genes involved in BR inactivation. In seedlings the mRNA level of *BAS1*, which was identified in a screen for suppressors of phytochrome B-deficient long hypocotyl, is downregulated by light. RT-PCR analyses showed that the amount of the *BAS1* transcript decreased following white, red or blue illumination, but was enhanced by far-red light and dark treatments. Histochemical assays revealed that the accumulation of the GUS::BAS1 fusion protein, expressed under the control of the *BAS1* promoter, has similar light dependence. Light conditions, however, did not influence BAS1 accumulation in the roots (Turk *et al.*, 2003). The *CYP72C1* gene, which encodes another BR inactivating P450, is also repressed by light (Nakamura *et al.*, 2005; Takahashi *et al.*, 2005). Deficiency of the BAS1 and CYP72C1 activities decreased, whereas their overexpression enhanced light responsiveness of hypocotyl elongation, indicating a role for these enzymes in the BR-mediated control of photomorphogenesis (Nakamura *et al.*, 2005; Turk *et al.*, 2005). In contrast to *BAS1* and *CYP72C1*, the gene that encodes the dihydroflavonol reductase homolog BEN1 was found positively regulated by light (Yuan *et al.*, 2007).

Microarray hybridization analyses indicate that most genes of BR removal are under both diurnal and circadian regulation. *BAS1* and *CYP72C1* were identified as genes having diurnal and circadian cycling profiles (Michael *et al.*, 2008; Pan *et al.*, 2009), and expression data (available at http://diurnal.cgrb.oregonstate.edu/) suggest that the activities of *UGT73C5*, *BEN1* and *AtST1* also show daily cycling. Remarkably, the circadian cycles of the BR inactivating genes have similar, or only slightly delayed phasing compared to those of the ones involved in biosynthesis. While, in accordance with their opposite roles, BR inactivating and biosynthetic genes tend to have oppositely acting homeostatic and light regulation, the daily rhythms of their expression are mostly overlapping (Michael *et al.*, 2008; Pan *et al.*, 2009).

4. COORDINATION OF BIOSYNTHETIC AND INACTIVATING FUNCTIONS

The similarities observed between the BR contents in several species of higher plants suggest that not only the functions of biosynthesis, but likely also those of inactivation are conserved. Though we have limited knowledge of the BR inactivating enzymes in *Arabidopsis* and, in some cases, also of their functional significance, some of their orthologs have already been identified and characterized in other species. Orthologs of BAS1 are known from pea (Nomura *et al.*, 2007), tomato (Ohnishi *et al.*, 2006a) and rice (Park *et al.*, 2009), whereas one of AtST1, BNST4 was described from *Brassica napus* (Marsolais *et al.*, 2004). These few data seem to be consistent with the proposed evolutionary coherence of BR metabolism in angiosperm species (Fujioka and Yokota, 2003; Szekeres and Bishop, 2006).

BR levels and light exert concerted, opposite effects on BR synthesis and removal. Whereas biosynthetic genes are upregulated by light and low BR levels, inactivating genes are repressed under the same conditions (Figure 2). Such opposing effects of environmental stimuli can be crucial for coordinated adjustments of the positive and negative metabolic impacts on the pools of active BRs, that is, the proper homeostatic control of the hormone. A different type of coordination can be observed when developmental induction of biosynthetic genes results in sudden increase of the active BR content. In such cases, during seed development or germination, both biosynthetic and inactivating genes are upregulated (Nomura *et al.*, 2007), thereby ensuring efficient enzymatic control over the increase of bioactive BR content.

Shoots preferentially accumulate 6-deoxoCS and bioactive BRs, whereas upstream intermediates are overrepresented in roots. Shoot accumulation of CS and BL results from the preferential *CYP85A2* expression in the aerial parts of the plant. On the other hand, primarily root-localized *BAS1*, *CYP72C1* and *BEN1* activities (Turk *et al.*, 2003; Takahashi *et al.*, 2005; Yuan *et al.*, 2007) may also be required for maintaining optimally low concentrations of the active hormone in this organ.

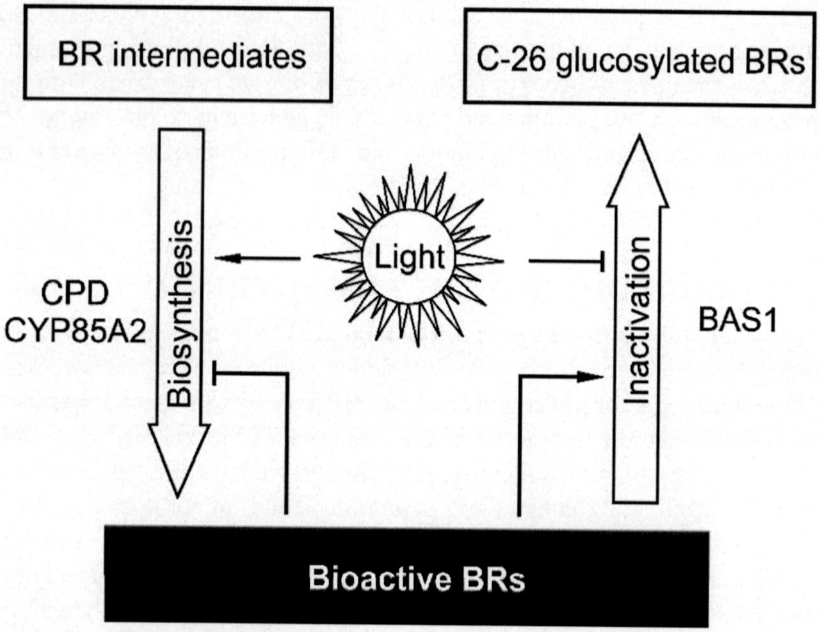

Figure 2. Coordinated hormonal and light regulation of key BR biosynthetic and inactivating genes in *Arabidopsis*.

5. CONCLUDING REMARKS

While our knowledge of the processes and physiology of BR synthesis is quite well established, the biological roles and impacts of the inactivating reactions are far less understood. Conjugated BRs, such as glycosylated or acylated compounds that are well known from several species (Fujioka and Yokota, 2003; Bajguz, 2007), may serve as reversibly inactivated storage forms, which can be mobilized enzymatically when hormone synthesis and accumulation are required. Such a reversible reaction is C-3 myristylation of teasterone that was described in lily (*Lilium longiflorum*), and was suggested to support BL synthesis during pollen development (Asakawa *et al.*, 1996; Soeno *et al.*, 2000). Conjugative reactions, rather than catabolic breakdown, can be responsible for the rapid decrease of the total BR content in fully developed pea seeds (Nomura *et al.*, 2007). These examples highlight the potential importance of reversible inactivation processes of BR metabolism, which still need to be elucidated.

BR metabolism is involved in complex, self-regulatory networks controlled by environmental and endogenous stimuli. Light influences BR levels through the transcriptional induction of biosynthetic, and repression of inactivating genes. Metabolic genes are also controlled by the endogenous circadian clock, entrained by light signals, through gating their light responsive expression (Robertson *et al.*, 2009). BRs modulate their own synthesis and removal via oppositely acting regulatory loops. Feedback regulation of the biosynthetic genes is mediated by the BZR1 transcription factor, which functions in a light-regulated manner. As light can influence BR sensitivity (He *et al.*, 2005; Bancos *et al.*, 2006; Jeong *et al.*, 2007), BR levels can also modulate light-responsiveness (Luccioni *et al.*, 2002; Turk *et al.*, 2003; Nakamura *et al.*, 2005). Therefore, interpretations of all experimental data relating to the control of BR metabolism need due consideration of these intricate regulatory relationships.

The lack of BR transport requires precise local coordination of biosynthetic and inactivating functions in order to optimize the level of the hormone. This regulation seems to act primarily through the transcriptional control of metabolic genes. It is not known, however, how the observed changes in the transcript levels influence abundances of the encoded enzymes, and how the availability of these enzymes determines biosynthetic or inactivating capacities in the relevant pathways. In several cases, however, good correlation was observed between the upregulation of key biosynthetic enzymes and the accumulation of bioactive BRs (Montoya *et al.*, 2005; Bancos *et al.*, 2006; Symons *et al.*, 2006; Yamamoto *et al.*, 2007), and the efficiency of homeostatic feedback control also suggests a tight dependence of BR synthesis on the activities of biosynthetic genes (Bancos *et al.*, 2002; Tanaka *et al.*, 2005). But while hormone levels critically influence the onset of BR-mediated physiological responses, these are also dependent on the local responsiveness of target tissues, which can be determined both by endogenous and environmental factors (He *et al.*, 2005; Bancos *et al.*, 2006). Defining the extents to which BR levels and differential sensitivities contribute to hormonal responses will require more precise knowledge of BR distribution, and its developmental changes, within the plant. This may be achieved by combining sensitive analytical techniques with the development and use of specific BR-responsive *in vivo* reporter constructs.

6. ACKNOWLEDGEMENTS

Research in the authors' laboratory has been supported by the Hungarian Scientific Research Fund (Grant T 68201) and the 'BRAVISSIMO' Marie Curie Initial Training Grant of the European Union.

7. REFERENCES

Asakawa, S., Abe, H., Nishikawa, N., Natsume, M., and Koshioka, M. 1996. Purification and identification of new acyl-conjugated teasterones in lily pollen. *Biosci. Biotech. Biochem.*, **60**: 1416–1420.

Azpiroz, R., Wu, Y.W., LoCascio, J.C., and Feldmann, K.A. 1998. An *Arabidopsis* brassinosteroid-dependent mutant is blocked in cell elongation. *Plant Cell*, **10**: 219–230.

Bai, M.Y., Zhang, L.Y., Gampala, S.S., Zhu, S.W., Song, W.Y., Chong, K., and Wang, Z.Y. 2007. Functions of OsBZR1 and 14-3-3 proteins in brassinosteroid signaling in rice. *Proc. Natl. Acad. Sci. USA*, **104**: 13839–13844.

Bajguz, A. 2007. Metabolism of brassinosteroids in plants. *Plant Pysiol. Biochem.*, **45**: 95–107.

Bajguz, A., and Tretyn, A. 2003. The chemical characteristic and distribution of brassinosteroids in plants. *Phytochemistry*, **62**: 1023–1046.

Bancos, S., Nomura, T., Sato, T., Molnár, G., Bishop, G.J., Koncz, C., Yokota, T., Nagy, F., and Szekeres, M. 2002. Regulation of transcript levels of the *Arabidopsis* cytochrome P450 genes involved in brassinosteroid biosynthesis. *Plant Physiol.*, **130**: 504–513.

Bancos, S., Szatmári, A.M., Castle, J., Kozma-Bognár, L., Shibata, K., Yokota, T., Bishop, G.J., Nagy, and F., Szekeres, M. 2006. Diurnal regulation of the brassinosteroid-biosynthetic *CPD* gene in *Arabidopsis*. *Plant Physiol.*, **141**: 299–309.

Bishop, G.J. 2007. Refining the plant steroid hormone biosynthesis pathway. *Trends Plant Sci.*, **12**: 377–380.

Bishop, G.J., Nomura, T., Yokota, T., Harrison, K., Noguchi, T., Fujioka, S., Takatsuto, S., Jones, J.D.G., and Kamiya, Y. 1999. The tomato DWARF enzyme catalyses C-6 oxidation in brassinosteroid biosynthesis. *Proc. Natl. Acad. Sci. USA*, **96**: 1761–1766.

Bishop, G.J., and Yokota, T. 2001. Plants steroid hormones, brassinosteroids: current highlights of molecular aspects on their synthesis/metabolism, transport, perception and response. *Plant Cell Physiol.*, **42**: 114–120.

Caño-Delgado, A., Yin, Y., Yu, C., Vafeados, D., Mora-Garcia, S., Cheng, J.C., Nam, K.H., Li, J., and Chory, J. 2004. BRL1 and BRL3 are novel brassinosteroid receptors that function in vascular differentiation in *Arabidopsis*. *Development*, **131**: 5341–5351.

Castle, J., Szekeres, M., Jenkins, G., and Bishop, G.J. 2005. Unique and overlapping expression patterns of *Arabidopsis CYP85* genes involved in brassinosteroid C-6 oxidation. *Plant Mol. Biol.*, **57**: 129–140.

Choe, S. 2006. Brassinosteroid biosynthesis and inactivation. *Physiol. Plant*, **126**: 539–548.

Choe, S. 2007. Signal-transduction pathways toward the regulation of brassinosteroid biosynthesis. *J. Plant Biol.*, **50**: 225–229.

Choe, S., Dilkes, B.P., Fujioka, S., Takatsuto, S., Sakurai, A., and Feldmann, K.A. 1998. The *DWF4* gene of *Arabidopsis* encodes a cytochrome-P450 that mediates multiple 22α-hydroxylation steps in brassinosteroid biosynthesis. *Plant Cell*, **10**: 231–243.

Choe, S., Fujioka, S., Noguchi, T., Takatsuto, S., Yoshida, S., and Feldmann, K. 2001. Overexpression of *DWARF4* in the brassinosteroid biosynthetic pathway results in increased vegetative growth and seed yield in *Arabidopsis*. *Plant J.*, **26**: 573–582.

Fu, F.Q., Mao, W.H., Shi, K., Zhou, Y.H., Asami, T., and Yu, J.Q. 2008. A role of brassinosteroids in early fruit development in cucumber. *J. Exp. Bot.*, **59**: 2299–2308.

Fujioka, S., and Yokota, T. 2003. Biosynthesis and metabolism of brassinosteroids. *Ann. Rev. Plant Biol.*, **54**: 137–164.

Fujita, S., Ohnishi, T., Watanabe, B., Yokota, T., Takatsuto, S., Fujioka, S., Yoshida, S., Sakata, K., and Mizutani, M. 2006. *Arabidopsis* CYP90B1 catalyses the early C-22 hydroxylation of C_{27}, C_{28} and C_{29}, sterols. *Plant J.*, **45**: 765–774.

Fujioka, S., Takatsuto, S., and Yoshida, S. 2002. An early C-22 oxidation branch in the brassinosteroid biosynthetic pathway. *Plant Physiol.*, **130**: 930–939.

Goda, H., Shimada, Y., Asami, T., Fujioka, S., and Yoshida, S. 2002. Microarray analysis of brassinosteroid-regulated genes in *Arabidopsis*. *Plant Physiol.*, **130**: 1319–1334.

Hardtke, C.S., Dorcey, E., Osmont, K.S., and Sibout, R. 2007. Phytohormone collaboration: zooming in on auxin-brassinosteroid interactions. *Trends Cell Biol.*, **17**: 485–492.

He, J.X., Gendron, J.M., Sun, Y., Gampala, S.S., Gendron, N., Sun, C.Q., and Wang, Z.Y. 2005. BZR1 is a transcriptional repressor with dual roles in brassinosteroid homeostasis and growth responses. *Science*, **307**: 1634–1638.

Hedden, P., and Phillips, A.L. 2000. Gibberellin metabolism: new insights revealed by the genes. *Trends Plant Sci.*, **5**: 523–530.

Hirano, K., Aya, K., Hobo, T., Sakakibara, H., Kojima, M., Shim, R.A., Hasegawa, Y., Ueguchi-Tanaka, M., and Matsuoka, M. 2008. Comprehensive transcriptome analysis of phytohormone biosynthesis and signaling genes in microspore/pollen and tapetum of rice. *Plant Cell Physiol.*, **49**: 1429–1450.

Hong, Z., Ueguchi-Tanaka, M., Shimizu-Sato, S., Inukai, Y., Fujioka, S., Shimada, Y., Takatsuto, S., Agetsuma, M., Yoshida, S., Watanabe, Y., Uozu, S., Kitano, H., Ashikari, M., and Matsuoka, M. 2002. Loss-of-function of a rice brassinosteroid biosynthetic enzyme, C-6 oxidase, prevents the organized arrangement and polar elongation of cells in the leaves and stem. *Plant J.*, **32**: 495–508.

Hwang, J.Y., Joo, S.H., Park, C.H., Lee, S.C., and Kim, S.K. 2009. Substrate specificity for cytochrome P450 85A1 and 85A2 in brassinosteroids biosynthesis. *Bull. Korean Chem. Soc.*, **30**: 293–294.

Jeong, D.H., Lee, S., Kim, S.L., Hwang, I., and An, G. 2007. Regulation of brassinosteroid responses by Phytochrome B in rice. *Plant Cell Environ.*, **30**: 590–599.

Kang, J.G., Yun, J., Kim, D.H., Chung, K.S., Fujioka, S., Kim, J.I., Dae, H.W., Yoshida, S., Takatsuto, S., Song, P.S., and Park, C.M. 2001. Light and brassinosteroid signals are integrated via a dark-induced small G protein in etiolated seedling growth. *Cell*, **105**: 625–636.

Katsumata, T., Hasegawa, A., Fujiwara, T., Komatsu, T., Notomi, M., Abe, H., Natsume, M., and Kawaide, H. 2008. *Arabidopsis* CYP85A2 catalyzes lactonization reactions in the biosynthesis of 2-deoxy-7-oxalactone brassinosteroids. *Biosci. Biotech. Biochem.*, **72**: 2110–2117.

Kim, B.K., Fujioka, S., Takatsuto, S., Tsujimoto, M., and Choe, S. 2008. Castasterone is a likely end product of brassinosteroid biosynthetic pathway in rice. *Biochem. Biophys. Res. Comm.*, **374**: 614–619.

Kim, G.T., Fujioka, S., Kozuka, T., Tax, F.E., Takatsuto, S., Yoshida, S., and Tsukaya, H., 2005. CYP90C1 and CYP90D1 are involved in different steps in the brassinosteroid biosynthesis pathway in *Arabidopsis thaliana*. *Plant J.*, **41**: 710–721.

Kim, G.T., Tsukaya, H., Saito, Y., and Uchimiya, H., 1999. Changes in the shapes of leaves and flowers upon overexpression of cytochrome P450 in *Arabidopsis*. *Proc. Natl. Acad. Sci. USA*, **96**: 9433–9437.

Kim, H.B., Kwon, M., Ryu, H., Fujioka, S., Takatsuto, S., Yoshida, S., An, C.S., Lee, I., Hwang, I., and Choe, S. 2006. The regulation of *DWARF4* expression is likely a critical mechanism in maintaining the homeostasis of bioactive brassinosteroids in *Arabidopsis*. *Plant Physiol.*, **140**: 548–557.

Kim, T.W., Hwang, J.Y., Kim, Y.S., Joo, S.H., Chang, S.C., Lee, J.S., Takatsuto, S., and Kim, S.K. 2005. *Arabidopsis* CYP85A2, a cytochrome P450, mediates the Baeyer-Villiger oxidation of castasterone to brassinolide in brassinosteroid biosynthesis. *Plant Cell*, **17**: 2397–2412.

Kwon, M., Fujioka, S., Jeon, J.H., Kim, H.B., Takatsuto, S., Yoshida, S., An, C.S., and Choe, S. 2005. A double mutant for the *CYP85A1* and *CYP85A2* genes of *Arabidopsis* exhibits a brassinosteroid dwarf phenotype. *J. Plant Biol.*, **48**: 237–244.

Li, J., Nagpal, P., Vitart, V., McMorris, T.C., and Chory, J. 1996. A role for brassinosteroids in light-dependent development of *Arabidopsis*. *Science*, **272**: 398–401.

Li, J., Nam, K.H., Vafeados, D., and Chory, J., 2001. *BIN2*, a new brassinosteroid-insensitive locus in *Arabidopsis*. *Plant Physiol.*, **127**: 14–22.

Luccioni, L.G., Oliviero, K.A., Yanofsky, M.J., Boccalandro, H.E., and Casal, J.J. 2002. Brassinosteroid mutants uncover fine tuning of phytochrome signaling. *Plant Physiol.*, **128**: 173–181.

Ma, L., Li, J., Qu, L., Hager, J., Chen, Z., Zhao, H., and Deng, X.W. 2001. Light control of *Arabidopsis* development entails coordinated regulation of genome expression and cellular pathways. *Plant Cell*, **13**: 2589–2607.

Marsolais, F., Boyd, J., Paredes, Y., Schinas. A.M., Garcia, M., Elzein, S., and Varin, L. 2007. Molecular and biochemical characterization of two brassinosteroid sulfotransferases from *Arabidopsis*, AtST4a (At2g14920) and AtST1 (At2g03760). *Planta*, **225**: 1233–1244.

Marsolais, F., Sebastia, C.H., Rousseau, A., and Varin, L. 2004. Molecular and biochemical characterization of BNST4, an ethanol-inducible steroid sulfotransferase from *Brassica napus*, and regulation of *BNST* genes by chemical stress and during development. *Plant Sci.*, **166**: 1359–1370.

Mathur, J., Molnár, G., Fujioka, S., Takatsuto, S., Sakurai, A., Yokota, T., Adam, G., Voigt, B., Nagy, F., Maas, C., Schell, J., Koncz, C., and Szekeres, M. 1998. Transcription of the *Arabidopsis CPD* gene, encoding a steroidogenic cytochrome P450, is negatively controlled by brassinosteroids. *Plant J.*, **14**: 593–602.

Michael, T.P., Breton, G., Hazen, S.P., Priest, H., Mockler, T.C., Kay, S.A., and Chory, J., 2008. A morning-specific phytohormone gene expression program underlying rhythmic plant growth. *PLOS Biol*, **6**: e225.

Mockaitis, K., and Estelle, M. 2004. Integrating transcriptional controls for plant cell expansion. *Genome Biol.*, **5**: 245.

Montoya, T., Nomura, T., Yokota, T., Farrar, K., Harrison, K., Jones, J.G.D., Kaneta, T., Kamiya, Y., Szekeres, M., and Bishop, G.J. 2005. Patterns of *Dwarf* expression and brassinosteroid accumulation in tomato reveal the importance of brassinosteroid synthesis during fruit development. *Plant J.*, **42**: 262–269.

Mouchel, C.F., Osmont, K.S., and Hardtke, C.S. 2006. *BRX* mediates feedback between brassinosteroid levels and auxin signaling in root growth. *Nature*, **443**: 458–461.

Nakamura, M., Satoh, T., Tanaka, S.I., Mochizuki, N., Yokota, T., and Nagatani, A. 2005. Activation of the cytochrome P450 gene, *CYP72C1*, reduces the levels of active brassinosteroids *in vivo*. *J. Exp. Bot.*, **56**: 833–840.

Neff, M.M., Nguyen, S.M., Malancharuvil, E.J., Fujioka, S., Noguchi, T., Seto, H., Tsubuki, M., Honda, T., Takatsuto, S., Yoshida, S., and Chory, J. 1999. *BAS1*: a gene regulating brassinosteroid levels and light responsiveness in *Arabidopsis*. *Proc. Natl. Acad. Sci. USA*, **96**: 15316–15323.

Nemhauser, J.L. 2008. Dawning of a new era: photomorphogenesis as an integrated molecular network. *Curr. Opin. Plant Biol.*, **11**: 4–8.

Nemhauser, J.L., Mockler, T.C., and Chory, J. 2004. Interdependency of brassinosteroid and auxin signaling in *Arabidopsis*. *PLOS Biol.*, **2**: E258.

Noguchi, T., Fujioka, S., Choe, S., Takatsuto, S., Yoshida, S., Yuan, H., Feldmann, K.A., and Tax, F.E., 1999. Brassinosteroid-insensitive dwarf mutants of *Arabidopsis* accumulate brassinosteroids. *Plant Physiol.*, **121**: 743–752.

Nomura, T., and Bishop, G.J. 2006. Cytochrome P450s in plant steroid hormone synthesis and metabolism. *Phytochem. Rev.*, **5**: 421–432.

Nomura, T., Kushiro, T., Yokota, T., Kamiya, Y., Bishop, G.J., and Yamaguchi, S. 2005. The last reaction producing brassinolide is catalyzed by cytochrome P450s, CYP85A3 in tomato and CYP85A2 in *Arabidopsis*. *J. Biol. Chem.*, **280**: 17873–17879.

Nomura, T., Sato, T., Bishop, G.J., Kamiya, Y., Takatsuto, S., and Yokota, T. 2001. Accumulation of 6-deoxocathasterone and 6-deoxocastasterone in *Arabidopsis*, pea and tomato is suggestive of common rate-limiting steps in brassinosteroid biosynthesis. *Phytochemistry*, **57**: 171–178.

Nomura, T., Ueno, M., Yamada, Y., Takatsuto, S., Takeuchi, Y., and Yokota, T. 2007. Roles of brassinosteroids and related mRNAs in pea seed growth and germination. *Plant Physiol.*, **143**: 1680–1688.

Ohnishi, T., Bancos, S., Watanabe, B., Fujioka, S., Yokota, T., Sakata, K., Szekeres, M., and Mizutani, M. 2007. Biochemical characterization of brassinosteroid C-3 oxidase. *Plant Cell Physiol.*, **48**: S200.

Ohnishi, Y., Nomura, T., Watanabe, B., Ohta, D., Yokota, T., Miyagawa, H., Sakata, K., and Mizutani, M. 2006a. Tomato cytochrome P450 CYP734A7 functions in brassinosteroid catabolism. *Phytochemistry*, **67**: 1895–1906.

Ohnishi, T., Szatmári, A.M., Watanabe, B., Fujita, S., Bancos, S., Koncz, C., Lafos, M., Shibata, K., Yokota, T., Sakata, K., Szekeres, M., and Mizutani, M. 2006b. C-23 hydroxylation by *Arabidopsis* CYP90C1 and CYP90D1 reveals a novel shortcut in brassinosteroid biosynthesis. *Plant Cell*, **18**: 3275–3288.

Ohnishi, T., Yokota, T., and Mizutani, M. 2009. Insights into the function and evolution of P450s in plant steroid metabolism. *Phytochemistry*, **70**: 1918–1929.

Pan, Y., Michael, T.P., Hudson, M.E., Kay, S.A., Chory, J., and Schuler, M.A. 2009. Cytochrome P450s as reporters for circadian-regulated pathways. *Plant Physiol.*, **150**: 858–878.

Paponov, I.A., Paponov, M., Teale, W., Menges, M., Chakrabortee, S., Murray, J.A.H., and Palme, K., 2008. Comprehensive transcriptome analysis of auxin responses in *Arabidopsis*. *Molec. Plant*, **2**: 321–337.

Park, W., Kim, H.B., Kim, W.T., Park, P.B., An, G., and Choe, S., 2009. Rice *bending lamina 2 (bla2)* mutants are defective in a cytochrome P450 (CYP734A6) gene predicted to mediate brassinosteroid catabolism. *J. Plant Biol.*, **49**: 469–476.

Pien, S., Wyrzykowska, J., and Fleming, A.J. 2001. Novel marker genes for early leaf development indicate spatial regulation of carbohydrate metabolism within the apical meristem. *Plant J.*, **25**: 663–674.

Pilati, S., Perazzolli, M., Malossini, A., Cestaro, A., Demattè, L., Fontana, P., Dal Ri, A., Viola, R., Velasco, R., and Moser, C., 2007. Genome-wide transcriptional analysis of grapevine berry ripening reveals a set of genes similarly modulated during three seasons and the occurrence of an oxidative burst at vèraison. *BMC Genomics*, **8**: 428.

Poppenberger, B., Fujioka, S., Soeno, K., George, G.L., Vaistij, F.E., Hiranuma, S., Seto, H., Takatsuto, S., Adam, G., Yoshida, S., and Bowles, D. 2005. The UGT73C5 of *Arabidopsis thaliana* glucosylates brassinosteroids. *Proc. Natl. Acad. Sci. USA*, **102**: 15253–15258.

Robertson, F.C., Skeffington, A.W., Gardner, M.J., and Webb, A.A.R. 2009. Interactions between circadian and hormonal signaling in plants. *Plant Mol. Biol.*, **69**: 419–427.

Rouleau, M., Marsolais, F., Richard, M., Nicolle, L., Voigt, B., Adam, G., and Varin, L. 1999. Inactivation of brassinosteroid biological activity by a salicylate-inducible steroid sulfotransferase from *Brassica napus*. *J. Biol. Chem.*, **274**: 20925–20930.

Scacchi, E., Osmont, K.S., Beuchat, J., Salinas, P., Navarrete-Gómez, M., Trigueros, M., Ferrándiz, C., and Hardtke, C.S., 2009. Dynamic, auxin-responsive plasma membrane-to-nucleus movement of *Arabidopsis* BRX. *Development*, **136**: 2059–2067.

Shimada, Y., Fujioka, S., Miyauchi, N., Kushiro, M., Takatsuto, S., Nomura, T., Yokota, T., Kamiya, Y., Bishop, and G.J., Yoshida, S., 2001. Brassinosteroid-6-oxidases from *Arabidopsis* and tomato catalyze multiple C-6 oxidations in brassinosteroid biosynthesis. *Plant Physiol.*, **126**: 770–779.

Shimada, Y., Goda, H., Nakamura, A., Takatsuto, S., Fujioka, S., and Yoshida, S., 2003. Organ-specific expression of brassinosteroid-biosynthetic genes and distribution of endogenous brassinosteroids in *Arabidopsis*. *Plant Physiol.*, **131**: 287–297.

Sibout, R., Sukumar, P., Hettiarachchi, C., Holm, M., Muday, and G.K., Hardtke, C.S., 2006. Opposite root growth phenotypes of *hy5* versus *hy5 hyh* mutants correlate with increased constitutive auxin signaling. *PLoS Genet.*, **2**: 1898–1911.

Soeno, K., Asakawa, S., Natsume, M., and Abe, H., 2000. Reversible conversion between teasterone and its ester conjugates in lily cell cultures. *J. Pesticide Sci.*, **25**: 117–122.

Symons, G.M., and Reid, J.B. 2004. Brassinosteroids do not undergo long-distance transport in pea. Implications for the regulation of endogenous brassinosteroid levels. *Plant Physiol.*, **135**: 2196–2206.

Symons, G.M., Davies, C., Shavrukov, Y., Dry, I.B., Reid, J.B., and Thomas, M.R. 2006. Grapes on steroids. Brassinosteroids are involved in grape berry ripening. *Plant Physiol.*, **140**: 150–158.

Symons, G.M., Schultz, L., Kerckhoffs, L.H., Davies, N.W., Gregory, D., and Reid, J.B., 2002. Uncoupling brassinosteroid levels and de-etiolation in pea. *Physiol. Plant.*, **115**: 311–319.

Szekeres, M., and Bishop, G.J. 2006. Integration of brassinosteroid biosynthesis and signaling. *In*: P Hedden, S Thomas (Eds), Plant hormone signaling. *Annual Plant Reviews*, 24. Blackwell, Oxford: pp. 67–92.

Szekeres, M., Németh, K., Koncz-Kálmán, Z., Mathur, J., Kauschmann, A., Altmann, T., Rédei, G., Nagy, F., Schell, J., and Koncz, C. 1996. Brassinosteroids rescue the deficiency of CYP90, a cytochrome P450, controlling cell elongation and de-etiolation in *Arabidopsis*. *Cell*, **85**: 171–182.

Takahashi, N., Nakazawa, M., Shibata, K., Yokota, T., Ishikawa, A., Suzuki, K., Kawashima, M., Ichikawa, T., Shimada, H., and Matsui, M. 2005. *shk1-D*, a dwarf *Arabidopsis* mutant caused by activation of the *CYP72C1* gene has altered brassinosteroid levels. *Plant J.*, **42**: 13–22.

Tanaka, K., Asami, T., Yoshida, S., Nakamura, Y., Matsuo, T., and Okamoto, S., 2005. Brassinosteroid homeostasis in *Arabidopsis* is ensured by feedback expressions of multiple genes involved in its metabolism. *Plant Physiol.*, **138**: 1117–1125.

Turk, E.M., Fujioka, S., Seto, H., Shimada, Y., Takatsuto, S., Yoshida, S., Denzel, M.A., Torres, Q.I., and Neff, M.M., 2003. CYP72B1 inactivates brassinosteroid hormones: an intersection between photomorphogenesis and plant steroid signal transduction. *Plant Physiol.*, **133**: 1643–1653.

Turk, E.M., Fujioka, S., Seto, H., Shimada, Y., Takatsuto, S., Yoshida, S., Wang, H., Torres, Q.I., Ward, J.M., Murthy, G., Zhang, J., Walker, J.C., and Neff, M.M. 2005. *BAS1* and *SOB7* act redundantly to modulate *Arabidopsis* photomorphogenesis via unique brassinosteroid inactivation mechanisms. *Plant J.*, **42**: 23–34.

Vert, G., Nemhauser, J.L., Geldner, N., Hong, F., and Chory, J. 2005. Molecular mechanisms of steroid hormone signaling in plants. *Annu. Rev. Cell Dev. Biol.*, **21**: 177–201.

Wang, Z.Y., Nakano, T., Gendron, J., He, J., Chen, M., Vafeados, D., Yang, Y., Fujioka, S., Yoshida, S., Asami, T., and Chory, J. 2002. Nuclear-localized BZR1 mediates brassinosteroid-induced growth and feedback suppression of brassinosteroid biosynthesis. *Dev. Cell*, **2**: 505–513.

Yamaguchi, S., and Kamiya, Y., 2000. Gibberellin biosynthesis: its regulation by endogenous and environmental signals. *Plant Cell Physiol.*, **41**: 251–257.

Yamamoto, R., Fujioka, S., Iwamoto, K., Demura, T., Takatsuto, S., Yoshida, S., and Fukuda, H. 2007. Co-regulation of brassinosteroid biosynthesis-related genes during xylem cell differentiation. *Plant Cell Physiol.*, **48**: 74–83.

Yokota, T., Sato, T., Takeuchi, Y., Nomura, T., Uno, K., Watanabe, T., and Takatsuto, S. 2001. Roots and shoots of tomato produce 6-deoxo-28-norcastasterone, 6-deoxo-28-nortyphasterol and 6-deoxo-28-norcastasterone, possible precursors of 28-norcastasterone. *Phytochemistry*, **58**: 233–238.

Yuan, T., Fujioka, S., Takatsuto, S., Matsumoto, S., Guo, X., He, K., Russell, S.D., and Li, J. 2007. *BEN1*, a gene encoding a dihydroflavonol 4-reductase (DFR)-like protein, regulates the levels of brassinosteroids in *Arabidopsis thaliana*. *Plant J.*, **51**: 220–233.

Chapter 4

BRASSINOSTEROID SIGNALING IN RICE

H. NAKAGAWA, A. TANAKA AND M. MORI

Division of Plant Sciences, National Institute of Agrobiological Sciences, 2-1-2 Kannondai, Tsukuba, Ibaraki 305-8602, Japan

Abstract: Brassinosteroids (BRs) are steroidal hormones that have various physiological and morphological effects on plants. The mechanisms of perception and signal transduction of BRs have been clarified mainly in *Arabidopsis* using BR insensitive mutants. In contrast, little is known about BR signaling in monocots. Recently, analyses of rice mutants and BR-regulated genes have provided new insights into BR signaling in rice. In addition, engineering BR signaling genes improved rice architecture and increased grain yield. This chapter summarizes recent advances in BR signaling research in rice.

Key words: rice, brassinosteroids, signal transduction

1. INTRODUCTION: UTILITY OF RICE FOR BR SIGNALING STUDIES

In *Arabidopsis*, several key factors for BR signaling have been cloned by chemical screening approaches using BR (Li and Chory, 1997; Li *et al.*, 2001), BR-synthesis inhibitor, Brassinazole (Wang *et al.*, 2002) or suppressor screening in the BR receptor mutant, *bri1* (Yin *et al.*, 2002). Such chemical screening approaches are powerful systems in *Arabidopsis*, and these have been successfully used to identify many important signaling factors for other phytohormones such as ABA (*ABI1 to -5*), ethylene (*EIN1 to -5, ETR1*), cytokinin (*CKI1, CKI2, CRE1, AHK*) and GA (*GAI/RGA*). Such large-scale screening approaches have not been successfully performed in rice because of its larger plant size and longer life cycle.

S. Hayat and A. Ahmad (eds.), *Brassinosteroids: A Class of Plant Hormone*,
DOI 10.1007/978-94-007-0189-2_4,

However, rice has at least three great advantages for research in BR signaling. The first advantage is the phenomenon known as "Lamina inclination in rice" or, the degree of bending between the leaf blade and leaf sheath, because this is very responsive to external signals, light, chemicals, and phytohormones especially to IAA and BRs. Lamina joint bending to exogenous BR is extremely sensitive, and it has become a commonly used bioassay for BR activity (Fujioka *et al.*, 1998). Lamina joint bending of rice also reflects endogenous BR levels and internal BR signaling. In fact, rice BR biosynthesis mutants with moderately reduced active BR levels (*d2*, *d11*, *osdwarf4*) share a common, semi-dwarf and erect leaf (less lamina joint bending) phenotype (Hong *et al.*, 2003; Tanabe *et al.*, 2005; Sakamoto *et al.*, 2006) that is distinguishable from dwarf mutants ascribed to disturbances in other plant hormones. Two major genes for BR signal perception in rice, *OsBRI1* and *RGA1* have been initially assigned to BR signaling and the corresponding mutants (*d61* and *d1*, respectively) show semi-dwarf and erect leaf phenotypes. In addition, some BR signaling factors have been identified using the altered lamina bending phenotype and several genes show increased lamina bending or erect leaf phenotype, when they are overexpressed or repressed (Table 1). Recently, our team and another group have independently identified helix-loop-helix transcription factors that control bending of the lamina-joint in rice; overexpression of these factors induce the extreme lamina-joint bending phenotype (Tanaka *et al.*, 2009; Zhang *et al.*, 2009).

The second advantage for rice is that it has been extensively studied as the primary monocot model in plant molecular biology as well as a model crop since the mid 1990s. Therefore, a lot of genetic information, genome sequences and gene annotation, expression data, resources such as gene-knockout mutant population, activation-tagging lines, promoter trap lines, collection of cultivars and wild populations are available for rice research. The physiology and development process of growing rice have been extensively studied at the molecular and anatomical levels (Hoshikawa, 1989; Itoh *et al.*, 2005).

The third advantage for rice is that it is one of the major crops cultivated in the world and it provides sustenance to more than two billion people. Therefore, useful BR-related traits can be directly applied for breeding of rice and food production. Mild mutations of BR-signaling or BR-biosynthetic gene lead to characteristic semi-dwarf phenotype with erect leaves and shorter panicles. This phenotype has been shown to improve grain yield in high density plantings (Sakamoto *et al.*, 2006; Morinaka *et al.*, 2006; Wu *et al.*, 2008). In addition, modification of BR expression using a vegetative-tissue promoter can be used to improve filling of rice grains in field conditions (Li *et al.*, 2009). Moreover, BR has been shown to enhance tolerance to abiotic stresses, such as salt, high temperature, cold and drought (reviewed by Bajguz and Hayat, 2009). Therefore, engineering of BR-signaling or its biosynthetic pathway may be useful for the breeding of new cultivars that can be grown in unfavorable climates and field conditions.

Table 1. BR signaling genes in rice

Gene	*Arabidopsis* homologues	Protein	Loss of function	Gain of function	References
BR signaling genes described in this review					
OsBRI1	*BRI1*	BR receptor	erect leaves, semidwarf, BR insensitive		Yamamuro *et al.* (2000)
OsBRL1, OsBRL3	*BRL1, BRL3*	BR receptor			Nakamura *et al.* (2006)
OsBAK1	*BAK1*	LRR-kinase	erect leaves, BR insensitive	semidwarf, enhanced lamina joint inclination	Li *et al.* (2009)
OsGSK1	*BIN2*	GSK3-like kinase	enhanced BR-response		Koh *et al.* (2007)
OsBZR1	*BZR1*	Transcription Factor	erect leaves, semidwarf, enhanced lamina joint inclination	enhanced lamina joint inclination	Bai *et al.* (2007)
14-3-3	*14-3-3*	14-3-3			Bai *et al.* (2007)
DLT		GRAS family	erect leaves, semidwarf,		Tong *et al.* (2009)
RGA1	*GPA1*	G-protein α-subunit	semidwarf (*d1*), erect leaves, small grain	Long grain, enhanced coleoptile response to BR	Wang *et al.* (2006), Oki *et al.* (2009b and 2009c)
BU1	*ATBS1, PRE1, KIDARI*	HLH transcription factor	erect leaves	enhanced lamina joint inclination	Tanaka *et al.* (2009), this review
ILI1	*ATBS1, PRE1, KIDARI*	HLH transcription factor	erect leaves	enhanced lamina joint inclination	Zhang *et al.* (2009)
IBH1	*AtIBH1*	bHLH transcription factor	enhanced lamina joint inclination	erect leaves	Zhang *et al.* (2009)
Other BR signal related genes					
*OsCKI*1		Casein kinase 1	BR-insensitive		Liu *et al.* (2003)
OsBLE3		Unknown protein	BR insensitive lamina joint, semi-dwarf		Yang *et al.* (2006)

(*continued*)

(*continued Table 1.*)

Gene	*Arabidopsis* homologues	Protein	Loss of function	Gain of function	References
OsMDP1		MADS box protein	enhanced BR-response		Duan *et al.* (2006)
OsMADS22, OsMADS55	*SVP, AGL24*	MADS box protein (SVP-group)	enhanced lamina joint inclination, reduced stem elongation	short panicle, affect stem elongation	Lee *et al.* (2008)
OsGSR1		GAST-family			Wang *et al.* (2008a)
OsIAA1		IAA family		enhanced lamina joint inclination	Song *et al.* (2009)
OsARF1	*ARF2*	ARF family	altered BR response		Song *et al.* (2009)

2. *Os*BRI1 PATHWAY AND OTHER SIGNALING FACTORS

In *Arabidopsis*, BR signals perceived by the receptor-kinase BRI1 (Brassinosteroid insensitive 1, Li and Chory, 1997; Wang *et al.*, 2001) are transduced by downstream factors that control transcription of BR regulated genes (reviewed by Kim and Wang, 2010). Upon sensing BRs, BRI1 auto-phosphorylates itself, and releases an inhibitor for BRI1, BKI (BRI1-kinase inhibitor 1). BRI1 is heterodimerized with BAK1 (BRI1 associated kinase 1) and transphosphorylates BAK1. BRI1 is activated *via* interphosphorylation by BAK1 and it subsequently phosphorylates BSKs (Brassinosteroid signaling kinases). Phosphorylated BSKs bind to BSU1 (BRI1 suppressor 1) in turn BSU1 subsequently inactivates BIN2 (Brassinosteroid insensitive 2), a negative regulator of the nuclear transcription factors, BZR1 (Brassinazole resistant 1) and BES1 (BRI1 EMS suppressor 1). Through the inactivation of BIN2, BZR1 and BES1 can control transcription of BR-regulated genes (BRI1-dependent BR signaling pathway). Under low BR condition, BKI (BRI1-kinase inhibitor 1) and 14-3-3 proteins form inactive complexes with BRI1 and BZR1, respectively, to repress BR signal transduction and response.

The rice orthologue of BRI1 (OsBRI1) and some downstream factors (OsBAK1, OsGSK1, OsBZR1, 14-3-3) have already been shown to be involved in BR signaling (Table 1) so the BRI1-dependent BR signaling pathway is likely to be conserved in rice (Figure 1). However, some of the essential downstream factors (BKI1, BSKs, and BSU1) in rice have not been characterized. The rice genome encodes putative counterparts of these *Arabidopsis* factors. For example, for BSKs, BIN2 and BES1, highly homologous sequences (80–90% sequence identity) are encoded in the rice genome.

For BKI1 and BSU1, moderately similar sequences (50–60% identity) are found in the rice genome. Functional analyses of these homologous factors in rice will be required to understand the role of OsBRI1-mediated BR signaling pathway in rice.

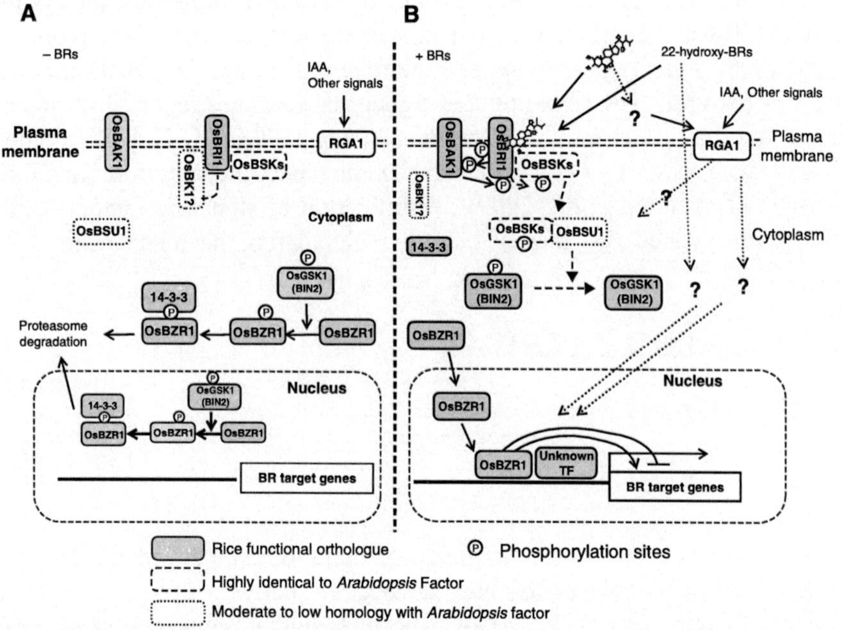

Figure 1. Brassinosteroid signal transduction pathway in rice. Gray rectangles enclosed by solid lines indicate rice factors that function in BR signaling. Dashed rectangles indicate putative rice homologues showing high sequence identity with *Arabidopsis* factors; however, these functions in BR signaling have not yet been verified in rice. Dotted rectangles indicate putative rice homologues showing moderate or low sequence identity with *Arabidopsis* factors. (A) Low BR condition. OsBRI1 exists as homo-oligomer and keeps OsBRI1 in an inactive state through interaction with OsBKI1 (BRI1 kinase inhibitor). On the other hand, the nuclear transcription factor OsBZR1 is phosphorylated by OsGSK1, then the phosphorylated OsBZR1 is captured by a 14-3-3 protein, and subsequently degraded by the proteasome. (B) High BR condition. Upon binding of BRs, OsBRI1 autophosphorylates its kinase domain and initiates the basal BR response. Activated OsBRI1 releases OsBKI1, allowing binding of OsBAK1 (formation of OsBRI1/OsBAK1 complex). Interphosphorylation between OsBRI1 and OsBAK1 fully activates the BR signaling complex. OsBRI1 phosphorylates its substrates (OsBSKs etc.). The functions of OsBSK and OsBSU1 homologues have not been verified in rice. However, it has been shown in *Arabidopsis* that activated BRI1 phosphorylates and releases BSKs. The phosphorylated BSK1 binds and activates BSU1. After that, BSU1 inactivates BIN2 through dephosphorylation. Under high BR condition, OsBZR1 is not phosphorylated by OsGSK1. The non-phosphorylated OsBZR1 is readily localized into the nucleus and controls transcription of BR-regulated genes both negatively and positively. In addition, RGA1 is also involved in BR signaling or response.

In addition to the OsBRI1 pathway, RGA1 (Rice heterotrimeric G protein alfa 1) is also involved in BR-signaling (Wang *et al.*, 2006; Oki *et al.*, 2009c). G protein α plays an important role in many physiological processes in rice and *Arabidopsis* (reviewed by Perfus-Barbeoch *et al.*, 2004). In *Arabidopsis*, the orthologous protein GPA1 (G protein alfa 1) has been shown to be involved in BR-mediated elongation of hypocotyl and BR-controlled GA sensitivity of seed germination (Ullah *et al.*, 2001; Ullah *et al.*, 2002; Gao *et al.*, 2008). To date, RGA1/GPA1 upstream factors that perceive BR signals (unknown receptor or BRI1?), downstream factors that mediate the signaling pathway, and the relationship with the BRI1 pathway are still not clear. BR signaling and signal-related genes reported in rice are listed in Table 1. Recently, Nakamura *et al.* reported that C-22-hydroxylated-BRs mediate auxin-induced lamina joint bending independently from functional OsBRI1 (Nakamura *et al.*, 2009). Identification of signaling factors for the C-22-hydroxylated-BRs may be helpful in elucidating this mechanism.

3. ***Os*****BRI1 PATHWAY**

3.1 ***OsBRI1***

3.1.1 OsBRI1

OsBRI1 gene, the rice homologue of membrane-localized BR receptor kinase, *BRI1* of *Arabidopsis*, was isolated by analyses of rice *d61* dwarf mutants (Yamamuro *et al.*, 2000). The *d61* mutant showed a semi-dwarf phenotype due to the reduced elongation of leaf sheath and of specific culm internode. In addition, *d61* mutants had erect leaves because of impaired lamina-joint bending. Since BRs enhance lamina-joint bending in rice, erect leaves of *d61* mutants indicate that the *D61* gene may be involved in the biosynthesis or signal transduction of BRs. The gene for *D61*was cloned and it appeared to be the rice homologue of *Arabidopsis BRI1,* and so it was named *OsBRI1* (*Oryza sativa BRI1*). OsBRI1 protein has extensive similarity to BRI1, and it has all of the functional domains of the BRI1 protein (Figure 2A).

OsBRI1 protein belongs to the Leucine Rich Repeat Receptor-like kinase (LRR-RLK) family and contains four functional domains:

(i) Extracellular 22 LRRs (Leucine-Rich Repeats) in its N-terminal region,
(ii) 70-amino acids of the island domain (ID) that is important for BR-binding activity,
(iii) juxta-membrane (JM) region,
(iv) intracellular kinase domain (KD) in C-terminus region.

The functions of these BRI1 domains have been deduced by extensive studies in *Arabidopsis* (Wang *et al.*, 2001; Russinova *et al.*, 2004; Kinoshita *et al.*, 2005; Wang *et al.*, 2008). The corresponding functions of these domains in OsBRI1 are still not well-elucidated.

The LRR region of BRI1 is a protein-protein interaction domain that functions in the homo-/ hetero-dimerization with itself or its co-receptor BAK1 (Figure 2B, Chinchilla *et al.*, 2009). The island domain (ID) and its downstream 24 amino acids are required for steroid-binding activity and are highly conserved across species (Caño-Delgado *et al.*, 2004; Kinoshita *et al.*, 2005). Two active BR-receptors in *Arabidopsis*, BRL1, BRL3 (BRI1 like 1 and -3) also show similarity in this region. In contrast, BRL2, a more distant homologue of BRI1 which does not show BR binding activity, has a less similar island domain. The JM region and KD are required for phosphorylation, homodimerization, interaction with co-receptor BAK1, inhibitor BKI, and BSKs (reviewed by Kim and Wang, 2010).

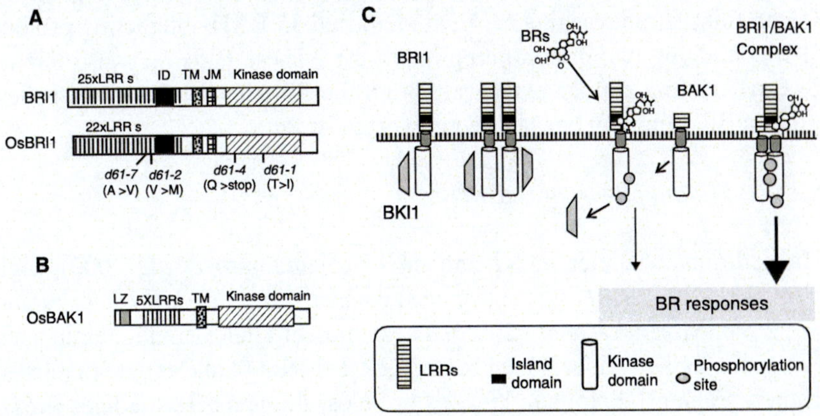

Figure 2. Structures of OsBRI1, OsBAK1 and a model for BR signal perception mediated by BRI1/BAK1 complex. (A) Structure of BRI1 and OsBRI1. BRI1 and OsBRI1 are composed of 25 or 22 leucine rich repeats (LRRs), 70-amino acids island domain (ID), transmembrane domain (TM), juxtamembrane domain (JM) and kinase domain. The positions of four mutation sites in *OsBRI1* in the *d61* mutants, whose phenotypes are mentioned in this review, are indicated. (B) Structure of OsBAK1. OsBAK1 is composed of 4 leucine-zipper domains (LZ), 5 leucine rich repeats (LRRs), a transmembrane domain (TM) and a kinase domain. (C) Proposed model for BR signal perception and transduction mediated by BRI1/BAK1 complex. Without BR signal, BRI1 is kept as an inactive monomer or homodimer *via* interaction with its inhibitor BKI1 (BRI1 kinase inhibitor 1). Upon binding of a brassinosteroid molecule to its island domain, BRI1 activates itself by autophosphorylation. The activated BRI1 releases BKI1, in turn it interacts with BAK1. BRI1 activates BAK1 through transphosphorylation. BRI1 is fully activated through interphoshorylations between BRI1 and BAK1, after which the fully activated BRI1 phosphorylates its signaling targets (BSKs *etc.*).

The activation of BRI1 by BR is shown in Figure 2C. Under low BR condition, BRI1 exists as a homo-oligomer. The BKI inhibitor of BRI binds the intracellular kinase domain of BRI1 and keeps BRI1 in an inactive state. Upon binding with BRs, BRI1 autophosphorylates its kinase domain, initiating the basal BR response. Activated BRI1 phosphorylates BKI, causes the release of BKI, and allows the binding of BAK1 to form the BRI1/BAK1 complex. Interphosphorylation between BRI1 and BAK1 causes full activation of the BR signaling complex. BRI1 subsequently phosphorylates its substrates (BSKs etc.) and the BR signal transduces other downstream factors. The rice genome encodes orthologues of most of the aforementioned genes so homologous mechanisms are likely to mediate BR-signaling in rice.

Rice has at least two putative proteins that are homologous (~80% identity) to BSK that might be direct targets of OsBRI1. However, the function of rice BSK homologues in BR signaling and its interaction with OsBRI1 have not been studied so far. Direct phosphorylation screening of rice cDNA library was performed as an alternative approach to identify BRI1 targets in rice (Hirabayashi *et al.*, 2004). Using the recombinant kinase domain of *Arabidopsis* BRI1 protein, Hirabayashi *et al.* have identified 35 BRI1-interacting proteins (BIPs) containing two proton-pump interactors. These BIPs may also interact with OsBRI1 to a similar extent, although the functional relevance of these factors in BR signaling has not been reported so far.

3.1.2 Mutant alleles of OsBRI1

In addition to alleles *d61-1* and *d61-2* (Yamamuro *et al.*, 2000), eight mutant alleles of *OsBRI1* gene have been isolated (*d61-3* to *d61-10*, Nakamura *et al.*, 2006; Morinaka *et al.*, 2006). These mutant alleles contain amino acid substitutions in the various domains of OsBRI1: LRRs (4 alleles), ID (2 alleles), transmembrane (1 allele) and kinase (3 alleles) domains. Two alleles (*d61-4* and *d61-6*) have premature stop codons that are predicted to produce truncated OsBRI1 proteins. The *d61* mutant alleles are classified into three groups according to fertility and plant height at heading stages. The weakest group (*d61-1*, *d61-7* to *d61-10*, mild-phenotype) of mutants is fertile and approximately 80–90% of the height of wild-type plants. The intermediate mutant for *d61-2* is fertile and 50–60% of the height of wild-type plants. Plants with mild or intermediate alleles of *OsBRI1* gene have dark green and erect leaves. In contrast to these two groups, the most abnormal mutant group (*d61-3* to *d61-6*) are sterile and extremely dwarf (only 5 cm in height after 6 months) and show multiple abnormalities similar to a knockout mutant of the BR biosynthetic BR C-6 oxidase gene, the *brd1* mutant (Mori *et al.*, 2002; Hong *et al.*, 2002).

3.1.3 d61-4, a putative null mutant of OsBRI1

The *d61-4* mutant of the *OsBRI1* gene has a premature stop codon in its kinase domain (Figure 2A). Because of the loss of catalytic domain and the consequent extreme mutant phenotype, *d61-4* is considered to be a null mutant of the *OsBRI1* gene. The phenotype of the *d61-4* mutant throughout most of its growth and development has been described in detail (Nakamura *et al.*, 2006). Embryos of the *d61-4* mutant appeared to be normal except for their size and longitudinal cell elongation. After germination, the *d61-4* mutant seedlings were severely stunted, and produced rolled and twisted dark green leaves. In addition to abnormality in leaf appearance, the elongation of the leaf sheath was extremely repressed, resulting in alteration of the ratio between leaf blade to leaf sheath. In contrast to severe defects in *d61-4* shoots, roots of the *d61-4* mutant are relatively normal in their appearance, size and number. The difference in severity between aerial organs and roots of the *d61-4* mutant phenotype will be discussed in a later section on functional divergence between OsBRI1 and other members of OsBRI1 family (OsBRL1, OsBRL3).

3.1.4 Agronomic use of d61-7, the weakest allele of OsBRI1, and OsBRI1 knock-down plants

In modern agriculture, large amounts of fertilizer are essential to increase grain yield. However fertilizer overuse also causes excessive plant height, leading to an increased risk of yield loss through lodging. Therefore, the dwarf phenotype, such as that used in the "Green revolution", is one of the most valuable traits in crop breeding. On the other hand, the erect-leaf phenotype, by enhancing light capture efficiency, has the potential to increase grain yield under dense planting. Mild mutations in BR-biosynthetic or signaling genes can confer both of these agronomically important traits (semi-dwarf stature and erect leaves). These mutations have been used to increase grain yield and biomass production in rice (Morinaka *et al.*, 2006; Sakamoto *et al.*, 2006). In East Asia, the spontaneous mutant of the barley homologue of *BRI1* (*HvBRI1*), "*uzu*", a dwarf phenotype, has been widely used in barley breeding to improve canopy and lodging resistance (Saisho *et al.*, 2004).

OsBRI1 gene has been demonstrated as a potential target for rice breeding to increase biomass and grain yield under dense planting condition (Morinaka *et al.*, 2006). The *d61-1* and *d61-2* alleles also show a semi-dwarf and erect-leaf phenotype but they are not used for breeding because their negative pleiotropic effect on the reproductive organs reduce grain yield. Among the other eight alleles, the weakest allele, *d61-7*, was tested for its effect on grain and biomass yield in paddy field conditions. Similar to *d61-1* and *d61-2*, the *d61-7* mutant shows a semi-dwarf stature and erect leaves but its panicles are significantly longer and bear more grains than the wild-type plant. However, it produces smaller grains and shows decreased fertility compared to the wild-type. Consequently, at the conventional planting density of 22.2 plants m^{-2}, the yield of the *d61-7* mutant was only 80% of the wild-type. However, the grain yield of the *d61-7* mutant increased with planting density, whereas the grain yield of wild-type plants reached a maximum at mid-density. At a high planting density of 66.7 plants m^{-2}, the *d61-7* mutant and the wild-type had similar grain yields but the former produced more aboveground biomass (dry matter weight) than the wild-types. The vegetative biomass of the *d61-7* mutant was 35% higher than that of wild-type at high planting density.

Subsequent studies using *OsBRI1* knockdown (KD) lines suggested the possibility that a slight reduction of *OsBRI1* expression may lead to a higher yield than wild-type. The *OsBRI1-KD* lines, produced by the introduction of truncated *OsBRI1* cDNA, showed an erect-leaf phenotype, but its plant height, morphology of panicles and grains were similar to wild-type. The estimated grain yield of these transgenic lines was about 30% higher than that of wild-type at high density (Morinaka *et al.*, 2006). To date, the yield potential of these *OsBRI1-KD* lines was only estimated in experiments under greenhouse condition and not tested in open paddy field. In the same way, a mild BR biosynthetic mutant, *osdwarf4-1*, was shown to increase both biomass production and grain yield in paddy field (Sakamoto *et al.*, 2006).

3.2 *OsBRL1* and *OsBRL3*

According to sequence similarity, three homologous genes for *BRI1*, *BRL1* to *-3* (*BRI1-Like 1* to *-3*) had been isolated from *Arabidopsis* (Caño-Delgado *et al.*, 2004; Ceserani *et al.*, 2009). BRI1 and these BRL proteins belong to BRI1-family, a small subfamily of LRR-RLK class proteins, which are characterized by the presence of a conserved 70-amino acid island domain that serves as BR binding motif.

In *Arabidopsis*, BRI1 is expressed ubiquitously and acts as a main BR receptor in many organs and developmental processes. BRL1 and BRL3, which can bind to BR with high affinity as well as BRI1, are predominantly expressed in vascular tissues and play important roles in vascular differentiation. Despite their differential expression, BRL1 and BRL3 are functionally similar to BRI1. *BRL1* and *BRL3* can complement weak *bri1*-mutation when expressed under the *BRI1* promoter (Caño-Delgado *et al.*, 2004). BRL2/VH1, the most distant homologue of BRI1, has a distinct island domain, but it cannot bind to BR *in vivo* so it does not seem to be a BR receptor. *BRL2* was previously isolated as a pro-vascular cell-specific LRR-kinase gene, *VH1* (*Vascular Highway 1*), and it is responsible for vascular patterning in *Arabidopsis* (Clay and Nelson, 2002).

Three *BRL1* homologues, *OsBRL1* to *-3* are found in the rice genome and the structure and expression of *OsBRL1* and *OsBRL3* have been reported by Nakamura *et al.* (2006). Members of the BRI1-family in rice, *Arabidopsis* and various species can be subdivided into three categories. *OsBRL1* and *OsBRL3* belong to the same class with *BRL1* and *BRL3*. Likewise, *OsBRI1* belongs to the same class as *BRI1,* and *OsBRL2* does as *BRL2*, respectively. Similar to their *Arabidopsis* counterparts, OsBRI1, OsBRL1 and OsBRL3 seem to share a common protein function as BR receptors, but they differ in expression pattern. *OsBRI1* is mainly expressed in shoots, whereas *OsBRL1* and *OsBRL3* are preferentially expressed in root-tissues. Functional analyses of *OsBRL1* and *OsBRL3* by loss-of-function approaches have not been performed, however their functions can be elucidated from the phenotype of *d61-4*, a putative null mutant of the *OsBRI1* gene. The abnormal phenotype of *d61-4* mutant resembles the overall phenotype of the *brd1* mutant, however the severity of the abnormalities varied among the organs. In shoots, dwarfness and defects in cell elongation and organization are more pronounced than those of *brd1*. In contrast, the defects are milder than *brd1* in roots where *OsBRL1* and *OsBRL3* are expressed. It is likely that root-specific expression of *OsBRL1* and *OsBRL3* can replace the function of *OsBRI1* in roots of the *d61-4* mutant. In contrast, loss of *OsBRI1* function causes severe defects in shoots because *OsBRL1* and *OsBRL3* are also poorly expressed in the aerial tissues.

However, the phenotype of *d61-4* mutant cannot be completely explained by organ-specific expression of the *OsBRI1* family. For example, even though the *d61-4* mutant shoots have more severe defects than that of the *brd1* mutant, its vascular patterning, plastochron index, phyllotaxy, and embryo development appeared to be normal. On the other hand, although the roots of *d61-4* show defects in cell elongation, their appearance, size and vascular patterning are normal. One possibility is cell or tissue specificity of *OsBRL1* and *OsBRL3* might differ from that of *OsBRI1*, just as *BRL1*, *BRL3* are preferentially expressed in vascular tissues. It is also possible that the protein functions of each member differ in some other aspects. Loss-of function analysis of *OsBRL1*, *OsBRL3* or detailed analyses of expression of the *OsBRI1* family by *GUS*-reporter genes, or *in situ* hybridization may help elucidate the functional divergence in the OsBRI1 family.

3.3 *OsBAK1*

BAK1 is a member of a membrane localized LRR-RLK protein family that plays an essential role in BR signaling as a co-receptor of the BR-signaling complex (Figure 1B). BAK1 belongs to the <u>S</u>omatic <u>E</u>mbryogenesis <u>R</u>eceptor

Kinase (SERK) family characterized by shorter (~5 repeats of) extracellular LRR domains and intracellular kinase domain. The BAK/SERK family of LRR-RLKs is involved in several signaling pathways, somatic embryogenesis, male sporogenesis, cell death, pathogen resistance as well as BR signaling (reviewed by Chinchilla *et al.*, 2009).

Rice has four LRR-RLK proteins showing homology with BAK1. Among them, OsBAK1 (*Oryza sativa* BAK1, Figure 2B), with the highest homology to BAK1, is involved in BR signaling (Li *et al.*, 2009). *OsBAK1* overexpression partially suppressed the *Arabidopsis bri1* mutant phenotype, therefore BAK1 function is conserved in rice and *Arabidopsis*. *OsBAK1* overexpression in rice led to increased bending of the lamina-joint, hypersensitivity to exogenous BR and development of a dwarf phenotype (about 60% of WT in height). In addition, repression of *OsBAK1* expression by the introduction of a truncated cDNA encoding the extracellular domain of OsBAK1 produced a dwarf phenotype with erect leaves similar to that of a typical BR-insensitive mutant of rice. These results suggest the essential role of *OsBAK1* in BR signaling. Antisense suppression of *OsBAK1* generated an erect leaf phenotype, but it had no effect on plant height, reproductive development, grain weight. The phenotype of *OsBAK1-AS* rice may also be suitable for the breeding of rice cultivars with improved grain yield in high density planting (Li *et al.*, 2009).

3.4 *OsGSK1* (A *BIN2*-orthologue)

GSK3 (Glycogen synthase kinase3)-like kinases are non-receptor serine/threonine protein kinases involved in various biological processes in various organisms ranging from yeast to animals. In mammals, two enzymes, GSK α and GSK β are involved in insulin signaling, cytoskeletal stability, and oncogenesis. By contrast, plants have a large set of GSK3-like kinases forming a multigene family involved in diverse biological functions, such as hormone signaling, organ development and stress responses (reviewed by Jonak and Hirt, 2002). Plant GSKs are grouped into four classes (I-IV) based on sequence homology (Yoo *et al.*, 2006). Functional analyses of plant GSKs revealed that different GSKs are involved in various processes. Class I factors, AtSK11 and 12, are implicated in flower patterning. On the other hand, WIG, a class III factor of *Medicago*, is involved in wounding response.

In the BR signaling pathway, the class II GSK kinases in *Arabidopsis*, BIN2 (Brassinosteroid insensitive 2) and two BIN2 homologues, play essential roles as negative regulators of BR response. BIN2 was first identified as a dominant mutant *bin2-1* by screening for BR insensitive mutants (Li *et al.*, 2001). The *bin2-1* mutant is a gain-of-function allele with increasing *BIN2* kinase activity and it shows a BR-insensitive dwarf phenotype similar to the *bri1* mutant. On the other hand, loss of function of *BIN2* and its two paralogues lead to a constitutive BR-response phenotype (Vert and Chory, 2006). In the absence of BR, BIN2 inactivates BZR1 and BES1 *via* phosphorylation and subsequent recruitment by 14-3-3 (reviewed by Li and Jin, 2007). Upon perception of the BR signal by BRI1, BSU1 dephosphorylates and inactivates BIN2. As a consequence of BIN2 inactivation, dephosphorylated BZR1 and BES1 control the transcription of BR-regulated genes.

3.4.1 OsGSK1

In rice, a putative *BIN2* orthologue, *OsGSK1* (*Oryza sativa GSK1*) was isolated and its knockout phenotype and expression pattern were characterized (Koh *et al.*, 2007). OsGSK1 protein shows 96% homology to BIN2. In addition to *OsGSK1*, three other rice GSK kinases showing high sequence identity with BIN2, were also grouped as class II GSK kinases (Yoo *et al.*, 2006). Therefore, it is likely that OsGSK1 and some class II rice GSK kinases have redundant roles in BR signaling in rice, just like the functional redundancy between *BIN2* and its two homologues in *Arabidopsis* (Vert and Chory, 2006).

Rice *OsGSK1* knockout (KO) mutant was isolated from a *GUS* enhancer tagging population by screening for salt-responsiveness. The *OsGSK1* KO mutant had a T-DNA insertion in the 11th exon of the *OsGSK1* gene, resulting in the deletion of 35 amino acids at the C-terminal variable domain. The *GUS* gene of the *OsGSK1* KO line was fused in frame with *OsGSK1 ORF* in the same orientation, so its GUS activity likely reflected the activity of the *OsGSK1* promoter (Koh *et al.*, 2007).

3.4.2 Developmental role of OsGSK1

Organ specificity of *OsGSK1* expression was studied by RNA blot analysis using various organs of wild-type rice and GUS histochemical staining. *OsGSK1* mRNA is highly expressed in callus and young panicles, in roots and shoots of seedlings but not in adult leaves, suggesting that OsGSK1 may play an important role in developing tissues. *OsGSK1*-driven *GUS* activity was high in root tips and root hairs and weak in shoots, and vascular bundle of coleoptiles in seedlings. In later stages, GUS staining was restricted to the lamina-joint, suggesting the role of OsGSK1 in BR-related lamina-joint bending. *OsGSK1* KO line did not show any morphological alteration until flowering stage. It is likely that other rice GSK kinases have redundant roles with OsGSK1 during this stage.

In the reproductive stage, GUS activity was detected in entire young panicles, awn, vascular bundles of lemma and palea of spikelet. In comparison with control plant, *OsGSK1* KO line had longer awns and larger grains. Therefore OsGSK1 may play an important role in controlling the sizes of these organs. In floral organs, GUS activity was detected strongly in the stigma and rachilla but it was quite faint in the anther (Koh *et al.*, 2007). In *Arabidopsis*, two GSK kinases, AtSK11 and AtSK12, are expressed in the flower meristem and control gynoecium patterning. *OsGSK1* may play a similar role in the development of flower organs such as the stigma.

3.4.3 *Role of OsGSK1 in abiotic stresses tolerance*

The potential role of BR to increase plant tolerance to environmental stress was reviewed recently (Bajguz and Hayat 2009). *OsGSK1-KO* was isolated because the corresponding GUS-fusion protein was produced in response to salinity. Therefore, the response of *OsGSK1* expression to various abiotic stresses was evaluated. *OsGSK1* mRNA was induced by cold and reached maximum transcription at 3h. In contrast, drought stress decreased *OsGSK1* expression after a 24h exposure, and ABA did not affect *OsGSK1* expression. Since BIN2 acts as a negative regulator of BR response, OsGSK1 (a rice BIN2-homologue) may repress acquisition of abiotic stress tolerances *via* repression of the BR response. In agreement with this idea, *OsGSK1-KO* mutant is more tolerant to salinity, drought, cold and heat stresses than wild-type plant. In addition, the expression of stress-responsive genes, *SalT*, *lip5*, and *OsDhn1* under salt or drought stress are affected in the *OsGSK1-KO* mutant. Therefore OsGSK1 may control abiotic stress tolerance *via* repression of stress responsive genes (Koh *et al.*, 2007).

3.4.4 *Role of OsGSK1 in BR signaling*

The role of *OsGSK1* in BR signaling is an important issue but it has not been studied as much as *OsBRI1*. BR sensitivity and expression of BR inducible genes have been reported in the *OsGSK1*-KO mutant. *OsGSK1*-KO mutant seedling showed more coleoptile elongation than wild-type seedling after treatment with 10^{-8} M BL (Brassinolide), implying that the *OsGSK1*-KO mutant is more sensitive to BL than the wild-type plant. In addition, expression of known BR-inducible genes, *OsXTR3* and *SalT*, were upregulated in *OsGSK1*-KO mutant. Regarding coleoptile elongation and expression of BR-inducible genes, *OsGSK1* is likely to function as a negative regulator of BR response in rice. Considering the redundancy of rice BIN2-like GSK kinases, simultaneous disruption of OsGSK1 and other class II GSK kinases may cause a clear BR hypersensitive phenotype.

Overexpression of *OsGSK1* in *Arabidopsis* (*OsGSK1:OX*) led to a stunted growth phenotype (1/3 of wild type in height) but not as severe as the dwarfism observed in *BIN2:OX* plants. The results imply that *OsGSK1* can negatively regulate BR signaling in *Arabidopsis* (Koh *et al.*, 2007). Overexpression of *OsGSK1* in rice may help to elucidate the role of *OsGSK1* in BR signaling in more detail.

3.5 *OsBZR1 and 14-3-3*

In *Arabidopsis,* two nuclear transcription factors, BZR1 (Brassinazole resistant 1) and its closest homologue BES1, are key components of BR signaling because they regulate the transcription of many BR-responsive genes. BZR1 consists of 335 amino acids containing an alanine-rich region, NLS and DNA-binding domain in its N-terminal region, 14-3-3 binding motif in its center, and a PEST domain which serves for rapid protein turnover, near its C-terminal region.

Through its N-terminal DNA binding domain, BZR1 directly binds to the CGTGT/CG sequence, also known as the BR-Response Element (BRRE), which is required for transcriptional control by BZR1. BZR1 acts as a repressor of transcription in protein function, but it plays a dual role as a repressor of BR synthetic genes for feedback regulation and as an activator for growth response (Wang *et al.*, 2002; He *et al.*, 2005). BRRE was first identified as a BZR1 binding sequence in the promoter of *CPD*, a BR-biosynthetic gene and is required for repression by BZR1 and BR. Subsequently, BRRE was found to be present at a high frequency in the promoters of a large number of BR-repressed genes, including four feedback-regulated BR synthetic genes (*CPD*, *DWF4*, *ROT3* and *BR6ox*). Therefore BZR1 is likely to act as a repressor of these BR-repressed genes. In addition, BZR1 is also likely to activate transcription of BR-inducible genes through interaction with other factors. In contrast to the repressor function of BZR1, BES1 is as an activator of a BR-inducible gene, *SAUR-AC1,* through interaction with a bHLH transcription factor, BIM1 (Yin *et al.*, 2002, 2005). BES1/BIM1 binds the E-box (CANNTG) sequence in the *SAUR-AC1* promoter and activates the expression of *SAUR-AC1*.

Transcription activity of BZR1 and BES1 is mainly controlled by their nuclear localization and degradation by the proteasome. A GSK-like kinase, BIN2, and 14-3-3 proteins are involved in the control of BZR1 and BES1 (Gampala *et al.*, 2007). Under a low BR level, BIN2 phosphorylates multiple serine residues, which are distributed from the central to the C-terminal region of BZR1 protein. Due to phoshorylation of the 14-3-3 binding motif, the 14-3-3 protein binds to BZR1 and enhances cytoplasmic retention of BZR1. As a result, BIN2 represses activity of BZR1. BZR1 protein is rapidly degraded by the proteasome through its PEST domain, thereby enabling rapid response of transcriptional control to BR signal. A dominant mutation in the BZR1 gene, *bzr1-1D*, has a P234L mutation in its PEST domain that stabilizes the BZR1 protein and causes constitutive BR-responsive growth in the dark.

3.5.1 OsBZR1

The rice genome encodes four BZR1 homologous proteins, OsBZR1 to 4. Among them, OsBZR1 shows the highest sequence identity with BZR1 (53.4%) and BES1 (53.2%). Phylogenetic analysis revealed that OsBZR1 shows lower homology with BZR1 and BES1 than that between BZR1 and BES1. Despite moderate sequence identity with BZR1 and BES1, the putative functional domains (DNA binding domain, 14-3-3 binding motif and PEST domain) of OsBZR1 are highly similar to those of BZR1 and BES1.

The role of *OsBZR1* in BR signaling was studied using RNAi suppression and a gain of function approach. *OsBZR1* RNAi plants had significantly depressed *OsBZR1* expression, whereas the transcript levels of *OsBZR2* and *OsBZR3* were only slightly suppressed. The *OsBZR1* RNAi plants showed a semi-dwarf phenotype with erect leaves, the typical BR-insensitive phenotype in rice. The *OsBZR1* RNAi plants showed various degrees of reduced culm elongation. Lamina joint bending of the RNAi plants was reduced and it was not affected by BL treatment. *BZR1* mediates feedback inhibition of BR-biosynthetic genes in *Arabidopsis*, therefore the response of BR-biosynthetic genes to BR was analyzed using the *OsBZR1* RNAi plants. In the absence of BR, expression of BR-biosynthetic genes (*D2*, *D12* and *BRD1*), in the *OsBZR1* RNAi plants were similar to WT. The expression levels of *D2*, *D12* and *BRD1* were reduced by BR-treatment in wild-type plants. In contrast, the expression of BR-biosynthetic genes was not affected by BR-treatment in the *OsBZR1* RNAi plants. Thus OsBZR1 (and homologues) is responsible for BR-regulated growth response and feedback inhibition of BR biosynthetic genes (Bai *et al.*, 2007).

In *Arabidopsis* BZR1, the P234L dominant mutation in the PEST domain increases BZR1 stability and causes constitutive BR-responsive growth phenotype. Consistent with the observation in *Arabidopsis*, overexpression of *OsBZR1P206L*, which contains the same dominant mutation in PEST domain, increased bending of the lamina-joint and reduced expression of BR-biosynthetic genes in rice. In addition, overexpression of *OsBZR1P206L* produced a *bzr1-1D*-like phenotype in *Arabidopsis*. Therefore the roles of BZR1/OsBZR1 in BR responsive gene expression and growth response are conserved in rice and *Arabidopsis* (Bai *et al.*, 2007).

3.5.2 *14-3-3 proteins: OsBZR1-interacting proteins*

14-3-3 proteins are highly conserved phosphopeptide-binding proteins, which are involved in various signaling processes in all eukaryotes. 14-3-3 proteins bind to specific sequences (RXXXpSXP and RXXpSXP) of phosphorylated proteins and control activity of interaction-partners. In plants, 14-3-3 proteins are known to control the function of various proteins such as transcription factors, metabolic enzymes, ion channels and kinases. In *Arabidopsis*, they also play an essential role in BR-mediated transcription through the control of subcellular localization of BZR1. As described above, under low BR condition, BZR1 protein is phosphorylated by BIN2. 14-3-3 proteins bind to the phosphorylated BZR1 and enhance cytoplasmic retention of BZR1. Thereby 14-3-3 proteins repress BZR1-mediated controls of transcription.

Consistent with the observations in *Arabidopsis*, eight isoforms of rice 14-3-3 proteins (GF14a-h) are shown to interact with OsBZR1 by yeast two-hybrid screening for OsBZR1-interaction proteins (Bai *et al.*, 2007). Therefore, it is likely that rice 14-3-3 proteins are also involved in OsBZR1-mediated transcriptional control in a manner similar to that in *Arabidopsis*. This possibility was tested by a series of experiments in tobacco and in *Arabidopsis* but not in rice.

Among rice 14-3-3 proteins, GF14c most frequently interacted with OsBZR1 in the yeast two-hybrid screening. Therefore GF14c was used to verify the interaction between OsBZR1 and 14-3-3 proteins *in vivo*. BiFC (Bi-molecular Fluorescence Complementation) assay using nYFP-OsBZR1 and cCFP-GF14c demonstrated that GF14c specifically interacts with OsBZR1. In addition, the S156G mutation in the 14-3-3 binding motif of OsBZR1 abolishes the interaction. Specific interaction between GF14c and OsBZR1 through 14-3-3 binding motif was also confirmed by Co-IP (Co-immuno precipitation) experiments.

The functions of the 14-3-3 binding motif of OsBZR1 in BR signaling and in gene expression were analyzed in detail in *Arabidopsis*. Overproduction of OsBZR1S156G (OsBZR1 mutant protein abolishing 14-3-3 binding) in *bri1-5 Arabidopsis* suppressed its mutant phenotype. Heterologous expression of wild-type OsBZR1 in *bri1-5 Arabidopsis* had no effect on the mutant phenotype. In addition, the expression of BR-biosynthetic genes, *DWF4* and *CPD* were down-regulated but the expression of *SAUR-AC1*, a BR-inducible gene, was increased in the gain-of-function mutant OsBZR1S156G. Therefore, the function of OsBZR1/BZR1 in BR responses and importance of 14-3-3 binding are conserved in *Arabidopsis* and rice.

Furthermore, BR treatment reduced interaction between OsBZR1 and 14-3-3 protein and increased nuclear localization of OsBZR1 in tobacco and in *Arabidopsis*, respectively. Although there are no direct evidences in rice, these results suggest that the essential roles of OsBZR1 and 14-3-3 proteins in BR-regulated expression are likely to be conserved in rice (Bai *et al.*, 2007).

3.6 *DLT*

The rice *dwarf and low-tillering* (*dlt*) mutant, displays a semi-dwarf phenotype, with reduced tiller number and erect and dark-green leaves that were wider, and shorter than wild type. In addition, it has a shorter second internode because of decreased cell length. These phenotypes of *dlt* are similar to those of BR-deficient or signaling mutants in rice. The *dlt* mutant is less sensitive to BL in terms of coleoptile elongation and lamina bending, suggesting that DLT is involved in BR signaling. Moreover, BR biosynthetic genes were highly expressed in *dlt* mutant, suggesting that DLT is also required for feedback inhibition of BR-biosynthetic genes.

DLT was isolated *via* map-based cloning. It encodes a new member of the plant-specific GRAS family (Tong *et al.*, 2009). The GRAS gene family is an important plant-specific gene family of putative transcription factors and the name was coined from the first three functionally characterized members (GAI, RGA, SCR). Many *GRAS* genes have been found to be involved in many developmental processes, such as gibberellin signal transduction, axillary meristem initiation, shoot meristem maintenance, and phytochrome A signal transduction, etc. (Tian *et al.*, 2004).

DLT is negatively regulated by either exogenous or endogenous BRs because BL treatment of wild type plants reduced *DLT* transcripts whereas BR deficient mutants accumulated *DLT* transcripts in the absence of BL treatment. Consistent with this, a gel mobility shift assay revealed that OsBZR1 can bind to the *DLT* promoter through the known BR-response element, CGTG(C/T)G. These results suggest that OsBZR1 represses *DLT* expression at the transcriptional level (Tong *et al.*, 2009).

4. OTHER BR SIGNALING FACTORS IN RICE

In addition to the OsBRI1-mediated BR-signaling pathway, *RGA1* (Wang *et al.*, 2006; Oki *et al.*, 2009a) is also known to be involved in BR-response in rice. Recently, Nakamura *et al.* reported that C22-hydroxylated-BRs are involved in IAA-mediated lamina joint bending (Nakamura *et al.*, 2009).

4.1 *RGA1*

Heterotrimeric G-proteins, highly conserved signal transducers that are composed of three subunits, Gα, Gβ, Gγ, play essential roles in various biological processes in many eukaryotes. Heterotrimeric G-proteins transduce signals from G-protein coupled receptors (GPCRs) to downstream effector molecules (Figure 3A–E). In animals, each subunit of heterotrimeric G-proteins and GPCRs compose large gene families. By contrast, plants have only a few genes for each subunit of G-protein and GPCR. For example, rice and *Arabidopsis* have only one gene for each of Gα, Gβ and two genes for Gγ. Despite fewer members in plants, plant heterotrimeric G proteins are involved in various signal responses (ABA, GA, BR, D-Glucose etc.) and many physiological processes (reviewed by Perfus-Barbeoch *et al.*, 2004).

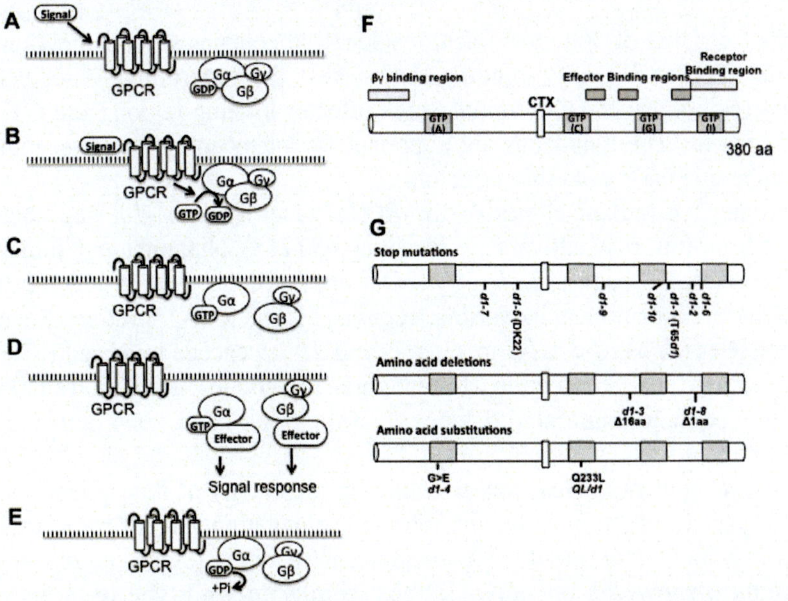

Figure 3. A model for signaling mechanism mediated by heterotrimeric G-protein, structure of RGA1 and its mutation alleles.

(A) To (E) Model for mechanism of G-protein mediated signal transduction.

(A) Signal perception by GPCR (G-protein coupled receptor) causes conformational change of GPCR. (B) The conformational change allows activation of its GTP/GDP exchange activity. GPCR exchanges GDP for GTP from the Gα subunit. (C) This exchange triggers dissociation of $G\alpha\text{-}_{GTP}$ subunit and Gβγ dimer. (D) Both of $G\alpha\text{-}_{GTP}$ and Gβγ can activate separate downstream signaling pathways *via* interaction with different effectors. (E) $G\alpha_{\text{-}GTP}$ subunit inactivates itself by its GTPase activity, then $G\alpha\text{-}_{GDP}$ subunit associates with Gβγ to form the inactive heterotrimeric complex shown in (A).

(F) Structure of RGA1 protein. G βγ subunits binding region, three effector binding regions, receptor binding region, cholera toxin binding site (CTX) and four GTP-binding regions [GTP(A), GTP(C), GTP(G) and GTP(I)] are indicated.

(G) Diagram showing mutation sites in RGA1 protein of *d1* mutant alleles. Premature stop mutations, amino acid deletions and amino acid substitutions are separately drawn.

4.1.1 RGA1

Rice heterotrimeric G-protein alfa 1(RGA1) is involved in many developmental processes (germination, cell division, cell elongation, morphogenesis) and signaling processes (GA, BR, pathogen, oxidative stress, etc.).

The loss of function mutant of *RGA1* gene, *dwarf1* (*d1*) was first identified as a spontaneous dwarf rice mutant, Daikoku dwarf. The *d1* mutant has short, round, broad, dark green leaves, compact panicle and short, round grains. RNAi-based suppression of *RGA1,* sequence analysis and map-based cloning of the *d1* locus revealed that the *D1* gene encodes RGA1 (Figure 3, Fujisawa *et al*., 1999; Ashikari *et al*., 1999).

4.1.2 Structure and mutation alleles of RGA1

RGA1 consists of 380 amino acid residues and contains several functional domains (Figure 3F), a βγ-binding region near the N-terminal, a receptor binding region near the C-terminal, three effector binding regions and GTP-binding motifs. These domains are important for interaction with its receptor, other subunits, GTP and other effectors.

To date, ten mutant alleles of the *RGA1* gene, *d1-1* to *-10*, have been isolated from four rice cultivars cv. Taichung 65 (T65), Nipponbare, Kinmaze and Blue Rose (Fujisawa *et al*., 1999; Oki *et al*., 2009a, Figure 3G). These mutations are positioned in various regions of the RGA1 protein. Seven mutant alleles (*d1-1*, *d1-2*, *d1-5* to *-7*, *d1-9* and *d1-10*) encode truncated RGA1 proteins due to the introduction of premature stop codons. Two alleles (*d1-3* and *d1-8*) contain amino acid deletions, on the other hand, *d1-4* contains a single amino acid substitution.

These *d1* mutant alleles showed variation in severity of their phenotype, seed size, seed weight, plant height, internode elongation and BR response in the lamina inclination assay. For example, *d1-4* allele (single amino acid substitution) showed a phenotype almost similar to wild-type in terms of grain size and BR response. On the other hand, all of the nonsense and deletion alleles showed extreme dwarf and short round grain phenotypes and were less sensitive to BR. Their extreme phenotypes were explained by a lack of Gα accumulation in plasma membrane. However, the reason for the difference in severity in the same genetic background could not be completely explained. For example, the *d1-6* allele encodes a larger truncated Gα (37.7 kDa) showed a more extreme phenotype than the *d1-7* allele that encodes a smaller protein (10.7 kDa) in the same Kinmaze background (Oki *et al*., 2009a).

4.1.3 Constitutive active forms of RGA1

In addition to loss of function *d1* mutants, Oki *et al.* (2005) have generated transgenic rice plants designated QL/*d1*, that express the constitutive active form of RGA1 under the control of *RGA1*-promoter in a *d1* null mutant background. In the G-protein signaling process, Gα is activated through GTP-binding and inactivated through hydrolysis of GTP by its GTPase activity. In mammalian Gα proteins, a single amino acid substitution in their GTP binding region is sufficient for conversion to constitutive form. Therefore, Oki *et al.* (2005) designed four site-directed mutant RGA1 proteins that had single amino acid substitutions in their GTP binding regions. Two of the mutated RGA proteins, R191C and Q233L, were chosen as the strong candidates of constitutive active forms of RGA, since they retained GTP binding ability but abolished GTPase activity *in vitro*. One of the candidates, Q233L, was fused downstream of the *RGA1*-promoter and used to complement the *d1* mutant. The transgenic plants that harbored Q233L were designated as QL/*d1* and they showed some interesting phenotypes that were opposite to those exhibited by the *d1* mutant. In contrast to the *d1* mutant, the grains of QL/*d1* were larger than that of wild-type plant. In addition, QL/*d1* was more sensitive to BR than the wild-type plant based on the coleoptile elongation response (Oki *et al.*, 2009b). However, QL/*d1* was indistinguishable from wild-type plants in terms of plant height and internode elongation.

4.1.4 Role of RGA1 in BR signaling

The role of Gα protein in BR signaling was first described for *Arabidopsis* GPA1, a gene that controls germination and vegetative growth (Ullah *et al.*, 2001, 2002; Gao *et al.*, 2008). In wild-type *Arabidopsis*, BR can rescue GA-deficient inhibition of seed germination. In contrast, both *gpa1* and BR-insensitive *bri1* mutants failed to rescue seed germination. The *gpa1* mutant was less sensitive to BR than the wild-type plant in the BR-mediated control of hypocotyl and root elongation. Therefore Gα protein is involved in BR response in *Arabidopsis*.

Rice *RGA1* is involved in various BR-mediated responses, such as lamina joint inclination, coleoptile elongation and root growth inhibition (Wang *et al.*, 2006; Oki *et al.*, 2009c). Wang *et al.* reported the reduced response of the *d1* mutant (*d1-5*, DK22 cultivar Nipponbare background) to BR. In comparison with wild-type plant, the *d1* mutant showed reduced BR responses in lamina inclination, coleoptile elongation and root inhibition. In rice, BR plays important roles in skotomorphogenesis (dark-adapted morphogenesis) such as mesocotyl elongation and coleoptile elongation. Similar to the *d61-1* mutant, the *d1* mutant showed less coleoptile elongation than wild-type plant in darkness. However, the *d1* mutant showed normal mesocotyl elongation in the dark, which was different from the non-elongated mesocotyl of the *d61-1* mutant. The *d1* mutant was less sensitive to BR, but BR-insensitive phenotype of *d1* was somewhat distinct from *d61-1* in root inhibition and mesocotyl elongation.

Oki *et al.* have also reported on the BR response of the T65*d1* mutant (*d1-1*, Oki *et al.*, 2009c). Consistent with the report on the *d1* mutant (Wang *et al.*, 2006), the T65*d1* mutant showed reduced responses to BR in lamina joint bending response, in inhibiting the growth of root and that of the aerial parts, and in the elongation of coleoptile and second leaf sheath. However, the T65*d1* mutant was different from the *d1* mutant in skotomorphogenesis. The T65*d1* mutant did not elongate internodes in the dark, whereas the *d1* mutant had elongated internodes like the wild-type plant (Wang *et al.*, 2006). This discrepancy may be due to differences in genetic background or allelic differences between the T65*d1* and *d1* mutants.

The T65*d1* mutant had a lower degree of lamina joint bending and it was less sensitive to BR than the wild-type plant. Histological observation of the lamina joint region revealed that reduction of lamina joint bending in the T65*d1* mutant was caused by a reduction in cell elongation. In addition, the T65*d1* mutant showed specific patterns of internodal elongation. In the T65*d1* mutant, the second internode was extremely shortened relative to the other internodes. This pattern was classified as a dm-type elongation pattern (Takeda, 1977) that was commonly observed in other BR-related mutants, such as those for the weak alleles of *osbri1* (*d61-1*, *d61-7*) and the mild BR-biosynthetic mutant *d2-1* (Oki *et al.*, 2009c).

4.1.5 Interaction between RGA1 and OsBRI1 pathway

The interaction between *RGA1* and *OsBRI1* was studied using a double mutant T65*d1*/*d61-7*, that was obtained by crossing a *RGA1* null mutant, T65*d1*, and a mild mutant of *OsBRI1*, *d61-7* (Nakamura *et al.*, 2006; Oki *et al.*, 2009c). At the reproductive stage, the plant height of the T65*d1*/*d61-7* double mutant was shorter than those of the single mutants because of greater suppression of internodal elongation. Histological observation showed that the reduction of leaf sheath elongation in the T65*d1*/*d61-7* double mutant was mainly caused by the reduction in cell number. Thus RGA1 and OsBRI1 promote organ elongation by increasing cell number in an additive manner. The double mutant showed a d6-type internode elongation pattern that is also found in a more severely BR-insensitive mutant, *d61-2*. The angle of flag leaf bending in the T65*d1*/*d61-7* double mutant was much smaller than those of both single mutants. In addition, the T65*d1*/*d61-7* double mutant had corrugated leaf blades, which have been found in severe BR-deficient mutants as well (Hong *et al.*, 2005). Thus, the T65*d1*/*d61-7* double mutant showed an additive phenotype of both the single mutants in the aspects of plant height, internodal elongation, abnormal leaf morphology and erectness of leaves. Therefore, RGA1 and OsBRI1 are likely to act in parallel pathways to regulate these processes.

Rice BR-deficient mutants, *d2*, *d11* as well as *d61*, have characteristic long panicles (Hong *et al.*, 2003; Tanabe *et al.*, 2005; Yamamuro *et al.*, 2000). In contrast, the panicle of the T65*d1* mutant was shorter than that of the wild-type plant. The panicle of the T65*d1*/*d61-7* double mutant was slightly shorter than wild-type plant. Therefore, panicle length of rice is likely to be regulated by OsBRI1 and RGA1 in an additive manner. Regarding seed size, the *d61-7* mutant was slightly smaller than the wild-type plant in both longitudinal and horizontal axes. In contrast, the grain of the T65*d1* mutant was significantly shorter (*i.e.* at the longitudinal axis) than wild-type grain. Seed size of the T65*d1*/*d61-7* double mutant was almost comparable to that of the T65*d1* mutant. Therefore T65*d1* may be epistatic to *d61-7* in the control of seed size (Oki *et al.*, 2009c). The effect of *RGA1* and *OsBRI1* on the control of seed size was also studied by using the constitutive active RGA1 line, QL/T65*d1* (Oki *et al.*, 2009b). Interestingly, expression of constitutive active form of RGA, Q233L, increased seed size in the absence of *OsBRI1*. Therefore RGA1 is likely to promote growth of seed size independent of OsBRI1.

4.1.6 RGA1 is not involved in feedback regulation of BR biosynthesis

In rice, BR controls expression of BR-biosynthetic genes, *D2*, *D11* and *OsDWARF (BRD1)* by negative-feedback regulation (Hong *et al.*, 2002, 2003; Tanabe *et al.*, 2005). This feedback regulation is mediated by the Os*BRI1* pathway. Therefore a mutation in the Os*BRI1* pathway, such as *d61*, abolishes the feedback regulation and causes hyper accumulation of BR intermediates, such as castasterone and typhasterol (Yamamuro *et al.*, 2000).

To determine whether *RGA1* was involved in the feedback regulation of BR biosynthesis, the effect of BL on the expression of biosynthetic genes was evaluated in the T65*d1* mutant. Treatment with BL reduced the expression of *D2*, *D11* and *OsDWARF* (*BRD1*) in both T65*d1* mutant and wild-type plants. In contrast, expression levels of these genes were not affected by BL in the *d61-2* mutant. Despite its BR-insensitive phenotype, the amounts of BR intermediates in the T65*d1* mutant were similar to those of the wild-type plant. Therefore, unlike OsBRI1, RGA1 is not involved in the feedback regulation of BR-biosynthetic pathway (Oki *et al.*, 2009c).

4.1.7 *RGA1 functions in other regulation, GA, etc.*

Because of its dwarf phenotype, the *d1* mutant was first classified as a Gibberellin (GA)-insensitive mutant (Ashikari *et al.*, 1999). In fact, the *d1* mutant expressed GA-responses in some aspects, such as GA-inducible gene expression (*GAMYB*, *Ca^{2+}-ATPase* and *α-Amylase*), internode elongation and leaf sheath elongation (Ueguchi-Tanaka *et al.*, 2000). As mentioned above, internodes of the *d1* mutant were shorter than those of the wild type plant, specifically, the second internode of *d1* was preferentially shortened. Expression of *GA20ox* (GA biosynthetic gene) and active GA levels in the second internode were significantly increased in the *d1* mutant, may be due to a defect in feedback inhibition of GA biosynthesis. The specific suppression of the second internode is a common characteristic of BR-related mutants. Therefore, the preferential suppression of the second internode and deregulation of GA-biosynthesis in *d1* indicates that RGA1 is likely to be involved in signal crosstalk between GA and BR signaling.

In addition, RGA1 also plays an important role in epidermal cell death induced by H_2O_2 and ethylene (Steffens and Sauter, 2009). The *d1* mutant showed a defect in epidermal cell death and subsequent adventitious root elongation.

4.1.8 *Relationship between OsBRI1 pathway, RGA1 and other signaling factors*

Relationship between RGA1 and OsBRI1 pathway in BR signaling is somewhat analogous to that of RGA1 and the GA-GID-SLR1 pathways (reviewed by Ueguchi-Tanaka *et al.*, 2007) in GA signaling. In both cases, RGA1 plays important roles in hormone signaling. However in both cases, the upstream signal perception factors and downstream signaling components have not been identified so far. In both cases, relationship between RGA1 and main signaling pathways are also not clear.

Since RGA1 is involved in multiple signaling processes, such as those associated with ABA, GA and Auxin, the role of RGA1 in BR signaling may be to modulate the BR signal response and coordinate with other signaling pathways. Regarding organ size, mutations in *d1* and *d61* additively control the size of organs as if OsBRI1 and RGA1 independently regulate organ growth. OsBRI1 is specialized for BR signaling, in contrast, RGA1 also mediates other signals, such as GA and IAA as well as BRs. Therefore, the apparent independence of RGA1 from OsBRI1 may be due to effects of other signals mediated by RGA1.

Currently, the presence of unidentified receptors for BR cannot be ruled out. It is possible that C-22-hydroxylated BRs are involved in auxin-mediated lamina joint inclination through the OsBRI1-independent pathway (Nakamura *et al.*, 2009). Moss (*Physcomitrella patens*) has BRs, but not CS nor BL, even though there are no *BRI1*-homologous receptor genes in the moss genome. Hence they proposed that another ancient receptor mediated system, which operates BRs signaling pathway in moss, might be involved in the OsBRI1-independent C-22-hydroxylated BR signaling pathway in rice (Nakamura *et al.*, 2009).

Undoubtedly, BR plays essential roles in many physiological and developmental processes in rice. Therefore severe mutations in BR-signaling and biosynthetic pathways cause severe developmental defects, such as severe growth retardation, malformed organ formation and sterility in rice (Mori *et al.*, 2002; Hong *et al.*, 2002; Nakamura *et al.*, 2006). Because of the aberrant phenotype of these severe mutants, most analysis of BR signaling in rice have been performed by using mild or intermediately weak mutants such as *d61-1*, *d61-2* and *brd1-3*. Therefore, the possibility that residual BR-sensitivity of mild-mutant OsBRI1 protein may be responsible for such "OsBRI1-independent" BR-signaling in rice cannot be ruled out.

5. BR RESPONSIVE HELIX LOOP HELIX (HLH) PROTEINS

In *Arabidopsis*, *BEE1* (*BR enhanced expression1*) to *BEE3* are early BR-responsive genes encoding redundant bHLH proteins that are necessary for the full BR response (Friedrichsen *et al.*, 2002). A bHLH protein has two functional domains, the basic region and the HLH region. The former is required for binding to DNA to regulate the expression of genes, while the latter is required for interaction with other bHLH proteins for hetero- or homo-dimerization (Murre *et al.*, 1989). BIM1 (BES1-interacting MYC-like 1), encoding a MYC-like bHLH protein, interacts with transcription factor BES1 and both proteins together bind to a promoter of the target gene to regulate BR-induced expression (Yin *et al.*, 2005). In rice, little is known about the role of HLH proteins in BR signaling.

5.1 *BU1*

In order to identify novel BR signal related genes in rice, we used microarray to search for BL responsive genes in the *brd1* mutant that was defective in BR biosynthesis (Mori *et al.*, 2002). One of the genes, designated as *BU1* (*Brassinosteroid upregulated1*) containing a HLH motif, was analyzed in detail (Tanaka *et al.*, 2009). Since BU1 lacks the basic region necessary for binding to DNA, it is classified into Group D of bHLH proteins, the non-DNA binding protein family (Li *et al.*, 2006). In rice, there are three proteins which have high identity with BU1 (Figure 4A).

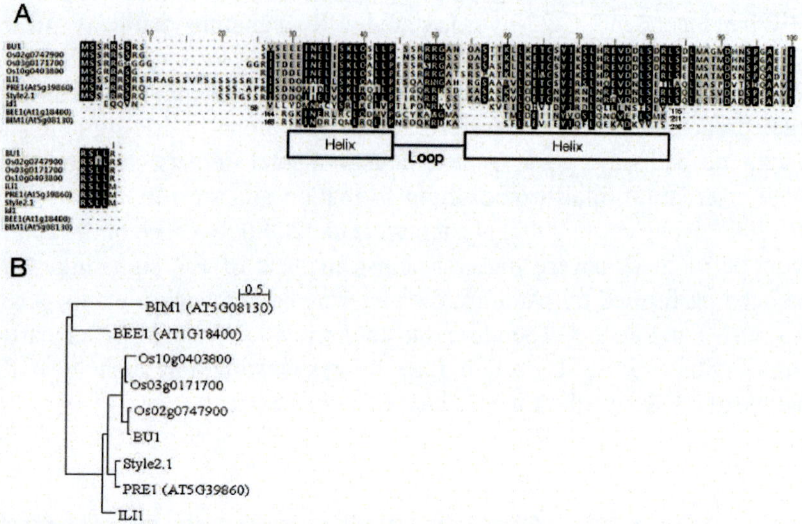

Figure 4. Protein sequence analysis of BU1. Alignment (A) and phylogenic tree (B) of Helix-Loop-Helix (HLH) domains for BU1 (Os06g0226500), its three homologous proteins, ILI1, PRE1, Style2.1, and two *Arabidopsis* bHLH proteins associated with BR signaling, BEE1 and BIM1. Style2.1 is a PRE1 homolog in tomato and regulates cell elongation in developing styles (Chen *et al.*, 2007). The upper seven protein sequences are full-length. HLH domains are indicated below the sequences. Black and gray backgrounds indicate identical and similar amino acids, respectively. The bar corresponds to 0.5 amino acid substitutions per site.

BU1 and these proteins have a high similarity to PRE1 (Paclobutrazol resistance 1) of *Arabidopsis*, a HLH protein that lacks the basic region which had been reported to be involved in gibberellin (GA) signaling (Lee *et al.*, 2006). Phylogenetic analysis indicates that the similarity between BU1 and BEE1 or BIM1, which are bHLH proteins related to BR signaling in *Arabidopsis*, is very low (Figure 4B). Therefore, BU1 and its homologs are thought to belong to a novel protein family.

Rice plants overexpressing *BU1* (*BU1*:OX) showed BR characteristic phenotypes. *BU1*:OX displayed increased bending of the lamina joint and produced larger seeds than WT (Figure 5A and B). Since bending of the lamina joint and seed size are regulated by BR (Yamamuro *et al.*, 2000; Mori *et al.*, 2002; Hong *et al.*, 2003), these results suggest that *BU1* may be involved in BR signaling or biosynthesis. *BU1*:OX also had some characteristic abnormalities in the pattern of internode elongation: it had elongated fifth and sixth internodes (Figure 5C) and shortened first internode. Moreover, development of tillers was observed at the nodes of aerial parts (Figure 5D).

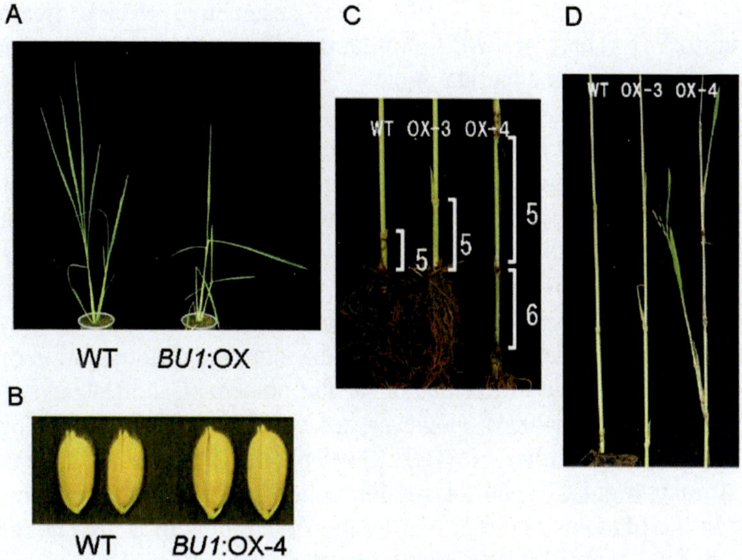

Figure 5. Phenotypes of rice plants overexpressing *BU1* (*BU1*:OX). (A) Gross morphology of wild type (WT) and *BU1*:OX at vegetative phase. (B) Seed morphology of WT and *BU1*:OX-4. *BU1*:OX-4 has larger seed than that of WT. (C) Lower internodes of WT and *BU1*:OX-3,4. Numbers 5 and 6 represent the fifth and sixth internodes. (D) Upper internodes of WT and *BU1*:OX-3,4. Development of tillers in nodes of aerial part was observed in *BU1*:OX-3,4.

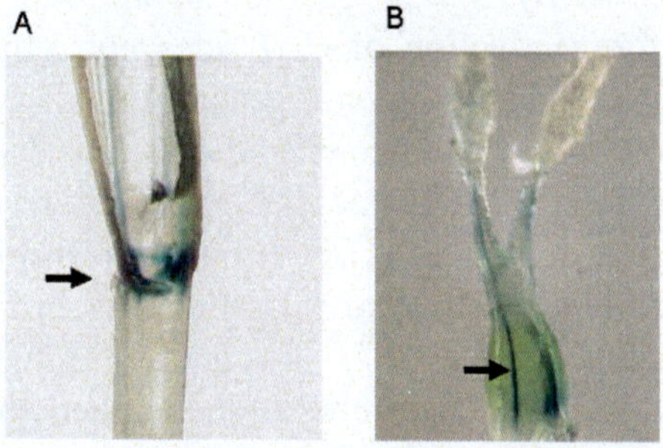

Figure 6. Histochemical GUS staining of rice plants expressing GUS under the control of the BU1 promoter. (A) Lamina joint region. (B) Stigma and ovule. Arrows indicate staining tissues.

There were no obvious differences between *BU1*:OX and WT in the amounts of internal BRs, so BU1 is not involved in BR biosynthesis but it may be involved in BR signaling in rice. We performed complementation analysis using the loss of function mutant of *OsBRI1* (BR receptor gene), *d61,* which has erect leaves. Overexpression of *BU1* in the *d61* background led to increased bending of the lamina joint, indicating that *BU1* is involved in BR signaling. Although loss-of-function transgenic rice that suppressed *BU1* did not show an obvious phenotype, transgenic plants suppressing *BU1* and its homologs (*BU1*F:RNAi) had erect leaves. In addition, the lamina joint test revealed that *BU1*F:RNAi was insensitive to BL treatment compared to WT. Therefore, *BU1* and the three homologues play redundant roles in the bending of the lamina joint.

BU1 is highly expressed in the lamina joint in vegetative organs and in the panicle especially at heading stage. BU1 promoter:GUS analyses revealed that the lamina joint region (Figure 6A), phloem in leaves, vascular bundles of the floral organs (Figure 6B) and epithelium in embryos were stained. The expression of *BU1* in the lamina joint strongly suggests that BU1 is involved in bending of the lamina joint. The expression of *BU1* in the vascular bundle suggests a new function of BR. BR is known to play an important role in vascular differentiation, enhancing the differentiation of the xylem and inhibiting the differentiation of the phloem (Caño-Delgado *et al.*, 2004; Nakamura *et al.*, 2006). However, we could not detect morphological changes in vascular bundles of *BU1*F:RNAi plants. It is possible, therefore, that BU1 may regulate gene expression related to transport or unknown function in the phloem. The role of BU1 in the epithelium is also unknown but we suppose that BU1 is involved in the expression of homeostatic-related genes.

BU1 was not induced by IAA or by GA_3 but it was repressed by ABA, a known BR antagonist. *BU1* was greatly up-regulated by exogenous BL in WT plants as well as in *brd1* plants. In addition, induction of *BU1* by exogenous BL was not affected by cycloheximide, an inhibitor of *de novo* protein synthesis. Therefore, induction of *BU1* does not require *de novo* protein synthesis and *BU1* is an early response gene to BR signaling. Induction levels of *BU1* by exogenous BL were compared between WT and two BR signal related mutants, *d61* and *d1*. The *d61* mutant has a defective OsBRI1 whereas the *d1* mutant has a defective RGA1 (rice heterotrimeric G protein alpha subunit). *BU1* was less sensitive to exogenous BL in both *d61* and *d1* than the wild type, indicating that *BU1* is induced through both OsBRI1 and RGA1 pathway by BRs. Therefore we concluded that *BU1* is an early response gene to BL and acts in two pathways associated with OsBRI1 and RGA1 (Figure 7).

As described above, BU1 protein is categorized as a putative non-DNA binding bHLH protein, since it lacks a basic region (Figure 4A). The eGFP-BU1 fusion protein was dispersed throughout the cells and not localized in the nucleus, suggesting that BU1 is not a transcription factor. Therefore, downstream genes of BU1 may be regulated by more complex mechanisms in addition to transcription factors. The molecular mechanism of non-DNA binding HLH proteins is understood well in human Inhibitor of DNA binding (Id) proteins. Id proteins heterodimerize with other DNA binding bHLH protein partners via HLH motif and abolish their functions as transcription factors, binding to the cis-element of their target genes (Benezra *et al.*, 1990; Sun *et al.*, 1991). As a result, Id proteins inhibit expression of their target genes by partners of Id proteins. Likewise, BU1 may also interact or form a complex with putative BR-negative regulators (maybe bHLH proteins) and inhibit their functions as transcription factors (Figures 7 and 8).

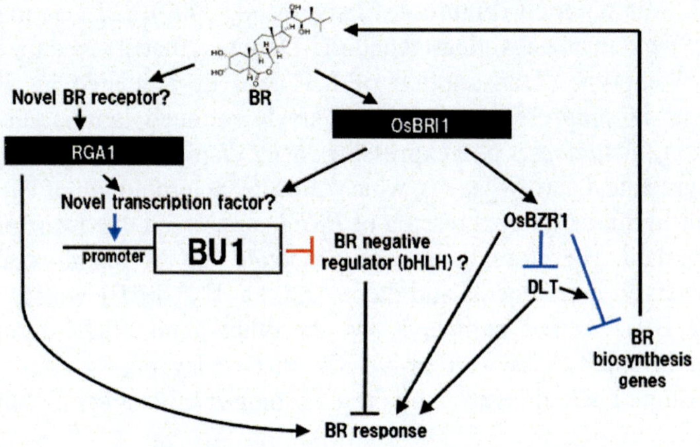

Figure 7. BR signaling model in rice.

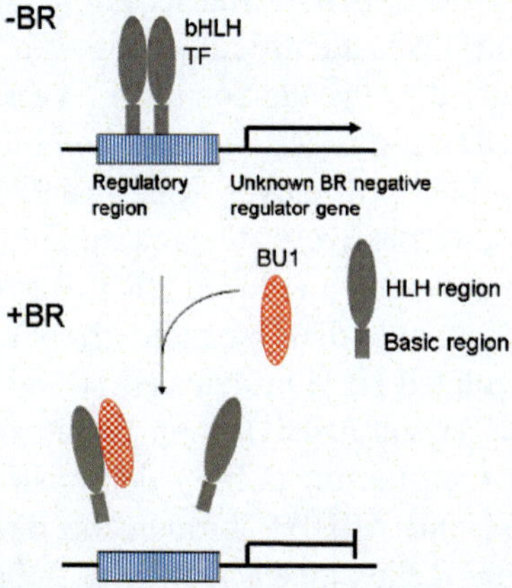

Figure 8. Possible molecular function of BU1 in the interaction with bHLH protein. Without BR, homo (or hetero) – dimer of unknown bHLH transcription factors bind to the cis-element of their target BR negative regulator genes. With BR, BU1 proteins heterodimerize with the bHLH protein partners via HLH motif and abolish their functions as transcription factors, binding to the cis-element of their target genes. As a result, expression of the target genes regulated by the bHLH transcription factors is abolished.

5.2 *ILI1* and *IBH1*

5.2.1 ILI1

After our report on BU1, another rice HLH protein, ILI1 (Increased leaf inclination 1), which has a similar sequence to both BU1 and *Arabidopsis* PRE1, was reported (Figure 4, Zhang *et al.*, 2009). *ILI* was identified as a causal gene in an activation-tagged *ili1-D* mutant that showed the bent lamina joint phenotype. Transcription of ILI1 was up-regulated by BL addition. Anti-sense suppression of *ILI1* reversely reduced lamina inclination. In addition, *Arabidopsis* overexpressing *PRE1* displayed elongated petioles and light-green and narrow leaves, which indicate the promotion of BR responses.

Furthermore, overexpression of *PRE1* suppressed the dwarf phenotype of *bri1* mutant. Therefore, these results indicate that the function of ILI1/PRE1 is conserved in monocots and dicots and that ILI1/PRE1 works downstream of the BRI1 or BR pathways. On the other hand, PRE1 is up-regulated by auxin and GA as well as by BR. PRE1 may play a common role for regulation of cell elongation by these hormones (Zhang *et al.*, 2009).

5.2.2 IBH1

ILI1 interacting proteins were screened by yeast two-hybrid system and a cDNA encoding a typical bHLH transcription factor called IBH1 (ILI1 binding bHLH protein1) was identified (Zhang *et al.*, 2009). Transcription of IBH1 was down-regulated by the addition of BL. Yeast two-hybrid experiments indicated that both IBH1 and its *Arabidopsis* homolog, AtIBH1, also interact with PRE1, a HLH protein. Moreover, ILI1 and PRE1 interacted with IBH1 and AtIBH1 *in vivo*, respectively. Overexpression of IBH1 generated an erect-leaf phenotype in rice and reduced BR sensitivity in *Arabidopsis* root, whereas overexpression of AtIBH1 caused dwarfism in *Arabidopsis*. It is likely that BR up-regulated HLH proteins (ILI1 and PRE1) inactivate bHLH negative regulators (IBH1 and AtIBH1) through heterodimerization. In addition, chromatin immunoprecipitation (ChIP) demonstrated that BR represses transcription of IBH1 and AtIBH1 through the direct binding of OsBZR1 and BZR1, respectively. Taken together, BR may inhibit the function of IBH1 at both RNA and protein levels. These results demonstrated a conserved mechanism of BR regulation through an interacting pair of HLH/bHLH factors, ILI1 and IBH1 in rice and PRE1 and AtIBH1 in *Arabidopsis*.

6. CONCLUSION

Recently, after our report on BU1, two families of HLH proteins capable of modulating BR signaling, ATBS (Activation-tagged bri1 suppressor) and AIF (ATBS1-interaction factor), were reported in *Arabidopsis* (Wang *et al.*, 2009) in addition to PRE1 and AtIBH1. ATBS1 is homologous to PRE1 and positively regulates BR signaling. In contrast, AIF1 negatively regulates BR signaling. AIF family is able to interact with ATBS1. These results indicate that plants use a lot of BU1-like or PRE1-like HLH proteins to positively regulate, and IBH1-like or AIF-like HLH proteins to negatively regulate BR signaling. In conclusion, primary BR perception, signal transduction and early BR responsive pathways are mostly conserved in dicots and monocots.

7. ACKNOWLEDGEMENT

The authors are grateful to the Program for Promotion of Basic Research Activities for Innovative Biosciences (PROBRAIN) for financial support.

8. REFERENCES

Ashikari, M., Wu, J., Yano, M., Sasaki, T. and Yoshimura A. 1999. Rice gibberellin-insensitive dwarf mutant gene *Dwarf 1* encodes the α-subunit of GTP-binding protein. *Proc. Natl. Acad. Sci. USA*, **96**: 10284–10289.

Bai, M.Y., Zhang, L.Y., Gampala, S.S., Zhu, S.W., Song, W.Y., Chong, K. and Wang, Z.Y. 2007. Functions of OsBZR1 and 14-3-3 proteins in brassinosteroid signaling in rice. *Proc. Natl. Acad. Sci. USA.*, **104**: 13839–13844.

Bajguz, A. and Hayat, S. 2009. Effects of brassinosteroids on the plant responses to environmental stresses. *Plant Physiol Biochem.*, **47**: 1–8.

Benezra, R., Davis, R.L., Lockshon, D., Turner, D.L. and Weintraub, H. 1990. The protein Id: a negative regulator of helix-loop-helix DNA binding proteins. *Cell*, **61**: 49–59.

Caño-Delgado, A., Yin, Y., Yu, C., Vafeados, D., Mora-Garćia, S., Cheng, J.-C., Nam, K.H., Li, J. and Chory, J. 2004. BRL1 and BRL3 are novel brassinosteroid receptors that function in vascular differentiation in *Arabidopsis*. *Development*, **131**: 5341–5351.

Ceserani, T., Trofka, A., Gandotra, N. and Nelson, T. 2009. VH1/BRL2 receptor-like kinase interacts with vascular-specific adaptor proteins VIT and VIK to influence leaf venation. *Plant J.*, **57:** 1000–1014.

Chen, K.Y., Cong, B., Wing, R., Vrebalov, J. and Tanksley, S.D. 2007. Changes in regulation of a transcription factor lead to autogamy in cultivated tomatoes. *Science*, **318**: 643–645.

Chinchilla, D., Shan, L., He, P., de Vries, S. and Kemmerling, B. 2009. One for all: the receptor-associated kinase BAK1. *Trends Plant Sci.*, **14**: 535–541.

Clay, N.K. and Nelson, T. 2002. VH1, a provascular cell-specific receptor kinase that influences leaf cell patterns in *Arabidopsis*. *Plant Cell*, **14**: 2707–2722.

Duan, K., Li, L., Hu, P., Xu, S.P., Xu, Z.H. and Xue, H.W. 2006. A brassinolide-suppressed rice MADS-box transcription factor, OsMDP1, has a negative regulatory role in BR signaling. *Plant J.*, **47**: 519–531.

Friedrichsen, D.M., Nemhauser, J., Muramitsu, T., Maloof, J.N., Alonso, J., Ecker, J.R., Furuya, M. and Chory, J. 2002. Three redundant brassinosteroid early response genes encode putative bHLH transcription factors required for normal growth. *Genetics*, **162:** 1445–1456.

Fujioka, S., Noguchi, T., Takatsuto, S. and Yoshida, S. 1998. Activity of brassinosteroids in the dwarf rice lamina inclination bioassay. *Phytochemistry*, **49**: 1841–1848.

Fujisawa, Y., Kato, T., Ohki, S., Ishikawa, A., Kitano, H., Sasaki, T., Asahi, T. and Iwasaki, Y. 1999. Suppression of the heterotrimeric G protein causes abnormal morphology, including dwarfism, in rice. *Proc. Natl. Acad. Sci. USA*, **96**: 7575–7580.

Gampala, S.S., Kim, T.W., He, J.X., Tang, W., Deng, Z., Bai, M.Y., Guan, S., Lalonde, S., Sun, Y., Gendron, J.M., Chen, H., Shibagaki, N., Ferl, R.J., Ehrhardt, D., Chong, K., Burlingame, A.L. and Wang, Z.Y. 2007. An essential role for 14-3-3 proteins in brassinosteroid signal transduction in *Arabidopsis*. *Dev. Cell*, **13**: 177–189.

Gao, Y., Wang, S., Asami, T. and Chen, J.G. 2008. Loss-of-function mutations in the Arabidopsis heterotrimeric G-protein alpha subunit enhance the developmental defects of brassinosteroid signaling and biosynthesis mutants. *Plant Cell Physiol.*, **49**: 1013–1024.

He, J.X., Gendron, J.M., Sun, Y., Gampala, S.S., Gendron, N., Sun, C.Q. and Wang, Z.Y. 2005. BZR1 is a transcriptional repressor with dual roles in brassinosteroid homeostasis and growth responses. *Science*, **307**: 1634–1638.

Hirabayashi, S., Matsushita, Y., Sato, M., Oh-i, R., Kasahara, M., Abe, H. and Nyunoya, H. 2004. Two proton pump interactors identified from direct phosphorylation screening of a rice cDNA library by using a recombinant BRI1 receptor kinase. *Plant Biotechnol.*, **21**: 35–45.

Hong, Z., Ueguchi-Tanaka, M., Shimizu-Sato, S., Inukai, Y., Fujioka, S., Shimada, Y., Takatsuto, S., Agetsuma, M., Yoshida, S., Watanabe, Y., Uozu, S., Kitano, H., Ashikari, M. and Matsuoka, M. 2002. Loss-of-function of a rice brassinosteroid biosynthetic enzyme, C-6 oxidase, prevents the organized arrangement and polar elongation of cells in the leaves and stem. *Plant J.*, **32**: 495–508.

Hong, Z., Ueguchi-Tanaka, M., Umemura, K., Uozu, S., Fujioka, S., Takatsuto, S., Yoshida, S., Ashikari, M., Kitano, H. and Matsuoka, M. 2003. A rice brassinosteroid-deficient mutant, ebisu dwarf (*d2*), is caused by a loss of function of a new member of cytochrome P450. *Plant Cell*, **15**: 2900–2910.

Hong, Z., Ueguchi-Tanaka, M., Fujioka, S., Takatsuto, S., Yoshida, S., Hasegawa, Y., Ashikari, M., Kitano, H. and Matsuoka, M. 2005. The Rice brassinosteroid-deficient dwarf2 mutant, defective in the rice homolog of *Arabidopsis* DIMINUTO/DWARF1, is rescued by the endogenously accumulated alternative bioactive brassinosteroid, dolichosterone. *Plant Cell*, **17**: 2243–2254.

Hoshikawa, K. 1989. The growing rice plant, an anatomical monograph. Nobunkyo, Tokyo.

Itoh, J., Nonomura, K., Ikeda, K., Yamaki, S., Inukai, Y., Yamagishi, H., Kitano, H. and Nagato, Y. 2005. Rice plant development: from zygote to spikelet. *Plant Cell Physiol.*, **46**: 23–47.

Jonak, C. and Hirt, H. 2002. Glycogen synthase kinase 3/SHAGGY-like kinases in plants: an emerging family with novel functions. *Trends Plant Sci.*, **7**: 457–461.

Kim, T.-W. and Wang, Z.-Y. 2010. Brassinosteroid signal transduction from receptor kinases to transcription factors. *Annual Rev. Plant. Biol.*, **61**: published online review.

Kinoshita, T., Caño-Delgado, A., Seto, H., Hiranuma, S., Fujioka, S., Yoshida, S. and Chory, J. 2005. Binding of brassinosteroids to the extracellular domain of plant receptor kinase *BRI1*. *Nature*, **433**: 167–171.

Koh, S., Lee, S.C., Kim, M.K., Koh, J.H., Lee, S., An, G., Choe, S. and Kim, S.R. 2007. T-DNA tagged knockout mutation of rice OsGSK1, an orthologue of *Arabidopsis* BIN2, with enhanced tolerance to various abiotic stresses. *Plant Mol Biol.*, **65**: 453–466.

Lee, S., Lee, S., Yang, K.Y., Kim, Y.M., Park, S.Y., Kim, S.Y. and Soh, M.S. 2006. Overexpression of PRE1 and its homologous genes activates Gibberellin-dependent responses in *Arabidopsis thaliana*. *Plant Cell Physiol.*, **47**: 591–600.

Lee, S., Choi, S.C. and An, G. 2008. Rice SVP-group MADS-box proteins, OsMADS22 and OsMADS55, are negative regulators of brassinosteroid responses. *Plant J.*, **54**: 93–105.

Li, D., Wang, L., Wang, M., Xu, Y.Y., Luo, W., Liu, Y.J., Xu, Z.H., Li, J. and Chong, K. 2009. Engineering *OsBAK1* gene as a molecular tool to improve rice architecture for high yield. *Plant Biotechnol.*, **7**: 791–806.

Li, J. and Chory, J. 1997. A putative leucine-rich repeat receptor kinase involved in brassinosteroid signal transduction. *Cell*, **90**: 929–938.

Li, J. and Jin, H. 2007. Regulation of brassinosteroid signaling. *Trends in Plant Sci.*, **12**: 37–41.

Li, J., Nam, K.H., Vafeados, D. and Chory, J. 2001. *BIN2*, a new brassinosteroid-insensitive locus in *Arabidopsis*. *Plant Physiol.*, **127**: 14–22.

Li, X., Duan, X., Jiang, H., Sun, Y., Tang, Y., Yuan, Z., Guo, J., Liang, W., Chen, L., Yin, J., Ma, H., Wang, J. and Zhang, D. 2006. Genome-wide analysis of basic/helix-loop-helix transcription factor family in rice and Arabidopsis. *Plant Physiol.*, **141**: 1167–1184.

Liu, W., Xu, Z.-H., Luo, D. and Xue H.-W. 2003. Roles of *OsCKI*1, a rice casein kinase I, in root development and plant hormone sensitivity. *Plant J.*, **36**: 189–202.

Mori, M., Nomura, T., Ooka, H., Ishizaka, M., Yokota, T., Sugimoto, K., Okabe, K., Kajiwara, H., Satoh, K., Yamamoto, K., Hirochika, H. and Kikuchi, S. 2002. Isolation and characterization of a rice dwarf mutant with a defect in brassinosteroid biosynthesis. *Plant Physiol.*, **130**: 1152–1161.

Morinaka, Y., Sakamoto, T., Inukai, Y., Agetsuma, M., Kitano, H., Ashikari, M. and Matsuoka, M. 2006. Morphological alteration caused by brassinosteroid insensitivity increases the biomass and grain production of rice. *Plant Physiol.*, **141**: 924–931.

Murre, C., McCaw, P.S., Vaessin, H., Caudy, M., Jan, L.Y., Jan, Y.N., Cabrera, C.V., Buskin, J.N., Hauschka, S.D., Lassar, A.B. *et al.* 1989. Interactions between heterologous helix-loop-helix proteins generate complexes that bind specifically to a common DNA sequence. *Cell*, **58**: 537–544.

Nakamura, A., Fujioka, S., Sunohara, H., Kamiya, N., Hong, Z., Inukai, Y., Miura, K., Takatsuto, S., Yoshida, S., Ueguchi-Tanaka, M., Hasegawa, Y., Kitano, H. and Matsuoka, M. 2006. The role of *OsBRI1* and its homologous genes, *OsBRL1* and *OsBRL3*, in rice. *Plant Physiol.*, **140**: 580–590.

Nakamura, A., Fujioka, S., Takatsuto, S., Tsujimoto, M., Kitano, H., Yoshida, S., Asami, T. and Nakano, T. 2009. Involvement of C-22-hydroxylated brassinosteroids in auxin-induced lamina joint bending in rice. *Plant Cell Physiol.*, **50**: 1627–1635.

Oki, K., Fujisawa, Y., Kato, H. and Iwasaki, Y. 2005. Study of the constitutive active form of the α subunit of rice heterotrimeric G proteins. *Plant Cell Physiol.*, **46**: 381–386.

Oki, K., Inaba, N., Kitano, H., Takahashi, S., Fujisawa, Y., Kato, H. and Iwasaki, Y. 2009a. Study of novel *d1* alleles, defective mutants of the α subunit of heterotrimeric G-protein in rice. *Genes Genet. Syst.*, **84**: 35–42.

Oki, K., Kitagawa, K., Fujisawa, Y., Kato, H. and Iwasaki, Y. 2009b. Function of α-subunit of heterotrimeric G protein in brassinosteroid response in rice plants. *Plant Signal. Behav.*, **4**: 126–128.

Oki, K., Inaba, N., Kitagawa, K., Fujioka, S., Kitano, H., Fujisawa, Y., Kato, H. and Iwasaki, Y. 2009c. Function of the alpha subunit of rice heterotrimeric G protein in brassinosteroid signaling. *Plant Cell. Physiol.*, **50**: 161–172.

Perfus-Barbeoch, L., Jones, A.M. and Assmann, S.M., 2004. Plant heterotrimeric G protein function: insights from *Arabidopsis* and rice mutants. *Curr. Opin. Plant Biol.*, **7**: 719–731.

Russinova, E., Borst, J.W., Kwaaitaal, M., Caño-Delgado, A., Yin, Y., Chory, J. and de Vries, S.C. 2004. Heterodimerization and endocytosis of *Arabidopsis* brassinosteroid receptors BRI1 and AtSERK3 (BAK1). *Plant Cell*, **16**: 3216–3229.

Saisho, D., Tanno, K., Chono, M., Honda, I., Kitano, H. and Takeda, K. 2004. Spontaneous brassinolide-insensitive barley mutants "uzu" adapted to East Asia. *Breed. Sci.*, **54**: 409–416.

Sakamoto, T., Morinaka, Y., Ohnishi, T., Sunohara, H., Fujioka, S., Ueguchi-Tanaka, M., Mizutani, M., Sakata, K., Takatsuto, S., Yoshida, S., Tanaka, H., Kitano, H. and Matsuoka, M. 2006. Erect leaves caused by brassinosteroid deficiency increase biomass production and grain yield in rice. *Nat. Biotechnol.*, **24**: 105–109.

Song, Y., You, J. and Xiong, L. 2009. Characterization of *OsIAA1* gene, a member of rice *Aux/IAA* family involved in auxin and brassinosteroid hormone responses and plant morphogenesis. *Plant Mol. Biol.*, **70**: 297–309.

Steffens, B. and Sauter, M. 2009. Heterotrimeric G protein signaling is required for epidermal cell death in rice. *Plant Physiol.*, **151**: 732–740.

Sun, X.H., Copeland, N.G., Jenkins, N.A. and Baltimore, D. 1991. Id proteins Id1 and Id2 selectively inhibit DNA binding by one class of helix-loop-helix proteins. *Mol. Cell. Biol.*, **11**: 5603–5611.

Takeda, K. 1977. Internode elongation and dwarfism in some gramineous plants. *Gamma Field Symp.*, **16**: 1–18.

Tanabe, S., Ashikari, M., Fujioka, S., Takatsuto, S., Yoshida, S., Yano, M., Yoshimura, A., Kitano, H., Matsuoka, M., Fujisawa, Y., Kato, H. and Iwasaki, Y. 2005. A novel cytochrome P450 is implicated in brassinosteroid biosynthesis via the characterization of a rice dwarf mutant, dwarf11, with reduced seed length. *Plant Cell*, **17**: 776–790.

Tanaka, A., Nakagawa, H., Tomita, C., Shimatani, Z., Ohtake, M., Nomura, T., Jiang, C.J., Dubouzet, J.G., Kikuchi, S., Sekimoto, H., Yokota, T., Asami, T., Kamakura, T. and Mori, M. 2009. *BRASSINOSTEROID UPREGULATED1*, encoding a helix-loop-helix protein, is a novel gene involved in brassinosteroid signaling and controls bending of the lamina joint in rice. *Plant Phys.*, **151**: 669–680.

Tian, C., Wan, P., Sun, S., Li, J. and Chen, M. 2004. Genome-wide analysis of the GRAS gene family in rice and *Arabidopsis*. *Plant Mol. Biol.*, **54**: 519–532.

Tong, H., Jin, Y., Liu, W., Li, F., Fang, J., Yin, Y., Qian, Q., Zhu, L. and Chu, C. 2009. Dwarf and low-tillering, a new member of the GRAS family, plays positive roles in brassinosteroid signaling in rice. *Plant J.*, **58**: 803–816.

Ueguchi-Tanaka, M., Fujisawa, Y., Kobayashi, M, Ashikari, M., Iwasaki, Y., Kitano, H. and Matsuoka, M. 2000. Rice dwarf mutant *d1*, which is defective in the α subunit of heterotrimeric G protein, affects gibberellin signal transduction. *Proc. Natl. Acad. Sci. USA*, **97**: 11638–11643.

Ueguchi-Tanaka, M., Nakajima, M., Ashikari, M. and Matsuoka, M. 2007. Gibberellin receptor and its role in gibberellin signaling in plants. *Ann. Rev. Plant Biol.*, **58**: 183–198.

Ullah, H., Chen, J.G., Young, J.C., Im, K.H., Sussman, M.R. and Jones, A.M. 2001. Modulation of cell proliferation by heterotrimeric G protein in *Arabidopsis*. *Science*, **292**: 2066–2069.

Ullah, H., Chen, J.G., Wang, S. and Jones, A.M. 2002. Role of a Heterotrimeric G protein in regulation of *Arabidopsis* seed germination. *Plant Physiol.*, **129**: 897–907.

Vert, G. and Chory, J., 2006 Downstream nuclear events in brassinosteroid signaling. *Nature*, **441**: 96–100.

Wang, H., Zhu, Y., Fujioka, S., Asami, T., Li, J. and Li, J. 2009. Regulation of *Arabidopsis* brassinosteroid signaling by atypical basic helix-loop-helix proteins. *Plant Cell*, **21**: 3781–3791.

Wang, L., Xu, Y-Y., Ma, Q.-B., Li, D., Xu, Z.-H. and Chong, K. 2006. Heterotrimeric G protein α subunit is involved in rice brassinosteroid response. *Cell Res.*, **16**: 916–922.

Wang, L., Wang, Z., Xu, Y., Joo, S.-H., Kim, S.-K., Xue, Z., Xu, Z., Wang, Z. and Chong, K. 2008a. *OsGSR1* is involved in crosstalk between gibberellins and brassinosteroids in rice. *Plant J.*, **57**: 498–510.

Wang, X., Kota, U., He, K., Blackburn, K., Li, J., Goshe, M.B., Huber, S.C. and Clouse, S.D. 2008b. Sequential transphosphorylation of the BRI1/BAK1 receptor kinase complex impacts early events in brassinosteroid signaling. *Dev. Cell*, **15**: 220–235.

Wang, Z.-Y., Seto, H., Fujioka, S., Yoshida, S. and Chory, J. 2001. BRI1 is a critical component of a plasma-membrane receptor for plant steroids. *Nature*, **410**: 380–383.

Wang, Z.Y., Nakano, T., Gendron, J., He, J., Chen, M., Vafeados, D., Yang, Y., Fujioka, S., Yoshida, S., Asami, T. and Chory, J. 2002. Nuclear-localized BZR1 mediates brassinosteroid-induced growth and feedback suppression of brassinosteroid biosynthesis. *Dev. Cell.*, **2**: 505–513.

Wu, C.Y., Trieu, A., Radhakrishnan, P., Kwok, S.F., Harris, S., Zhang, K., Wang, J., Wan, J., Zhai, H., Takatsuto, S., Matsumoto, S., Fujioka, S., Feldmann, K.A. and Pennell, R.I. 2008. Brassinosteroids regulate grain filling in rice. *Plant Cell*, **20**: 2130–2145.

Yamamuro, C., Ihara, Y., Wu, X., Noguchi, T., Fujioka, S., Takatsuto, S., Ashikari, M., Kitano, H. and Matsuoka, M. 2000. Loss of function of a rice brassinosteroid insensitive1 homolog prevents internode elongation and bending of the lamina joint. *Plant Cell*, **12**:1591–1605.

Yang, G., Nakamura, H., Ichikawa, H., Kitano, H. and Komatsu, S. 2006. *OsBLE3*, a brassinosteroid-enhanced gene, is involved in the growth of rice. *Phytochemistry*, **67**: 1442–1454.

Yin, Y., Wang, Z.Y., Mora-Garcia, S., Li, J., Yoshida, S., Asami, T. and Chory, J. 2002. BES1 accumulates in the nucleus in response to brassinosteroids to regulate gene expression and promote stem elongation. *Cell*, **109**: 188–191.

Yin, Y., Vafeados, D., Tao, Y., Yoshida, S., Asami, T. and Chory, J. 2005. A new class of transcription factors mediates Brassinosteroid-regulated gene expression in *Arabidopsis*. *Cell*, **120**: 249–259.

Yoo, M.J., Albert, V.A., Soltis, P.S. and Soltis, D.E. 2006. Phylogenetic diversification of glycogen synthase kinase 3/SHAGGY-like kinase genes in plants. *BMC Plant Biol*, **6**:3.

Zhang, L.Y., Bai, M.Y., Wu, J., Zhu, J.Y., Wang, H., Zhang, Z., Wang, W., Sun, Y., Zhao, J., Sun, X., Yang, H., Xu, Y., Kim, S.H., Fujioka, S., Lin, W.H., Chong, K., Lu, T. and Wang, Z.Y. 2009. Antagonistic HLH/bHLH transcription factors mediate brassinosteroid regulation of cell elongation and plant development in rice and *Arabidopsis*. *Plant Cell*, **21**: 3767–3780.

Chapter 5

BRASSINOSTEROIDS AND LIGHT – REGULATORY FACTORS OF GROWTH AND DEVELOPMENT OF PLANTS

I.F. GOLOVATSKAYA
Plant Physiology and Biotechnology Department, Tomsk State University, Lenin Avenue, 36, 634050, Tomsk, Russia

Abstract: High biological activity of brassinosteroids determines their importance in regulating different processes in a plant. At present there is no general agreement on BRs role in regulating the morphogenesis and the light signal transduction. One group of researchers considers the BRs to be negative regulators of plant photomorphogenesis which reduces the effects caused by photoreceptors. In other researchers' opinion the BRs are positive photomorphogenesis regulators imitating the light effects on a plant. The existing chapter views physiological action of brassinosteroids which depends on their concentration in a plant and the medium, on the method of application, on light quality and the mechanism of regulation of the photomorphogenesis by brassinosteroids.

Key words: brassinosteroids, hormone interactions, photomorphogenesis

Abbreviations: BL – brassinolide; BlL – blue light; BRs – brassinosteroids; CRY – cryptochrome; CYT – Cytokinins; EBL – 24-epibrassinolide; GA – gibberellin; GBL – 28-gomobrassinolide; GL – green light; phy – phytochrome; RL – red light; ZR – zeatin riboside.

1. INTRODUCTION

The plants conducting the attached way of life, are compelled to be especially plastic in answers to influences of an environment. Light concerns to the most important external factors. Light is not only an energy source for photosynthesis, but also a signal, activating and changing the program of

S. Hayat and A. Ahmad (eds.), *Brassinosteroids: A Class of Plant Hormone*,
DOI 10.1007/978-94-007-0189-2_5,

development of plant. Green light occupies a special place in the regulatory phenomenon.

Realization of plant developmental programs largely depends on the light, which acts on various physiological processes during the entire plant life, including seed germination, seedling growth, determination of growth direction (phototropism, shadow avoidance), flowering, etc. On the other hand, the integrity of the plant organism is provided by the endogenous regulatory systems, by the hormonal system in particular (Beevers *et al.*, 1970; Kulaeva, 1982; Polevoi, 2001).

The plants absorb light via its photoreceptors, specifically regulatory pigments among them. They control all stages of plant development by the adaptation of physiological processes to changing conditions of illumination, thus playing a key role in plant transition from scotomorphogenesis (of a Greek word skotos-darkness) to photomorphogenesis (Konev and Volotovsky, 1974; Cosgrove, 1986; Voskresenskaya, 1987). Spectral specialization of the regulatory pigments is shown: phytochromes (PHYA–PHYE) absorb red light (RL) and far-red light (FRL); phototropins (PHOT1 and PHOT2) absorb blue light (BlL); cryptochromes (CRY1–CRY5), BlL and UV-A; and superchrome (PHY3), RL and BlL (Ahmad and Cashmore, 1993, 1997; Imaizumi *et al.*, 2000; Briggs and Olney, 2001; Casal, 2001). Regulatory pigments for green light (GL) are yet unknown. Based on the assessment of the extent of changes of physiological processes per the amount of a pigment in the biological system, some authors supposed the occurrence of light signal amplification. G-proteins, cGMP, phospholipase D, Ca^{2+}, Ca^{2+}-binding proteins (calmodulin and others), and kinases are believed to transduce a light signal inside the cell; phytohormones operate at the tissue level (Sternberg 1965; Neuhaus *et al.*, 1993; Dubovskaya *et al.*, 2001; Malec *et al.*, 2002; Kabachevskaya *et al.*, 2004; Golovatskaya, 2005). It was shown that, in *hy4 Arabidopsis* seedlings displaying disturbed CRY1 synthesis, the levels of phytohormones changed under both BlL and GL (Karnachuk *et al.*, 2001, 2002a and b). A phytochrome-mediated control of cereal growth by RL is believed to occur via the change in the gibberellins (GA) level (Brien *et al.*, 1985). An information appeared that signal transduction cascades induced by cytokinins (CYT) and brassinosteroids modulate signal transduction from phytochrome at the gene and cytoplasmic levels (Hutchison and Kieber, 2002; Zubo *et al.*, 2005; Neff *et al.*, 2000). Light was also shown to affect hormonal signal transduction, controlling expression/activity of key enzymes, involved in hormone biosynthesis (Kang *et al.*, 2001). Phytochrome-dependent changes in GA activity were demonstrated mainly under RL, whereas the spectrum of phytochrome action embraces the entire region of PAR (Mohr, 1970). Therefore, it is of interest to study the involvement of phytochromes in the control of hormonal balance in plant leaves, illuminated with GL.

At present, the mechanisms of scoto- and photomorphogenesis are widely investigated (Whitelam and Devlin, 1998; Turk *et al.*, 2003). According to the opinion of some researchers, phytohormones are involved in light signal transduction because many light-induced plant responses are also under hormonal control. Thus, cytokinins mediate photomorphogenesis, whereas GA, scotomorphogenesis (Chory *et al.*, 1994; Alabady *et al.*, 2004). In its turn, light changes GA, IAA and ABA metabolism and plant sensitivity to GA. *Arabidopsis* mutants *phyb* harboring a disturbed photoreceptor of red light display an increased sensitivity to GA, whereas tobacco mutants *pew1* defective in phytochrome A have an increased amounts of ABA (Bandurski *et al.*, 1977; Kraepiel *et al.*, 1994; Reed *et al.*, 1996; Weatherwax *et al.*, 1996). Both short-term and long-term illumination with the light of different quality was shown to affect growth and hormonal status of plant seedlings (Karnachuk *et al.*, 1990; Karnachuk and Golovatskaya, 1998).

High biological activity of brassinosteroids (BRs) determines their importance in regulating different processes in a plant (Grove *et al.*, 1979; Khripach *et al.*, 1993, 1999, 2003; Fujioka and Sakurai, 1997; Bajguz and Czerpak, 1998; Clouse and Feldman, 1999). It is known that they change the membranous potentials and the enzyme activity, promote the protein synthesis, nucleic and fatty acids, alter the balance of the other endogenic hormones, stimulating in this way, the cell elongation and division (Szekeres *et al.*, 1996; Vardhini and Ram Rao, 1998; Bajguz, 2000; Karnachuk *et al.*, 2002a and b; Müssig *et al.*, 2002; Lisso *et al.*, 2005). Alterations at a cell level are reflected in a whole plant by growth intensification, productivity rise and stress stability (Clouse and Sasse, 1998; Sasse, 1999; Khripach *et al.*, 2003; Shakirova, 2003; Tanaka *et al.*, 2003: Symons and Reid, 2004; Yu *et al.*, 2004; Golovatskaya and Vinnikova, 2007; Hayat *et al.*, 2007; Golovatskaya and Nikonorova, 2008). A tissue-specific expression of genes regulating the BRs biosynthesis has been observed. In the dark the expression of the *CPD* gene is limited by cotyledons and primordials of leaf seedlings, and in the light it is found in the parenchyma of plant leaves. Brassinolide reduces the *CPD* gene transcription both in dark and in the light, but the effects of other phytohormones (such as auxin, ethylene, gibberellins, cytokinins, jasmine and salicylic acids do not alter it (Mathur *et al.*, 1998). It is supposed that the inhibition of hypocotyl growth in the light is caused by decreasing the BRs synthesis (Li *et al.*, 1996; Clouse, 2001), and that the BRs control the plant photomorphogenesis (Chory *et al.*, 1991; Karnachuk *et al.*, 2002a; Tanaka *et al.*, 2003; Nemhauser *et al.*, 2003; Nemhauser and Chory, 2004).

At present there is no general agreement on the BRs role in regulating the morphogenesis and the light signal transduction. One group of researchers considers the BRs to be negative regulators of plant photomorphogenesis which reduces the effects caused by photoreceptors (Bishop *et al.*, 1999; Kim *et al.*, 2002; Asami *et al.*, 2004). In other researchers' opinion the BRs are positive photomorphogenesis regulators imitating the light effects on a plant (Nagata *et al.*, 2000; Karnachuk *et al.*, 2001, 2002a; Symons *et al.*, 2002).

2. THE PHYSIOLOGICAL ACTION OF BRASSINOSTEROIDS DEPENDS ON THEIR CONCENTRATION IN A PLANT AND THE MEDIUM

2.1 The BRs role in regulating the scotomorphogenesis of *Arabidopsis thaliana* seedlings

The studies on BRs role in regulating the scotomorphogenesis of *Arabidopsis thaliana* seedlings were made on the Col and L*er* parental lines and the *det2* and *hy4* (Koornneef *et al.*, 1980) mutants corresponding to them. The shape – size dependence of hypocotyl and cotyledons on the concentration of the exogenous and endogenic BRs have been shown (Karnachuk *et al.*, 2002a; Golovatskaya, 2004a, 2004b; Efimova *et al.*, 2008).

Under the stimulating effects of all the concentrations studied, the largest hypocotyl elongation was observed under the action of 10^{-7} M *28-homo-brassinolide* (GBL) in the etiolated seedlings of *det2* mutant, while in case of Col wild type GBL (10^{-6} and 10^{-5} M) inhibited the hypocotyl elongation. High mutant sensitivity to exogenous GBL was associated with the deficiency of the endogenic BRs as well as with the stimulating effect of BRs on cell elongation (Piechowski, 1997). The maximum elongation of *det2* cotyledons occurred at a higher concentration of the exogenous hormone (10^{-6} M) as compared with the hypocotyl. The seedling growth of the two other lines (L*er* and *hy4*) was sustained by means of lower GBL concentration (10^{-7} M) than in case of Col. However, the smaller absolute value of growth parameters of the etiolated *hy4* seedlings, as compared with the wild type L*er*, was determined by *HY4* gene mutation which is likely to be conjugated with the activity of the genes controlling the growth.

The more active *24-epibrassinolide* (EBL) and *brassinolide* (BL) brassinosteroids had a stimulating effect on the hypocotyl growth and *det2* cotyledons at low concentrations 10^{-10} and 10^{-9} M (Golovatskaya, 2004a; Efimova *et al.*, 2007, 2008), while EBL and BL (10^{-8} M) inhibited the elongation of Col and L*er* hypocotyl, which might be caused by an increase in the endogenic ethylene pool (Smalle *et al.*, 1997) or by the negative BRs regulation of their characteristic biosynthesis according to the feedback principle.

The elongation of hypocotyl and cotyledons in the seedlings of *hy3* mutant (which is RL *phyB* photoreceptor-deficient) was greater under the action of endogenic EBL (10^{-8} M) in the dark than in the wild type L*er*. One can expect an increase in sensitivity to EBL (similar to the reaction of the BR-deficient mutant *det2*) due to a decrease in the endogenic BRs level on the background of the *PHYB* gene mutation.

The differences in the growth responses of the cotyledons and hypocotyl to the exogenous BRs effects might be due to their different competence to these hormones or to controlling their amount along the transport line on the part of the buffer or metabolic activity of the transport systems (primarily the root and the hypocotyl). Our results on a high stimulating effect of the exogenous hormone on the cotyledon growth of the *Arabidopsis* seedlings (with the roots removed as compared with the intact seedlings) may serve as a proof of regulating the BRs level along the_transport line from root to cotyledons (the data are not presented here).

Comparison of EBL effects on elongation of the *Arabidopsis* hypocotyl and the *Triticum vulgare* coleoptile established a higher sensitivity of the growth processes of the dicotyledons (10^{-9} and 10^{-8} M) as compared with the monocotyledons (10^{-8} and 10^{-6} M) (Golovatskaya, 2004b). These differences are likely to be associated with a different level of the endogenic BRs and the tissue-specific regulation of expression of the definite genes.

2.2 Interaction of 24-epibrassinolide and gibberellic acid in the regulation of *Arabidopsis thaliana* seedlings scotomorphogenesis

Gibberellins (GAs) and brassinosteroids (BRs) are multifunctional phytohormones that control size of cells, organs, and whole plants. Mutations affecting the biosynthesis of these hormones lead to dwarfism of plants. GAs and BRs are believed to play an important role in the control of morphogenetic programs of plants, and in the first place, in the regulation of plant morphogenesis in darkness. It is considered that the most important hormones regulating growth of *Pisum sativum* in the dark are only GAs, whereas for *Arabidopsis* these are both GAs and BRs (Alabady *et al*., 2004). Low level of GAs induces complete de-etiolation of *P. sativum*, which is not reversed by the BR treatment.

ga1-3, an *Arabidopsis* mutant with the first step of GA biosynthesis blocked (Sun, 2000) is characterized by the absence of the apical hook, clearly manifested dwarfism, open cotyledons, and derepression of *CAB2* and *RbcS*, the lightregulated genes, in darkness. BRs could partially elongate short hypocotyls of *ga1-3* plants and mediate the regulatory effect of gibberellins on *CAB2* and *RbcS* gene expression in darkness (Azpiroz *et al*., 1998; Ephritikhine *et al*., 1999; Bouquin *et al*., 2001).

The studies of hypocotyl elongation, seed germination, and gene expression (Bouquin *et al*., 2001; Steber and McCourt, 2001) demonstrated interaction between GAs and BRs. These hormones have similar effects on many processes, but their effects on the expression of *GASA1* and *GA5* genes are opposite (Bouquin *et al.*, 2001; Steber and McCourt, 2001). An intersection of BR and GA biosynthesis pathways is suggested (Kasahara *et al.*, 2002), and the endogenous GA level could be changed by EBL treatment (Tishchenko *et al.*, 2001). However, the hierarchy of BR and GA roles in the regulation of plant scotomorphogenesis is not yet clear.

The effects of GA_3, 24-epibrassinolide (EBL), and their combination on morphogenesis of *Arabidopsis thaliana* (L.) Heynh 7-day-old seedlings were studied (Golovatskaya, 2008). Four plant lines were analyzed: wild type L*er* and *ga4-1* mutant (Talon *et al.*, 1990), belonging to the Landsberg *erecta* ecotype and wild type Col and *det2* mutant, both of the Columbia ecotype. In *ga4-1* and *det2*, $GA_{4/1}$- and brassinosteroid-deficient mutants, the highest hypocotyl growth response to the lack of hormones was noted. The cotyledon shape and size were dependent on EBL, but the root length was both GA_3- and EBL-regulated, indicating organ specificities in the responses to these hormones.

Simultaneous treatment of dark-grown plants with GA_3 and EBL exerted an additive stimulatory effect on the root growth of *det2*, reduced the inhibitory effect of EBL on hypocotyl elongation of *ga4-1*, and enhanced the effect of EBL on hypocotyl and cotyledon elongation of *det2*.

2.3 Interaction of 24-epibrassinolide and ecdysterone in the regulation of elongation of wheat coleoptiles

24-epibrassinolide affected the elongation of wheat coleoptiles in the same concentration range as IAA (10^{-7} and 10^{-6} M) (Golovatskaya, 2004b). A similarity between auxins and BS in their growth effects was also demonstrated in other bio-assays, such as the elongation of hook segments of dwarf pea, coleoptiles and mesocotyls of maize, epicotyls of azuki bean, as well as the hook opening in dwarf kidney bean (He *et al.*, 1991; Moor, 1989).

An enhancement of coleoptile growth after the addition of *ecdysterone* (ECD, 20-hydroxyecdysone) in the presence of low, inefficient EBL concentrations was the same as at 10^{-6} M ECD. At the same time, after adding ECD at high, efficient EBL concentrations, the effect of EBL on cell elongation decreased, that is, EBL displayed an anti-brassinosteroid effect, which might be determined by EBL and ECD competition for a common receptor (Golovatskaya, 2004b). This suggestion is supported by the structural similarity between ecdysteroids and brassinosteroids (Lafont and Wilson, 1996) and a decrease in the effect of a simultaneous action of ECD and EBL on the development of insects, that is, an anti-ecdysteroid effect of BS (Akhrem and Kovganko, 1989).

ECD increased the cotyledon area in *det2* mutant seedlings due to elongation of petiole by 37%, while, in the case of EBL, the increase was caused by an increase in the blade area. In the wild type (Col), blade increase was accompanied by the shortening of the petiole by 19%, whereas EBL induced the elongation of the cotyledon petiole (Golovatskaya, 2004b).

ECD is similar to EBL and GA_3 in its effect on some growth and metabolic processes. Therefore, it might be suggested that the mechanism of ECD activity involves its interaction with EBL and GA_3 receptors.

3. THE PHYSIOLOGICAL ACTION OF BRASSINOSTEROIDS DEPENDS ON THE METHOD OF TREATMENT OF THE EXOGENOUS HORMONE

The studies of the vegetation process show that there is a dependence between morphogenesis and the periods of plant's development on the activity of the regulatory photoreceptors and the method of plants treatment with brassinolide (Golovatskaya and Nikonorova, 2008). The plants of the *hy4* mutant (40- and 64-days old) with the disturbed regulatory photoreceptor blue light (BlL) are characterized by smaller leaf and shoot sizes, wider branching, a greater number of small seedpods with lower seed filling and their later ripening (low male fertility), lower seed productivity compared with wild type. The deficiency of cry1 photoreceptor increases the duration of separate ontogene phases (seed germination, cotyledons, first leaves, rosette, formation of the leading shoot and budding, flowering and fructification) of the mutant plants by retarding the transition from the vegetative to reproductive stage, and lowers the rate of the growth reactions.

The use of exogenous BL for *Arabidopsis* growth and development provides a stimulating effect dependent on the cultivation method (Golovatskaya and Nikonorova, 2008). The presowing treatment of L*er and hy4* seeds with 10^{-9} M BL (+BL) provides an earlier completion of the rosette formation and an active elongation of the *leading shoot* (LSh). Therefore, as the plants enter into the reproductive stage, the hormone enlarges the area and the biomass of the rosette leaves, the LSh length, stimulates shoot branching, increases the quantity of pods and their seed filling, causing the rise of the seed productivity in *Arabidopsis*, as compared with the control plants. Other scientists point out to speedy seed germination and rapid leaf elongation after BL and EBL treatment, performed before the planting-out (Prusakova and Chizhova, 1996), and to increase the productivity of crops by applying the exogenous GBL (Ramraj *et al.*, 1997).

It should be noted that the lack of cry1 photoreceptor rises the sensitivity of the growth reactions of *hy4* mutant *Arabidopsis* to the action of the exogenous BL, which leads to the restoration of some growth parameters in mutants (the LSh length, biomasses in the rosette and the LSh) and the developmental time (flowering and fructification) of the L*er* initial line.

The leaf treatment with brassinolide (BL*) 10^{-11} M, as in the case with presowing treatment, increases the growth parameters of a shoot and the number of the reproductive organs (RO). However, the leaf treatment with BL* is less effective to change the LSh sizes, compared with the presowing one, which is likely to be due to divertion of the assimilators to grow the leaf surface. Flowering in $hy4_{BL*}$ mutant is accelerated as compared with $hy4_{+BL}$, while at the initial line Ler_{+BL} and Ler_{BL*} flowering takes place simultaneously.

The distinctions of the growth responses to BL in both the *Arabidopsis* lines are sure to be associated with the features of the light signal transduction or with the endogenic brassinosteroids level conditioned by the tissue-specific regulation of genes expression of BRs biosynthesis in the ontogenesis (Bancos *et al.*, 2006) or by regulating the BRs level during their biosynthesis on the feedback principle (Jefferson, 1987)

The most promising effect is exerted by a double BL (BL+BL*) treatment which supports a definite endogenic hormone level in the ontogenesis process and ensures the increase in all the parameters. The combination of presowing and leaf treatments with BL speeds up fructification and the second-order shoot formation (side shoots – SSh).

In case of wild type is BL+BL* synergetic effect was observed on the LSh length, the additive effect – on the area and the biomass of rosette leaves and the RO amount, an antagonistic influence – on formation of meromes of Ler_{BL+BL*} SSh. Taken together these observations point to an important role of BL dynamics on the growth processes and plant development, much as its diurnal dynamics (Bancos *et al.*, 2006; Golovatskaya and Karnachuk, 2007). Double BL treatment of cry1 mutant reveals a synergetic effect while forming LSh and SSh $hy4_{BL+BL*}$ meromes, however, the enlargement of area and the biomass of $hy4_{BL+BL*}$ rosette leaves corresponds to the effect of a single BL treatment ($hy4_{+BL}$ or $hy4_{BL*}$).

The action effect of presowing hormone treatment is decreased in the process of the plant development while the subsequent BL leaf treatment (a double treatment) resumes the growth of Ler_{BL+BL*} and $hy4_{BL+BL*}$ rosette leaves and enlarges their area.

A comparative analysis of different methods of BL treatment for the *Arabidopsis* productivity showed that increasing the quantity of seeds in a mutant's pod is similar after the presowing and leaf treatments and it is summed up after a double treatment. The treatment with +BL and BL+BL* essentially influences the quantity of both the number of pods on a plant and the seeds in a pod of the wild type, improved L*er* seed productivity calculated per one plant (Golovatskaya and Nikonorova, 2008).

Joint action of active groups of gibberellins (GA) and brassinolide (BL) on the growth and development of plants of *Arabidopsis thaliana*, ecotype Landsberg *erecta* has been studied (Golovatskaya and Vinnikova, 2007). It has been observed that the presowing treatment of seeds with 10^{-9} M BL reduced the duration of phases of ontogenesis of plants, accelerated growth of vegetative and reproductive organs and increased seed production in plants of wild-type L*er* and of mutant *ga4-1*. Exogenous BL, in part compensated lack of endogenic $GA_{1/4}$ at mutant, enlarging common length of shoots and restoring seed production of mutant up to a level of wild-type. This effect might be the augmentation of quantity of side shoots, quantities of pods, length of pods and seed number per pod.

4. THE PHYSIOLOGICAL ACTION OF BRASSINOSTEROIDS DEPENDS ON LIGHT OF DIFFERENT SPECTRAL COMPOSITION

Light is an environmental signal that induces the expression of light-dependent genes and triggers photomorphogenetic programs in plants. Such genes are regulated by various types of photoreceptors, including phytochromes, cryptochromes, and possibly phototropins (Quail *et al*., 1995; Casal, 2001; Lin and Shalitin, 2003).

Multiple signal systems of plant cells triggered by both internal and external factors are involved in complicated interaction and employ various intermediates, of which some may be common. Most likely, phytohormones are involved in light signal transduction and thus in the regulation of the photomorphogenesis (Karnachuk *et al*., 1990). It was demonstrated, for example, that the presence of cytokinins was required for light regulation of wheat protein kinase homologous genes (*WPK4*) (Sano and Youssefian, 1994), whereas gibberellins, cytokinins, and brassinosteroids could compensate for the absence of light in seedlings germinated in darkness (Steber and McCourt, 2001). The studies of photomorphogenetic mutants showed that light signal changed the endogenous auxin level (Kraepiel *et al*., 1994; Karnachuk and Golovatskaya, 1998). Different parts of light spectrum changed the endogenous hormone balance by changing the levels of active gibberellins, auxins, ABA, and cytokinins (Karnachuk *et al*., 1988, 1990, 2001, 2002a; Golovatskaya *et al*., 2001).

We have shown the dependence of photomorphogenesis on the endogenic BRs level during a long period of cultivating the *A. thaliana* plants of the Columbia ecotype in GL. The deficiency of BL (*det2*) results in restricting elongation of the axial elements of the shoot (hypocotyl and stem) and the leaf surface, increasing the specific surface density of the rosette leaves and LSh, as compared with the Col wild type. The leaf surface and their number in a rosette and on the LSh in 30-day old mutant plants decreased, resulting in the loss of their dry biomass.

The GL action inhibits the elongation of the organs and increases the duration of the vegetative phase of *Arabidopsis* development as compared with the BlL action (Figure 1). The deficiency of the endogenic pool in BR enhances this GL effect. The *Arabidopsis* mutant plants begin flowering late in GL, on the 30th day they form a smaller number of buds by a factor of 2.

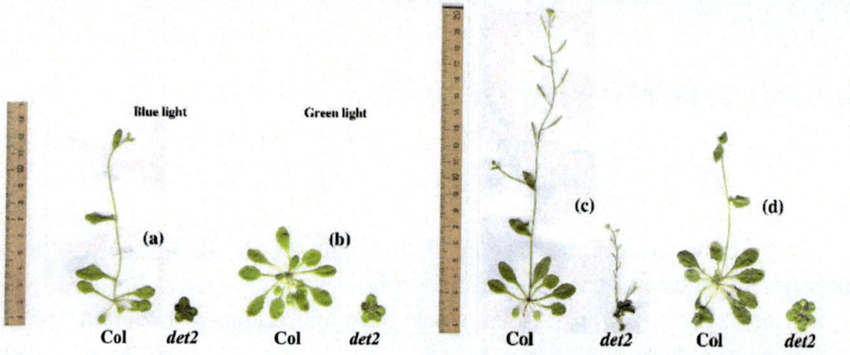

Figure 1. The 26- (a, b) and 37-day-old (c, d) plants of *A. thaliana* (L.) Heynh. Columbia ecotype grown in the blue (a, c) and green light (b, d): the lamps (produced by the "Philips" firm) of the blue TL-D 18W/18 and green TL-D 18W/17, 48 μmol quant / m^2s, photoperiod 16 h.

The formation of a photosynthetic system in GL occurs under the influence of endogenic regulation systems. The data presented in Table 1 shows the dependence of level of the photosynthetic pigments of *Arabidopsis* cotyledons upon the level of the endogenic BR (the Col and *det2* lines) and cry1 (the L*er* and *hy4* lines) in GL.

In the seedlings of *det2* mutant which is BR-deficient in carotenoids (*Car*) content in the dark is almost twice as much as *Car* content in the seedlings of Col initial line (Table 1) said the above observations allow one to associate the latter with the expression of light-sensitive genes which control *Car* biosynthesis in the mutants having the de-etiolated phenotype in the dark.

Table 1. The 28-gomobrassinolide effects (GBL, 10^{-6} M) on the photosynthetic pigment content in 7-day-old seedlings of the *A. thaliana* (L.) Heynh Columbia ecotype in green light (GL, 500–600 nm, 77 μmol quant / m^2s, photoperiod 14 h)

Line	Growth conditions	Pigment level, μg/cotyledon ×10^{-2}			
		Chl *a*	**Chl *b***	**Chl *a*/Chl *b***	**Carotenoids**
Col	Darkness	–	–	–	0.8±0.1
	darkness + GBL	–	–	–	1.4±0.2
	GL	7.5±0.6	4.0±1.2	1.90	4.7±0.3
	GL + GBL	4.7±0.5	2.0±0.4	2.41	3.0±0.4
det2	darkness	–	–	–	1.3±0.1
	darkness+ GBL	–	–	–	0.6±0.1
	GL	11.5±0.1	3.3±0.1	3.54	7.2±0.2
	GL + GBL	5.6±1.0	1.7±0.2	3.39	3.9±1.0

The GL effect increases the *Car* content by several orders of magnitude in the both line seedlings as compared with the controls in the dark, retaining all the advantages of the pigment content in *det2* (Table 1). The data obtained agree with the well-known data on the yellow pigment synthesis in the dark and on the essential increase in the *Car* pool, during the plastid development in light, which fits well into the pigment-protein complexes of photosynthetic membranes and transfers an additional energy to the reaction centers. In GL, total content of the green pigments is considerably increased and the Chl *a*/Chl *b* ratio also grows as compared with the wild type at the expense of Chl *a* fraction. A higher level of Chl *a* is attained on the background of *Car* high level, for which a protective function of the major photosynthesis pigment is shown.

The lack of cry1 in the *hy4* seedlings determines the reduction in the yellow and green pigments content with respect to L*er* in the GL with similar value of Chi *a*/Chl *b* (Table 2). The BL restores the level of *hy4* photosynthetic pigments to the level of the wild type. The observed change of the pigments' level allows one to suppose a possibility of crossing the transduction ways of the GL and BRs signals.

Table 2. The brassinolide effects (BL, 10^{-7} M) on the photosynthetic pigment content in 7-day-old seedlings of the *A. thaliana* (L.) Heynh. Landsberg *erecta* ecotype in green light (GL, photoperiod 16 h)

Line	Growth conditions	Pigment level, μg/cotyledon ×10^{-2}			
		Chl *a*	**Chl *b***	**Chl *a*/Chl *b***	**Carotenoids**
	Darkness	–	–	–	0.7±0.2
L*er*	GL	11.8±0.6	4.4±0.3	2.7±0.2	6.3±0.4
	GL + BL	12.5±0.3	4.7±0.2	2.7±0.2	6.7±0.2
	darkness	–	–	–	0.7±0.2
hy4	GL	8.6±0.9	3.1±0.2	2.8±0.1	4.7±0.4
	GL + BL	11.3±1.1	4.0±0.5	2.9±0.2	6.3±0.6

However, a partial preservation of the reverse response to green light (543 nm) by *hy4* seedlings enables one to propose participation of other photoreceptors, including CRY2, phytochromes and a non-identified green light receptor (Karnachuk *et al*., 1978; Tanada, 1984; Golovatskaya, 2005; Golovatskaya *et al*., 2007). A similar increment of cotyledon area in the case of L*er* and its *hy3* mutant during their de-etiolation in GL (524.5 nm) testifies the participation of a photoreceptor other than phyB. These very-low fluence responses (VLFR) are likely to have been caused either by a similar phyA level which is responsible for sensitivity to VLFR (Botto *et al*., 1996; Shinomura *et al*., 1996) or by some other photoreceptor.

The *det2* mutant of *Arabidopsis* possessing all the characteristics of a light phenotype when grown in dark is the model to investigate the hormone photomorphogenetic regulation (Chory *et al*., 1989). A de-etiolated phenotype of *det2* mutant seedlings growing in the dark are characterized by short and thickened hypocotyls, open and widened cotyledons as compared with the wild type Col (Figure 2). The ultimate growth cessation of the axis and the cotyledon development in the dark shows that the gene expression usually regulated by light is taking place in the mutants without any compelling necessity for it.

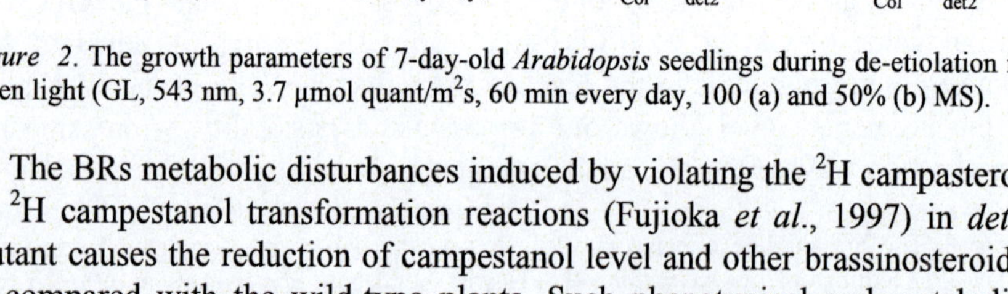

Figure 2. The growth parameters of 7-day-old *Arabidopsis* seedlings during de-etiolation in green light (GL, 543 nm, 3.7 μmol quant/m^{2}s, 60 min every day, 100 (a) and 50% (b) MS).

The BRs metabolic disturbances induced by violating the ^{2}H campasterol → ^{2}H campestanol transformation reactions (Fujioka *et al*., 1997) in *det2* mutant causes the reduction of campestanol level and other brassinosteroids as compared with the wild-type plants. Such phenotypical and metabolic differences among plants of *det2* mutant and wild type play an important part in *Arabidopsis* development.

The physiological action of brassinosteroids in regulating the *Arabidopsis* morphogenesis depends on the light qualities. The GL de-etiolation of *det2* and Col *Arabidopsis* seedlings differing in the level of the endogenic BRs was effective for the enlargement of cotyledon sizes but was ineffective for the inhibition of the hypocotyls growth in the both lines (Figure 3). A 60-minute GL action (543 nm) on the growth of Col cotyledons was two times as high as a 30-min BlL action (439 nm) having the same intensity;

this fact demonstrated a similar sensitivity of the *Arabidopsis* seedlings of the Columbia ecotype to the green and blue light unlike L*er*.

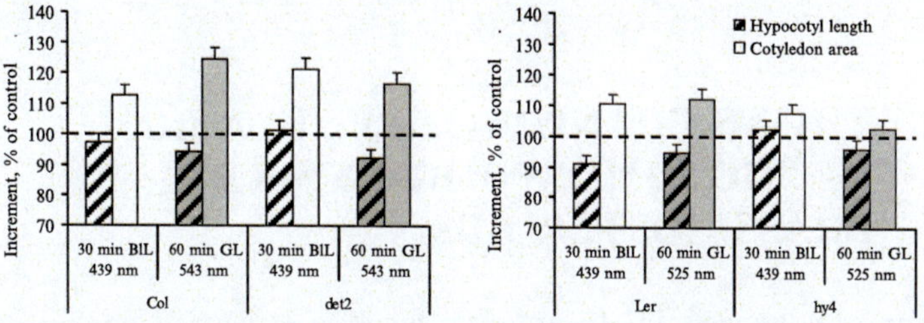

Figure 3. The growth parameters of 7-day-old *Arabidopsis* seedlings during de-etiolation in green (GL, 543 or 525 nm, 3 μmol quant/m^{2}s, 60 min every day) and blue light (BlL, 439 nm, 3 μmol quant/m^{2}s, 30 min every day).

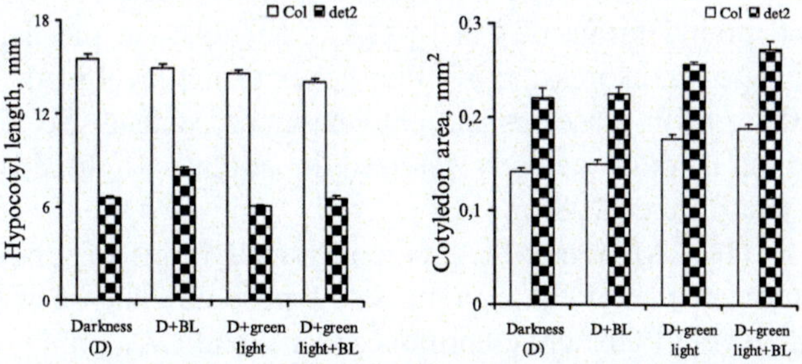

Figure 4. The brassinolide effects (BL, 10^{-7} M) on the morphogenesis of 7-day-old *Arabidopsis* seedlings in the darkness (D) and during de-etiolation in the green light (GL, 543 nm, 3 μmol quant/m^{2}s, 60 min every day).

The BL treatment (10^{-7} M) raised the growth response of cotyledon *det2* to the GL action (Figure 4).

BlL_{436} and BL both display antagonism under a joint influence on the hypocotyl growth of L*er* and Col *Arabidopsis* initial lines. This influence is lowered in BR-deficient mutant of *det2*, but it is lacking in *hy4* which has a disturbed photoreceptor BlL/UV-A cry1.

The factors (GL_{543}+ EBL and $GL_{524.5}$+EBL) additively influence the L*er* hypocotyl growth and antagonistically *hy3* growth. The latter is not observed in *hy4*. The obtained data point to the importance of *DET2* and *HY4* gene products in BlL signal transduction; the *PHYB* and *HY4* genes are also of fundamental importance for GL signal transduction. A joint action of light factors and the hormones (BlL_{436}+BL or GL_{543}+EBL) manifests an additive character, the *det2* a synergetic one, while the $GL_{524.5}$+EBL reveal an antagonistic action on the growth of *hy3* cotyledons.

An activating EBL action on the growth of hypocotyl and *hy3* cotyledon stick in the dark is removed by $GL_{524.5}$. This fact testifies a change in the

EBL signal transduction in GL under phyB deficit in *hy3*; it can also testify to a possible BRs involvement into the transfer of phyB signal to GL, as it was shown for the red light (Luccioni *et al.*, 2002).

5. THE MECHANISM OF REGULATION OF PHOTOMORPHOGENESIS BY BRASSINOSTEROIDS

The morphogenesis in *Arabidopsis* by BRs is carried out by means of variation of the contents and the correlation of free and bound forms of the endogenic hormones (gibberellins – GA, cytokinins – CYT, IAA and ABA) (Karnachuk *et al.*, 2002a; Golovatskaya and Karnachuk, 2010) and dependents on the light (GL and BlL). The *HY4* gene mutation causes the activity rise in the free and bound forms of GA_{1+3}, GA_{4+7} and GA_9 in the dark. Cry1 is likely to change the expression of other genes or the level of their products regulating the growth processes and phytohormone contents. So, $GA_{20} \rightarrow GA_1$ transformation is more actively realized in etiolated light dwarfs *Pisum* (Campbell and Bonner, 1986).

The EBL (10^{-6} M) raises the free and bound forms of GA_{4+7} and GA_9 content but lowers that of GA_{1+3} in the wild type, while in case of *hy4* mutant it lowers the level of the active hormones GA_{1+3} and GA_{4+7} and increases the contents of inactive GA_9 as compared with the plants grown in the dark without applying the hormone. In the absence of cry1, EBL reduces the activity of GA-3β-hydroxylase participating in the biosynthesis of active GA (GA_4) (Yamaguchi and Kamiya, 2000).

A joint action of GL and EBL on the level of free forms of GA_{4+7} in L*er* and GA_9 in *hy4* is antagonistic in character since the physiological effect of the joint action of the factors is lower than the action of every factor taken separately.

According to our data the de-etiolation in the GL (Table 3, +CRY1) lowers the level of CYT, i.e. the IAA antagonists (Chory *et al.*, 1994) participating in photomorphogenesis of L*er* *Arabidopsis*, but increases the level of IAA and ABA (Karnachuk *et al.*, 2002a). GL regulate the activity of IAA oxidase (Boll, 1965). One may propose that the realization of plant photomorphogenesis in GL depends on the activation level of the photoreceptors.

Table 3. The dynamics of the endogenic phytohormone contents under the action of hormone (EBL) and light signals (GL) on etiolated *A. thaliana* seedlings of Landsberg *erecta* ecotype (as compared with those grown in the dark)

Factor		GL	EBL	GL	EBL
Photoreceptor		+CRY1		−CRY1	
Endogenic phytohormones	zeatin	− −	o	−	o
	zeatin riboside	− −	o	o	−
	free IAA	+	+ +	o	+
	free ABA	+	+ +	−	+*

Note: activation (+), no action (o), inhibition (−), bound ABA (*)

It the absence of cry1 photoreceptor, *hy4* mutant shows an incomplete photoreceptor activation in GL and this fact determines the incomplete realization of the photomorphogenesis program, and first of all, this is observed at the phytohormone level. The GL action on *hy4* mutant seedlings of *Arabidopsis* (Table 3, −CRY1) lowered the ABA level and, to a lesser extent, the zeatin level, not touching the IAA and RZ contents. A different reaction of the hormonal balance of the two lines to GL action allows one to think of changing the GL signal transduction way in *Arabidopsis* plant, which is caused by *HY4* gene mutation.

The analysis of hormonal balance in the seedlings, treated with exogenous brassinosteroid (EBL) showed that the directivity of the exogenous hormone action on dynamics of ABA and IAA content in the L*er* seedlings (Table 3, +CRY1) was similar to that of GL.

A positive tendency (o) was observed in the change of CYT content, since no hormone content decreased due to the GL (− −). A similar tendency to change in the level of endogenic IAA and zeatin was also noted for the mutant (Table 1, −CRY1). The lack of cry1 modified the hormone response to EBL action with respect to RZ level and the associated forms of the ABA (+*).

BRs are called the negative regulators of photomorphogenesis, because of the shortage of their endogenic forms mediates the de-etiolated phenotype (short hypocotyls and large cotyledons) of 7-day seedlings of *det2* mutant in dark and the reduction of their growth reactions in response to GL action, as compared with Col. By increasing the light phenotype of the cotyledons the exogenous BL restores the etiolated hypocotyl phenotype of the initial line in the *det2* seedlings. The increase in the cotyledon area in response to the action of exogenous BL and the additive effect of GL points to the importance of BL in photomorphogenesis.

The deficit of endogenic BRs in the seedlings of *det2* mutant (Table 4, –BR) changes the amount of the hormone response to GL action, as compared with the wild type (Table 4, +BR), retaining the directivity of responses to GL action (Golovatskaya and Karnachuk, 2010).

Table 4. The hormone status of the etiolated seedlings of Col (+BRs) and *det2* (–BRs) *A. thaliana* ecotype Columbia under the action of the hormonal (BL) and light signals (GL) (as compared with those grown in the dark)

Factor		GL	BL	GL	BL
	BRs	**+BRs**		**–BRs**	
Endogenic phytohormones	zeatin	+	+	– –	– –
	zeatin riboside	–	+	– –	– –
	free IAA	–	o	– –	–
	free ABA	– –	–	–	+

The endogenic BR (Table 4, +BR) decreases the negative effect of GL on the hormone level (zeatin, RZ and IAA), which is one of the proofs of phytohormone entry into the chain of the light signal transduction after the photoreceptors. A partial recovery of BRs level at the expense of exogenous hormone increases ABA, preserving the level of endogenic CYT and IAA, as compared with the wild type. The exogenous BL changes the hormone balance of BR-deficient mutant with respect to the wild-type, which testifies the significance of achieving a definite level of the endogenic BRs in order to develop a hormone status of plants.

The diversity of BL action on the dynamics of CYT and IAA (*det2*, –BR) contents_is similar to that of GL.

Summing up the data on the dynamics of endogenic phytohormones level depending on GL and the BR, one should note that the disturbance of the hormonal balance of BR content (Table 4) is compensated to a lesser extent, as compared with cry1 photoreceptor deficiency (Table 3). The latter indicates in favour of BR participation as a link in the mechanism of GL signal transduction. It is likely that GL may use a transduction system lowering the deficit of one of the receptors by means of a compensatory activation of other photoreceptors with overlapping functions. The deficit of (BR) hormone lowers the compensatory reactions in which this hormone directly participates.

The data obtained allows the author to suppose that brassinosteroids participate in the green-light signal transduction in regulating the photomorphogenesis of *Arabidopsis* plants. The observed alterations of the growth reactions and the level of photosynthetic pigments allows the author to suppose that *HY4* gene mutation changes the transduction ways of GL signal, and first of all, the level of the endogenic phytohormones however, the exogenous BRs restore them according to the wild type.

6. CONCLUSION

The effect of light of different quality on seedlings and plant growth were found to correlate with the shift in the levels of endogenous hormones.

A suggestion can be made on the basis of data obtained that localization and quality of morphogenetic responses of *Arabidopsis* plants to exogenous GA_3 and BRs are linked to the dynamics and the ratio of endogenous hormone levels in the plant, inactivation and destruction of exogenous hormones during their transport in the plant, as well as on the degree of overlapping of the hormonal responses.

I have shown that both BRs and light (GL, BlL) participate in the regulation of different stages of *A. thaliana* morphogenesis. The results of my experiments suggest that regulation of *A. thaliana* seedlings morphogenesis by BRs and light is associated with the interaction of signal transduction systems triggered by these factors.

Action of BRs is tissue-specific, therefore it is very difficult to specify BRs as only negative or positive regulators the photomorphogenesis.

7. ACKNOWLEDGEMENT

This research was supported by Grant of Russian Foundation for Basic Research 08-04-90042-Bel_a, by the State program University of Russia, project UR no. 07.01.042, by the Federal target program "The scientific and scientific-pedagogical staff of innovational Russia" on 2009–2013 (contract P283). The author is thankful to Professor R.A. Karnachuk for consultation and to M.V. Efimova (Tomsk State University, Russia) for assistance in performing a number of experiments. The author is grateful to Professor V.A. Khripach (Institute of Bioorganic Chemistry, National Academy of Sciences of Belarus), who kindly placed at our disposal the sample of brassinosteroids. The author is also grateful to R. Scholl (Director of the *Arabidopsis* Biological Resource Center, Ohio State University, United States) for providing *A. thaliana* seeds.

8. REFERENCES

Ahmad, M., and Cashmore, A., 1993. *HY4* Gene of *A. thaliana* encodes a protein with characteristics of a blue-light photoreceptor. *Nature* **366:** 162–166.

Ahmad, M., and Cashmore, A., 1997. The blue-light receptor cryptochrome 1 shows functional dependence on phytochrome A or phytochrome B in *Arabidopsis thaliana*. *Plant J.* **11:** 421–427

Akhrem, A., and Kovganko, N., 1989. *Ekdisterony: khimiya i biotogicheskaya aktivnost* (Ecdysterones: chemistry and biological activity), Minsk: Nauka i tekhnika.

Alabady, D., Gil, J., Blazquez, M., and Garcya-Martynez, J., 2004. Gibberellins repress photomorphogenesis in darkness. *Plant Physiol.* **134**: 1050–1057.

Asami, T., Ohb, K., Jikumaru, Y., Shimada, Y., Kaneko, I., Nakanoa, I., Takatsuto, S., Fujioka, S., and Yoshida, S., 2004. A mammalian steroid action inhibitor spironolactone retards plant growth by inhibition of brassinosteroid action and induces light-induced gene expression in the dark. *J. Steroid Biochem. Mol. Biol.* **91**: 41–47.

Azpiroz, R., Wu, Y., LoCascio, J., and Feldmann, K., 1998. An *Arabidopsis* brassinosteroid-dependent mutant is blocked in cell elongation. *Plant Cell* **10**: 219–230.

Bajguz, A., 2000. Effect of brassinosteroids on nucleic acids and protein content in cultured cells of *Chlorella vulgaris*. *Plant Physiol. Biochem.* **38**: 209–215.

Bajguz, A., and Czerpak, R., 1998. Physiological and biochemical role of brassinosteroids and their structure – activity relationship in the green alga *Chlorella vulgaris*. *Plant Growth Regul.* **17**: 131–139.

Bancos, S., Szatmari, A.-M., Castle, J., Kozma-Bognar, L., Shibata, K., Yokota, T., Bishop, G., Nagy, F., and Szekeres, M., 2006. Diurnal regulation of the brassinosteroid-biosynthetic *CPD* gene in *Arabidopsis*. *Plant Physiol.* **141**: 299–309.

Bandurski, R., Schulze, A., and Cohen, J., 1977. Photoregulation of the ratio of ester to free indole-3-acetic acid. *Biochem. Biophys. Res. Commun.* **79**: 1219–1223.

Beevers, L., Loveys, B., Pearson, J., and Wareing, P., 1970. Phytochrome and hormonal control of expansion and greening of etiolated wheat leaves. *Planta* **90**: 286–294.

Bishop, G., Nomura, T., Yokota, T., Harrison, K., Noguchi, T., Fujioka, S., Takatsuto, S., Jones, J., and Kamiya, Y., 1999. The tomato DWARF enzyme catalyses C-6 oxidation in brassinosteroid biosynthesis. *Proc. Natl. Acad. Sci. USA* **96**: 1761–1766.

Boll, W., 1965. On the claimed "adaptive" nature of IAAOxidase and an effect of green light on IAA-oxidase activity. *Can. J. Bot.* **43**: 885–892.

Botto, J., Sa´nchez, R., Whitelam, G., and Casal, J., 1996. Phytochrome A mediates the promotion of seed germination by very low fluences of light and canopy shade light in *Arabidopsis*. *Plant Physiol.* **110**: 439–444.

Bouquin, T., Meier, C., Foster, R., Nielsen, M.E., and Mundy, J., 2001. Control of specific gene expression by gibberellin and brassinosteroid. *Plant Physiol.* **127**: 450–458.

Brien, T., Beall, F., and Smith, H., 1985. De-etiolation and plant hormones. *Encyclopedia of Plant Physiology. New Series*, Berlin: Springer-Verlag **11:** 282–307.

Briggs, W., and Olney, M., 2001. Photoreceptors in plant photomorphogenesis to date. Five phytochromes, two cryptochromes, one phototropin, and one superchrome. *Plant Physiol.* **125**: 85–88.

Campbell, B., and Bonner, B., 1986. Evidence for phytochrome regulation of gibberellin A20 3β-hydroxylation in shoots of dwarf (lele) *Pisum sativum* L. *Plant Physiol.* **82**: 909–915.

Casal, J., 2001. Phytochromes, cryptochromes, phototropin: photoreceptor interaction in plants. *Photochem. Photobiol.* **71**: 1–11.

Chory, J., Nagpal, P., and Peto, C., 1991. Phenotypic and genetic analysis of *det2*, a new mutant that affects light-regulated seedling development in *Arabidopsis*. *Plant Cell.* **3**: 445–459.

Chory, J., Peto, C., Feinbaum, R., Pratt, L., and Ausubel, F., 1989. *Arabidopsis thaliana* mutant that develops as a light-grown plant in the absence of light. *Cell* **5**: 991–999.

Chory, J., Reinecke, D., Sim, S., Washburn, T., and Brenner, M., 1994. A role for cytokinins in de-etiolation in *Arabidopsis: det* mutants have an altered response to cytokinins. *Plant Physiol.* **104**: 339–347.

Clouse, S., 2001. Integration of light and brassinosteroid signals in etiolated seedling growth. *Trends Plant Sci.* **6**: 443–445.

Clouse, S., and Feldmann, K., 1999. Molecular genetics of brassinosteroid action. In *Brassinosteroids: Steroidal Plant Hormones*, Sakurai, A., Yokota, T., and Clouse, S.D., Eds. Tokyo: Springer: 163–190.

Clouse, S., and Sasse, J., 1998. Brassinosteroids: essential regulators of plant growth and development. *Annu. Rev. Plant Physiol. Plant Mol. Biol.* **49**: 427–451.

Cosgrove, D., 1986. Photomodulation of growth. In *Photomotphogenesis in Plant*, Kendrick, R.E. and Kronenberg, G.H.M., Eds. Dordrecht: Martinus Nijhoff: 341–366.

Dubovskaya, L., Molchan, O., and Volotovsky, I., 2001. Photoregulation of the endogenous cGMP content in oat seedlings, *Fiziol. Rast.* (Moscow), **48**: 26–29 (*Russ. J. Plant Physiol.* Engl. Transl., pp. 19–22).

Efimova, M., Karnachuk, R., Golovatskaya, I., and Khripach, V., 2007. The role of brassinosteroids in transduction of green light signals. *Plant growth substances: intracellular hormonal signaling and applying in agriculture*: 2nd International Symposium (8–12 October, 2007 Kiev, Ukraine). Kiev, p. 19.

Efimova, M., Karnachuk, R., Golovatskaya, I., Khripach, V., and Kusnetsov, V., 2008. The role of brassinosteroids in transduction of light signals. *Physiol. Plant.* **133**: P01–P026. (XIV Congr. Feder. Eur. Soc. Plant Biol, 17–22 August, 2008. Tampere, Finland).

Ephritikhine, G., Fellner, M., Vannini, C., Lapous, D., and Barbier-Brygoo, H., 1999. The *sax1* dwarf mutant of *Arabidopsis thaliana* shows altered sensitivity of growth responses to abscisic acid, auxin, gibberellins and ethylene and is partially rescued by exogenous brassinosteroid. *Plant J.* **18**: 303–314.

Fujioka, S., and Sakurai, A., 1997. Brassinosteroids. *Nat. Prod. Rep.* **14**: 1–10.

Fujioka, S., Li, J., Choi, Y., Seto, H., Takatsuto, S., Noguchi, T., Watanabe, T., Kuriyama, H., Yokota, T., Chory, J., and Sakurai, A., 1997. The *Arabidopsis deetiolated2* mutant is blocked early in brassinosteroid biosynthesis. *Plant Cell* **9**: 1951–1962.

Golovatskaya, I., 2005. The role of cryptochrome 1 and phytochromes in the control of plant photomorphogenetic responses to green light. *Russ. J. Plant Physiol.* **52**: 724–730.

Golovatskaya, I., and Karnachuk, R., 2007. Dynamics of growth and the content of endogenous phytohormones during kidney bean scoto- and photomorphogenesis. *Russ. J. Plant Physiol.* **54**: 407–413.

Golovatskaya, I., and Karnachuk, R., 2010. The Effect of brassinolide on *Arabidopsis* seedlings growth and hormonal balance under green light. *Vestn. Tomsk. Gos. Univ. Biology* **1(9)**: 13–19.

Golovatskaya, I., Karnachuk, R., and Efimova, M., 2001. *Arabidopsis* growth and hormonal balance under green light. *Vestn. Bashkir. Gos. Univ.* **2**: 114–116.

Golovatskaya, I., Karnachuk, R., Efimova, M., Kopylova, T., and Svetlichnyi, V., 2007. The role of cryptochrome 1 and phytochromes A–E in the control of *Arabidopsis* growth under green light. *Vestn. Tomsk. Gos. Univ.* **297**: 184–187.

Golovatskaya, I.F., 2004a. Brassinosteroids and *Arabidopsis* morphogenesis. *Aktual'nye problemy meditsiny i biologii* (Actual problems in medicine and biology), NN Il'inskikh (Ed.). Tomsk: Sib. Gos. Med. Univ., **3**: 74–76.

Golovatskaya, I.F., 2004b. Effect of ecdysterone on morphological and physiological processes in plants. *Russ. J. Plant Physiol.* **51**: 407–413.

Golovatskaya, I.F., 2008. Interaction of gibberellic acid and 24-epibrassinolide in the regulation of *Arabidopsis thaliana* seedling scotomorphogenesis. *Russ. J. Plant Physiol.* **55**: 663–669.

Golovatskaya, I.F., and Nikonorova, N.M., 2008. Growth and productivity of plants as depending on their photosensitivity and brassinolide treatment. *Agrokhimiya* **1**: 46–51.

Golovatskaya, I.F., and Vinnikova, Ju. M., 2007. Action of gibberellins and brassinolide on growth and development of plants *Arabidopsis*. *Vestn. Tomsk. Gos. Pedagog. Univ.* **6(69)**: 48–53.

Grove, M., Spencer, G., Rohwedder, W., Mandava, N., Worley, J., Wartner, J., Steffens, G., Flippen-Anderson, J., and Cook J., 1979. Brassinolide, a plant growth-promoting steroid isolated from *Brassica napus* pollen. *Nature* **281**: 216–217.

Hayat, S., Ali B., Hasan, S.A., and Ahmad, A., 2007. Effect of 28-Homobrassinolide on salinity-induced changes in *Brassica juncea*. *Turk. J. Biol.* **31**: 141–146.

He, R.-Y, Wang, G.-J., and Wang, X.-S., 1991. Effects of brassinolide on growth and chilling resistance of maize seedlings. *Brassinosteroids – chemistry. bioactivity and applications (ACS Symp. Ser. 474)*, Cutler, H.G.C., Yokota, T., and Adam, G., Eds. Washington: Am. Chem. Soc., pp. 220–230.

Hutchison, C., and Kieber, J., 2002. Cytokinin signaling in *Arabidopsis*. *Plant Cell* **14**: 47–59.

Imaizumi, T., Kanegae, T., and Wada, M., 2000. Cryptochrome nucleocytoplasmic distribution and gene expression are regulated by light quality in the fern *Adiantum capillus-veneris*. *Plant Cell* **12**: 81–96.

Jefferson, A.R., 1987. Assaying chimeric genes in plants: The GUS gene fusion system. *Plant Mol. Biol. Reporter* **5**: 387–405.

Kabachevskaya, E., Lyakhnovich, G., and Volotovsky, I., 2004. Regulation of phospholipase D activity by light and phytohormones in oat seedlings. *Fiziol. Rast.* **51**: 855–859 (*Russ. J. Plant Physiol.* Engl. Transl. pp. 769–773).

Kang, J., Yan, J., Kim, D., Chung, K., Fuijoka, S., Kim, J., Dae, H., Yoshida, S., Takatsuto, S., Song, P., and Park, C., 2001. Light and brassinosteroid signals are integrated via a dark induced small G protein in etiolated seedling growth. *Cell* **105**: 625–636.

Karnachuk, R., and Golovatskaya, I., 1998. Effect of light spectral composition on the hormonal balance, growth, and photosynthesis in plant seedlings. *Russ. J. Plant Physiol.* **45**: 805–813.

Karnachuk, R., Postovalova, V., Belen'kaya, E., and Zhulanova, S., 1978. Phytochrome control of ^{14}C-carbohydrates in plants. *Fiziol. Rast.* (Moscow), **25**: 268–271 (*Sov. Plant Physiol.* Engl. Transl.).

Karnachuk, R., Protasova, N., and Golovatskaya, I., 1988. Dependence of plant srowth and hormonal content from light spectral composition. *Rost i ustoichivost rastenii* (Plant Growth and Tolerance), R.K. Salyaev and V.I. Kefeli (Eds.). Novosibirsk: Nauka, 71–81.

Karnachuk, R., Negretskii, V., and Golovatskaya, I., 1990. Hormonal balance in the leaves of plants under light of different spectral composition. *Fiziol. Rast.* (Moscow), **37**: 527–534 (*Sov. Plant Physiol.* Engl. Transl.).

Karnachuk, R., Tishchenko, S., and Golovatskaya, I., 2001. Endogenous phytohormones and regulation of morphogenesis of *Arabidopsis thaliana* by blue light. *Russ. J. Plant Physiol.* **48**: 226–230.

Karnachuk, R., Golovatskaya, I., Efimova, M., and Khripach, V., 2002a. The effect of epibrassinolide on *Arabidopsis* seedling morphogenesis and hormonal balance under green light. *Russ. J. Plant Physiol.*, **49**: 530–533.

Karnachuk, R., Khripach, V., Golovatskaya, I., and Efimova M., 2002b. Simultaneous effect of epybrassinolide and monochrome light on the level of growth substances in *Arabidopsis*. *13th Congr. Feder. Eur. Soc. Plant Biol.* (Hersonissos, Heraklion, Crete, Greece, 2–6 September 2002). Heraklion, p. 779.

Kasahara, H., Hanada, A., Kuzuyama, T., Takagi, M., Kamiya, Y., and Yamaguchi, S., 2002. Contribution of mevalonat and methylerythritol phosphate pathways to biosynthesis of gibberellins in *Arabidopsis*. *J. Biol. Chem.* **277**: 45188–45194.

Khripach, V., Lakhvich, F., and Zhabinskii, V., 1993. *Brassinosteroidy* (Brassinosteroids). Minsk: Nauka i tekhnika: 287.

Khripach, V., Zhabinskii, V., and Ae de Groot, 1999. *Brassinosteroids: A New Class of Plant Hormones*. San Diego: Academic Press: 456.

Khripach, V., Zhabinskii, V., and Karnachuk, R., 2003. Science at the interface of brassinosteroids: a new role of steroids as biosignaling molecules. Chemical probes in biology. M.P. Schneider. Ed. Netherlands: Kluwer Academic Publishers, **129**: 153–167.

Kim, T.-H., Kim B.-H., and Arnim A.G., 2002. Repressors of photomorphogenesis. *Cytology* **220**: 185–223.

Konev, S., and Volotovsky, I., 1974. *Fotobiologiya* (Photobiology), Minsk: Bel. Gos. Univ.

Koornneef, M., Rolff, E., and Spruit, C.J.P., 1980. Genetic control of light-inhibited hypocotyl elongation in *Arabidopsis thaliana* L. Heynh. *Z. Pflanzenphysiol.* **100**: 147–160.

Kraepiel, Y., Rousselin, P., Sotta, B., Kerhoas, L., Einhorn, J., Caboche, M., and Miginiac, E., 1994. Analysis of phytochrome- and ABA-deficient mutants suggests that ABA degradation is controlled by light in *Nicotiana plumbaginifolia. Plant J.* **6**: 665–672.

Kulaeva, O., 1982. Hormonal regulation of physiological processes in plants at the level of RNA and protein synthesis. *The 41st Timiryazev Lecture*, Moscow: Nauka: 84.

Lafont, R., and Wilson, I., 1996. *The Ecdysone Handbook*, Nottingham: The Chromatogr. Soc.

Li, J., Nagpal, P., Vitart, V., McMorris, T., and Chory, J., 1996. A role for brassinosteroids in light-dependent development of *Arabidopsis*. *Science* **272**: 398–401.

Lin, C., and Shalitin D., 2003. Cryptochrome structure and signal transduction. *Annu. Rev. Plant Biol.* **54**: 469–496.

Lisso, J., Steinhauser, D., Altmann, T., Kopka, J., and Müssig, C., 2005. Identification of brassinosteroid-related genes by means of transcript co-response analyses. *Nucl. Acids Res.* **33**: 2685–2696.

Luccioni, L., Oliverio, K., Yanovsky, M., Boccalandro, H., and Casal, J., 2002. Brassinosteroid mutants uncover fine tuning of phytochrome. *Plant Physiol.* **128**: 173–181.

Malec, P., Yahalom, A., and Chamovitz, A., 2002. Identification of a light-regulated protein kinase activity from seedlings of *Arabidopsis thaliana. Photochem. Photobiol.* **75**: 178–183.

Mathur, J., Molnar, G., Fujioka, S., Takatsuto, S., Sakurai, A., Yokota, T., Adam, G., Voigt, B., Nagy, F., Maas, C., Schell, J., Koncz, C., and Szekeres, M., 1998. Transcription of the *Arabidopsis CPD* gene, encoding a steroidogenic cytochrome P450, is negatively controlled by brassinosteroids. *Plant J.* **14**: 593–602.

Mohr, H., 1970. Regulation der enzymsynthese bei den hoheren pflanze. *Naturwiss. Rundschau.* **23**: 187–195.

Moor, T., 1989. *Biochemistry and Physiology of Plant Hormones.* New York: Springer-Verlag.

Müssig, C., Fischer, S., and Altmann, T., 2002. Brassinosteroid-regulated gene expression. *Plant Physiol.* **129**: 1241–1251.

Nagata, N., Min, Y.K., Nakano, T., Asami, T., and Yoshida, S., 2000. Treatment of dark-grown *Arabidopsis thaliana* with a brassinosteroid-biosynthesis inhibitor, brassinazole, induces some characteristics of light-grown plants. *Planta* **211**: 781–790.

Neff, M., Nguyen, S., Malancharuvil, E., Fujioka, S., Noguchi, T., Seto, H., Tsubuki, M., Honda, T., Takatsuto, S., Yoshida, S., and Chory, J., 2000. *BAS*1: a gene regulating brassinosteroid levels and light responsiveness in *Arabidopsis*. *Proc Natl. Acad. Sci. USA*, **96**: 15316–15323.

Nemhauser, J., and Chory, J., 2004. BRing it on: new insights into the mechanism of brassinosteroid action. *J. Exp. Bot.* **55**: 265–270.

Nemhauser, J., Maloof, J., and Chory, J., 2003. Building integrated models of plant growth and development. *Plant Physiol.* **132**: 436–439.

Neuhaus, G., Bowler, C., Kern, R., and Chua, N.-H., 1993. Calcium/calmodulin-dependent and independent phytochrome signal transduction pathways. *Cell*, **73**: 937–952.

Piechowski, K., 1997. Brassinosteroids. *Growth and development*, November 4, ABS798: 70–76.

Polevoi, V., 2001. The physiology of plant integrity. *Russ. J. Plant Physiol.* **48**: 545–555.

Prusakova, L., and Chizhova, S., 1996. The role of brassinosteroids in plant growth, resistance, and productivity. *Agrokhimiya* **11**: 137–150.

Quail, P., Boylan, M., Parks, B., Short, T., Xu, Y., and Wagner, D., 1995. Phytochrome: photosensory perception and signal transduction. *Science* **268**: 675–680.

Ramraj, V., Vyas, B., Godrej, N., Mistry, K., Swmai, B., and Singh, N., 1997. Effects of 28-homobrassinolide on yields of wheat, rice, groundnut, mustard, potato and cotton. *J. Agricultural Sci.* **128**: 405–413.

Reed, J., Foster, K., Morgan, P., and Chory, J., 1996. Phytochrome B affects responsiveness to gibberellins in *Arabidopsis. Plant Physiol.* **112**: 337–342.

Sano, J., and Youssefian, S., 1994. Light and nutritional regulation of transcripts encoding a wheat protein kinase homolog is mediated by cytokinins. *Proc. Natl. Acad. Sci. USA*, **91**: 2582–2586.

Sasse, J., 1999. Physiological actions of brassinosteroids. *Brassinosteroids: Steroidal Plant Hormones*, Springer: Tokyo: 137–161.

Shakirova, F., 2003. Nonspecific stability of plants to stressful factors and its regulation, Ufa: Gilem: 160.

Shinomura, T., Nagatani, A., Hanzawa, H., Kubota, M., Watanabe, M., and Furuya, M., 1996. Action spectra for phytochrome A- and phytochrome B-specific photoinduction of seed germination in *Arabidopsis thaliana. Proc. Natl. Acad. Sci. USA*, **93**: 8129–8133.

Smalle, J., Haegman, M., Kurepa, J., Van Montagu, M., Van Der Straeten, D., 1997. Ethylene can stimulate *Arabidopsis* hypocotyl elongation in the light. *Proc. Natl. Acad. Sci. USA*, **94**: 2756–2761.

Steber, C., and McCourt, P., 2001. A role for brassinosteroids in germination in *Arabidopsis. Plant Physiol.* **125**: 763–769.

Sternberg, M., 1965. Possible role of growth substances and nucleic acids in the mechanism of phytochrome action. *Regulyatory rosta rastenii i nukleinovyi obmen* (Regulators of Plant Growth and Nucleic Metabolism), Moscow: Nauka: 65–80.

Sun, T., 2000. Gibberellin signal transduction. *Curr. Opin. Plant Biol.* **3**: 374–380.

Symons, G., and Reid, J., 2004. Brassinosteroids do not undergo long-distance transport in pea. Implications for the regulation of endogenous brassinosteroid levels. *Plant Physiol.* **135**: 2196–2206.

Symons, G., Schultz, L., Kerckhoffs, L., Davies, N., Gregory, D., and Reid, J., 2002. Uncoupling brassinosteroids levels and de-etiolation in pea. *Physiol. Plant.* **115**: 311–319.

Szekeres, M., Nemeth, K., Konczkalman, Z., Mathur, J., Kauschmann, A., Altmann, T., Redei, G., Nagy, F., Schell, J., and Koncz, C., 1996. Brassinosteroids rescue the deficiency of CYP90, a cytochrome P450, controlling cell elongation and de-etiolation in *Arabidopsis. Cell* **85**: 171–182.

Talon, M., Koornneef, M., and Zeevaart, J., 1990. Endogenous gibberellins in *Arabidopsis thaliana* and possible steps blocked in the biosynthetic pathways of semidwarf *ga4* and *ga5* mutants. *Proc. Natl. Acad. Sci. USA* **87**: 7983–7987.

Tanada, T., 1984. Interaction of green or red light with blue light on the dark closure *Albizzia pinnules. Physiol. Plant* **61**: 35–37.

Tanaka, K., Nakamura, Y., Asami, T., Yoshida, S., Matsuo, T., and Okamoto, S., 2003. Physiological roles of brassinosteroids in early growth of *Arabidopsis*. Brassinosteroids have a synergistic relationship with gibberellin as well as auxin in light-grown hypocotyl elongation. *J. Growth Regul.* **22**: 259–271.

Tishchenko, S., Karnachuk, R., and Khripach, V., 2001. Participation of epibrassinolide in the photoregulation of growth and hormonal balance in *Arabidopsis* under blue light. *Vestn. Bashkir. Gos. Univ.* **2**: 166–167.

Turk, E., Fujioka, S., Seto, H., Shimada, Y., Takatsuto, S., Yoshida, S., Denzel, M., Torres, Q., and Neff, M., 2003. CYP72B1 inactivates brassinosteroid hormones: an intersection between photomorphogenesis and plant steroid signal transduction. *Plant Physiol.* **133**: 1643–1653.

Vardhini, B., and Ram Rao, S., 1998. Effect of brassinosteroids on growth, metabolite content and yield of arachis hypogaea. *Phytochemistry* **48**: 927–930.

Voskresenskaya, N., 1987. Photoregulatory reactions and activity of the photosynthetic apparatus. *Fiziol. Rast.* (Moscow), **34**: 669–684 (*Sov. Plant Physiol.* Engl. Transl.).

Weatherwax, S., Ong, M., Degenhardt, J., Bray, E., and Tobin, E., 1996. The interaction of light and abscisic acid in the regulation of plant gene expression. *Plant Physiol.*, **111**: 363–370.

Whitelam, G., and Devlin, R., 1998. Light signalling in *Arabidopsis. Plant Physiol. Biochem.*, **36**: 125–133.

Yamaguchi, S., and Kamiya, Y., 2000. Gibberellin biosynthesis: its regulation by endogenous and environmental signals. *Plant Cell Physiol.* **41**: 251–257.

Yu, J., Huang, L., Hu, W., Zhou, Y., Mao, W., Ye, S., and Nogués, S., 2004. A role for brassinosteroids in the regulation of photosynthesis in *Cucumis sativus*. *J. Exp. Bot.* **55**: 1135–1143.

Zubo, Ya., Selivankina, S., Yamburenko, M., Zubkova, N., Kulaeva, O., and Kusnetsov, V., 2005. Cytokinins activate transcription of chloroplast genes. *Dokl. Akad. Nauk* **400**: 396–399.

Chapter 6

BRASSINOSTEROIDS AND PHOTOSYNTHESIS

D. HOLÁ

Charles University in Prague, Faculty of Science, Department of Genetics and Microbiology, Viničná 5, Prague 2 – Nové Město, CZ 128 43, Czech Republic

Abstract: Brassinosteroids (BRs) are organic compounds with structure based on polyhydroxylated sterols, which show multiple effects on plant physiology, development and growth and are now included among main groups of phytohormones. One of the diverse functions of BRs in higher plants is their possible involvement in the regulation of photosynthesis. The exogenous application of BRs has been described to enhance the net photosynthetic rate in several plant species and this phenomenon is usually more pronounced in plants stressed by various abiotic stress factors (such as drought, high or low temperature, salinity or heavy metals). However, the exact causes of this enhancement are far from being clear. BRs could prevent *e.g.* loss of photosynthetic pigments by an activation of enzymes participating in chlorophyll biosynthesis or an induction of their synthesis. Another possibility is that BRs improve the efficiency of photosynthetic carbon fixation by overcoming stomatal limitations and, thus, increasing internal concentration of CO_2 available for photosynthetic enzymes. BRs have also been shown to induce the synthesis and/or activation of carbonic anhydrase, an enzyme that catalyses an interconversion of CO_2 and HCO_3^-, to increase the activation state of ribulose-1,5-carboxylase/oxygenase or to protect enzymes involved in the regeneration of ribulose-1,5-bisphosphate. The positive response of primary photochemistry to exogenously applied BRs (such as the increased efficiency of Photosystem II) has also been observed in some cases, but the role of BRs in the improvement of the efficiency of primary photosynthetic processes seems to be a rather secondary one. On the molecular genetics level, recent genomic or proteomic analyses of either BR-treated or BR-deficient plants have revealed several photosynthetic genes whose expression seems to be either up- or down-regulated by BRs. This chapter summarizes the current knowledge on BRs effects on photosynthetic organelles, *i.e.* chloroplasts, and on various components of photosynthetic apparatus.

S. Hayat and A. Ahmad (eds.), *Brassinosteroids: A Class of Plant Hormone*,
DOI 10.1007/978-94-007-0189-2_6, © Springer Science+Business Media B.V. 2011

Key words: Brassinosteroids, carbonic anhydrase, carotenoids, CO_2 fixation, chlorophylls, chloroplast structure and development, light-harvesting, photosynthesis, Photosystem I, Photosystem II, primary photochemistry, Rubisco, stomata, stress

1. INTRODUCTION

The existence of steroid compounds with hormonal function has been accepted for a long time in animals, whereas the idea that steroid hormones occur also in plants has not been experimentally supported until late 1970s. The first purification of such a substance was performed by Grove *et al.* (1979) from the pollen of *Brassica napus* L. and since then, almost one hundred structurally related compounds were isolated from various parts of many plant species. This group of plant steroid hormones is collectively termed brassinosteroids (BRs) and has been shown to play an important role in plant growth and development (Clouse and Sasse, 1998; Müssig, 2005). BRs participate in the regulation of many cellular and physiological processes occurring in plants, including cell division and elongation, biosynthesis of cell wall components, synthesis of DNA, RNA and various proteins, microtubule organization, nitrogen fixation, distribution of assimilates to plant organs, pollen tube growth, differentiation of plant vascular system, formation of adventitious roots, flowering and reproduction, germination of seeds, stress response, senescence *etc.* (reviewed *e.g.* by Khripach *et al.*, 1999, 2000; Sasse, 1999, 2003; Castle *et al.*, 2003; Hayat *et al.*, 2003; Krishna, 2003; Haubrick and Assman, 2006; Bajguz and Hayat, 2009; Divi and Krishna, 2009). Almost from the very beginning of BRs' research, the possible role for BRs in the regulation of photosynthetic processes has been also suggested; the study of Mandava *et al.* (1981) being probably the first one that mentioned the effect of these hormones on the loss of photosynthetic pigments during leaf senescence. Other studies dealing with the effect of BRs on various photosynthetic characteristics soon followed and today we have at our disposal relatively abundant data on the relationship between BRs and photosynthesis. Since the last comprehensive review on BRs' function in the regulation of photosynthetic efficiency has been published in 1999 (Khripach *et al.*, 1999), the time has come to update and summarize our current knowledge on this topic again.

To understand the influence any compound exerts on a particular physiological process enacted in plant, it is necessary to have a solid methodical and technical background for the evaluation of the respective process. Photosynthesis has now been studied for more than two centuries and a plentitude of methods for its measurement is available, which is partly due to the high complexity of this process. At present, the most frequently used methodology is based either on gas exchange measurements with portable infrared gas analysis systems (Long and Bernacchi, 2003), spectrophotometric, chromatographic or light transmitance analyses used for the determination of photosynthetic pigments' content (Richardson *et al.*, 2002; Gitelson *et al.*, 2003; Pocock *et al.*, 2004; Porra, 2006; Garrido and Zapata, 2006), and monitoring of both fast-transient and slow-transient phases of chlorophyll fluorescence induction kinetics or fluorescence emission spectra with pulse-modulated portable fluorometers or fluorescence imaging systems (Baker and Rosenqvist, 2004; Papageorgiou and Govindjee, 2004; Baker, 2008). This selection of methods offers an insight into various parts of photosynthetic processes including the capture of excitation energy by photosynthetic pigments in light-harvesting antennae, the electron transport chain performed by protein complexes and mobile electron carriers in chloroplast thylakoid membranes, the generation of ATP and NADPH, and the uptake of CO_2 through stomata and its assimilation by enzymes of photosynthetic carbon reduction cycle. For an analysis of chloroplast structure and development, an electron microscopy coupled with image analysis is usually employed (Staehelin, 2003). Other methods commonly used in cell and molecular biology provide various information about the expression of photosynthesis-related genes. From the application of all the above-mentioned methods to plants treated with exogenous BRs, BR-deficient or -insensitive mutants and transgenic plants, a picture of the possible role these steroid hormones play in the regulation of photosynthetic processes is at least slowly emerging, although various parts of this picture are still rather blurred or even completely blank.

2. EXOGENOUSLY APPLIED BRASSINOSTEROIDS AFFECT PHOTOSYNTHESIS AT VARIOUS LEVELS

Almost all studies examining the effect BRs exert on photosynthesis have been made with plants treated by exogenously applied BRs. However, a surprisingly great diversity exists among such studies. This diversity concerns the analyzed species, the mode of BRs' treatment, the time/developmental stage when BRs are applied and/or when the photosynthetic parameters are measured, the concentration of BRs used, the conditions experimental plants are cultivated in and, of course, the respective photosynthetic parameters measured. It is clearly discernible from Table 1 which attempts to sum up the majority of studies published in scientific journals or books, which deal with the effect of exogenously applied BRs on photosynthesis.

Actually, there is only one factor that seems to be rather constant in these studies and that is the type of BRs applied: most authors have worked with either 24-epibrassinolide (EBL), 28-homobrassinolide (HBL) or brassinolide (BL). The concentration of BRs ranges from 10^{-4} M to 10^{-14} M and studies examining two or more dosages of the respective BR are not infrequent. Further variability extends to the mode of BR treatment: although foliar application by spraying of the aboveground parts of plants (mostly once, but repeated spraying has also been tried by some authors) prevails, followed by seed soaking for time period ranging from 4 to 48 hours, other modes of BR treatment have been used as well, *e.g.* soaking of various parts of plant (roots, shoot, leaves or leaf segments) in BR solution or growth of plants in a liquid medium containing nutrients and BRs. As regards plant material, the greatest number of studies has been made with cucumber (*Cucumis sativus* L.), closely followed by wheat (*Triticum aestivum* L.), Indian mustard (*Brassica juncea* (L.) Czern.), mungbean (*Vigna radiata* (L.) Wilczek) and tomato (*Lycopersicon esculentum* Mill.); however, more "exotic" plant species have been also used (*e.g.* geranium (*Pelargonium graveolens* (L.) Herit), jack pine (*Pinus banksiana* Lamb.), banana (*Musa* sp.) or spotless watermeal (*Wolffia arrhiza* (L.) Wimm.)). The experimental plants have been grown both in field and fully- or partially-controlled conditions (growth chamber, greenhouse, net-house) and often subjected to unfavourable factors such as high or low temperature, drought, salinity, hypoxia, excess of heavy metals, herbicides, pesticides, fungicides or pathogen infection. The diversity in the time of BRs application or in the time of the determination of photosynthetic parameters is almost indescribable and the reader is referred to Table 1 and to the original studies for these details.

2.1 Net photosynthetic rate

Measurement of the net rate of CO_2 uptake expressed per unit of leaf area (the net photosynthetic rate, P_N) is commonly used for the assessment of the effect of BRs on photosynthesis. One of the first studies evaluating this parameter in BR-treated plants was work of Braun and Wild (1984) who measured the *in vivo* CO_2 fixation rate in young wheat plants sprayed with 10^{-6} M solution of BL and reported increased values of this parameter in BL-treated plants, compared to non-treated ones. Sairam (1994a) applied HBL solution in two concentrations by foliar spray to wheat plants as well and observed that plants treated with this BR showed significantly improved P_N (measured in the flag leaf at anthesis stage, *i.e.* not immediately after BR treatment), particularly when lower concentration of HBL was used.

Table 1. Studies dealing with the effect of exogenously applied brassinosteroids (BRs) on various photosynthetic parameters of plants. Abbreviations used: BL, brassinolide; CA, carbonic anhydrase; Car, carotenoids; Chl, chlorophyll; C_i, intercellular CO_2 concentration; C_a, ambient CO_2 concentration; DAS, days after sowing; EBL, 24-epibrassinolide; ETR, relative rate of photosynthetic electron transport; F_v/F_m, maximum quantum efficiency of Photosystem II; F'_v/F'_m, maximum efficiency of Photosystem II; Φ_{PSII}, effective quantum yield of Photosystem II; g_S, stomatal conductance; HBL, 28-homobrassinolide; J_{max}, maximum rate of ribulose-1,5-bisphosphate regeneration; *l*, stomatal limitation to CO_2 uptake; NPQ, nonphotochemical quenching of chlorophyll fluorescence; NS, not stated; P_N, net photosynthetic rate; PS I, Photosystem I; q_P, photochemical quenching of chlorophyll fluorescence; Rubisco, ribulose-1,5-bisphosphate carboxylase/oxygenase; V_{cmax}, maximum carboxylation rate of Rubisco

Plant species	**Growth conditions**	**BRs type**	**BRs dosage / concentration**	**BRs mode of treatment**	**Time of BRs application**	**Time of measurements**	**Photosynthetic parameters measured**	**References**
Rumex obtusifolius, Xanthium strumarium	Greenhouse → laboratory conditions (leaf segments)	BL	5×10^{-6} M, 5×10^{-5} M	Leaf segments soaking for 4 days	NS	After the end of BRs treatment	Chl content	Mandava *et al.* (1981)
Phaseolus vulgaris	Growth chamber; cool white fluorescent, incandescent and far-red fluorescent light	BL	5 µg / 200 µg	Application of lanolin with BRs to the second internode	NS	6 days after BRs treatment	Chl content	Krizek and Mandava (1983)

(*continued*)

(*continued Table 1.*)

Plant species	Growth conditions	BRs type	BRs dosage / concentration	BRs mode of treatment	Time of BRs application	Time of measurements	Photosynthetic parameters measured	References
Triticum aestivum, Leucosinapis alba	NS (hydroponic growth in nutrient solution)	BL	10^{-6} M	Foliar spray (repeated every other day)	10 DAS	12–20 DAS	P_N, Chl *a*, Chl *b* and Car content, Chl *a/b* ratio, Rubisco activity and content	Braun and Wild (1984)
Oryza sativa	Greenhouse and growth chamber (pots with soil)	BL	10^{-3} ppm, 10^{-2} ppm	Foliar spray	10 days before heading and at heading stage	3 days after heading stage	$^{14}CO_2$ assimilation and partitioning	Fujii *et al.* (1991)
Zea mays	NS (hydroponic growth in water); cold stress / non-stress	BL	10^{-3} ppm, 10^{-2} ppm, 10^{-1} ppm	Incubation in water with BRs	Plants with 1 fully developed leaf	After the end of BRs treatment	Chl content	He *et al.* (1991)
Cucumis sativus	NS (detached cotyledons)	BL	10^{-4} ppm, 10^{-3} ppm, 10^{-2} ppm, 10^{-1} ppm	Cotyledons soaking for 12 h	Plants with cotyledons	After the end of BRs treatment	Chl content	He *et al.* (1991)
Cucumis sativus	NS; cold stress	BL	10^{-7} M, 10^{-6} M, 10^{-5} M	Seed soaking	Seeds	12 DAS	Chl content	Katsumi (1991)

Hordeum vulgare	NS (leaf segments - incubation with water); salinity stress	EBL	10^{-8} M	Leaf segments soaking for 2 h	10 DAS	1 d after BRs treatment	Chloroplast structure	Kulaeva *et al.* (1991)
Triticum aestivum	NS (pots with soil); drought stress / recovery / non-stress	HBL	0.1 ppm, 1 ppm	Foliar spray (repeated 2×)	30 and 32 DAS (repeated treatment)	Plants in anthesis stage	P_N, Chl content	Sairam (1994a)
Triticum aestivum	Field (pots with soil); drought stress / non-stress	HBL	0.01 ppm, 0.05 ppm	Seed soaking for 6 h / Foliar spray	Seeds / Plants 25 DAS	Plants in anthesis stage	P_N, Chl content	Sairam (1994b)
Cucumis sativus	Greenhouse (pots with vermiculite) → growth chamber (hydroponic growth in nutrient solution)	EBL	0.1 mg/L (*approx.* 2×10^{-7} *M*)	Root incubation in BRs solution for 4 h	Approx. 10-15 DAS	After the end of BRs treatment	$^{14}CO_2$ assimilation and partitioning	Nakajima and Toyama (1995)
Vigna radiata	Field (natural conditions)	HBL	0.1 mg/L, 0.5 mg/L (*approx.* 2×10^{-7} *M*, 10^{-6} *M*)	Foliar spray (repeated 2×)	15 and 30 DAS (repeated treatment)	22, 36 or 50 DAS	Chl *a*, Chl *b* content, Chl *a*/*b* ratio, Hill reaction activity	Bhatia and Kaur (1997)

(*continued*)

(*continued Table 1.*)

Plant species	Growth conditions	BRs type	BRs dosage / concentration	BRs mode of treatment	Time of BRs application	Time of measurements	Photosynthetic parameters measured	References
Chlorella vulgaris	Controlled conditions (hydroponic growth in nutrient solution)	BL, EBL, others	10^{-12} M, 10^{-8} M	Incubation in nutrient solution with BRs for 24 h or 36 h	NS (cell culture)	After the end of BRs treatment	P_N, Chl *a*, Chl *b* content	Bajguz and Czerpak (1998)
Pinus banksiana	Greenhouse and growth chamber (pots with soil); drought stress	HBL	0.005 μg/L (*approx.* 10^{-11} *M*)	Xylem feeding with BRs solution for 7 d	18 months + 11 d	12 d after the end of BRs treatment	P_N, g_S, C_i/C_a	Rajasekaran and Blake (1999)
Brassica juncea	Net-house (pots with sand)	HBL	10^{-10} M, 10^{-8} M, 10^{-6} M	Foliar spray	30 DAS	60 DAS	P_N, CA activity	Hayat *et al.* (2000)
Triticum aestivum	Controlled conditions (pots with sand)	HBL	10^{-6} M, 3×10^{-6} M, 5×10^{-6} M	Seed soaking for 4, 8 or 12 h	Seeds	25 and 35 DAS	CA activity	Hayat *et al.* (2001a)
Brassica juncea	NS (pots with sand)	HBL	10^{-8} M	Foliar spray	30 DAS	60 DAS	P_N, Chl *a*, Chl *b* content, Chl *a/b* ratio, CA activity	Hayat *et al.* (2001b)

Lycopersicon esculentum	Field (plants - pots with soil) → laboratory conditions (leaf segments - flasks with water); heat stress / non-stress	Other	0.01 mg/L (*approx.* 2×10^{-8} *M*)	Leaf segments soaking for 24 h	15 DAS	After the end of BRs treatment	Chloroplast structure	Sam *et al.* (2001)
Phaseolus aureus	Growth chamber (pots with vermiculite → hydroponic growth in nutrient solution); Al stress / non-stress	BL	0.1 ng/L, 10 ng/L, 1 µg/L, 100 µg/L (*approx.* 2×10^{-13} *M*, 2×10^{-11} *M*, 2×10^{-9} *M*, 2×10^{-7} *M*)	Incubation in nutrient solution with BRs for 10 d	4 days-old plants	After the end of BRs treatment	Chl content	Abdullahi *et al.* (2003)
Oryza sativa	Growth chamber (pots with vermiculite); salinity stress	EBL, HBL	3×10^{-6} M	Seed soaking for 24 h	Seeds	20 DAS	Chl *a*, Chl *b* content	Anuradha and Rao (2003)

(continued)

(*continued Table 1.*)

Plant species	Growth conditions	BRs type	BRs dosage / concentration	BRs mode of treatment	Time of BRs application	Time of measurements	Photosynthetic parameters measured	References
Vigna radiata	NS (pots with soil)	HBL	10^{-8} M, 10^{-6} M, 10^{-4} M	Seed soaking for 4, 8 or 12 h	Seeds	30, 40, 50 DAS	P_N, g_S, carboxylation efficiency, Chl content, CA activity	Fariduddin *et al.* (2003)
Lycopersiccn esculentum	Greenhouse and growth chamber; pathogen infection / non-infection	Other	2 mg/L	Whole plant dipping in BRs solution (repeated 2×)	28 DAS	4 days after the end of BRs treatment	F_v/F_m, Φ_{PSII}, ETR, NPQ, *RbcS* expression	Berger *et al.* (2004)
Vigna radiata	Net-house (pots with soil)	HBL	10^{-10} M, 10^{-8} M, 10^{-6} M	Foliar spray	25 DAS	35 or 45 DAS	P_N, Chl content, CA activity	Fariduddin *et al.* (2004)
Cucumis sativus	Greenhouse (pots with soil → hydroponic growth in nutrient solution)	EBL	0.01 mg/L, 0.1 mg/L, 1 mg/L (*approx.* 2×10^{-8} *M*, 2×10^{-7} *M*, 2×10^{-6} *M*)	Foliar spray	Plants with 3 fully developed leaves	3, 6, 9, 24, 26 h, 3, 5, 7 days after BRs treatment	P_N, C_i, Chl content, V_{cmax}, J_{max}, *l*, total and initial Rubisco activity, Rubisco content, F_v/F_m, F_v'/F_m', Φ_{PSII}, q_P	Yu *et al.* (2004)

Wolffia arrhiza	Growth chamber (hydroponic growth in nutrient solution); with / without brassinazole	EBL	10^{-13} M, 10^{-12} M, 10^{-11} M, 10^{-10} M, 10^{-9} M, 10^{-8} M, 10^{-7} M, 10^{-6} M	Incubation in nutrient solution with BRs for 7 days	NS	After the end of BRs treatment	Chl *a*, Chl *b* and Car content	Bajguz and Asami (2005)
Musa sp.	*In vitro* culture followed by *ex vitro* (growth chamber); cold or heat stress / non-stress	Other	0.1 mg/L	Foliar spray	4 weeks after transfer of plants from *in vitro* to *ex vitro* conditions	4 days or 10 days after BRs treatment	P_N	Gonzáles-Olmedo *et al.* (2005)
Brassica napus	NS (vessels with nutrient medium); Cd stress / non-stress	EBL	10^{-7} M	Incubation in nutrient solution with BRs for 14 d	3 DAS	After the end of BRs treatment	Chl and Car content, fast transients of Chl fluorescence (OJIP test)	Janeczko *et al.* (2005)
Lycopersicon esculentum	Greenhouse and growth chamber (pots with sand); heat stress / non-stress	EBL	10^{-6} M, 10^{-5} M, 2×10^{-5} M	Foliar spray	28 DAS	42-44 DAS	P_N, C_i, g_S	Singh and Shono (2005)

(*continued*)

(continued Table 1.)

Plant species	Growth conditions	BRs type	BRs dosage / concentration	BRs mode of treatment	Time of BRs application	Time of measurements	Photosynthetic parameters measured	References
Lycopersicon esculentum	Net-house (pots with soil); natural conditions	HBL	10^{-8} M, 10^{-7} M, 10^{-6} M, 10^{-5} M	Root soaking for 15, 30 or 45 min	20 DAS	50 or 80 DAS	Chl content, CA activity	Ali *et al.* (2006)
Vigna radiata	NS (pots with soil)	HBL	10^{-10} M, 10^{-8} M, 10^{-6} M	Foliar spray (repeated 3×)	15 DAS	30 or 50 DAS	P_N, g_S, carboxylation efficiency, Chl content, CA activity	Fariduddin *et al.* (2006)
Cucumis sativus	Greenhouse (pots with soil → hydroponic growth in nutrient solution); herbicide, fungicide, insecticide stress / non-stress	EBL	0.1 mg/L (*approx.* 2×10^{-7} *M*)	Foliar spray	Plants with 4 fully developed leaves	2 d after BRs treatment	P_N, C_i, g_S, F_v/F_m, F_v'/F_m', Φ_{PSII}, q_P, NPQ	Xia *et al.* (2006)
Brassica juncea	NS (pots with sand); Ni stress / non-stress	HBL	10^{-8} M	Foliar spray	30 DAS	40 DAS	P_N, Chl content, CA activity	Alam *et al.* (2007)

Cicer arietinum	NS (pots with soil); salinity stress / non-stress	HBL	10^{-10} M, 10^{-8} M	Seed soaking for 4 or 8 h	Seeds	60, 90 or 120 DAS	Chl *a*, Chl *b* content, CA activity	Ali *et al.* (2007)
Brassica oleracea	Growth chamber (detached cotyledons - Petri dishes with water)	EBL	10^{-9} M, 10^{-7} M, 10^{-5} M	Cotyledons soaking for 3 days	8 DAS	After the end of BRs treatment	Chl content	Çağ *et al.* (2007)
Phaseolus vulgaris, Hordeum vulgare	Greenhouse (pots with vermiculite); salinity stress / non-stress	NS	5×10^{-6} M	Seed soaking for 6 h	Seeds	6, 8, 18 days after germination	Chl content	El-Fattah (2007)
Brassica juncea	Net-house (pots with sand); Cd stress / non-stress	HBL	10^{-8} M	Foliar spray	30 DAS	60 DAS	P_N, Chl content, CA activity	Hayat *et al.* (2007)

(*continued*)

(*continued Table 1.*)

Plant species	Growth conditions	BRs type	BRs dosage / concentration	BRs mode of treatment	Time of BRs application	Time of measurements	Photosynthetic parameters measured	References
Brassica napus	Greenhouse and growth chamber (plants - pots with soil; detached leaves – Petri dishes with water); cold stress / non-stress	EBL	5×10^{-7} M, 10^{-6} M	BRs injection into apoplast of cotyledons or primary leaves	21 DAS	25 DAS + 3 or 7 days after leaf cutting	Chl *a*, Chl *b* and Car content	Janeczko *et al.* (2007)
Triticum aestivum	NS (pots with sand); salinity stress / non-stress	EBL	0.0125 mg/L, 0.025 mg/L (*approx. 2.6×10^{-8} M, 5.2×10^{-8} M*)	Foliar spray	56 DAS	83 DAS	P_N, C_i, C_i/C_a, g_S, Chl *a*, Chl *b* content	Qayyum *et al.* (2007)
Triticum aestivum	Growth chamber (plants - pots with sawdust; detached leaves - Petri dishes with water)	EBL	10^{-9} M, 10^{-7} M, 10^{-5} M	Leaf segments soaking for 3, 5, 7 or 10 d	10 DAS	After the end of BRs treatment	Chl content	Sağlam-Çağ (2007)

Vigna radiata	Growth chamber (pots with sand); Al stress / non-stress	EBL, HBL	10^{-8} M	Foliar spray	14 DAS	21 DAS	P_N, C_i, g_S, Chl content, CA activity	Ali *et al.* (2008a)
Brassica juncea	Growth chamber (pots with sand); salinity or Ni stress / non-stress	EBL	10^{-6} M	Foliar spray	15 DAS	30 DAS	P_N, C_i, g_S, Chl and Car content, CA activity	Ali *et al.*, (2008b)
Triticum aestivum	Net-house (hydroponic growth in nutrient solution); salinity stress / non-stress	EBL	0.052×10^{-6} M, 0.104×10^{-6} M, 0.156×10^{-6} M,	Seed soaking followed by hydroponic growth in medium with BRs for 45 days	Seeds	After the end of BRs treatment	P_N, C_i, g_S, Chl *a* content, F_v/F_m	Ali *et al.* (2008c)
Glycine max	Growth chamber (pots with soil); light / dark	EBL	10^{-9} M, 10^{-7} M, 10^{-5} M	Foliar spray (repeated every other day for 12 days)	5 DAS	After the end of BRs treatment	Chl and Car content	Cevahir *et al.* (2008)

(*continued*)

(*continued Table 1.*)

Plant species	Growth conditions	BRs type	BRs dosage / concentration	BRs mode of treatment	Time of BRs application	Time of measurements	Photosynthetic parameters measured	References
Vigna radiata	Net-house (pots with soil)	HBL	10^{-8} M, 10^{-6} M	Seed soaking for 8 h / Foliar spray / Seed soaking for 8 h + Foliar spray	Seeds / Plants 15 DAS / Seeds + plants 15 DAS	30 or 50 DAS	P_N, g_S, carboxylation efficiency, Chl content, CA activity	Fariduddin *et al.* (2008)
Pisum sativum	NS (plants - hydroponic growth in nutrient solution; detached shoots – nutrient solution)	EBL	10^{-7} M	Shoot incubation in nutrient solution with BRs for 20 min	9 DAS	After the end of BRs treatment	Phosphorylation of Calvin cycle proteins	Fedina *et al.* (2008)
Cicer arietinum	Net-house (pots with soil); Cd stress / non-stress	HBL	10^{-8} M	Foliar spray	30 DAS	90 DAS	Chl content, CA activity	Hasan *et al.* (2008)
Capsicum annuum	Greenhouse (pots with soil); salinity stress / non-stress	EBL	0.01, 0.05, 0.1, 0.5 mg/L (*approx.* 2×10^{-8} *M*, 10^{-7} *M*, 2×10^{-7} *M*, 10^{-6} *M*)	Foliar spray	Plants with 2 fully developed leaves	12 days after BRs treatment	F_v/F_m	Houimli *et al.* (2008)

Lycopersicon esculentum	Greenhouse and growth chamber (pots with vermiculite); heat stress / non-stress	EBL	0.01 mg/L, 0.1 mg/L, 1 mg/L (*approx.* 2×10^{-8} *M*, 2×10^{-7} *M*, 2×10^{-6} *M*)	Foliar spray	42 DAS	0, 4, 8 or 12 days after BRs treatment	P_N, C_i, g_S, V_{cmax}, J_{max}, *l*, F_v/F_m, F_v'/F_m', Φ_{PSII}, q_P, NPQ	Ogweno *et al.* (2008)
Triticum aestivum	Net-house (pots with sand); salinity stress / non-stress	EBL	0.0125 mg/L, 0.025 mg/L, 0.0375 mg/L (*approx.* 2.6×10^{-8} *M*, 5.2×10^{-8} *M*, 7.2×10^{-8} *M*)	Foliar spray	43 DAS	88 DAS	P_N, C_i, C_i/C_a, g_S, Chl *a*, Chl *b* content, Chl *a/b* ratio, F_v/F_m	Shahbaz *et al.* (2008)
Pelargonium graveolens	Greenhouse (pots with soil)	HBL	0.5×10^{-6} M, 10^{-6} M, 3×10^{-6} M	Foliar spray	30, 60 or 90 d after cuttings transplantation into soil	120 days after cuttings transplantation into soil	P_N, Chl *a*, Chl *b* content	Swamy and Rao (2008)
Glycine max	Sheltered field (pots with soil); drought stress / non-stress	BL	0.1 mg/L (*approx.* 2×10^{-7} *M*)	Foliar spray	Plants in R1 stage	14 days after BRs treatment	$^{14}CO_2$ assimilation and partitioning, P_N, Chl content, F_v/F_m, Rubisco activity, phosphoenol-pyruvate carboxylase activity	Zhang *et al.* (2008)

(*continued*)

(*continued Table 1.*)

Plant species	Growth conditions	BRs type	BRs dosage / concentration	BRs mode of treatment	Time of BRs application	Time of measurements	Photosynthetic parameters measured	References
Raphanus sativus	Growth chamber (pots with vermiculite); Cd stress / non-stress	EBL	10^{-6} M, 2×10^{-6} M	Seed soaking for 24 h	Seeds	20 DAS	P_N, g_S, Chl and Car content, CA activity	Anuradha and Rao (2009)
Chlorella vulgaris	Controlled conditions (hydroponic growth in nutrient solution); heat stress / non-stress	BL	10^{-8} M	Incubation in nutrient solution with BRs for 1 or 3 h	NS (cell culture)	After the end of BRs treatment	Chl content	Bajguz (2009)
Lycopersicon esculentum	Greenhouse and growth chamber (pots with sand and soil); drought stress / non-stress	EBL	10^{-8} M, 10^{-6} M	Foliar spray (repeated 3× during 3 d)	Plants with 3 fully developed leaves	3 or 5 days after the end of BRs treatment	Chl *a*, Chl *b* and Car content	Behnamnia *et al.* (2009)

Zea mays	NS (pots with soil); salinity stress / non-stress	BL	0.25 ppm	Seed soaking for 12 h / Seed soaking for 12 h + Foliar spray (repeated 2×)	Seeds / Seeds + plants 14 and 28 DAS (repeated treatment)	21 or 35 DAS	Chl *a*, Chl *b* and Car content	El-Khallal *et al.* (2009)
Brassica juncea	Net-house (pots with soil); drought stress / non-stress	HBL	10^{-8} M	Foliar spray	30 DAS	60 DAS	P_N, C_i, g_S, Chl content, CA activity	Fariduddin *et al.* (2009a)
Brassica juncea	Growth chamber (pots with sand); Cu stress / non-stress	HBL	10^{-10} M, 10^{-8} M, 10^{-6} M	Seed soaking for 8 h	Seeds	30 DAS	P_N, C_i, g_S, Chl content, CA activity	Fariduddin *et al.* (2009b)
Oryza sativa	Growth chamber (pots with soil); drought stress	EBL, HBL	10^{-8} M	Seed soaking for 2 d / Foliar spray	Seeds / Plants 28 DAS	35 DAS	P_N, C_i, g_S, *l*	Farooq *et al.* (2009)

(continued)

(*continued Table 1.*)

Plant species	Growth conditions	BRs type	BRs dosage / concentration	BRs mode of treatment	Time of BRs application	Time of measurements	Photosynthetic parameters measured	References
Beta vulgaris	Field (natural conditions)	EBL, others	10^{-6} M	Foliar spray	Plants with at least 9 fully developed leaves	14 d after BRs treatment	Chl content, rapid Chl fluorescence induction parameters	Hradecká *et al.* (2009)
Cucumis sativus	Greenhouse (hydroponic growth in nutrient solution); hypoxia stress / non-stress	EBL	0.001 mg/L (*approx.* 2×10^{-9} *M*)	Incubation in nutrient solution with BRs for 2, 4, 6 or 8 d	Plants with 3 fully developed leaves	After the end of BRs treatment	P_N	Kang *et al.* (2009)
Vicia faba	Growth chamber (pots with soil); herbicide stress	EBL	2×10^{-9} M, 2×10^{-8} M	Seed soaking for 24 h	Seeds	6-30 DAS	P_N, F_v/F_m, F'_v/F_m', q_P, NPQ, ETR	Piñol and Simón, (2009)
Pelargonium graveolens	Greenhouse (pots with soil)	EBL	0.5×10^{-6} M, 10^{-6} M, 3×10^{-6} M	Foliar spray	30, 50 or 70 days after cuttings transplantation into soil	120 days after cuttings transplantation into soil	P_N, Chl *a*, Chl *b* content	Swamy and Rao (2009)

Cucumis sativus	Growth chamber (pots with soil)	EBL	10^{-7} M	Foliar spray (repeated every 5 days)	Plants with 4 fully developed leaves	1, 3, 5, 20 or 25 days after the first BRs treatment	P_N, C_i, g_S, Chl content, V_{cmax}, J_{max}, total and initial Rubisco activity, Rubisco activation state, F_v/F_m, F'_v/F'_m, Φ_{PSII}, q_P, expression of genes for Calvin cycle enzymes and Rubisco activase	Xia *et al.* (2009a)
Cucumis sativus	Growth chamber (pots with vermiculite followed by containers with nutrient medium); pesticide stress / non-stress	EBL	10^{-7} M	Foliar spray	Plants with 4 fully developed leaves	0 or 1 days after BRs treatment	P_N, F_v/F_m, Φ_{PSII}, q_P	Xia *et al.* (2009b)

(*continued*)

(*continued Table 1.*)

Plant species	Growth conditions	BRs type	BRs dosage / concentration	BRs mode of treatment	Time of BRs application	Time of measurements	Photosynthetic parameters measured	References
Zea mays	Greenhouse (pots with soil); cold stress / non-stress	EBL, other	10^{-14} M, 10^{-12} M, 10^{-10} M, 10^{-8} M	Foliar spray (repeated 2×)	11 and 25 DAS (repeated treatment)	39 DAS	Chl *a*, Chl *b* and Car content, Chl *a*/b and Chl/Car ratios, Hill reaction activity, PS I activity	Honnerová *et al.* (2010)
Zea mays	Field (natural conditions)	EBL, other	10^{-12} M	Foliar spray	41 DAS	72, 82 or 94 DAS	Chl *a*, Chl *b* and Car content, Chl *a*/b and Chl/Car ratios, Hill reaction activity, PS I activity	Kočová *et al.* (2010)

He further extended his analysis by examining the effect of not only foliar spray with HBL but seed soaking as well. The results of this second study again confirmed the positive effect of HBL on P_N with no significant differences between modes of BR treatment (Sairam, 1994b).

Other authors who worked with HBL usually obtained similarly positive results regarding the impact of this type of BR on P_N. A lot of this work has been done with Indian mustard and mungbean at Aligarh Muslim University, India. Spraying of 30 days-old Indian mustard plants with HBL solutions increased the values of P_N in leaves of those plants 30 days after BR treatment to about 170 % of the values recorded for control plants when 10^{-8} M or 10^{-6} M concentrations were used, and improved net photosynthesis even in case of lower HBL concentration, although not to such a great degree (Hayat *et al.*, 2000). Further studies made with the same species and the same type, mode and conditions of BR treatment used only 10^{-8} M concentration of HBL solution and confirmed a significant enhancement of P_N as the effect of HBL application to experimental plants (Hayat *et al.*, 2001b, 2007; Alam *et al.*, 2007; Fariduddin *et al.*, 2009a), although in some cases the values of this parameter increased only by about 30–40 % compared to BR-untreated plants. Likewise, soaking of Indian mustard seeds in solutions of HBL improved net photosynthesis in leaves 30 days after BR treatment; again, the most diluted solution appeared to be the least effective one (Fariduddin *et al.*, 2009b). The effect of foliar spray with HBL on P_N was examined also by Fariduddin *et al.* (2004, 2006) in mungbean with slightly different results: the most effective concentration in this case was 10^{-8} M while the more concentrated, *i.e.* 10^{-6} M solution showed the least effectivity for P_N enhancement (but still improved net photosynthesis relative to BR-untreated plants). Both these studies compared also two time-points P_N was measured; in the first case (Fariduddin *et al.*, 2004) 10 or 20 days after HBL treatment and in the second case (Fariduddin *et al.*, 2006) 15 or 45 days after HBL treatment, but the difference in the effect of BR on P_N was negligible. Another study made by the same authors examined how soaking of mungbean seeds in HBL solutions of three different concentrations for 4, 8 or 12 h affects net photosynthesis in leaves of plants at three different ages (Fariduddin *et al.*, 2003). In this case, 10^{-6} M concentration was slightly better than 10^{-8} M (and markedly better than 10^{-4} M), and the improvement in net photosynthesis again did not diminish with the prolongation of interval between HBL application and P_N measurement. To assess the response of P_N to different modes of HBL application (seed soaking, foliar spray or combination of both treatments), Fariduddin *et al.* (2008) made yet another study with mungbean plants, which revealed that combination of seed soaking and foliar spray has the best impact on this parameter. The list of papers dealing with HBL effect on P_N should not be complete without mentioning two other studies: Ali *et al.* (2008a) examination of P_N in mungbean plants and Swamy and Rao (2008) work with geranium, both of which confirmed the positive effect this BR had on net photosynthesis.

As regards other type of BR that is commonly used in physiological studies, *i.e.* EBL, the situation here is somewhat more controversial. Comparison of the effect of EBL and HBL in very young mungbean plants treated by foliar spray with these BRs slightly favoured EBL over HBL in respect to net photosynthesis improvement (Ali *et al.*, 2008a). Spraying of 15 days-old Indian mustard plants with EBL also increased the values of P_N in comparison to control plants (Ali *et al.*, 2008b) and similar treatment of tomato (Singh and Shono, 2005) or geranium (Swamy and Rao, 2009) resulted in the improvement of P_N as well. On the other hand, other authors working with EBL-sprayed tomato (Ogweno *et al.*, 2008) or wheat (Qayyum *et al.*, 2007; Shahbaz *et al.*, 2008) did not usually observe any significant effect of BR treatment on net photosynthesis when plants were not subjected to some stress factor. Two studies that examined P_N in wheat or radish (*Raphanus sativus* L.) plants, grown from seeds soaked in EBL solutions brought again positive results with respect to the effect this BR exerts on net photosynthesis (Ali *et al.*, 2008c; Anuradha and Rao, 2009).

Rather detailed analysis of an impact the treatment of plants with EBL has on photosynthetic CO_2 assimilation has been made by Chinese scientists working with cucumber. Yu *et al.* (2004) measured P_N in EBL-sprayed cucumber plants both at light-saturated conditions and at growing light intensity and demonstrated that the effect of this BR on net photosynthesis lasted approximately 7 days with the highest effectivity between Day 1 to 3 after EBL treatment. The effective concentrations were apparently in the 10^{-8} M or 10^{-7} M range, whereas the application of more concentrated EBL solution had no significant effect on net photosynthesis. They also simulated photorespiratory (21% atmospheric O_2 content) and non-photorespiratory (2% atmospheric O_2 content) conditions but the effect of EBL on P_N was similar in both cases leading to the conclusion that EBL does not reduce photorespiration. The enhancement of P_N continuing for at least 8 days was observed also in leaves of EBL-treated cucumber plants (Kang *et al.*, 2009); in this case, plants were grown hydroponically in nutrient solution with/without EBL. Xia *et al.* (2009a) also examined the time-course of P_N measured under saturating irradiation in young cucumber plants that were repeatedly sprayed with EBL and found that this type of BR significantly increased values of P_N up to 5 days after treatment but at 20 days this positive influence disappeared. Moreover, they applied brassinazole (Brz) to their experimental plants and showed that treatment with this inhibitor of BRs biosynthesis leads to a decrease in P_N, which could be alleviated by EBL application. These authors also published two other papers confirming the stimulating effect EBL exerts on P_N in leaves of this plant species (Xia *et al.*, 2006, 2009b).

Curiously enough, other types of BRs have been rarely examined with the aim to assess their impact on net photosynthesis. Zhang *et al.* (2008) sprayed soybean (*Glycine max* (L.) Merrill) plants with solution of BL and found that it increased their rate of net CO_2 uptake. Trihydroxylated spirostane analogue of brassinosteroid was used in study made with banana plants, the results of which showed no effect of this BR on P_N early after its application but a stimulating effect later (González-Olmedo *et al.*, 2005). Bajguz and Czerpak (1998) examined the effect of various types of BRs (BL, EBL, HBL, castasterone, 24-epicastasterone, homocastasterone) applied to green alga *Chlorella vulgaris* Beijerinck. They measured net photosynthesis polarographically as the amount of oxygen evolved by algal cells and found that the effectivity of individual BRs decreased in the order of BL – EBL – HBL – castasterone – 24-epicastasterone – homocastasterone.

Although gas exchange analysis is currently the preferred method for the determination of P_N in plants, some other studies evaluated the efficiency of CO_2 assimilation by monitoring the uptake of labelled $^{14}CO_2$. Works of Fujii *et al.* (1991) and Nakajima and Toyama (1995) made with BL-treated rice (*Oryza sativa* L.) plants or EBL-treated cucumber plants, respectively, should be mentioned in this context. Whereas Fujii *et al.* (1991) observed BL-induced promotion of ^{14}C assimilation in rice leaves and its translocation to other parts of plant, the second group of authors did not find any effect of EBL on the amount of $^{14}CO_2$ uptake by primary leaves of cucumber plants but observed a distinctive shift in the transport of photosynthates toward epicotyl in EBL-treated plants. However, the mode of BR treatment differed between both studies which could cause this discrepancy: the foliar spray in case of Fujii *et al.* (1991) and the incubation of roots for 4 h in BR-containing solution in case of Nakajima and Toyama (1995). Increased translocation of ^{14}C assimilates after BR treatment was described also by Zhang *et al.* (2008) who worked with BL-sprayed soybean plants.

The results of all the above-mentioned studies made with BR-treated plants can be summarized into the simple statement that this group of plant hormones undoubtedly stimulates photosynthesis, although the exact level of such stimulation depends both on plant species and BR type, concentration or treatment mode, and can change during the time. How, then, does this positive effect occur? Which parts of photosynthetic processes, which components of primary or secondary photosynthetic reactions are affected and which are not? Are BR-induced changes of photosynthetic efficiency accompanied by changes in the structure of photosynthetic organelles, *i.e.* chloroplasts, as well? The following paragraphs attempt to further dissect our current knowledge on BRs influence on photosynthesis as obtained from studies with exogenously applied BRs.

2.2 Stomatal function

The determination of the net photosynthetic rate by gas analysis systems is usually accompanied by the evaluation of other parameters associated with leaf gas exchange, namely the stomatal conductance (g_S) and the intercellular CO_2 concentration (c_i). An observation of the response of P_N to c_i offers the possibility to separate stomatal and non-stomatal factors that can limit photosynthesis (Farquhar *et al.*, 1980; Long and Bernacchi, 2003). Several authors have examined changes in g_S and c_i associated with the exposure of plants to exogenous BRs, but the results of their studies are rather ambiguous.

Dahse *et al.* (1991) found that EBL and HBL significantly inhibited stomatal opening in Asiatic dayflower (*Commelina communis* L.) epidermal strips. Stomatal limitation of photosynthetic efficiency, determined by Farquhar and Sharkey (1982) method, was in no way affected by treatment of cucumber plants with EBL (Yu *et al.*, 2004). Xia *et al.* (2006) worked with the same species and determined values of g_S and c_i in leaves 2 days after foliar spray with EBL. No effect of EBL on g_S was revealed but the intercellular concentration of CO_2 slightly diminished due to the application of this BR. Later, they followed the time-course of changes in these two parameters in cucumber leaves treated with the same type of BR and/or with Brz and demonstrated that while treatment with Brz increases the intercellular CO_2 concentration from 1 day up to 20 days after its application, using EBL *per se* leads to the decrease of this parameter during early days but not later (Xia *et al.*, 2009a). Again, they found almost no effect of either EBL or Brz on the stomatal conductance with the exception of Day 20 when g_S in leaves of plants treated with Brz significantly decreased.

Imbibition of EBL by wheat seeds followed by hydroponic growth of plants in medium with EBL did not cause any significant changes in either g_S or c_i (Ali *et al.*, 2008c). Shahbaz *et al.* (2008) also found no significant response of g_S in two cultivars of wheat sprayed with EBL solution (with the exception of the highest concentration used, *i.e.* approx. 7×10^{-8} M). Relative internal CO_2 values (calculated as the ratio of intercellular and ambient CO_2 concentrations, c_i/c_a) in leaves of their experimental plants somewhat decreased due to EBL treatment, but only in one cultivar and at one EBL level. Similar effect was described by Qayyum *et al.* (2007) who observed slight (statistically insignificant) decrease of both c_i and c_i/c_a values 3 weeks after foliar application of EBL to wheat plants. However, they also found that g_S measured in leaves of the same plants at the same time diminished due to EBL treatment.

Ogweno *et al.* (2008) did not find any effect of EBL application on either g_S or c_i in tomato when determined at Day 0, 4, 8 or 12 after BR treatment. On the other hand, EBL application resulted in an increase of g_S but no changes in c_i approx. 14 days after foliar spray of the plants of the same species (Singh and Shono, 2005); however, these authors used higher concentrations of EBL and younger plants for their experiments. Foliar application of EBL or HBL to mungbean or Indian mustard positively affected both stomatal conductance and intercellular CO_2 concentration in the leaves of these species (Ali *et al.*, 2008a,b; Fariduddin *et al.*, 2009a), although HBL treatment influenced these parameters to a lesser extent compared to EBL (Ali *et al.*, 2008a). A significant increase in g_S was found also in mungbean plants sprayed with three concentrations of HBL solution when this parameter was measured 15, 35 or 45 days after spraying (Fariduddin *et al.*, 2006, 2008). Soaking of mungbean seeds or the combined treatment of seed soaking and foliar spray by HBL had a positive impact on the stomatal conductance as well (Fariduddin *et al.*, 2003, 2008). Increased values of g_S and c_i were observed by Fariduddin *et al.* (2009b) in leaves of Indian mustard plants, grown from seeds treated for 8 h with HBL when measured 30 days after BR treatment. Anuradha and Rao (2009) reported an increased stomatal conductance in leaves of radish plants grown from seeds soaked in EBL solutions of various concentrations.

To draw any conclusion about the possible role of BRs in stomatal regulation of photosynthesis from the studies mentioned above is extremely difficult. In some cases, BR-induced increase in the stomatal conductance closely correlated with the improvement of net photosynthesis, which could be caused by the greater amount of CO_2 available for its fixation by photosynthetic enzymes. In other cases, no such correlation existed, stomatal function was not the limiting factor for photosynthesis and any effect BRs exerted on the efficiency of photosynthetic processes could not be associated with stomatal regulation. It seems that the impact of these hormones on parameters associated with the performance of stomata depends mainly on plant species (and in some cases on cultivar and plant age as well), and secondly on finding the effective dosage of BRs. A more detailed analysis made in many plant species, with several BRs examined in different dosages, and using simultaneous measurements of various parameters associated with stomatal function and photosynthesis is clearly needed.

2.3 Enzymes of photosynthetic carbon reduction cycle

Among non-stomatal factors affecting photosynthesis, the efficiency of CO_2 fixation and the regeneration of ribulose-1,5-bisphosphate (RuBP) as CO_2 acceptor by enzymes of Calvin cycle play a very important role. A possibility that BRs could somehow affect this part of photosynthetic processes has always been an extremely attractive idea and has been examined in several studies. The main focus of these studies has been the enzyme ribulose-1,5-bisphosphate carboxylase/oxygenase (Rubisco). This multisubunit protein complex (composed of eight small – SSU – and eight large – LSU – subunits in the majority of photosynthetic organisms) is probably the most abundant protein at the Earth. Its carboxylase activity catalyzes the coupling of CO_2 with RuBP and the formation of 3-phosphoglycerate, whereas competing reaction catalyzed by its oxygenase activity leads subsequently to photorespiration (Roy and Andrews, 2000). An increase in both carboxylase activity and content of Rubisco in wheat and white mustard (*Leucosinapis alba* L.) treated with BL was described by Braun and Wild (1984). Yu *et al.* (2004) found that exogenous application of EBL highly increased carboxylation rate of Rubisco determined *in vivo* in cucumber leaves. Further dissection revealed that while neither total Rubisco activity nor the content of this enzyme (measured *in vitro* in the extracts made from these leaves) was affected by EBL application, the initial Rubisco activity (reflecting Rubisco activation state) increased as early as 3 h after BR treatment and continued to be significantly elevated compared to untreated plants up to 5 days after BR treatment (but not longer). Recently, this group of scientists published another study dealing with the promoting effects of EBL on photosynthesis in cucumber, where they advanced our knowledge about effects of BRs on the Calvin cycle enzymes even further (Xia *et al.*, 2009a). Again, they observed the statistically significant increase of both maximum Rubisco carboxylation rate and the initial Rubisco activity caused by EBL treatment of plants. On the other hand, plants treated with the inhibitor of BR biosynthesis, Brz, showed significantly lower values of these parameters and combination of EBL and Brz restored the values to the level comparable with BR-untreated plants. Moreover, they analyzed the expression of genes coding for SSU and LSU of Rubisco and demonstrated the elevated levels of transcripts from these genes in EBL treated plants and the diminished levels in Brz-treated plants. As regards the amount of proteins, Brz treatment had negative effect both on SSU and LSU whereas EBL application increased only SSU amount but not LSU. A role of BRs in the enhancement of the Rubisco activation state was supported further by finding the elevated expression of Rubisco activase (observed both on the mRNA and protein levels) in EBL-treated plants and the reduced expression of gene for this enzyme in Brz-treated plants (Xia *et al.*, 2009a). Rubisco activase is a protein necessary for maintaining Rubisco in an active conformation state (von Caemmerer and Quick, 2000). Its enhanced amount and/or activity could indeed be responsible for the stimulation of CO_2 fixation and, consequently, for the enhancement of net photosynthesis caused by BRs, as suggested by Xia *et al.* (2009a).

Yu *et al.* (2004) and Xia *et al.* (2009a) also observed that the maximum potential rate of RuBP regeneration was positively affected by EBL application to cucumber leaves and hypothesized that this could perhaps be associated with an increased activity/amount of other Calvin cycle enzymes. This idea was partly supported by their later study with the same plant species, where they analyzed the expression of genes coding for six enzymes involved in RuBP regeneration, *i.e.* triose-3-phosphate isomerase, fructose-1,6-bisphosphate aldolase, fructose-1,6-bisphosphatase, sedoheptulose-1,7-bisphosphatase, ribulose phosphate epimerase and ribulose-5-phosphate kinase (Xia *et al.*, 2009a). The steady-state level of transcripts from all these genes was significantly elevated in EBL-treated plants and (with the exception of fructose-1,6-bisphosphate aldolase and ribulose phosphate epimerase) significantly reduced in Brz-treated plants as compared to the control ones.

Besides these two studies, the positive effect of BRs on enzymes of photosynthetic carbon reduction cycle was observed by other authors as well. Zhang *et al.* (2008), who worked with BL-sprayed soybean plants, demonstrated that the application of BR can significantly increase Rubisco activity as well as the activity of phosphoenolpyruvate carboxylase. Interesting results were published by Fedina *et al.* (2008), who examined the effect of EBL on tyrosine phosphorylation of Calvin cycle enzymes and found that BR substantially increases phosphorylation of Rubisco SSU and fructose-1,6-bisphosphate aldolase precursors and – to a much lesser extent – of fructose-1,6-bisphosphate aldolase and three isoforms of Rubisco LSU. Application of BR also diminished phosphorylation of the α-subunit of Rubisco-binding protein precursor. Altered phosphorylation levels could be associated with changed activities of the Calvin cycle enzymes and, thus, with BR-induced improvement of photosynthesis.

Another protein related to photosynthetic CO_2 fixation, which is frequently studied in association with the possible effect of BRs on photosynthesis, is the carbonic anhydrase. This enzyme catalyzes the inter-conversion between CO_2 and HCO_3^- and can increase the availability of CO_2 for Rubisco (Coleman, 2000). Several studies demonstrated that treatment of plants with exogenously applied HBL or EBL (regardless of the mode of application, *i.e.* foliar spray or seed soaking) can significantly improve the activity of carbonic anhydrase in Indian mustard (Hayat *et al.*, 2000, 2001b, 2007; Alam *et al.*, 2007; Ali *et al.*, 2008b; Fariduddin *et al.*, 2009a, b), mungbean (Fariduddin *et al.*, 2003, 2004, 2006, 2008; Ali *et al.*, 2008a), chickpea (*Cicer arietinum* L.) (Ali *et al.*, 2007; Hasan *et al.*, 2008), radish (Anuradha and Rao, 2009), tomato (Ali *et al.*, 2006) or wheat (Hayat *et al.*, 2001a) and that this increased activity usually corresponded well with the enhanced photosynthetic efficiency.

So far, it would seem that the above mentioned results strongly suggest that the activities and/or levels of enzymes associated with photosynthetic carbon fixation are indeed regulated by BRs, although the exact mechanism of this regulation is unknown. However, the situation is more complex. Ogweno *et al.* (2008) did not find any significant changes in either maximum Rubisco carboxylation rate or maximum rate of RuBP regeneration in tomato plants treated with EBL and grown in optimum conditions. Application of BR-containing agens ComCat to tomato plants had no effect on the level of Rubisco SSU mRNA in case these plants were not infected by pathogens (Berger *et al.*, 2004). Asami *et al.* (2000), as well as Sekimata *et al.* (2001) found an increased steady-state level of Rubisco SSU mRNA in *Arabidopsis thaliana* (L.) Heynh. or tobacco (*Nicotiana tabacum* L.) plants treated with Brz and grown in the dark, whereas light-grown Brz-treated *Arabidopsis* plants did not differ in this respect from Brz-untreated ones (Asami *et al.*, 2000). Thus, although various components of photosynthetic carbon reduction cycle remain good candidates for the site of BR action on photosynthesis, there are still many questions which can be answered only by additional detailed analyses.

2.4 Photosynthetic pigments and light-harvesting

Almost from the beginning, analyses of various morphological, physiological and developmental traits in BR-treated plants were often accompanied by the determination of chlorophyll (Chl) content in leaves. Chls *a* and *b* constitute an important part of photosynthetic apparatus due to their role as main light-capturing molecules in photosynthetic light-harvesting antennae complexes (besides their function as electron-transfer cofactors in reaction centers of Photosystems I and II) (Melkozernov and Blankenship, 2006). A positive effect of BRs on their content could therefore lead to an increased efficiency of light capture resulting in the improvement of net photosynthesis. Quite a number of studies documented an elevated amounts of both total Chl and Chls *a* or *b* caused by BR application to plants. However, not so negligible number of studies also did not find any effect of BRs on these parameters at all or revealed their strong dependance on BR dosage in combination with plant species, so there still exists some doubt as to the possible influence BRs could have on these photosynthetic pigments. One thing, however, rather seems to stand out when comparing these studies: a manifest dependance of the effect BRs possibly have on Chl content on the type of BR. Almost all authors who described an increase in the amounts of these photosynthetic pigments worked with HBL whereas those that did not observe any such effect worked with EBL (although there are some exceptions to this rule as well).

Positive effect of either seed soaking or foliar spray with HBL on the content of total Chl in leaves was observed for various plant species: wheat (Sairam, 1994a, b), mungbean (Bhatia and Kaur, 1997; Fariduddin *et al.*, 2003, 2004, 2006, 2008; Ali *et al.*, 2008a), Indian mustard (Hayat *et al.*, 2001b, 2007; Alam *et al.*, 2007; Fariduddin *et al.*, 2009a, b), chickpea (Ali *et al.*, 2007; Hasan *et al.*, 2008), or geranium (Swamy and Rao, 2008). There are also some studies that describe an elevated amount of Chl by the application of BL to kidney bean (*Phaseolus vulgaris* L.) (Krizek and Mandava, 1983), cucumber (He *et al.*, 1991) or maize (*Zea mays* L.) (He *et al.*, 1991; El-Khallal *et al.*, 2009) plants. An examination of the effect of both 6-ketone (castasterone, 24-epicastasterone and homocastasterone) and 7-oxalactone (BL, EBL, HBL) types of BRs on *Chlorella vulgaris* cells made by Bajguz and Czerpak (1998) revealed that BRs of the 7-oxalactone type had more effect on the content of Chls, compared to the other type of compounds, with BL being the most effective one. As regards EBL, Ali *et al.* (2008a, b) showed that the application of this BR leads to the elevated total Chl content when sprayed to young mungbean or Indian mustard plants. Hradecká *et al.* (2009) reported an increased values of this parameter in the leaves of sugar beet (*Beta vulgaris* L.) treated with EBL as well, and similar phenomenon was observed for both Chl *a* and Chl *b* contents by Swamy and Rao (2009) who worked with geranium. Other synthetic derivatives of BRs were examined as well and it was found that these compounds are more effective than EBL in elevating Chl amounts in leaves of sugar beet or maize (Hradecká *et al.*, 2009, Honnerová *et al.*, 2010; Kočová *et al.*, 2010).

The above mentioned studies described positive effect of BRs on Chl content regardless of the concentration of these phytohormones. Some other authors, however, published results that were less unequivocal. Bajguz and Asami (2005) analyzed a wide range of EBL concentrations (10^{-13} M to 10^{-6} M) applied to growth medium of spotless watermeal plants and observed a distinctive dependance of Chl *a* and Chl *b* content on EBL dosage with peak at 10^{-9} M and both boundary concentrations almost totally ineffective. Qayyum *et al.* (2007) found that foliar application of EBL to wheat plants could both decrease and increase contents of Chls *a* and *b* depending on the concentration used; curiously enough, the concentration that exerted the stimulating effect on Chl *a* reduced the content of Chl *b* and *vice versa*. The application of EBL to soybean led to the increase in both Chl *a* and Chl *b* contents in cotyledons when plantlets were grown in light but the opposite was true for dark-grown plantlets; as regards hypocotyl, 10^{-5} M concentration of EBL reduced whereas 10^{-7} M concentration slightly elevated the content of these pigments under light conditions (Cevahir *et al.*, 2008). The effect of

EBL on total Chl content in leaves of radish also depended on the concentration used (Anuradha and Rao, 2009) and similar phenomenon was described for red cabbage (*Brassica oleracea* L.) grown from seeds soaked in EBL solutions of different concentrations (Çağ *et al.*, 2007), mungbean plants cultivated in growth medium containing BL (Abdullahi *et al.*, 2003) or tomato treated with different concentrations of HBL applied to its roots (Ali *et al.*, 2006).

Finally, a third group comprizes the studies that did not find any effect of BRs on Chl content at all. A majority of EBL applications belongs into this group, *e.g.* works of Yu *et al.* (2004) or Xia *et al.* (2009a) with cucumber, two papers of Janeczko *et al.* (2005, 2007) who worked with winter or spring rape (*Brassica napus* L.) or some studies made with wheat plants (Ali *et al.*, 2008c; Shahbaz *et al.*, 2008). Braun and Wild (1984) and Zhang *et al.* (2008) also did not observe any significant effect of treatment with BL on total Chl content in wheat, white mustard or soybean leaves and the same was true for BL applied to *Chlorella vulgaris* cell culture (Bajguz, 2009).

Carotenoids (Cars), a second group of photosynthetic pigments, have been less frequently examined in connection with BRs compared to Chls. Bajguz and Asami (2005) determined the content of total Cars in spotless watermeal after 7 days growth in medium with EBL and found that, similarly to the effect on the Chl content, EBL increased this parameter depending on its dosage (10^{-9} M being the most effective one). No significant effect of the addition of EBL to growth medium of winter rape plants on the content of total Cars was reported by Janeczko *et al.* (2005) and the same was true when EBL was injected into the apoplast of cotyledons or primary leaves of spring rape (Janeczko *et al.*, 2007). On the other hand, Ali *et al.* (2008b) demonstrated that foliar spray of Indian mustard with EBL solution caused the elevation of the content of these pigments. Examination of soybean plants grown under light or dark conditions and treated with three concentrations of EBL revealed no changes of total Cars' content in hypocotyl or the first internode but the values of this parameter increased in cotyledon when lower EBL concentrations were used (regardless of growth conditions) (Cevahir *et al.*, 2008). The content of Cars increased also in radish plants whose seeds imbibed EBL for 24 h (Anuradha and Rao, 2009) or in maize plants grown from seeds soaked in BL solution for 12 h and later sprayed with the same BR (El-Khallal *et al.*, 2009), but not in tomato or maize sprayed with EBL (Behnamnia *et al.*, 2009; Honnerová *et al.*, 2010; Kočová *et al.*, 2010).

If we acknowledge that at least *some* BRs in *specific* dosages and modes of application can actually positively affect the content of Chls and Cars in *some* plant species, what mechanism is employed? BRs could either increase biosynthesis or reduce degradation of these pigments. Alternatively, BRs could also affect the levels of proteins these pigments are associated with, particularly the proteins of light-harvesting complexes (LHC) in chloroplast thylakoid membranes or the early light-inducible proteins (ELIPs). Both these possibilities are supported by some, although extremely sparse data. Sağlam-Çağ (2007) examined the effect of EBL on senescence of wheat and found that the use of 10^{-9} M concentration of this BR somehow slowed down the loss of Chls from leaf segments. On the other hand, an incubation of segments cut from mature *Rumex obtusifolius* L. or *Xanthium strumarium* L. leaves for 4 d with BL in the dark actually promoted Chl loss compared to the control without BL (Mandava *et al.*, 1981). An examination of an expression of gene coding for one light-harvesting protein in cotyledons of dark-grown *Arabidopsis* or tobacco seedlings treated with Brz revealed higher steady-state levels of the respective mRNA compared to Brz-untreated seedlings; under light conditions, the levels of mRNA between Brz-treated and -untreated plants did not differ (Asami *et al.*, 2000; Sekimata *et al.*, 2001).

2.5 Primary photochemistry

Some effort has also been put into the examination of the possible influence of BRs on the efficiency of primary photochemical processes and on the protein complexes that participate in the electron transfer in thylakoid membranes of chloroplasts. *In situ* measurement of various parameters associated with Chl fluorescence is the most common method employed for this analysis, although some authors have used other approaches as well. Actually, one of the first studies aimed at the assessment of photosynthetic electron transport in BR-treated plants used the determination of the Hill reaction activity in chloroplasts isolated from mungbean leaves (Bhatia and Kaur, 1997). This parameter can express the efficiency of photosynthetic electron transport, although whether it represents the activity of Photosystem (PS) II, PS I or both depends on the specific conditions of its measurement. In any case, the results of this study were rather ambiguous: application of HBL decreased the Hill reaction activity early after treatment but enhanced it at later developmental stages, and the degree of these changes strongly depended on HBL dosage used. The results of analyses made in my own Laboratory with maize plants treated by foliar spray with either EBL or a synthetic analogue of castasterone also did not prove any relationship between BR treatment and the Hill reaction activity (Honnerová *et al.*, 2010; Kočová *et al.*, 2010).

The determination of minimal chlorophyll fluorescence (F_0), maximal fluorescence (F_m) and variable fluorescence ($F_v = F_m - F_0$) in dark-adapted leaves and the calculation of the maximum quantum efficiency of PS II (F_v/F_m) from these parameters has been frequently used in plants treated with exogenously applied BRs. Ali *et al.* (2008c) did not observe any significant changes of F_v/F_m values in two cultivars of wheat, grown hydroponically in EBL-containing medium. Similar results were obtained by Shahbaz *et al.* (2008) who worked with the same species but used a different mode of EBL treatment (foliar spray). No effect of spray with EBL on this parameter was found also in pepper (*Capsicum annuum* L.) (Houimli *et al.*, 2008) or tomato (Ogweno *et al.*, 2008) plants. The later study examined other parameters of chlorophyll fluorescence as well, *e.g.* the maximum efficiency of Photosystem II in light-adapted leaves (F'_v/F'_m), the effective quantum yield of PS II (Φ_{PSII}), which gives a measure of the rate of linear electron transport, or the photochemical (q_P) and non-photochemical (NPQ) quenching of chlorophyll fluorescence (q_P informs about the proportion of "open" PS II reaction centers whereas NPQ is associated with heat dissipation from PS II and its light harvesting complexes) (Maxwell and Johnson, 2000; Baker and Rosenqvist, 2004). None of these parameters was affected by the EBL treatment (Ogweno *et al.*, 2008). Yu *et al.* (2004) also did not note any significant effect of foliar spray with EBL on the values of either F_v/F_m or F_v'/F_m' in cucumber leaves; however, the efficiency of excitation energy capture by "open" PS II reaction centers (as expressed by Φ_{PSII}) as well as the proportion of "open" PS II reaction centers (as expressed by q_P) were significantly higher in EBL-treated plants compared to the control ones and this increase persisted up to 3 days after EBL treatment. Similar observations were made by Xia *et al.* (2006, 2009a, b) who worked with the same species and the same type of BR. On the other hand, treatment of cucumber plants with Brz led to the reduced efficiency of excitation energy capture by "open" PS II reaction centers (expressed by both Φ_{PSII} and F_v'/F_m' parameters) but had no efect on either q_P or F_v/F_m (Xia *et al.*, 2009a). Janeczko *et al.* (2005) measured fast transients of Chl *a* fluorescence (OJIP), which reflect the successive reduction of electron-acceptor pool of PS II, in winter rape grown in a medium containing EBL. This analysis revealed that EBL caused only a small and mostly insignificant increase of specific energy fluxes and flux ratios per reaction center of PS II. The activity of oxygen-evolving complexes and the number of active reaction centers were not affected by EBL application. Asami *et al.* (2000) and Sekimata *et al.* (2001) observed an increased accumulation of mRNA for D1 protein of PS II reaction center in dark-grown seedlings of *Arabidopsis* and tobacco treated with Brz. In light-grown *Arabidopsis*, there was no difference between Brz-treated and -untreated plants in the level of this mRNA (Asami *et al.*, 2000).

Components of photosynthetic electron-transport chain other than PS II have been seldom examined in BR-treated plants. My Laboratory analyzed the effect of EBL and some synthetic derivatives of BRs on the activity of PS I in suspensions of chloroplasts, isolated from BR-treated maize plants but did not find any effect of these compounds on this parameter (Honnerová *et al.*, 2010; Kočová *et al.*, 2010). However, as regards other parts of photosynthetic electron transport, no data on the possible regulation of their amounts or activities by BRs are currently available. Thus, similarly to what has been said about other parts of photosynthetic processes, a more exhaustive analysis is needed to reach a deeper understanding of the relationship between BRs and various components of thylakoid electron-transport chain.

2.6 Chloroplast ultrastructure

The effect of BRs on chloroplast structure has been investigated very rarely and this is certainly one aspect of BRs' research that deserves more attention. I have found only two studies that analyzed ultrastructure of chloroplasts in BR-treated plants and both were aimed at the analysis of whole cell ultrastructure, not chloroplasts *per se*. Kulaeva *et al.* (1991) reported that incubation of barley (*Hordeum vulgare* L.) leaf segments for 26 h in EBL solution had little or no effect on leaf cell structure including chloroplasts. Sam *et al.* (2001) made electron microscopic analysis of mesophyll cells in tomato leaf segments subjected to 24 h incubation in aquatic solution of a synthetic BR analogue with spirostanic structure. They reported that this BR causes structural changes in chloroplasts similar to symptoms of heat stress but did not pursue it further. Further analyses of ultrastructure of various plastid types in BR-treated plants would certainly shed more light on the role of these hormones in chloroplast development.

3. STRESS FACTORS ENHANCE THE EFFECT OF BRASSINOSTEROIDS ON PHOTOSYNTHESIS

Up to this point, the description of the current state of our knowledge on the effect of exogenously applied BRs on various aspects of photosynthesis given in this chapter dealt with the results obtained from plants cultivated in non-stress conditions. However, a great number of studies have addressed the application of BRs to plants exposed to some unfavourable environmental factors as well. BRs have long been known to alleviate stress symptoms in a variety of plant species subjected to both abiotic and biotic stressors (reviewed e.g. by Krishna, 2003; Bajguz and Hayat, 2009). As photosynthetic processes are usually among the first to be affected by unfavourable environment, it can only be expected that the measurement of photosynthetic parameters is rather a favorite way to assess the response of plants to stressful conditions and the impact BRs have on this response. What we did learn so far from such studies seems to point to the fact that exogenously applied BRs not only diminish the reduction of photosynthesis caused by various stressors, but that their influence on photosynthetic processes is actually more pronounced in stressed plants than in non-stressed ones, regardless of the stressor type.

Low temperature reduced synthesis of Chls in etiolated maize leaves exposed to light but incubation of those leaves with BL strongly diminished this reduction (He *et al.*, 1991). Similar effect on Chl content was observed in cucumber seedlings grown from seeds treated with BL and subjected to cold stress (Katsumi, 1991). Janeczko *et al.* (2007) described lower degradation of Chls in segments cut from primary spring rape leaves and incubated at low temperature (but not at optimum temperature) when plants were injected with EBL solutions. A decrease in the net photosynthetic rate caused by the exposure of banana plants to 7 days-long period of cold was not alleviated by treatment with synthetic BR analogue but when the plants were exposed to high temperatures, BR had significantly positive effect on this parameter (González-Olmedo *et al.*, 2005). EBL-treated plants of tomato also showed enhancement of P_N as well as g_S compared to the control when exposed to high temperatures for 24 h, and this enhancement was maintained after another 24 h of recovery in optimum temperature conditions as well (Singh and Shono, 2005). Another study with tomato subjected to high temperatures and sprayed with EBL solutions of different concentrations confirmed that this BR diminishes heat-induced depression of photosynthesis and showed that whereas photosynthetic parameters are not affected by BR treatment under optimum temperatures, the positive effect of BRs on photosynthesis manifests itself under stress conditions (Ogweno *et al.*, 2008).

Drought also strongly reduces photosynthetic efficiency and BRs have been shown to alleviate its negative effect in various plant species. Both the net photosynthetic rate and the content of Chls in leaves of wheat subjected to withholding water supply for 7 days were significantly improved by spraying of plants with HBL solutions or by imbibition of HBL by seeds (Sairam, 1994a, b). Several photosynthetic parameters (particularly the Chl content and the maximum quantum efficiency of PS II) measured in soybean treated with BL, which did not show any significant change when plants were grown in conditions with sufficient water supply, were improved by BL after exposure of plants to a period of drought (Zhang *et al.*, 2008). The content of Cars in the leaves of tomato also responded positively to BR treatment only under drought conditions but not in non-stressed plants (Behnamnia *et al.*, 2009). Both EBL and HBL treatment improved the net photosynthetic rate and the intercellular CO_2 concentration in drought-stressed rice plants compared to the stressed but BR-untreated ones (Farooq *et al.*, 2009). Spraying of Indian mustard with HBL alleviated the negative effect of drought on photosynthesis and restored the values of various photosynthetic parameters to the level observed in non-stressed plants (Fariduddin *et al.*,

2009a). However, no significant effect of xylem feeding with HBL on either P_N, g_S or c_i/c_a was observed in jack pine subjected to 12 days of withholding water supply (Rajasekaran and Blake, 1999).

Many studies that analyzed the effect of BRs on photosynthesis were made with salt-stressed plants. EBL significantly reduced the damage to leaf cell ultrastructure (including chloroplasts) caused by salt stress in barley (Kulaeva *et al.*, 1991). A reduction of P_N, g_S and c_i associated with the exposure to high levels of NaCl was partially overcome by spraying of Indian mustard plants with EBL (Ali *et al.*, 2008b). Negative effect of salinity on leaf gas exchange parameters in wheat plants was reported to be ameliorated by exogenous application of EBL as well, but this phenomenon strongly depended both on EBL concentration and cultivar examined (Ali *et al.*, 2008c; Shahbaz *et al.*, 2008). Salinity-caused loss of Chls *a* and *b* was prevented by the incubation of rice seeds in BL, EBL or HBL solutions (Anuradha and Rao, 2003) and similar effect was observed in barley, kidney bean, chickpea or maize (Ali *et al.*, 2007; El-Fattah, 2007; El-Khallal *et al.*, 2009). EBL also alleviated the negative effect of NaCl on the Chl content in Indian mustard when applied as a foliar spray (Ali *et al.*, 2008b). On the other hand, the Chl content in leaves of salt-stressed wheat plants was not significantly affected by EBL treatment (Qayyum *et al.*, 2007; Ali *et al.*, 2008c; Shahbaz *et al.*, 2008). The maximum quantum efficiency of PS II measured in leaves of wheat plants subjected to salinity conditions was not much influenced by EBL application (Ali *et al.*, 2008c), although Shahbaz *et al.* (2008) did observe a slight increase in F_v/F_m parameter due to EBL treatment. This parameter was measured also in salt-stressed pepper plants sprayed with EBL solutions of various concentrations and no changes of its values due to the BR application were found (Houimli *et al.*, 2008).

BRs can also reduce toxic effects of heavy metals on plants, which can be reflected in changes of the photosynthetic efficiency as well. Several authors have examined the effect of cadmium on photosynthetic parameters in BR-treated plants and found that the exogenous application of HBL can to some extent overcome Cd-induced loss of Chls and reduction in carbonic anhydrase activity in leaves of Indian mustard, chickpea and radish (Hayat *et al.*, 2007; Hasan *et al.*, 2008; Anuradha and Rao, 2009). BRs also diminished the damage to reaction centers and oxygen-evolving complexes of PS II, enabled the maintenaince of an efficient photosynthetic electron transport in Cd-stressed winter rape (Janeczko *et al.*, 2005) and alleviated the depression of net photosynthesis caused by cadmium in Indian mustard and radish (Hayat *et al.*, 2007; Anuradha and Rao, 2009). Toxic effects of aluminium on the content of Chls were mitigated in mungbean plants treated with BL, EBL or HBL (Abdullahi *et al.*, 2003; Ali *et al.*, 2008a); the second group of authors also found that both EBL and HBL improved the activity of carbonic anhydrase and the net photosynthetic rate in the leaves of mungbean plants grown in a medium containing Al (Ali *et al.*, 2008a). The reduction of photosynthesis caused by nickel treatment was partially neutralized by spraying of plants with HBL or EBL in plants of Indian mustard (Alam *et al.*, 2007; Ali *et al.*, 2008b) and soaking of seeds in HBL solutions improved photosynthetic parameters in the leaves of the same plant species subjected to copper stress (Fariduddin *et al.*, 2009b).

As regards other types of stressors, Kang *et al.* (2009) reported a positive effect of exogenously applied EBL on P_N in cucumber plants subjected to hypoxic conditions. The reduction in net CO_2 assimilation in the leaves of cucumber caused by herbicide paraquat was not affected much by the application of EBL and the same applied for the stomatal conductance and the maximum quantum efficiency of PS II in the dark- or light-adapted leaves, whereas the decline of the effective quantum yield of PS II was alleviated by EBL (Xia *et al.*, 2006). Another study made with faba bean (*Vicia faba* L.) described the positive effect of EBL on both CO_2 assimilation and chlorophyll fluorescence parameters in plants treated with herbicide terbutryn (Piñol and Simón, 2009). Adverse effects of two other herbicides (fluazifop-p-butyl and haloxyfop) as well as selected fungicides (flusilazole, cuproxat, cyazofamid) and insecticides (imidacloprid, chlorpyrifos and abamectin) on the net photosynthetic rate were alleviated by EBL pre-treatment of cucumber plants and similar phenomenon was observed for the effective quantum yield of PS II and the photochemical quenching of chlorophyll fluorescence (Xia *et al.*, 2006, 2009b). BRs-containing agens ComCat also counteracted down-regulation of the expression of gene coding for Rubisco SSU in tomato plants infected with *Botrytis cinerea* (Berger *et al.*, 2004).

4. PHOTOSYNTHESIS IN BRASSINOSTEROIDS-DEFICIENT MUTANTS OR TRANSGENIC PLANTS

An analysis of photosynthetic parameters in BR-deficient or BR-insensitive mutants or transgenic plants is less common than the work with plants treated with exogenously applied BRs. However, some studies utilizing such mutants brought interesting results relevant to the possible role of BRs in the development of photosynthetic organelles and the effectivity of photosynthetic processes.

BR-deficient mutants of *Arabidopsis thaliana* (L.) Heynh. usually show dwarf phenotype, often have dark-green leaves (due to the reduction of cell volume causing the closer location of chloroplasts) and sometimes exhibit photomorphogenetic development even in the absence of light (Bishop, 2003). From the photosynthetic point of view, the most interesting and the best characterized is the deetiolated2 (*det2*) mutant. DET2 protein catalyzes one of the early steps in BR biosynthesis, *i.e.* the conversion of campesterol to campestanol (Noguchi *et al.*, 1999). When grown in the dark, mutants in this gene display 10- to 20-fold higher levels of mRNA for several photosynthetic proteins (Rubisco LSU and SSU, LHC proteins, D1 protein of PS II reaction center and PSAB protein of PS I reaction center) compared to dark-grown *wild type* (*wt*) plants. An increase in the expression of genes coding for Rubisco LSU and SSU observed in *det2* mutant under dark conditions is reflected also in the amount of Rubisco protein, whereas for LHC proteins, no such relationship was found due to the fact that the accumulation of Chls in dark-grown *det2* plants is inhibited and LHC proteins show rapid turnover in the absence of these pigments (Chory *et al.*, 1991; Nagata *et al.*, 2000). Dark-grown *det2* mutant plants also have condensated and scattered nucleoids (plastid DNA) in their etioplasts, which is a state typical for a more advanced stage of plastid development during differentiation of etioplasts into chloroplasts (Nagata *et al.*, 2000). Under light conditions, the development of chloroplast inner structures in *det2* mutant seems to be delayed which is visible particularly during early stages of plant growth. *det2* mutant also shows a delayed senescence of leaves accompanied by the maintenance of high levels of both Chls and mRNA for LHC proteins during later stages of development (Chory *et al.*, 1991).

Another *Arabidopsis* mutant with a block in the early steps of BR biosynthesis is cabbage1/dwarf4 (*cbb1/dwf4*) (Klahre *et al.*, 1998). Contrary to *det2* mutant, this type of BR-deficient mutant does not differ from *wt* plants in the development of plastids (Azpiroz *et al.*, 1998). Schlüter *et al.* (2002) analyzed activities of several enzymes of Calvin cycle (glyceraldehyde-3-phosphate dehydrogenase, triose-3-phosphate isomerase, fructose-1,6-bisphosphate aldolase, fructose-1,6-bisphosphatase and phosphoribulokinase), as well as the mRNA levels for Rubisco SSU in this mutant and did not find any changes compared to *wt* plants. The same analysis was made also with transgenic plants carrying an antisense construct for the *CPD* gene that normally codes for cytochrome P450 monooxygenase participating in BR biosynthesis (Szekeres *et al.*, 1996). Neither in this case were any differences between transgenic and *wt* plants in the activities of Calvin cycle enzymes or the expression of gene coding for Rubisco SSU found. However, the antisense *CPD* plants were also characterized by their reduced capacity for CO_2 fixation (Schlüter *et al.*, 2002).

Some work has also been done with mutants or transgenic plants of rice. A study made with transgenic rice plants overexpressing *CYP* genes coding for sterol hydroxylases involved in BR biosynthesis reported both increased rate of CO_2 uptake and the maximum quantum efficiency of PS II compared to *wt* plants, but very few changes in the expression of genes involved with CO_2 assimilation (however, one exception was the gene coding for Rubisco SSU, the expression of which was induced in these transgenic plants) (Wu *et al.*, 2008). Some Japanese scientists suggested that very erect leaves of rice mutants defective in *CYP* genes or in the ortholog of *Arabidopsis BRI1* gene (which codes for a kinase functioning as a BR receptor) can increase an amount of light available for photosynthesis in lower leaves by diminishing their shading, thus improving the overall photosynthetic efficiency (Sakamoto *et al.*, 2005; Morinaka *et al.*, 2006). However, they did not actually measure any photosynthetic parameters in their mutants, nor the canopy photosynthesis, so this association between BR-induced changes of plant morphology and photosynthetic light capture still remains rather speculative.

5. THE AGE OF "OMICS": WHAT DID WE LEARN FROM LARGE-SCALE ANALYSES?

The adoption of transcriptomics and proteomics as routine methods of plant biology enabled scientists to further advance their knowledge on BRs effects on plants. Various studies of gene expression patterns based on microarray analyses were made in plants treated with exogenously applied BRs or in BR-deficient mutants and a large number of BR-responsive genes has been identified. Most of these analyses were made with *Arabidopsis*, although other plant species were sometimes examined as well. From the functional view, BR-responsive genes fall into various categories and their products participate in *e.g.* regulation of cell division and elongation, cell wall and cytoskeleton organization, signal transduction (including signaling pathways of other phytohormones), transcription, protein degradation, stress response, BR biosynthesis and metabolism, and other metabolic processes (reviewed *e.g.* by Müssig and Altmann, 2003). Are photosynthesis-related genes also induced or repressed by this group of phytohormones? Some evidence in favour of this does exist, other results offer a negative answer to this question.

Genome-wide analysis of the expression of BR-responsive genes made in EBL-treated *Arabidopsis* plants (Hu *et al.*, 2001) revealed the increased level of mRNA for CHLH subunit of Mg-protoporphyrin IX chelatase, an enzyme participating in Chl biosynthesis (Yaronskaya and Grimm, 2006) and offered the possibility that the expression of gene coding for a subunit of oxygen-evolving complex of PS II could also be affected by EBL application.

Down-regulation of the expression of genes coding for one photosynthetic light-harvesting protein and carbonic anhydrase was noted by Goda *et al.* (2002, 2004), who worked with *Arabidopsis* plants treated with BL. Exogenous application of BL to rice plants stimulated the degradation of Rubisco LSU (Konishi and Komatsu, 2003). Yang *et al.* (2004) identified a putative subunit of PS I (PSI-L) whose mRNA levels were down-regulated by BL. The greatest number of BR-responsive photosynthesis-related genes were identified in the recent study of Song *et al.* (2009) who worked either with BR-deficient *det2* mutant of *Arabidopsis* or with *wt* plants of the same species treated with exogenously applied BL. These authors found that the levels of mRNA representing genes coding for several LHC and ELIP proteins, as well as for PSBS protein (which is an integral part of PS II), one of smaller PS I subunits (PSI-L), ferredoxin (a protein that accepts electrons from PS I reaction center), and Rubisco activase were actually higher in the BR-deficient mutant compared to *wt* plants. Treatment of *wt Arabidopsis* plants with BL also significantly decreased mRNA levels for two LHC proteins and significantly elevated mRNA levels for lycopene ε-cyclase (an enzyme participating in biosynthesis of Cars; Hirschberg, 2001).

However, an almost equal number of large-scale studies aimed at the identification of BR-responsive genes did not reveal any genes that would code for proteins directly (or indirectly, as in the case of proteins participating in the biosynthesis of photosynthetic pigments) participating in photosynthesis (Müssig *et al.*, 2002; Lisso *et al.*, 2005; Deng *et al.*, 2007; Dhaubhaudel and Krishna, 2008; Tang *et al.*, 2008). It is, of course, possible that the rather conspicuous absence of photosynthetic genes in the list of BR-responsive genes presented in microarray or proteomics studies is due not to the actual unresponsiveness of photosynthetic genes to BRs, but to rather stringent selection criteria applied in such analyses. It is also possible that with new, recently developed high-throughput technologies, photosynthesis will be another functional category commonly appearing in these lists. Further dissection of complex signaling pathways BRs seem to participate in (particularly light-signaling and stress response), as well as the application of metabolomics and interactomics technologies would also provide a valuable insight into the role BRs play in the regulation of photosynthesis. Unfortunately, BRs' research is still somewhat lagging behind other groups of phytohormones in the implementation of new technologies for large-scale analyses and we can only hope that this situation will be improved in future. Until then, our knowledge on the association between BRs and photosynthesis will remain mainly at the physiological level.

6. CONCLUSIONS AND FUTURE CHALLENGES

The end of the 20th and particularly the beginning of the 21st centuries brought along an unquestionable evidence that BRs can, and do, affect the photosynthetic efficiency in various plant species. However, although several attempts to resolve the actual relationship between these phytohormones and the functioning of various parts of photosynthetic apparatus have been made, our knowledge on the mode of BRs' action in the regulation of photosynthetic processes is still far from being complete. So far, it seems that the main site of BRs' impact on photosynthesis is probably the photosynthetic carbon reduction cycle and that these compounds could perhaps somehow affect the activation state of Rubisco, the main CO_2 fixing enzyme. We can also speculate about the possible effect of BRs on the activity of carbonic anhydrase which modulates the ratio of inorganic carbon species (CO_2/HCO_3^-) and thus influences the availability of CO_2 for Rubisco. On the other hand, the actual process of CO_2 uptake through stomata makes an unlikely target for BRs' action. The role of BRs in the improvement of the efficiency of photosynthetic light capture and primary photosynthetic processes appears to be a secondary one as well and the same applies for the possible BRs' function in the regulation of the expression of photosynthetic genes. However, all these statements must be taken *cum grano salis*, as the current data on the role of BRs in various parts of photosynthetic processes are still insufficient.

What are the challenges that we should face in order to better elucidate the relationship beteween BRs and photosynthesis? First, a detailed examination of the participation of these phytohormones in the development of photosynthetically active chloroplasts is sorely needed. Second, an analysis of their role in photosynthetic electron transfer, aimed at the dissection of components other than Photosystem II, should be made as well, as almost no data on this topic are currently available. Third, the determination of BRs' influence on *all* enzymes participating in photosynthetic CO_2 fixation (not only in plants with C_3 pathway of carbon fixation but C_4 and perhaps CAM pathway as well) would not be amiss to further clarify whether this part of photosynthetic process really serves as the main target for the action of these hormones. Fourth, a more comprehensive BR structure-photosynthetic activity studies could perhaps provide some reasons why some photosynthetic characteristics (*e.g.* photosynthetic pigments' content) seem to be affected only by specific types of BRs. Fifth, a question why the effect of BRs on photosynthesis is more pronounced in plants subjected to some unfavourable environmental factor demands an answer as well. Sixth, a more frequent utilization of modern methods of molecular and cell biology including various "omics" technologies should enhance our knowledge on BRs' role in the regulation of photosynthesis at the sub-organellar level. A new information on the interconnection between BRs, photosynthesis and light signalling is also necessary to improve our understanding of the complex network of regulationary pathways these phytohormones participate in.

7. ACKNOWLEDGEMENT

I am very grateful to Ladislav Kohout from the Institute of Organic Chemistry and Biochemistry, Academy of Sciences of the Czech Republic, who introduced me to the fascinating world of brassinosteroids, and to Shamsul Hayat from the Aligarth Muslim University, India, for the opportunity to participate in the creation of this book. Part of my own research on the role of brassinosteroids in photosynthesis was supported by grant No. MSM0021620858 from the Ministry of Education, Youth and Sports of the Czech Republic. I also apologize to all scientists working in the field of brassinosteroids and photosynthesis, whose results I have unintentionally missed during writing of this chapter.

8. REFERENCES

Abdullahi, B.A., Gu, X.G., Gan, Q.L., and Yang, Y.H., 2003. Brassinolide amelioration of aluminium toxicity in mungbean seedling growth. *J. Plant Nutr.* **26**: 1725–1734.

Alam, M.M., Hayat, S., Ali, B., and Ahmad, A., 2007. Effect of 28-homobrassinolide treatment on nickel toxicity in *Brassica juncea*. *Photosynthetica* **45**: 139–142.

Ali, B., Hasan, S.A., Hayat, S., Hayat, Q., Yadav, S., Fariduddin, Q., and Ahmad, A., 2008a. A role for brassinosteroids in the amelioration of aluminium stress through antioxidant system in mung bean (*Vigna radiata* L. Wilczek). *Environ. Exp. Bot.* **62**: 153–159.

Ali, B., Hayat, S., and Ahmad, A., 2007. 28-Homobrassinolide ameliorates the saline stress in chickpea (*Cicer arietinum* L.). *Environ. Exp. Bot.* **59**: 217–223.

Ali, B., Hayat, S., Fariduddin, Q., and Ahmad, A., 2008b. 28-Epibrassinolide protects against the stress generated by salinity and nickel in *Brassica juncea*. *Chemosphere* **72**: 1387–1392.

Ali, B., Hayat, S., Hasan, S.A., and Ahmad, A., 2006. Effect of root applied 28-homobrassinolide on the performance of *Lycopersicon esculentum*. *Sci. Hortic.* **110**: 267–273.

Ali, Q., Athar, H.R., and Ashraf, M., 2008c. Modulation of growth, photosynthetic capacity and water relations in salt stressed wheat plants by exogenously applied 24-epibrassinolide. *Plant Growth Regul.* **56**: 107–116.

Anuradha, A., and Rao, S.S.R., 2003. Application of brassinosteroids to rice seeds (*Oryza sativa* L.) reduced the impact of salt stress on growth, prevented photosynthetic pigment loss and increased nitrate reductase activity. *Plant Growth Regul.* **40**: 29–32.

Anuradha, S., and Rao, S.S.R., 2009. Effect of 24-epibrassinolide on the photosynthetic activity of radish plants under cadmium stress. *Photosynthetica* **47**: 317–320.

Asami, T., Mina, Y.K., Nagata, N., Yamagishi, K., Takatsuto, S., Fujioka, S., Murofushi, N., Yamaguchi, I., and Yoshida, S., 2000. Characterization of brassinazole, a triazole-type brassinosteroid biosynthesis inhibitor. *Plant Physiol.* **123**: 93–99.

Azpiroz, R., Wu, Y., LoCascio, J.C., and Feldmann, K.A., 1998. An *Arabidopsis* brassinosteroid-dependent mutant is blocked in cell elongation. *Plant Cell* **10**: 219–230.

Bajguz A., 2009. Brassinosteroid enhanced the level of abscisic acid in *Chlorella vulgaris* subjected to short-term heat stress. *J. Plant Physiol.* **166**: 882–886.

Bajguz, A., and Asami, T., 2005. Suppression of *Wolffia arrhiza* growth by brassinazole, an inhibitor of brassinosteroid biosynthesis and its restoration by endogenous 24-epibrassinolide. *Phytochemistry* **66**: 1787–1796.

Bajguz, A., and Czerpak, R., 1998. Physiological and biochemical role of brassinosteroids and their structure-activity relationship in the green alga *Chlorella vulgaris* Beijerinck (*Chlorophyceae*). *J. Plant Growth Regul.* **17**: 131–139.

Bajguz, A., and Hayat, S., 2009. Effect of brassinosteroids on the plant responses to environmental stresses. *Plant Physiol. Biochem.* **47**: 1–8.

Baker, N.R., 2008. Chlorophyll fluorescence: A probe of photosynthesis *in vivo. Annu. Rev. Plant Biol.* **59**: 89–113.

Baker, N.R., and Rosenqvist, E., 2004. Applications of chlorophyll fluorescence can improve crop production strategies: An examination of future possibilities. *J. Exp. Bot.* **55**: 1607–1621.

Behnamnia, M., Kalantari, K.M., and Rezanejad, F., 2009. Exogenous application of brassinosteroid alleviates drought-induced oxidative stress in *Lycopersicon esculentum* L. *Gen. Appl. Plant Physiol.* **35**: 22–34.

Berger, S., Papadopoulos, M., Schreiber, U., Kaiser, W., and Roitsch, T., 2004. Complex regulation of gene expression, photosynthesis and sugar levels by pathogen infection in tomato. *Physiol. Plant.* **122**: 419–428.

Bishop, G.J., 2003. Brassinosteroid mutants of crops. *J. Plant Growth Regul.* **22**: 325–335.

Bhatia, D.S., and Kaur, J., 1997. Effect of homobrassinolide and humicil on chlorophyll content, Hill activity and yield components in mungbean (*Vigna radiata* (L.) Wilczek). *Phytomorphology* **47**: 421–426.

Braun, P., and Wild, A., 1984. The influence of brassinosteroid on growth and parameters of photosynthesis of wheat and mustard plants. *J. Plant Physiol.* **116**: 189–196.

Çağ, S., Gören-Sağlam, N., Çıngıl-Barış Ç., and Kaplan, E., 2007. The effect of different concentration of epibrassinolide on chlorophyll, protein and anthocyanin content and peroxidase activity in excised red cabbage (*Brassica oleracea* L.) cotyledons. *Biotechnol. Biotechnol. Eq.* **21**: 422–425.

Castle, J., Montoya, T., and Bishop, G.J., 2003. Selected physiological responses of brassinosteroids: a historical approach. In: S Hayat and A Ahmad (Eds.) Brassinosteroids. Bioactivity and Crop Productivity. *Kluwer Academic Publishers, Dordrecht-Boston-London*. pp. 45–68.

Cevahir, G., Yentür, S., Eryilmaz, F., and Yilmazer, N., 2008. Influence of brassinosteroids on pigment content of *Glycine max* L. (soybean) grown in dark and light. *J. Appl. Biol. Sci.* **2**: 23–28.

Chory, J., Nagpal, P., and Peto, C.A., 1991. Phenotypic and genetic analysis of *det2*, a new mutant that affects light-regulated seedling development in *Arabidopsis*. *Plant Cell* **3**: 445–459.

Clouse, S.D., and Sasse, J.M., 1998. Brassinosteroids: Essential regulators of plant growth and development. *Annu Rev. Plant Physiol. Plant Mol. Biol.* **49**: 427–451.

Coleman, J.R., 2000. Carbonic anhydrase and its role in photosynthesis. In: RC Leegood, TD Sharkey, and S von Caemmerer (Eds) Photosynthesis: Physiology and Metabolism. *Kluwer Academic Publishers, Dordrecht*. pp. 353–367.

Dahse, I., Petzold, U., Willmer, C.M., and Grimm, E., 1991. Brassinosteroid-induced changes of plasmalemma energization and transport and of assimilate uptake by plant tissues. In: HG Cutler, T Yokota, and G. Adam (Eds.) Brassinosteroids. Chemistry, Bioactivity, and Applications. *American Chemical Society, Washington, DC*. pp. 167–175.

Deng, Z., Zhang, X., Tang, W., Oses-Prieto, J.A., Suzuki, N., Gendron, J.M., Chen, H., Guan, S., Chalkley, R.J., Peterman, T.K., Burlingame, A.L., and Wang, Z.Y., 2007. A proteomics study of brassinosteroid response in *Arabidopsis*. *Mol. Cell. Proteomics* **6**: 2058–2071.

Dhaubhaudel, S., and Krishna, P., 2008. Identification of differentially expressed genes in brassinosteroid-treated *Brassica napus* seedlings. *J. Plant Growth Regul.* **27**: 297–308.

Divi, U.K., and Krishna, P., 2009. Brassinosteroids: A biotechnological target for enhancing crop yield and stress tolerance. *New Biotechnol.* **26**: 131–136.

El-Fattah, R.I.A., 2007. Osmolytes-antioxidant behaviour in *Phaseolus vulgaris* and *Hordeum vulgare* with brassinosteroid under salt stress. *Am-Euras. J. Agric. Environ. Sci.* **2**: 639–647.

El-Khallal, S.M., Hathout, T.A., Ashour, A.A., and Kerrit, A.A., 2009. Brassinolide and salicylic acid induced growth, biochemical activities and productivity of maize plants grown under salt stress. *Res. J. Agric. Biol. Sci.* **5**: 380–390.

Fariduddin, Q., Ahmad, A., and Hayat, S., 2003. Photosynthetic response of *Vigna radiata* to pre-sowing seed treatment with 28-homobrassinolide. *Photosynthetica* **41**: 307–310.

Fariduddin, Q., Ahmad, A., and Hayat, S., 2004. Responses of *Vigna radiata* to foliar application of 28-homobrassinolide and kinetin. *Biol. Plant.* **48**: 465–468.

Fariduddin, Q., Hasan, S.A., Ali, B., Hayat, S., and Ahmad, A., 2008. Effect of modes of application of 28-homobrassinolide on mung bean. *Turk. J. Biol.* **32**: 17–21.

Fariduddin, Q., Hayat, S., Ali, B., and Ahmad, A., 2006. Effect of 28-homobrassinolide on the nitrate reductase, carbonic anhydrase activities and net photosynthetic rate in *Vigna radiata*. *Acta Bot. Croat.* **65**: 19–23.

Fariduddin, Q., Khanam, S., Hasan, S.A., Ali, B., Hayat, S., and Ahmad, A., 2009a. Effect of 28-homobrassinolide on the drought stress-induced changes in photosynthesis and antioxidant system of *Brassica juncea* L. *Acta Physiol. Plant.* **31**: 889–897.

Fariduddin, Q., Yusuf, M., Hayat, S., and Ahmad, A., 2009b. Effect of 28-homobrassinolide on antioxidant capacity and photosynthesis in *Brassica juncea* plants exposed to different levels of copper. *Environ. Exp. Bot.* **66**: 418–424.

Farooq, M., Wahid, A., Basra, S.M.A., and Din, I., 2009. Improving water relations and gas exchange with brassinosteroids in rice under drought stress. *J. Agron. Crop Sci.* **195**: 262–269.

Farquhar, G.D., and Sharkey, T.D., 1982. Stomatal conductance and photosynthesis. *Annu. Rev. Plant Physiol.* **33**: 317–345.

Farquhar, G.D., von Caemmerer, S., and Berry, J.A., 1980. A biochemical model of photosynthetic CO_2 assimilation in leaves of C3 species. *Planta* **149**: 78–90.

Fedina, E.O., Karimova, F.G., Tarchevsky, I.A., Toropygin, I.Y., and Khripach, V.A., 2008. Effect of epibrassinolide on tyrosine phosphorylation of the Calvin cycle enzymes. *Russ. J. Plant Physiol.* **55**: 193–200.

Fujii, S., Hirai, K., and Saka, H., 1991. Growth-regulating action of brassinolide in rice plants. In: HG Cutler, T Yokota, and G. Adam (Eds.) Brassinosteroids. Chemistry, Bioactivity, and Applications. *American Chemical Society, Washington, DC*. pp. 306–311.

Garrido, J.L., and Zapata, M., 2006. Chlorophyll analysis by new high performance liquid chromatography methods. In: B Grimm, RJ Porra, W Rüdiger, and H Scheer (Eds). Chlorophylls and Bacteriochlorophylls. Biochemistry, Biophysics, Functions and Applications. *Springer, Dordrecht*. pp. 109–121.

Gitelson, A.A., Gritz, Y., and Merzlyak, M.N., 2003. Relationships between leaf chlorophyll content and spectral reflectance and algorithms for non-destructive chlorophyll assessment in higher plant leaves. *J. Plant Physiol.* **160**: 271–282.

Goda, H., Sawa, S., Asami, A., Fujioka, S., Shimada, Y., and Yoshida, S., 2004. Comprehensive comparison of auxin-regulated and brassinosteroid-regulated genes in *Arabidopsis*. *Plant Physiol.* **134**: 1555–1573.

Goda, H., Shimada, Y., Asami, T., Fujioka, S., and Yoshida, S., 2002. Microarray analysis of brassinosteroid-regulated genes in *Arabidopsis*. *Plant Physiol.* **130**: 1319–1334.

Gonzáles-Olmedo, J.L., Córdova, A., Aragón, C.E., Pina, D., Rivas, M., and Rodríguez, R., 2005. Effect of an analogue of brassinosteroid on FHIA-18 plantlets exposed to thermal stress. *InfoMusa* **14**: 18–20.

Grove, M.D., Spencer, G.F., Rohwedder, W.K., Mandava, N., Worley, J.F., Warthen, J.D., Jr., Steffens, G.L., Flippen-Anderson, J.L., and Cook, J.C., Jr., 1979. Brassinolide, a plant growth-promoting steroid isolated from *Brassica napus* pollen. *Nature* **281**: 216–217.

Hasan, S.A., Hayat, S., Ali, B., and Ahmad, A., 2008. 28-Homobrassinolide protects chickpea (*Cicer arietinum*) from cadmium toxicity by stimulating antioxidants. *Environ. Pollut.* **151**: 60–66.

Haubrick, L.L., and Assmann, S.M., 2006. Brassinosteroids and plant function: some clues, more puzzles. *Plant Cell Environ.* **29**: 446–457.

Hayat, S., Ahmad, A., and Fariduddin, Q., 2003. Brassinosteroids: a regulator of 21st century. *In:* S Hayat and A Ahmad (Eds.) Brassinosteroids. Bioactivity and Crop Productivity. *Kluwer Academic Publishers, Dordrecht-Boston-London*. pp. 231–246.

Hayat, S., Ahmad, A., Hussain, A., and Mobin, M., 2001a. Growth of wheat seedlings raised from grains treated with 28-homobrassinolide. *Acta Physiol. Plant.* **23**: 27–30.

Hayat, S., Ahmad, A., Mobin, M., Fariduddin, Q., and Azam, Z.M., 2001b. Carbonic anhydrase, photosynthesis, and seed yield in mustard plants treated with phytohormones. *Photosynthetica* **39**: 111–114.

Hayat, S., Ahmad, A., Mobin, M., Hussain, A., and Fariduddin, Q., 2000. Photosynthetic rate, growth, and yield of mustard plants sprayed with 28-homobrassinolide. *Photosynthetica* **38**: 469–471.

Hayat, S., Ali, B., Hasan, S.A., and Ahmad, A., 2007. Brassinosteroid enhanced the level of antioxidants under cadmium stress in *Brassica juncea*. *Environ. Exp. Bot.* **60**: 33–41.

He, R., Wang, G.J., and Wang, X.S., 1991. Effects of brassinolide on growth and chilling resistance of maize seedlings. In*:* HG Cutler, T Yokota, and G. Adam (Eds.) Brassinosteroids. Chemistry, Bioactivity, and Applications. *American Chemical Society, Washington, DC*. pp. 220–230.

Hirschberg, J., 2001. Carotenoid biosynthesis in flowering plants. *Curr. Opin. Plant Biol.* **4**: 210–218.

Honnerová, J., Rothová O., Holá, D., Kočová, M., Kohout, L., and Kvasnica, M., 2010 The exogenous application of brassinosteroids to *Zea mays* (L.) stressed by long-term chilling does not affect the activities of photosystem 1 or 2. *J. Plant Growth Regul.* doi.10.1007/s00344-010-9153-0.

Houimli, S.M., Denden, M., and El Hadj, S.B., 2008. Induction of salt tolerance in pepper (*Capsicum annuum*) by 24-epibrassinolide. *Eur. Asia J. Bio. Sci.* **2**: 83–90.

Hradecká, D., Urban, J., Kohout, L., Pulkrábek, J., and Hnilička, R., 2009. Utilization of brassinosteroids to stress control during growth and yield formation of sugar beet. *Listy cukrovarnické a řepařské* **125**: 271–273. [*In Czech; English abstract*].

Hu, Y., Wang, Z., Wang, Y., Bao, F., Li, N., Peng, Z., and Li, J., 2001. Identification of brassinosteroid responsive genes in *Arabidopsis* by cDNA array. *Science in China (Series C)* **44**: 637–643.

Janeczko, A., Gullner, G., Skoczowski, A., Dubert, F., and Barna, B., 2007. Effect of brassinosteroid infiltration prior to cold treatment on ion leakage and pigment contents in rape leaves. *Biol. Plant.* **51**: 355–358.

Janeczko, A., Kościelniak, J., Pilipowicz, M., Szarek-Łukaszewska, G., and Skoczowski, A., 2005. Protection of winter rape photosystem 2 by 24-epibrassinolide under cadmium stress. *Photosynthetica* **43**: 293–298.

Kang, Y.Y., Guo, S.R., Li, J., and Duan, J.J., 2009. Effect of root applied 24-epibrassinolide on carbohydrate status and fermentative enzyme activities in cucumber (*Cucumis sativus* L.) seedlings under hypoxia. *Plant Growth Regul.* **57**: 259–269.

Katsumi, M., 1991. Physiological modes of brassinolide action in cucumber hypocotyl growth. In*:* HG Cutler, T Yokota, and G. Adam (Eds.) Brassinosteroids. Chemistry, Bioactivity, and Applications. *American Chemical Society, Washington, DC.* pp. 246–254.

Khripach, V.A., Zhabinskii, V.N., and de Groot, A.E., 1999. Physiological mode of action of BS. *In*: VA Khripach, VN Zhabinskii, and AE de Groot (Eds.) Brassinosteroids. A New Class of Plant Hormones. *Academic Press, San Diego-London-Boston-New York-Sydney-Tokyo-Toronto*. pp. 219–299.

Khripach, V., Zhabinskii, V., and de Groot, A., 2000. Twenty years of brassinosteroids: Steroidal plant hormones warrant better crops for the XXI century. *Ann. Bot.* **86**: 441–447.

Klahre, U., Noguchi, T., Fujioka, S., Takatsuto, S., Yokota, T., Nomura, T., Yoshida, S., and Chua N.H., 1998. The *Arabidopsis DIMINUTO/DWARF1* gene encodes a protein involved in steroid synthesis. *Plant Cell* **10**: 1677–1690.

Kočová, M., Rothová, O., Holá, D., Kvasnica, M., Kohout, L., 2010. The exogenous application of brassinosteroids does not affect photosynthetic parameters in leaves of two field-grown maize inbred lines or their F1 hybrid. *Biol. Plant.*, **54**: 785–788.

Konishi, H., and Komatsu, S., 2003. A proteomics approach to investigating promotive effects of brassinolide on lamina inclination and root growth in rice seedlings. *Biol. Pharm. Bull.* **26**: 401–408.

Krishna, P., 2003. Brassinosteroid-mediated stress responses. *J. Plant Growth Regul.* **22**: 289–297.

Krizek, D.T., and Mandava, N.B., 1983. Influence of spectral quality on the growth response of intact bean plants to brassinosteroid, a growth-promoting steroidal lactone. II. Chlorophyll content and partitioning of assimilate. *Physiol. Plant.* **57**: 324–329.

Kulaeva, O.N., Burkhanova, E.A., Fedina, A.B., Khokhlova, V.A., Bokebayeva, G.A., Vorbrodt, H.M., and Adam, G., 1991. Effect of brassinosteroide on protein synthesis and plant-cell ultrastructure under stress conditions. *In:* HG Cutler, T Yokota, and G. Adam (Eds.) Brassinosteroids. Chemistry, Bioactivity, and Applications. *American Chemical Society, Washington, DC.* pp. 141–155.

Lisso, J., Steinhauser, D., Altmann, T., Kopka J., and Müssig, C., 2005. Identification of brassinosteroid-related genes by means of transcript co-response analyses. *Nucleic Acids Res.* **33**: 2685–2696.

Long, S.P., and Bernacchi, C.J., 2003. Gas exchange measurements, what can they tell us about the underlying limitations to photosynthesis? Procedures and sources of error. *J. Exp. Bot.* **54**: 2393–2401.

Mandava, N.B., Sasse, J.M., and Yopp, J.H., 1981. Brassinolide, a growth-promoting steroidal lactone. II. Activity in selected gibberellin and cytokinin bioassays. *Physiol. Plant.* **53**: 453–461.

Maxwell, K., and Johnson, G.N., 2000. Chlorophyll fluorescence – a practical guide. *J. Exp. Bot.* **51**: 659–668.

Melkozernov, A.N., and Blankenship, R.E., 2006. Photosynthetic functions of chlorophylls. *In*: B Grimm, RJ Porra, W Rüdiger, and H Scheer (Eds). Chlorophylls and Bacteriochlorophylls. Biochemistry, Biophysics, Functions and Applications. *Springer, Dordrecht.* pp. 397–412.

Morinaka, Y., Sakamoto, T., Inukai, Y., Agetsuma, M., Kitano, H., Ashikari, M., and Matsuoka, M., 2006. Morphological alteration caused by brassinosteroid insensitivity increases the biomass and grain production of rice. *Plant Physiol.* **141**: 924–931.

Müssig, C., 2005. Brassinosteroid-promoted growth. *Plant Biol.* **7**: 110–117.

Müssig, C., and Altmann, T., 2003. Genomic brassinosteroid effects. *J. Plant Growth Regul.* **22**: 313–324.

Müssig, C., Fischer, S., and Altmann, T., 2002. Brassinosteroid-regulated gene expression. *Plant Physiol.* **129**: 1241–1251.

Nagata, N., Min, Y.K., Nakano, T., Asami, T., and Yoshida, S., 2000. Treatment of dark-grown *Arabidopsis thaliana* with a brassinosteroid-biosynthesis inhibitor, brassinazole, induces some characteristics of light-grown plants. *Planta* **211**: 781–790.

Nakajima N., and Toyama, S., 1995. Study on brassinosteroid-enhanced sugar accumulation in cucumber epicotyls. *Jpn. J. Crop. Sci.* **64**: 616–621.

Noguchi, T., Fujioka, S., Takatsuto, S., Sakurai, A., Yoshida, S., Li, J., and Chory, J., 1999. *Arabidopsis det2* is defective in the conversion of (24 R)-24-methylcholest-4-en-3-one to (24 R)-24-methyl-5a-cholestan-3-one in brassinosteroid biosynthesis. *Plant Physiol.* **120**: 833–839.

Ogweno, J.O., Song, X.S., Shi, K., Hu, W.H., Mao, W.H., Zhou, Y.H., You, J.Q., and Nogués, S., 2008. Brassinosteroids alleviate heat-induced inhibition of photosynthesis by increasing carboxylation efficiency and enhancing antioxidant systems in *Lycopersicon esculentum*. *J. Plant Growth Regul.* **27**: 49–57.

Papageorgiou, G.C., and Govindjee (Eds), 2004. Chlorophyll *a* Fluorescence. A Signature of Photosynthesis. *Springer, Dordrecht.*

Piñol, R., and Simón, E., 2009. Effect of 24-epibrassinolide on chlorophyll fluorescence and photosynthetic CO_2 assimilation in *Vicia faba* plants treated with the photosynthesis-inhibiting herbicide terbutryn. *J. Plant Growth Regul.* **28**: 97–105.

Pocock, T., Król, M., and Huner, N.P.A., 2004. The determination and quantification of photosynthetic pigments by reverse phase high-performance liquid chromatography, thin-layer chromatography, and spectrophotometry. *In*: R Carpentier (Ed.) Photosynthesis Research Protocols (Methods in Molecular Biology, Vol. 274). *Humana Press, Totowa NJ.* pp. 137–148.

Porra, R.J., 2006. Spectrometric assays for plant, algal and bacterial chlorophylls. *In*: B Grimm, RJ Porra, W Rüdiger, and H Scheer (Eds). Chlorophylls and Bacteriochlorophylls. Biochemistry, Biophysics, Functions and Applications. *Springer, Dordrecht.* pp. 95–107.

Qayyum, B., Shahbaz, M., and Akram, N.A., 2007. Interactive effect of foliar application of 24-epibrassinolide and root zone salinity on morpho-physiological attributes of wheat (*Triticum aestivum* L.). *Int. J. Agri. Biol.* **9**: 584–589.

Rajasekaran, L.R., and Blake, T.J., 1999. New plant growth regulators protect photosynthesis and enhance growth under drought of jack pine seedlings. *J. Plant Growth Regul.* **18**: 175–181.

Richardson, A.D., Duigan, S.P., and Berlyn, G.P., 2002. An evaluation of noninvasive methods to estimate foliar chlorophyll content. *New Phytol.* **153**: 185–194.

Roy, H., and Andrews, T.J., 2000. Rubisco: Assembly and mechanism. *In*: RC Leegood, TD Sharkey, and S von Caemmerer (Eds) Photosynthesis: Physiology and Metabolism. *Kluwer Academic Publishers., Dordrecht.* pp. 53–83.

Sağlam-Çağ, S., 2007. The effect of epibrassinolide on senescence in wheat leaves. *Biotechnol. Biotechnol. Eq.* **21**: 63–65.

Sairam, R.K., 1994a. Effect of homobrassinolide application on metabolic activity and grain yield of wheat under normal and water-stress condition. *J. Agron. Crop Sci.* **173**: 11–16.

Sairam, R.K., 1994b. Effects of homobrassinolide application on plant metabolism and grain yield under irrigated and moisture-stress conditions of two wheat varieties. *Plant Growth Regul.* **14**: 173–181.

Sakamoto, T., Morinaka, Y., Ohnishi, T., Sunohara, H., Fujioka, S., Ueguchi-Tanaka, M., Mizutani, M., Sakata, K., Takatsuto, S., Yoshida, S., Tanaka, H., Kitano, H., and Matsuoka, M., 2005. Erect leaves caused by brassinosteroid deficiency increase biomass production and grain yield in rice. *Nature Biotechnol.* **24**: 105–109.

Sam, O., Núñez, M., Ruiz-Sánchez, M.C., Dell'Amico, J., Falcón, V., De La Rosa, M.C., and Seoane, J., 2001. Effect of a brassinosteroid analogue and high temperature stress on leaf ultrastructure of *Lycopersicon esculentum*. *Biol. Plant.* **44**: 213–218.

Sasse, J., 1999. Physiological actions of brassinosteroids. *In:* A Sakurai, T Yokota, and SD Clouse (Eds.) Brassinosteroids. Steroidal Plant Hormones. *Springer, Tokyo-Berlin-Heidelberg-New York.* pp. 137–161.

Sasse, J.M., 2003. Physiological actions of brassinosteroids: An update. *J. Plant Growth Regul.* **22**: 276–288.

Schlüter, U., Köpke, D., Altmann, T., and Müssig, C., 2002. Analysis of carbohydrate metabolism of *CPD* antisense plants and the brassinosteroid-deficient *cbb1* mutant. *Plant Cell Environ.* **25**: 783–791.

Sekimata, K., Kimura, T., Kaneko, I., Nakano, T., Yoneyama, K., Takeuchi, Y., Yoshida, S., and Asami, T., 2001. A specific brassinosteroid biosynthesis inhibitor, Brz2001: evaluation of its effects on *Arabidopsis*, cress, tobacco, and rice. *Planta* **213**: 716–721.

Shahbaz, M., Ashraf, M., and Athar, H.R., 2008. Does exogenous application of 24-epibrassinolide ameliorate salt induced growth inhibition in wheat (*Triticum aestivum* L.)? *Plant Growth Regul.* **55**: 51–61.

Singh, I., and Shono, M., 2005. Physiological and molecular effects of 24-epibrassinolide, a brassinosteroid on thermotolerance of tomato. *Plant Growth Regul.* **47**: 111–119.

Song, L., Zhou, X.Y., Li, L., Xue, L.J., Yang, X., and Xue, H.W., 2009. Genome-wide analysis revealed the complex regulatory network of brassinosteroid effects in photo-morphogenesis. *Mol. Plant* **2**: 755–772.

Staehelin, L.A., 2003. Chloroplast structure: From chlorophyll granules to supra-molecular architecture of thylakoid membranes. *Photosynth. Res.* **76**: 185–196.

Swamy, K.N., and Rao, S.S.R., 2008. Influence of 28-homobrassinolide on growth, photosynthesis metabolite and essential oil content of geranium (*Pelargonium graveolens* (L.) Herit). *Amer. J. Plant Physiol.* **3**: 173–179.

Swamy, K.N., and Rao, S.S.R., 2009. Effect of 24-epibrassinolide on growth, photosynthesis, and essential oil content of *Pelargonium graveolens* (L.) Herit. *Russ. J. Plant Physiol.* **56**: 616–620.

Szekeres, M., Németh, K., Koncz-Kálmán, Z., Mathur, J., Kauschmann, A., Altmann, T., Rédei, G.P., Nagy, F., Schell J., and Koncz C., 1996. Brassinosteroids rescue the deficiency of

CYP90, a cytochrome P450, controlling cell elongation and de-etiolation in *Arabidopsis*. *Cell* **85**: 171–182.

Tang, W., Deng, Z., Oses-Prieto, J.A., Suzuki, N., Zhu, S., Zhang, X., Burlingame, A.L., and Wang, Z.Y., 2008. Proteomics studies of brassinosteroid signal transduction using prefractionation and two-dimensional DIGE. *Mol. Cell. Proteomics* **7**: 728–738.

von Caemmerer, S., and Quick, W.P., 2000. Rubisco: Physiology *in vivo*. *In*: RC Leegood, TD Sharkey, and S von Caemmerer (Eds) Photosynthesis: Physiology and Metabolism. *Kluwer Academic Publishers, Dordrecht*. pp. 85–113.

Wu, C., Trieu, A., Radhakrishnan, P., Kwok, S.F., Harris, S., Zhang, K., Wang, J., Wan, J., Zhai, H., Takatsuto, S., Matsumoto, S., Fujioka, S., Feldmann, K.A., and Pennell, R.I., 2008. Brassinosteroids regulate grain filling in rice. *Plant Cell* **20**: 2130–2145.

Xia, X.J., Huang, L.F., Zhou, Y.H., Mao, W.H., Shi, K., Wu, J.X., Asami, T., Chen, Z., and Yu, J.Q., 2009a. Brassinosteroids promote photosynthesis and growth by enhancing activation of Rubisco and expression of photosynthetic genes in *Cucumis sativus*. *Planta* **230**: 1185–1196.

Xia, X.J., Huang, Y.Y., Wang, L., Huang, L.F., Yu, Y.L., Zhou, Y.H., and Yu, J.Q., 2006. Pesticides-induced depression of photosynthesis was alleviated by 24-epibrassinolide pretreatment in *Cucumis sativus* L. *Pesticide Biochem. Physiol.* **86**: 42–48.

Xia, X.J., Zhang, Y., Wu, J.X., Wang, J.T., Zhou, Y.H., Shi, K., Yu, Y.L., and Yu, J.Q., 2009b. Brassinosteroids promote metabolism of pesticides in cucumber. *J. Agric. Food Chem.* **57**: 8406–8413.

Yang, G.X., Jan, A., Shen, S.H., Yazaki, J., Ishikawa, M., Shimatani, Z., Kishimoto, N., Kikuchi, S., Matsumoto, H., and Komatsu, S., 2004. Microarray analysis of brassinosteroids- and gibberellin-regulated gene expression in rice seedlings. *Mol. Gen. Genomics* **271**: 468–478.

Yaronskaya, E., and Grimm, B., 2006. The pathway from 5-aminolevulinic acid to protochlorophyllide and protoheme. *In*: B Grimm, RJ Porra, W Rüdiger, and H Scheer (Eds). Chlorophylls and Bacteriochlorophylls. Biochemistry, Biophysics, Functions and Applications. *Springer, Dordrecht*. pp. 173–188.

Yu, J.Q., Huang, L.F., Hu, W.H., Zhou, Y.H., Mao, W.H., Ye, S.F., and Nogués, S., 2004. A role for brassinosteroids in the regulation of photosynthesis in *Cucumis sativus*. *J. Exp. Bot.* **55**: 1135–1143.

Zhang, M., Zhai, Z., Tian, X., Duan, L., and Li, Z., 2008. Brassinolide alleviated the adverse effect of water deficits on photosynthesis and the antioxidant of soybean (*Glycine max* L.). *Plant Growth Regul.* **56**: 257–264.

Chapter 7

PHYSIOLOGICAL EFFECTS RELATED TO BRASSINOSTEROID APPLICATION IN PLANTS

M.M.A. GOMES

Instituto Superior de Tecnologia em, Ciências Agrárias/Fundação de Apoio à Escola, Técnica do Estado do Rio de Janeiro (FAETEC), Av. Wilson Batista, s/n, Parque Aldeia, Campos dos Goytacazes-RJ, CEP28070-620, Brazil*; e-mail: maramag@ig.com.br*

Abstract: Brassinosteroids are plant hormones whose functions have been discovered in the past years. In order to confirm scientifically the biological effects caused exclusively by these compounds, different tools can be used, such as BR-deficient or BR-perceptive mutants, molecular studies, biological assays, application of brassinosteroid biosynthesis inhibitors, endogenous quantification and exogenous application. This work aims at relating the physiological effects in plants when exposed to different dosages and analogues of brassinosteroids during different phases of development (germination, flowering, fructification) and when submitted to biotic and abiotic stress (pathogens, water stress, saline stress, hypoxia, temperature, heavy metals and pesticides) as well as the particularities related to tropisms, circadian rhythms and interactions with other plant hormones. The use of brassinosteroids with the objective of increasing crop yield in the field and to improve the quality of the seedlings has also received attention in recent papers. The main objective of this chapter is to discuss the physiological effects that occur in cells, tissue or whole plants when submitted to brassinosteroid applications, taking into account the possible mechanism of action of these compounds and their practical use in agriculture, describing the analogues and the dosages used in field and laboratory experiments during the last 10 years.

Key words: brassinosteroid concentration, brassinosteroid analogue, germination, flowering, plant tissue culture, plant stress

Abbreviations: BR-brassinosteroid, BL-brassinolide, EBL-epibrassinolide, HBL-homobrassinolide, POD-peroxidase, CAT-catalase, SOD-superoxide dismutase

S. Hayat and A. Ahmad (eds.), *Brassinosteroids: A Class of Plant Hormone*,
DOI 10.1007/978-94-007-0189-2_7,

1. INTRODUCTION

This chapter aims at providing information for researchers that work with exogenous brassinosteroid application. The use of this compound has to be evaluated in order to establish how plants function when BRs are present or to discover the metabolic route involved in a plant's response to hormones or even to prevent toxic effects in plants caused by a variety of stressing situations. Sometimes, the BR concentration used in a plant does not apply to other plants or the developing stage varies the responsiveness dose. When this compound is applied exogenously, it should have a practical use in agriculture as well as in plant propagation for instance, as a constituent of plant culture medium or to establish seedlings in the field or seed priming. The use of exogenous BRs to alleviate plant stress can not be forgotten as it can be used as an antistress agent in a wide range of biotical and abiotic stress conditions (heavy metals, herbicides, saline stress, and drought stress). There is also some information on the increase of plant yield by brassinosteroids application but this is not yet commercially viable.

The use of exogenous BRs to study plant metabolism is desirable but it has to be taken into account that the concentrations used are not always compatible with the endogenous concentrations that can actually be found in plants, so the effects may be confusing. The use of BR plant mutants or endogenous quantification are more appropriate to verify gene expression or changes in plant metabolism caused by BRs because the concentrations applied to explain biological effects may suppress, repress or super express some genes.

There are also problems relating to hormone volatilization if it is spray-applied, the use of surfactants also should be considered to promote best adherence to plant surface. The time of the day and the quality of light should be taken into account as BRs sometimes act in darkness but not under light conditions or, on the contrary, it works in light but not in dark conditions depending on the expected response, so it is very important to decide which time of the day it will be applied. The moment of application is also fundamental because studies have proved that there are significant differences if BR is applied before, after or at the same time of the stressing agent. If it is applied before the stressing agent, it can protect the plant from this agent but, on the other hand, when applied after this external factor, the effects can be deleterious or vice-versa. Besides, BRs endogenous concentrations vary during the day and also according to the plant ontogeny, which causes significant differences in endogenous concentrations among plant organs and plant species. Few plants in Plant Kingdom were tested for BRs, the most common results being for plants of the *Brassica* species, *Arabidopsis thaliana* and some crops with agricultural interests.

We have selected here a series of studies that took place in the last decade in which BRs application was used as a tool to explain plant's behavior to this hormone. Although the BRs concentrations used are commonly very low, there were detected slight differences among the responsiveness even when a varied series of low concentrations get the same response. So, it is important to take into account the concentrations and the mode of application that other researchers have used to improve the desirable results. This work is dedicated to researchers that believe that this plant hormone is responsible for a lot of plant responses and that it can be used in crops in several situations but, at the same time, they know that there are still gaps to be filled in.

2. BRASSINOSTEROID APPLICATION TRIGGERS SEVERAL RESPONSES RELATED TO PLANT GROWTH AND DEVELOPMENT

Brassinosteroids have biological effects at low concentrations and they are found in gymnosperms, alga, monocotyledons (Liliopsida) and dicotyledons (Magnoliopsida) plants in different organs, such as leaves, floral buds, seeds, fruits, stems and roots. It can affect a great variety of developmental process during plant growth and development (Table 1).

Table 1. Effects and concentrations of brassinosteroid and brassinosteroid analogues on plant growth and development in different plant species

Species	Concentrations	BR	Organ	Mode of application	Physiological effects analyzed	Author
Arabidopsis thaliana	10^{-10}, 10^{-8} and 10^{-6} M	BL	Seed	Medium constituent	Growth	Tanaka *et al.* (2003)
Arabidopsis thaliana	0.1 and 0.5 nM 10 nM	24-EBL 24-ECAS	Seed	Medium constituent	Root growth	Müssig *et al.* (2003)
Arabidopsis thaliana	1–100 nM	BL	Seedling	Medium constituent	Lateral root formation	Bao *et al.* (2004)
Arabidopsis thaliana	100 nM	EBL	Seed	Medium constituent	Hypocotyl growth, apical hook	De Grauwe *et al.* (2005)
Arabidopsis thaliana	0.05 and 0.1 μM	24-EBL	Seed	Medium constituent	Root growth	Golovatskaya (2008)
Arabidopsis thaliana	2 μM	BL	Inflorescence stalk	Medium constituent	Ethylene production	Arteca and Arteca (2008)
Vigna radiata	10^{-8}, 10^{-6}, 10^{-4} M	28 HBL	Seed	Seed soaking	Photosynthesis, Enzyme activities	Fariduddin *et al.* (2003)
Vigna radiata	0.0001, 0.01, 1 μM	28-HBL	Leaves (25 day-age)	Spray	Enzymes activities, Photosynthesis, protein and chlorophyll content	Fariduddin *et al.* (2004)
Vigna radiata	1 μM 0.01 μM	28HBL	Seed Leaves	Seed soaking Spray	Photosynthesis, enzymes activities, chlorophyll content, yield	Fariduddin *et al.* (2008)
Pisum sativum	0.1 μM	EBL	Detached pea shoot	Medium constituent (incubation for 20 minutes)	Protein	Fedina *et al.* (2008)

(*continued*)

(*continued Table 1.*)

Species	Concentrations	BR	Organ	Mode of application	Physiological effects analyzed	Author
Cucumis sativus	0.1 mg/L	24-EBL	Leaves	Spray application	Photosynthesis	Yu *et al.* (2004)
Brassica oleraceae	0.001 µM 10 µM	EBL	Cotyledon	Cotyledon incubation for 3 days	Growth, chlorophyll content, anthocyanin acontent	Çag *et al.* (2007)
Lupinus albus *Lupinus angustifolius* *Lupinus luteus*	10^{-9} M 10^{-9} M	EBL HBL	Seed	Seed soaking	Protein and amino acid content	Kandenlinskaya *et al.* (2007)
Oryza sativa	10^{-5}, 10^{-6}, 10^{-7}, 10^{-8}, 10^{-9} M	BL	Coleoptile	Medium constituent	Lamina joint inclination and coleoptile elongation	Jeong *et al.* (2007)
Wolffia arriza	10^{-9} M	24-EBL	Culture cells	Addition to cells	Growth, photosynthetic pigments, protein and sugar content	Bajguz and Asami (2005)
Chlorella vulgaris	10 nM	BL	Culture cells	Addition to cells	Growth	Bajguz and Asami (2004)
Glycine max	0.21, 2, 1, 21 µM 0.22, 2.2, 22 µM 0.22, 2.2, 22 µM	24-EBL MH-5 BB-6	Seedling	Medium constituent	Growth	Mazorra *et al.* (2004)
Ananas comosus	0.1, 0.3, 0.5 1 mg/L	BB-16	Micropropagated seedling	Spray	Growth	Catunda *et al.* (2008)
Citrus reshni	0.1, 0.5, 1.0 mg/L	BB-16	Seedling	Spray	Stem diameter	Altoé *et al.* (2008)
Passiflora edulis	0.1 mg/L (successive applications)	BB-16	Plant (6 months)	Spray	Yield (number of fruits)	Gomes *et al.* (2006)
Opuntia ficus-indica	0.1 and 10 mg/L (BB-6) 0.001 and 10 mg/L (BB-16)	BB-6 BB-16	Cladodes	Spray	Yield (number of vegetative buds) and growth	Cortes *et al.* (2003)
Lycopersicon esculentum	10^{-8} M	28-HBL	Roots	Roots dipped	Enzymes, chlorophyll content, yield (number of fruits)	Ali *et al.* (2006)
Allium cepa	0.005 ppm, 0.05 and 0.5 ppm	24-EBL	Bulb	Medium constituent	Root length, number of mitoses	Howell *et al.* (2007)
Hordeum vulgare	0.1, 0.5 and 1.0 µM	28-HBL	Seeds	HBL-supplemented distilled water	Primary root growth, mitoses activity, protein content, antioxidant enzymes activities	Kartal *et al.* (2009)

There are many studies relating the effects of brassinosteroid application on *Arabidopsis thaliana* plant development, so they will be related together in the next paragraphs.

Exogenous BL (10^{-10}, 10^{-8} and 10^{-6} M) remarkably promoted the growth of hypocotyls and cotyledonous leaf-blades in a dose-dependent manner (Tanaka *et al.*, 2003). The BL-induced hypocotyls elongation was shown only in conditions of light. Hypocotyls seedlings that were treated with 10^{-6} M of BL became 2.4 times longer than untreated seedlings. In darkness, BL did not promote but rather inhibited hypocotyls elongation at concentrations higher than 10^{-8} M. On the other hand, BL-induced elongation of cotyledonous leaf blade was found in both dark and light, although BL was less effective in darkness. Taproot elongation was severely inhibited by exogenously applied BL in a dose dependent fashion in both dark and light conditions. The inhibition of taproot growth in the light was observed when seedlings were treated with BL concentrations greater than 10^{-10} M. Conversely, in darkness, taproot growth was inhibited when treated with 100-fold higher concentrations of BL (10^{-8} M) at a minimum. The authors suggest that the results indicate that responsiveness of organs to BL differs among organs. Besides, cytological observation disclosed that BL-induced hypocotyls elongation was achieved through cell enlargement rather than cell division. Furthermore, a serial experiment with hormone inhibitors showed that BL induced hypocotyl elongation not through gibberellins and auxin actions. However a synergistic relationship of BL with gibberellins A_3 and indole-3-acetic acid (IAA) was observed on elongation growth in light-grown hypocotyls, even though gibberellins have been reported to be additive to BR action in other plants. These authors showed that BRs act on light-grown hypocotyl elongation independent of, but cooperatively with, gibberellins and auxin (Tanaka *et al.*, 2003). Müssig *et al.* (2003) showed that low concentrations of 24-epicastasterone (10 nM) and 24-EBL (0.1 and 0.5 nM) promoted root elongation in *A. thaliana* wild-type plants up to 50% and in BR-deficient mutants up to 150%. The growth stimulating effect of exogenous BRs is not reduced by the auxin transport inhibitor 2,3,5,-triiodobenzoic acid. BR-deficient mutants show normal gravitropism and 2,3,5-triiodobenzoic acid or higher concentrations of 2,4-dichlorophenoxyacetic acid (2,4D) and naphthalene acetic acid (NAA) inhibit root growth in the mutants in the same extent as in wild-type plants. They verified that simultaneous administration of EBR and 2,4D results in largely additive effects but exogenous gibberellins do not promote root elongation in the BR-deficient mutants and the sensitivity to the ethylene precursor 1-aminocyclopropane-1-carboxylic acid is not altered. Thus, the root growth stimulating effects of BRs appears to be largely independent of auxin and gibberellins action. Bao *et al.*(2004) verified that BRs are required for lateral root development in *Arabidopsis* and that BRs acts synergistically with auxin to promote lateral root formation. The authors showed that the number of lateral roots increased in response to 1–100 nM BL with 10 nM being optimal, inducing nearly an eightfold increase in lateral root formation but the elongation of primary roots was inhibited by 1–100 nM BL. Auxin is necessary for lateral root development and lateral root emergence can be blocked by the auxin transport inhibitor N-(1-naphthyl) phthalamic acid (NPA). The authors found that 2 μM NPA not only inhibited lateral root formation in wild-type seedlings but also reduced BL promotion of lateral root formation, so these results indicate that BRs promote lateral root development by increasing acropetal auxin transport.

Dark-grown *Arabidopsis* seedlings develop an apical hook by differential cell elongation and division, a process driven by cross-talk between multiple hormones like auxins, ethylene and gibberellins as they interact in the formation of the apical hook. In the light, a similar complexity of hormonal regulation has been revealed at the level of hypocotyls elongation (De Grauwe *et al.*, 2005). These authors analyzed the involvement of brassinosteroids in auxin-and ethylene controlled process in the hypocotyls of both light-and-dark-grown seedlings. They showed that BR biosynthesis is necessary for the formation of an exaggerated apical hook and that either application of BRs or disruption of BR synthesis alters auxin response, presumably by affecting auxin transport, eventually resulting in the disappearance of the apical hook. When EBL (100 nM) was applied to light-grown hookless mutants or light-grown PIN mutants, EBL increased hypocotyls length even when ACC was supplied together with EBL. When treated with EBR, the relative hypocotyl elongation in the mutants was comparable with that of the wild type. However, when ACC and EBR were applied simultaneously, a small synergistic effect was observed in all hookless mutants, whereas this was not the case for the wild type. Thus, these components may act in a signal transduction route either upstream or independently of BRs.

The effects of GA_3, 24-EBL and their combination on morphogenesis of *Arabidopsis thaliana* (L.) Heynh (7-day-old seedlings) were studied by Golovatskaya (2008). The cotyledons shape and size were dependent on 24-EBL and the root length was both GA_3 and EBL regulated, indicating organ specificities in the responses to these hormones. Simultaneous treatment of dark-grown plants with GA_3 (0.01–1.0 μM) and EBL (0.05–0.1 μM) exerted an additive stimulatory effect on the root growth of *det-2* (BR-deficient mutant), reduced the inhibitory effect of EBL on hypocotyl elongation of *ga4-1* (GA deficient mutant) and enhanced the effect of EBL on hypocotyls and cotyledon elongation of *det-2*. The authors suggest that localization and quality of morphogenetic responses of *Arabidopsis* plants to exogenous GA_3 and EBL are linked to the dynamics and the ratio of endogenous hormone levels in the plant, inactivation and destruction of exogenous hormones during their transport in the plant, as well as on the degree of overlapping of the hormone responses.

The effects of varying concentrations of IAA alone or in combination with BL or 6-benzylaminopurine, a cytokinin, on ethylene production were evaluated in order to determine the relationship between IAA (0–100 μM) and BL (2 μM) or BAP (10 μM) in *A. thaliana* inflorescences (Arteca and Arteca, 2008). Inflorescences treated with BL alone had no effect on ethylene production. However, when BL was used in combination with IAA there was a dramatic increase in ethylene production above the induction promoted by IAA alone so, there was a synergistic effect due to the co-application of these hormones in ethylene production.

Mung bean (*Vigna radiata* L. Wilczek cv. T-44) seeds soaked in 28-HBL (10^{-8}, 10^{-6}, 10^{-4} M) for 4, 8 or 12 hours enhanced net photosynthetic rate, leaf chlorophyll content, carbonic anhydrase activity, carboxylation efficiency, stomatal conductance and seed yield at harvest being the best combination the concentration of 10^{-6} M for 8 hours, the others concentrations were either very low or supra optimal for most of the parameters (Fariduddin *et al.*, 2003). When 28-HBL (0.0001, 0.01, 1 μM) and kinetin (0.01, 1, 100 μM) were applied to the leaves of 25-day old plants of mung bean (*Vigna radiata* L. Wilczek), the activities of nitrate reductase and carbonic anhydrase, chlorophyll and total protein contents and net photosynthetic rate in the leaves and pod number and seed yield increased at harvest (Fariduddin *et al.*, 2004). Other study with this plant tested the effects of 28-HBL supplied to the seeds (1.0 μM) and/or to the foliage (0.01 μM) of mung bean (*Vigna radiata* L.Wilczek) plants (Fariduddin *et al.*, 2008). It was observed that the activities of carbonic anhydrase and nitrate reductase, leaf chlorophyll content, net photosynthetic rate, stomatal conductance, carboxylation efficiency, dry mass, pod number and seed yield at harvest increased significantly over the control, irrespective of the mode of application. However the plants raised by seed soaking and also received foliar application of HBL had most successful results. Detached pea (*Pisum sativum* L.) shoot was incubated in a medium supplemented with 0.1 μM EBL for 20 minutes and then submitted to a protein quantitative assay (Fedina *et al.*, 2008). This analysis revealed that the phosphorylation of PY20 phosphotyrosine polypeptides was changed under the action of EBL. The results indicate that eight of these proteins belong to the Calvin Cycle enzymes so the observed changes in phosphorylation of these proteins may partly explain the effects of BRs on photosynthesis. The effects of 24-EBL spray application on gas exchange, chlorophyll fluorescence characteristics, Rubisco activity and carbohydrate metabolism were investigated in cucumber (*Cucumis sativus* L. cv. Jinchun No. 3) plants (Yu *et al.*, 2004). EBR significantly increased the light-saturated net CO_2 assimilation rate from 3 hours to 7 days after spraying, with 0.1

mg.L^{-1} EBR proving most effective. EBR-treated leaves also had a higher quantum yield of PSII electron transport than the controls, which was mainly due to a significant increase in the photochemical quenching (q_p), with no change in the efficiency of energy capture by open PSII reaction centers (F'_v/F'_m). It was concluded that EBR increases the capacity of CO_2 assimilation in the Calvin Cycle, which was mainly attributed to an increase in the initial activity of RUBISCO. Different concentrations of EBL (0.001, 0.1 and 10 μM) were tested in excised red cabbage (*Brassica oleraceae* L.) by incubating the cotyledons on those solutions for 3 days (Çag *et al.*, 2007). There was a significant increase in chlorophyll content of cotyledons incubated in 0.001 μM compared to the control and the concentration of 10 μM showed the lowest peroxidase activity and the optimal concentration for anthocyanin concentration was 10 μM. The concentration of 0.001 μM EBL promotes growth on cotyledons, whereas 10 μM inhibits it and 0.01 μM causes no significant effect on cotyledon growth. Presowing treatment of seeds of lupine (*Lupinus angustiflolius*, *Lupinus luteus* L., *Lupinus albus* L.) of various species and cultivars with EBL (10^{-9} M) and HBL (10^{-9} M) caused an increase in protein content and a change in the proportion of some amino acids (Kandenlinskaya *et al.*, 2007). These changes in protein metabolism correlated with an increase in the concentration of indole acetic acid and a decrease in the content of abscisic acid.

Jeong *et al.* (2007) verified the responses of lamina joint inclination and coleoptiles elongation of rice (*Oryza sativa* L.) to exogenous BL under light or dark conditions. Both responses were more pronounced under darkness, implying that BR signalling is inhibited by light. The authors suggest that phytochrome B acts as a negative regulator of BL-regulated growth and development processes in rice.

The use of brassinazole, an inhibitor of BR biosynthesis, to prove the involvement of BRs on plant development is well known and exogenous BR is used to restore the effects caused by brassinazole. The application of 24-EBL (10^{-13}–10^{-6} M) to *Wolffia arrhiza* cultures, an aquatic monocotyledon, stimulated the growth and increased the content of photosynthetic pigments, sugar and protein and the concentration of 10^{-9} M showed the greatest effect (Bajguz and Asami, 2005). Addition of Brz2001 (10^{-6}–10^{-4} M) to *Wolffia arrhiza* cultures, a kind of brassinazole, inhibits growth after 7 days of cultivation and this inhibition could be reversed by the addition of EBL so BR is important for *Wolffia arrhiza* growth and development. Cultured *Chlorella vulgaris* Beijerinck cells with 0.1–10 μM Brz2001 inhibits their growth during the first 48 hours of cultivation in the light and this inhibition is prevented by the co-application of BL (10 nM) (Bajguz and Asami, 2004). This result suggests that the presence of endogenous BRs during the initial steps of the *C. vulgaris* cell cycle is indispensable for their normal growth in light. In darkness, a treatment with 10 nM BL promotes growth through the first 24 hours of culture but during the following 24 hours the cells undergo complete stagnation. Brz2001, was also used to block the growth of roots, hypocotyls and epicotyls of soybean (*Glycine max* cv. Cubasoy 27) seedlings producing a dwarf phenotype (Mazorra *et al.*, 2004). The application of 24-EBL (0.21, 2.1 and 21 μM) completely reversed the inhibitory effects caused by brassinazole and two growth-promoting brassinosteroid analogues tested (MH-5 and BB-6, 0.22, 2.2 and 22μM of each) partially overcame the Brz2001-induced growth defects but MH-5 proved to be more effective and the largest reversible growth reduction was obtained with 0.22 μM MH-5 and 2.2 μM BB-6. BB-6 and MH-5 have biological activity despite presenting the spirostane side chain instead of the 22α,23α-dihydroxicholestane characteristic of the natural brassinosteroids.

BB-16 (0.1, 0.3, 0.5 and 1.0 mg.l), a spirostanic analogue of brassinosteroid, was applied to micropropagated seedlings of Imperial pineapple (*Ananas comosus* L. Merrill) to evaluate the development under two substrates (Plantmax® and a mix of composting sugar cane bagasse and filter cake, CC) (Catunda *et al.*, 2008). The plants that were cultivated on CC substrate and sprayed with 0.1mgL^{-1} BB-16 produced 2.8 times more dry matter than the control cultivated in Plantmax substrate. Besides, this treatment showed higher growth of shoots with greater number of leaves, rosette diameter, leaf width, fresh and dry matter production at 150 days after planting, the authors suggest that BR would influence carbohydrate translocation from root to shoot because CC is richer in organic compound when compared to Plantmax. This analogue (BB-16), at 0.1, 0.5 and 1.0 mg.L^{-1}, was also used to verify the development of "Cleopatra" orange rootstock (*Citrus reshni* Hort ex Tanaka) mycorrhized and non-mycorrhized, BB-16 promoted an enhancement in the diameter of the stem which is an important parameter for plant grafting (Altoé *et al.*, 2008). Successive applications of BB-16 (0.1 mg.L^{-1}) to yellow passion fruit plants (*Passiflora edulis* f. *flavicarpa*) after the appearance of the first flowers enhanced yield when applied during three successive weeks (81.5 fruits/plant) when compared to the control (53.5 fruits/plant) (Gomes *et al.*, 2006). BB-6 and BB-16 applied to the cladodes of *Opuntia ficus-indica* L. Mill. var. lutea stimulated larger number of vegetative buds under both greenhouse and field conditions and promoted precocity, accelerating growth during the first stages of vegetative bud development, but did not alter the morphology of the harvested cladodes (Cortes *et al.*, 2003).

The quantities of nitrate reductase, carbonic anhydrase and content of chlorophyll in leaves were significantly higher than control in 30 and 60 days old plants of tomato (*Lycopersicon esculentum* Mill.) whose roots (20 days old seedling) were dipped in 28-HBL (10^{-8}, 10^{-7}, 10^{-6}, 10^{-5} M), there were more fruits in BR-treated plants than in control (Ali *et al.*, 2006).

Besides, the fruits at ripening had higher levels of lycopene and β-carotene, being the treatment 15 min feeding of 10^{-8} M the best of all them (Ali *et al.*, 2006).

Low doses of 24-EBL (0.005ppm) nearly doubled the mean root length and the number of mitoses over that of controls of onion (*Allium cepa*) root tips (Howell *et al.*, 2007). Intermediate doses of EBL (0.5 ppm) also produced mean root lengths and number of mitoses that were significantly greater than those of the controls. Seeds of barley (*Hordeum vulgare* L.) germinated between filter papers in 0.1, 0.5 and 1.0 μM HBL-supplemented distilled water, the HBR-seedlings showed significant increases in the primary root growth (twofold increase in 1.0 μM HBR), the roots treated with HBR showed more mitotic activity, mitotic abnormalities and significant enlargements at the root tips when compared with control material (Kartal *et al.*, 2009). HBR application to barley seeds decreased total soluble protein content, superoxide dismutase, catalase and peroxidase activities significantly at 1.0 μM HBR concentration.

2.1 Brassinosteroid interactions with other plant growth substances

Plant metabolism involves responses that are regulated mainly by hormones and environment. The involvement of a unique substance to explain such intricate processes does not exist, so it is a consequence of a complex mechanism that involves hormones, phytochromes, DNA synthesis, gene expression. The next paragraphs will relate the influence of applied BR and other applied plant growth regulators in plant physiology (Table 2).

When auxin (10^{-5} M) and BL (10^{-7} M) were applied exogenously to *Zea mays* L. cv. Golden Cross Bantam plants, both increased ethylene production (Yun *et al.*, 2009). When these hormones were tested simultaneously, the increase in the level of ethylene was greater than the sum of effects by each one. Such a positive interaction was also recorded for changes in the activity of ACC synthase and the expression of its gene. For ACC oxidase, however, the two hormones had no apparent influence. When applied separately neither affected root elongation nor proton extrusion. However, when given in combination, both phenomena occurred. These results suggest that BL interacts with IAA to promote ethylene biosynthesis and elongation in roots. Therefore it is possible that BL acts by inducing auxin, which then stimulates both ethylene production (at the early stage) and root development.

A high concentration of BL induced abnormal shoot shape in rice (*Oryza sativa* L. cv. Tai Nguyen) seedlings, that is, the newly developed leaf sheath

Table 2. Effects and concentrations of brassinosteroids and other hormones on hormone interactions

Species	**Hormones and concentrations applied**	**Organ which received BR**	**Mode of application**	**Physiological effects analyzed**	**Author**
Zea mays	Auxin (10^{-5} M) BL (10^{-7} M)	Root tissue	Growth medium constituent	Ethylene production	Yun *et al.* (2009)
Oryza sativa *Echinochloa crus-galli*	BL (0.1, 1.10 μM) Ethephon (1.2, 5.5 mM)	Root	Root soaking	Shoot shape Leaf sheath length Number of leaves	Chon *et al.* (2008)
Vicia faba	BL (10 mM) ABA (50 mM)	Detached leaf	Incubation solution constituent	movement	Haubrick *et al.* (2006)
Tabebuia alba	BL (0.052, 0.104, 0.208 mM) GA3 (0.072, 0.1443 and 0.2887 mM) IAA (0.1427, 0.2854, 0.5708 mM)	Seedling (104 d-age)	Spray	Development and leaf anatomy	Ono *et al.* (2000)

was shorter than the old sheath and increased the number of leaves under light (Chon *et al.*, 2008). In this same study, the number of leaves of barnyard grass (*Echinochloa crus-galli* L.) increased at 0.1 μM BL while the number of leaves of rice increased at 1.0 μM. BL strongly stimulated ethylene production and the amount of ethylene and the third leaf sheath length in rice seedlings were negatively correlated. The effect of ethephon on leaf sheath was very similar to that of BL. The results indicate that BL-induced abnormal shoot growth of rice seedlings was probably mediated by ethylene production.

When analyzed the regulation of stomatal movement of *Vicia faba* L. cv. Longpod, BL did not oppose the effect of ABA (Haubrick *et al.*, 2006). On the contrary, BL (10 mM) modulated stomatal aperture by promoting stomatal closure and inhibiting stomatal opening. BL inhibited inwardly rectifying K^+ currents of *Vicia faba* guard cell protoplasts in a manner similar to ABA. The effects of BL and ABA (50 mM stock in 95% ethanol) applied together were not additive, suggesting that these two hormones may function in interacting pathways to regulate stomatal apertures and guard cell physiology.

The effects of gibberellin, GA_3 (0.072 mM, 0.1443 mM and 0.2887 mM), auxin (0.1427 mM, 0.2854 mM, 0.5708 mM) and BL (0.052 mM, 0.104 mM, 0.208 mM) in different combinations were evaluated on the development and leaf anatomy of *Tabebuia alba* (Cham.) Sandow seedlings (Ono *et al.*, 2000). The results showed that GA_3 plus BL produced the highest stem and petiole growth rates and also produced a significant development of lateral buds but BL application alone stimulated petiole growth but not stem growth, thus indicating the influence of both hormones to promote stem growth and lateral bud development in *Tabebuia alba*.

2.2 Does exogenous BR change the endogenous concentration of other plant hormones?

It is expected that brassinosteroid application may change other endogenous plant hormones because it is well known that the level of a hormone in a tissue will alter the plant response by regulating endogenous levels of other hormones. Some researchers have studied how BR can affect other endogenous hormone levels in organs or cells (Table 3).

Table 3. Effects of brassinosteroids on the endogenous concentrations of other plant hormones

Species	Hormones and some concentrations applied	Organ which received BR	Mode of application	Endogenous hormone quantified	Author
Triticum aestivum	24-EBL (0.4 nM and 0.4 μM)	Root seedling	Root incubation medium	Cytokinin Auxin Abscisic acid	Aval'baev *et al.* (2003)
Pisum sativum	BL (200 ng/2μL)	The oldest unexpanded internode of 21 day-old plants	2μL applied to the internode	GA_{20} BRs IAA	Jager *et al.* (2005)

Aval'baev *et al.* (2003) studied the effects of 24-EBL (0.4 nM and 0.4 μM) on the dynamics of the concentration of auxins, cytokinins and abscisic acid in the seedlings of wheat (*Triticum aestivum* L. cv. Moskovskaya 35). The results showed that neither of the EBL-concentrations induced changes in the concentration of IAA or ABA in seedlings roots. On the other hand, even 1 hour after application of EBL, there was an almost twofold increase in the rate of accumulation of cytokinins and such an enhanced level was verified through the period of experiment and they suggested that the growth stimulation activity of EBL in wheat seedlings was primarily due to the effects of this agent on the cytokinin metabolism in plants.

The application of BL (200 ng/2 μl 100% ethanol) to *lkb* (BR-deficient mutants) pea plants (*Pisum sativum* L.) reduced GA_{20} levels and metabolism studies revealed a reduced conversion of GA_{19} to GA_{20} in EBL-treated (1 μM) lkb plants (Jager *et al.*, 2005). These results indicate that BRs actually negatively regulate GA_{20} levels in pea. Although GA_{20} levels are affected by BR levels, this does not result in consistent changes in the level of the bioactive GA, GA_1. It appears that the BR growth response is not mediated by changes in bioactive GA levels, thus providing further evidence that BRs are important regulators of stem elongation.

2.3 Seed germination is enhanced by exogenous brassinosteroids

It is supported that brassinosteroids have the ability to induce seed germination as they are naturally occurring substances in seeds and are probably as important as gibberellins and abscisic acid in the control of this process (Table 4).

Seed germination of *Nicotiana tabacum* L. cv. Havana 425 is determined by the balance of forces between the growth potential of the embryo and the mechanical restraint of the micropylar endosperm (Leubner-Metzger, 2001). In contrast to the gibberellin (GA_4), the BL did not release photodormancy of dark-imbibed photodormant seeds. BL and GA_4 promoted endosperm rupture of dark-imbibed non-photodormant seeds, but did not appreciably affect the induction of class I β-1,3-glucanase (βGLU I) in the micropylar endosperm. Promotion of endosperm rupture by BL was dose-dependent and 0.01 μM was most effective. It is proposed that BRs promote seed germination by directly enhancing the growth potential of the emerging embryo in a GA and βGLU I-independent manner.

Wheat seeds (*Triticum aestivum* L. cv. HD2204) were soaked in aqueous solutions of 28-HBL (10^{-10}, 10^{-8} and 10^{-6} M) for 8 hours and the α-amilase levels were increased in HBR-treated seeds, the most effectives dosages were 10^{-10} and 10^{-8} M, the levels of catalase and peroxidases also increased in seedlings whose seeds were treated with HBR (Hayat and Ahmad, 2003a). These authors used the same concentrations of 28-HBL in seeds of *Lens culinaris* cv. Pusa-6 and they verified that HBR-treated plants decreased root length and nodule number per plant but increased nitrate reductase activity and the most effective concentration was 10^{-8} M (Hayat and Ahmad, 2003b).

Broomrapes (*Orobanche spp.*) are serious root parasitic weeds that cause great damage to crop production. Different plant growth regulators were used to verify the potential of germination of this species (Song *et al.*, 2005, 2006). Exogenous gibberellins, BL (1 mg.L^{-1}) and fluridone, inhibitor of carotenoid biosynthesis, significantly increased the broomrape seed response to a germination stimulant (Gr24, 10^{-6} M) even when seeds were first conditioned at a suboptimal temperature and under water stress. Exogenous GA_3 and BL could restore the germination of *Orobanche spp*. seeds. This may be due to breaking of the secondary dormancy, which is induced by the suboptimal temperature and by water stress.

Table 4. Effects of brassinosteroids in different developmental stage of plants

Species	Developmental stage	BR and concentrations applied*	Organ	Mode of application	Physiological effects analyzed	Author
Nicotiana tabacum	Seed germination	BL (0.01 μM)	Seed	Filter paper wetted with BL and other compounds	Endosperm rupture of dark-imbibed non-photodormant seeds	Leubner-Metzger (2001)
Triticum aestivum	Seed germination	28-HBL (10^{-10}, 10^{-8}, 10^{-6} M)	Seed	Seed soaking	Increase in α-amylase levels	Hayat and Ahmad (2003a)
Lens culinaris	Seed germination	28-HBL (10^{-10}, 10^{-8}, 10^{-6} M)	Seed	Seed soaking	Decrease in root length, nodule number and increase in nitrate reductase activity	Hayat and Ahmad (2003b)
Orobanche spp.	Seed germination	BL (0.5–1.0 mg/L)	Seed	Petri dishes wetted with BL and other compounds	Increased seed germination	Song *et al.* (2005) Song *et al.* (2006)
Vitis vinifera	Fructification	BL (200 ng/5 μL)	Berries	Application of 5μL to each berry	Accelerated ripening, increase on total soluble solids	Symons *et al.* (2006)
Cucurbitaceae family (zucchini, melon, cucumber)	Flowering and fructification	EBL (0.1, 1, 10 μM) EBL (0.1, 10 μM)	Plants Seedling	Pipetting 250μL onto the apical meristem and developing leaf 20μL applied	Differences in the appearance of female and male flowers, ethylene production	Papadopoulou and Grumet (2005)
Cucumis sativus	Flowering and fructification	24-EBL (0.02, 0.2, 2 μM)	Unpollinated ovaries	Spray on to unpollinated ovaries at anthesis	Parthenocarpic growth, expression of cell-cycle related genes	Fu *et al.* (2004)
Pharbitis nil	Flowering	BL (0.01 and 1.0 μM) CAS (0.01 and 1.0 μM) BL (0.1–10 μM)	Cotyledon Shoot apices (*in vitro*)	Soft paintbrush Medium constituent	Flowering inhibition	Kesy *et al.* (2003)
Oryza sativa	Flowering and fructification	BL (2.1×10^{-9} M and 2.1×10^{-8} M)	Whole rice plants	Spray application	Panicle ripening, endogenous hormone quantification, growth	Saka *et al.* (2003)
Litchi sinensis	Flowering and Fructification	BL (0.5, 0.75 and 1.0 mg/L)	Trees (6-y-old)	Spray before anthesis and at early fruit stage	Enzymes activities, pectin content, Ca level	Peng *et al.* (2004)
Lycopersicon esculentum	Flowering and fructification	28-HBL (0.1, 1.0 and 3.0 μM) 24-EBL (0.1, 1.0 and 3.0 μM)	Pericarp discs	Petriplates supplied with BRs	Carbohydrates, lycopene, chlorophyll and ascorbic acid content	Vardhini and Rao (2002)

Triticum aestivum	Senescence	EBL (0.001, 0.1, 10.0 μM)	Leaves segments	Petri dishes containing EBL	Enzymes activities, chlorophyll content	Çag et al. (2007)
Zea mays	Tropism	BL (10^{-10}, 10^{-9}, 10^{-8} and 10^{-7} M)	Primary roots	Immersion of root caps	Gravitropic curvature	Chang *et al.* (2004)
Arabidopsis thaliana	Tropism	BL (10^{-10},10^{-9}, 10^{-8} and 10^{-7} M)	Seed	Medium constituent	Gravitropic curvature, root length	Kim *et al.* (2007)
Zea mays	Tropism	BL (10^{-9}, 10^{-8}, 10^{-7}, 10^{-6} and 10^{-5} M)	Root	Immersion of root caps	Gravitropic curvature	Kim *et al.* (2000)
Brassica napus	Tropism	BL (0.1 μM)	Root	Medium constituent	Polar auxin transport, endogenous IAA quantification	Li *et al.* (2005)
Pisum sativum	Tropism	BL (10^{-13} to 10^{-8} M)	Root	Root solution	Gravitropic curvature	Amzallag and Vaisman (2006)
Arabidopsis thaliana	Tropism	BL (1.0 nM)	Seedling	Growth medium constituent	Hypocotyl curvature Phototropism	Whippo *et al.* (2005)
Arabidopsis thaliana	Circadian rhythms	HBL (20 μM)	Seedling	Medium constituent (imaging microtiter plates)	Modulation of circadian periodicity	Hanano *et al.* (2006)

2.4 How exogenous BR application may regulate flowering and fructification?

Brassinosteroid spraying at flowering generally leads to a significant increase in the production of various crops and it is reported in several article papers (Table 4).

Exogenous EBL application (200 ng dissolved in 100% ethanol) in grape (*Vitis vinifera* L. cv. Cabernet Sauvignon) significantly promoted ripening, while brassinazole, significantly delayed fruit ripening (Symons *et al*., 2006). Using the first appearance of coloring (anthocyanin production) in the berry skin as an indicator for the onset of ripening, the authors showed that EBR significantly promoted *véraison* while brassinazole delayed *véraison*. Total soluble solids measured in berries 28 days after the first treatment (13.4°Brix) was greater than in control plants (12.7°Brix) and brassinazole treated plants (11.7°Brix) indicating that BRs also stimulate sugar accumulation. These results provide evidence that changes in endogenous BRs levels influence this key developmental process.

BR application in species of the Cucurbitaceae family was tested in order to identify different phenotypes in relation to sexual expression. In this family a phase of male flowers precedes either female or bisexual flower production. In this study EBL application (0.1, 1, 10 μM) caused a significant decrease in time of appearance of the first female flowers in monoecious cucumber plants and increased the total number of female flowers on the main stem. Increasing concentrations had a stronger effect (Papadopoulou and Grumet, 2005). EBL application in cucumber, melon and zucchini caused an increase in ethylene production suggesting that BR effect can be mediated by ethylene. The concentration of 10 μM epi-BL leads to an ethylene production comparable to that induced by 5 ppm ethefon. The treatment with 5 ppm ethefon was sufficient to increase femaleness of cucumber plants but not zucchini plants suggesting that the difference in response to EBL treatment may reflect differences in the sensitivity to ethylene (Papadopoulou and Grumet, 2005).

In a work conducted by Fu *et al*. (2008) 24-EBL application (0.02, 0.2 and 2.0 μM) in cucumber (*Cucumis sativus* L.) unpollinated ovaries induced parthenocarpic growth accompanied by active cell division in Jinchun No. 4, a cultivar without parthenocarpic capacity, whereas brassinazole treatment inhibited fruit set and, subsequently, fruit growth in Jinchun No.2, a cultivar with natural parthenocarpic capacity, and this inhibitory effect could be rescued by the application of EBR. RT-PCR analysis showed both pollination and EBR induced expression of cell-cycle related genes (CycA, CycB, CycD3.1, CycD3.2 and CDKB) after anthesis. BRs triggered active cell division associated with increased transcripts of cell cycle-related genes, especially that of cyclin D3 genes. These results indicate that BRs play a regulatory role in early fruit development of cucumber plants. It is noteworthy that fruit development is a complex process and BRs could cross talk with other hormones such as auxins and gibberellins.

In order to verify the effects of exogenous BRs on the flowering induction of *Pharbitis nil Chois* cv. Violet, a short-day plant, BL and castasterone in the concentrations of 0.01 and 1.0 μM were applied to the cotyledons (Kesy *et al.*, 2003). Both BRs used inhibit flowering, forming less number of flowers in relation to control plants and flowering inhibition was depended on the concentration and the method of BR application as well as the length of the inductive dark period. In plants regenerated from sub-induced apices treated with BL (1 and 10 μM), the flower formation was inhibited completely. These authors suggest that BR can be acting similar as auxin because this hormone is proved to inhibit flowering of this short-day plant when applied exogenously.

BL applied to rice (*Oryza sativa* L. cv. Nippon bare) promoted panicle ripening (Saka *et al.*, 2003). These authors analyzed if exogenous BR application at the meiosis and flowering stages affected the endogenous levels of abscisic acid, auxin or ethylene. When brassinolide (2.1×10^{-9} and 2.1×10^{-8} M) were applied to the whole rice plants by spraying twice, 10 days before heading and on the day of heading, the free-IAA content slightly increased and greatly increased the bound-IAA content at the milk ripe stage in field conditions (22–33°C) and in low temperature (17–22°C). BL slightly decreased the ABA content of the spikelet at the milk-ripe stage in field conditions and slightly increased it in the low temperature condition. The rate of ethylene production was markedly high at the milk-ripe stage and low at the dough-ripe stage (21 days after heading) in field conditions. BL treatment clearly increased the rate of ethylene production from the panicles under both light and dark conditions at the milk-ripe stage. BL treatment also increased panicle weight and grain weight. BL promotes the assimilate translocation and accumulation of carbohydrates in the panicle and it may promote ripening by regulating the amounts of endogenous hormones such as auxins, abscisic acid and ethylene, not only in field conditions but also at low temperature conditions. Under low temperature condition, BL may maintain or rescue the sites of action of the other endogenous hormones mentioned as auxin and abscisic acid to promote grain filling after anthesis in rice plants.

Litchi trees (*Litchi chinensis* cv. nuomoci) were sprayed with 0.5, 0.75 and 1.0 $mg.L^{-1}$ BL at full blossom and at early fruit stage (Peng *et al.*, 2004). The enzyme activities of pectin methylesterase (PME) and polygalacturonase (PG) increased and showed the same trend under BR treatments. The content of water soluble pectin remained at a higher concentration during early stages of development of treated fruit compared to control fruit (water-prayed). Calcium concentration of fruit pericarp was higher in treated fruit than control fruit and showed significant dosage response. The cellulase activity was inhibited by BL treatment and BL reduced fruit cracking compared to control. The rise of PME and PG activities in fruit from trees treated with BL might reflect the rise of pectin metabolism, which may be related to cell division and cell elongation resulting in fruit growth. Meanwhile, the increase in calcium during early stages of fruit development could provide a good basis of fruit pericarp development and the final increase in protopectin content in the pericarp might thus guarantee the good quality of the fruit pericarp. The results showed that BL sprayed before anthesis may play an important role in increasing the commercial value of litchi fruits (Peng *et al.*, 2004).

The application of 28-HBL (0.1, 1.0 and 3.0 μM) and 24-EBL (0.1, 1.0 and 3.0 μM) to tomato (*Lycopersicon esculentum* Mill.) pericarp discs resulted in increased levels of lycopene and lowered chlorophyll levels (Vardhini and Rao, 2002). Brassinosteroid-treated pericarp discs exhibited decreased ascorbic acid and increased carbohydrate contents. Fruit ripening as induced by brassinosteroids was associated with an increase in ethylene production. This study revealed the ability of BRs in accelerating fruit-senescence.

2.5 Senescence is also a process regulated by brassinosteroid

Senescence is controlled by phytohormones and the involvement of auxins, ethylene and cytokinins is well documented but it is likely that BR influence this process. Saglam-Çag (2007) showed that EBL accelerated senescence in wheat (*Triticum aestivum* L.) leaves segments especially at high concentration. An increase in peroxidase activity (at 0.1 μM) and a decrease in protease activity (at 10 μM) were detected. Application of 0.1 and 10 μM was effective in accelerating chlorophyll breakdown while 0.001 μM EBL treatment showed the highest chlorophyll content of leaves inhibiting the chlorophyll loss.

2.6 New insights related to tropism and circadian rhythms when exogenous brassinosteroid is applied in plants

When BL (10^{-10}, 10^{-9}, 10^{-8}, 10^{-7}M) was applied exogenously to maize (*Zea* (*Zea mays* L.cv. Golden Cross Bantam) primary roots, it reduced the presentation time and lag period for the gravitropic response, whereas ethylene increased them (Chang *et al.*, 2004). Ethylene increases the rate of gravitropic curvature in a dose-dependent manner. If AVG, a specific action inhibitor of ACC synthase, is applied to the primary roots, it reduces the gravitropic curvature in the presence and absence of BL. The authors suggested that BL is involved in the gravitropic response in maize primary roots via ethylene production, but it acts in a way that differs somewhat from that of ethylene. It is possible that BL affects protein kinase activity, since the protein kinase inhibitors, staurosporine and H89, reduce BL-increased gravitropic response. The effect of brassinosteroids on plant gravitropism was verified by Kim *et al.* (2007) in the primary roots of *Arabidopsis thaliana* ecotype Columbia and Wassilewskija and BR-related mutants. Exogenously applied BL (10^{-10} M) increased both gravitropic curvature and length of primary roots, whereas at higher concentrations (10^{-9}, 10^{-8}, 10^{-7} M), BL further increased gravitropic curvature while it inhibited primary root growth. IAA is the primary hormone involved in the gravitropic response of plants. BL may activate the gravitropic response through a combination of its effect on polar IAA transport and IAA biosynthesis, resulting in the modulation of endogenous IAA levels in both upper and lower side of roots. Kim *et al.* (2000) showed that the stimulatory effect of BL is more pronounced in the presence of IAA, suggesting that BL increases the sensitivity of maize (*Zea mays* L.) roots to IAA but in *A. thaliana* the dose-dependent increase in the root gravitropic curvature was not as clearly evident in the presence of BL than the control upon IAA treatment. Kim *et al.* (2007) suggested that BL interacts negatively with IAA in the regulation of plant gravitropic response and root growth, and its regulation is achieved partly by modulating biosynthetic pathways of the counterpart hormone. Kim *et al.* (2000) also demonstrated the occurrence of endogenous BR (castasterone) in the primary roots which provides the first evidence of BR in plant roots. Analysis of the polar auxin transport capacities were analyzed in response to BL (0.1 µM) treatment to explore the potential interactions between them in *Brassica napus* ecotype Huyou 15 (Li *et al.*, 2005). Analysis of the polar auxin transport (PAT) activities of *Brassica napus* seedlings using [^{14}C]IAA showed that BL treatment strongly promoted shoot basipetal IAA transport and exogenous BL treatment (0.01–0.1 µM) changed the IAA concentrations in different organs.

In roots of pea (*Pisum sativum* L. cv. Alaska) seedlings, the average lag-time required for initiation of the gravitropic response was reduced proportionally to the concentration of 24-EBL added to the root solution (10^{-13}–10^{-8} M) (Amzallag and Vaisman, 2006). A treatment with clotrimazole, an inhibitor of steroid synthesis, prevents the initiation of gravitropic response and this effect was partly reverted by EBR application. They suggested that BR stimulates the root curvature through a gravitropic-induced change in sensitivity to the hormones regulating cell elongation.

Whippo and Hangarter (2005) suggested that brassinosteroids, which are hormonal repressors of photomorphogenesis, are involved in the repression of very-low-light phototropism, given by hypocotyl curvature, and the enhancement of high-light phototropism as addition of BL (1.0 nM) resulted in a strongly enhanced high-light response in *Arabidopsis thaliana* plants.

Studies with *Arabidopsis thaliana* ecotype Wassilewskija and mutants were used to test the effect of HBL (20 μM) in circadian clock (Hanano *et al.*, 2006). These authors showed that cytokinins delayed circadian phase, auxins regulated clock precision and brassinosteroid and abscisic acid modulated circadian periodicity. As a result of HBL application, circadian periodicity was shortened for CCR2, CAB2 and CCA1 mutants rhythms (1.0–2.7 hours) under both constant-light or constant darkness conditions.

3. BRASSINOSTEROID-APPLIED AMELIORATIVE EFFECTS TO A WIDE RANGE OF PLANT STRESS

Since the discovery and isolation of brassinosteroid, there is a continuous effort to discover how this compound acts and how it can be used in plants. The most related articles focus on the role to diminish toxic or non-desirable effects caused by biotical and abiotical stress. In the next topics, some information has been collected about the positive action in protecting plants from reactive oxygen species, pigments destruction and the ability to help plants to synthesize protective substances and expression of genes involved in defense responses as well as biosynthesis of other plant hormones.

3.1 Disease stress

Recent studies try to elucidate the possible role of BR-applied on plant tolerance and resistance to pathogen attack (Table 5).

BL induced resistance in rice to rice blast and bacterial blight diseases caused by *Magnaporthe grisea* and *Xanthomonas oryzae* pv. oryzae, respectively (Nakashita *et al.*, 2003). They observed that the application of BL (100 or 10 μg/pot) reduced disease symptoms caused by infection with the virulent pathogen *Xanthomonas oryzae* pv. oryzae race 003. These authors also verified that wild-type tobacco treated with BL exhibited enhanced resistance to the viral pathogen tobacco mosaic virus (TMV), to the bacterial pathogen *Pseudomonas syringae* pv. tabaci (Pst) and

Table 5. Effects of brassinosteroids on disease stress

Species	Pathogen	Hormone and concentrations applied	Organ which received BR	Mode of application	Physiological effects analyzed	Author
Oryza sativa	*Magnaporthe grisea*	BL (2, 20 and 100 μg/pot)	Root	Soil drench application (pre-treatment)	Reduction disease symptoms	Nakashita, *et al.* (2003)
Oryza sativa	*Xanthomonas oryzae* pv. oryzae	BL (2, 20 and 100 μg/pot)	Root	Soil drench application (pre-treatment)	Reduction disease symptoms	Nakashita *et al.* (2003)
Nicotiana tabacum	Tobacco Mosaic Virus (TMV)	BL (20, 40 and 200 μM)	Selected leaves	Foliar spraying (pre-treatment)	Enhanced resistance	Nakashita *et al.* (2003)
Nicotiana tabacum	*Pseudomonas syringae* pv. tabaci	BL (20 μM)	Whole plant (5-week-old)	Foliar spraying (pre-treatment)	Enhanced resistance	Nakashita *et al.* (2003)
Nicotiana tabacum	*Oidium* sp.	BL (20 μM)	Whole plant (5-week-old)	Foliar spraying (pre-treatment)	Enhanced resistance	Nakashita *et al.* (2003)
Cucumis sativus	*Fusarium*	24-EBL (0.1 μM) 24-EBL (0.2 μM–10 mL plant)	Root and leaves	Addition to nutrient solution Spraying on leaves	Reduction in pathogen-induced accumulation of reactive oxygen species, reduction disease severity	Ding *et al.* (2009a, b)
Cucumis sativus	Cucumber mosaic virus (CMV)	24-EBL (0.1 μM)	Seedling	Whole plant spraying	Expression of genes involved in defense response	Xia *et al.* (2009)

to the fungal pathogen *Oidium* sp (Nakashita *et al.*, 2003). BL-treatment did not induce either acidic or basic pathogenesis-related (PR) gene expression, suggesting that BL-induced resistance is distinct from systemic acquired resistance (SAR) and wound inducible disease resistance. They suggested that BL functions as one of the common signaling molecules in the innate immunity system of higher plants. It seems that BR is also involved in plant defense, by regulating thionin protein which is a low-molecular-weight, basic cysteine-rich antimicrobial protein and is expressed in a broad range of plant species. Transcripts of thionin genes encoding antimicrobial peptides were present at a high level in rice coleoptiles just after germination and decreased to an undetectable level after about 3 days but this decline was suppressed by co-treatment with gibberellic acid and brassinosteroid (Kitanaga *et al.*, 2006). Rice plants (*Oryza sativa* L. cv Nipponbare) were treated with a 1 μl solution containing 10 ng of GA_3 or 10 ng of BL at the base of the coleoptiles in the light. The results indicate an action of sequential regulation between the biosynthesis of GA/BR and JA jasmonic acid in a light dependent manner, mediated by a kind of collaborative cross-signaling process from GA and BR, leading to control of thionin transcript levels.

Root and foliar applications of 24-EBL were evaluated for their effects on reducing *Fusarium* wilt and their influence on antioxidant and phenolic metabolism in roots of cucumber plants (*Cucumis sativus* L. cv. Jinyan No. 4) (Ding *et al.*, 2009a, b). EBR treatments significantly reduced pathogen-induced accumulation of reactive oxygen species (ROS), flavonoids and phenolic compounds, activities of defense-related and ROS-scavenging enzymes (superoxide dismutase, ascorbate peroxidase, guaiacol peroxidase, catalase as well as phenylalanine ammonia-lyase and polyphenoloxidase). Xia *et al.* (2009) verified the effects of 24-EBL (0.1 μM) in cucumber (*Cucumis sativus* L. cv. Jinyan No. 4) seedlings submitted to cucumber mosaic virus (CMV) inoculation. Water-treated plants developed typical CMV symptoms 10 days post-inoculation. CMV-disease severity and malondialdehyde content in EBR-treated plants were lower than those of water-treated plants, the authors showed that EBR treatment induced expression of genes involved in defense response.

3.2 Water Stress

In plants submitted to stresses, one of the first symptoms is the synthesis of enzymes whose functions are to maintain the cell membrane integrity, the antioxidant enzymes like catalase, peroxidase and superoxide dismutase, the ABA content increases and, as a consequence, stomata close in order to diminish loss of water by transpiration (Table 6).

In soybean (*Glycine max* L.), BL (0.1 $mg.L^{-1}$) was applied by foliar spray at the beginning of bloom in plants submitted to two levels of soil moisture (80% field capacity for well watered control and 35% for drought stress treatment) (Zhang *et al.*, 2008). BR increased biomass accumulation and seed yield for both treatments. Drought stress inhibited translocation of assimilated ^{14}C from the labeled leaf, but BR increased the translocation for both treatments. BR treatment increased maximum quantum yield of PSII, the activity of ribulose-1,5-bisphosphate carboxylase and the leaf water potential of drought stressed plants. BR also increased the concentration of soluble sugars and proline and the activities of peroxidase and superoxide dismutase of soybean leaves when drought-stressed. However, it decreased the malondialdehyde concentrations and electrical conductivity of leaves under drought stress. This study shows that BR can be used as a plant growth regulator to enhance drought tolerance and minimize the yield loss of soybean caused by water deficits.

Table 6. Effects of brassinosteroids on water stress

Species	Hormone and concentrations applied	Organ which received BR	Mode of application*	Physiological effects analyzed	Author
Glycine max	BL (0.1 mg/L)	Leaves	Foliar Spraying at the beginning of bloom	Biomass accumulation, grain yield, quantum yield of PSII, enzymes activities, soluble sugar and proline content	Zhang *et al.* (2008)
Robinia pseudoacacia	BL (0 to 0.4 mg/L) BL (0 to 0.4 mg/L)	Roots and leaves of seedlings (1-y-old)	Root soaking before planting Foliar spraying	Soluble sugar and proline content, antioxidant enzymes activities, gas exchanges, seedling survival	Li *et al.* (2008)
Sorghum vulgare	28-HBL (2 and 3 μM) 24-EBL (2 and 3 μM)	Seeds	Petriplates provided with filter paper supplied with BRs	Antioxidant enzymes activities, soluble proteins and proline content	Vardhini and Rao (2003)
Phaseolus vulgaris	28-HBL (1 and 5 μM) 24-EBL (1 and 5 μM)	Seedling at flowering stage	Foliar spraying	Root nodulation, endogenous ABA and cytokinin, nitrogenase activity	Upreti and Murti (2004)
Arabidopsis thaliana *Brassica napus*	24-EBL (1 μM)	Seed	Addition to nutrient solution	Growth, morphological changes	Kagale *et al.* (2007)

* All the BR-treatments were done before water stress

In *Robinia pseudoacacia* seedlings, soaking roots in BL (0–0.4 mg.L^{-1}) prior to planting increased the activities of POD, SOD and catalase in water stressed plants when compared to the control plants and the survival and growth of seedlings (Li *et al.*, 2008). The best results were in 0.2 mg.L^{-1} BL treatment, because plants decreased the transpiration rate, stomatal conductance and malondialdehyde content. In sorghum (*Sorghum vulgare* Pers), 28-HBL and 24-EBL application in susceptible and resistant varieties reduced peroxidase and ascorbic acid oxidase activities while catalase activity was increased in plants submitted to osmotic stress (Vardhini and Rao, 2003). The higher activity of catalase in brassinosteroid treated seedlings of sorghum might have resulted in increased oxidation of harmful substrates, leading to increased seedling growth. A crucial mechanism in the adaptation process of plants to water stress is the osmotic adjustment because it might support the tissue metabolic activity and provide *de novo* growth after rewatering. The osmotic compounds include proteins and amino acids. Under osmotic stress (polyethylene glycol alone), there was increase in the levels of free proline in seedlings as compared to unstressed control seedlings. The application of brassinosteroids further increased the levels of free proline under osmotic stressed conditions. The free proline levels were higher in 3 μM brassinosteroid treatments, where the stress alleviation was also found to be maximum.

Application of 1 and 5 μM EBL or HBL prior to water stress induction in the nodulated roots of *Phaseolus vulgaris* L. cv Arka Suvidha was studied (Upreti and Murti, 2004). Brassinosteroids in the unstressed plants increased root nodulation, zeatin contents and nitrogenase activity and also ameliorated their stress-induced decline in the nodulated roots, the response was more prominent at 5 μM concentration in 4-days stressed plants. Among the brassinosteroids, EBL was relatively more effective than HBL (Upreti and Murti, 2004).

Arabidopsis thaliana and *Brassica napus* seedlings grown on nutrient solution containing 1 μM 24-EBL and then transplanted to sand were subjected to drought stress by withholding water for 96 or 60 hours (Kagale *et al.*, 2007). Visible morphological changes in response to drought stress, such as leaf wilting, reduction in growth and complete drying of some seedlings were frequently observed in untreated but were considerably reduced in EBR-treated seedlings.

3.3 Saline Stress

Proline accumulation is an indicator of saline stress; this compound is accumulated in order to maintain plant water relations. BR application enhances proline accumulation (Table 7).

The application of 24-EBL and 28-HBL to rice seeds (*Oryza sativa* L.) reduced the impact of salt stress on growth, prevented photosynthetic pigment loss and increased nitrate reductase activity (Anuradha and Rao, 2003). The seeds treated with 3.0 μM of brassinosteroid solution considerably reduced the growth inhibitory effect of salt stress as reflected in the growth of the plants.

Osmotic stress-induced accumulation of proline, an important protective osmolyte in higher plants, is dependent on the expression of Δ^1-pyrroline-5-carboxylate synthase (P5CS) and proline dehydrogenase (PDH) enzymes that catalyze the rate-limiting steps of proline biosynthesis and degradation, respectively. Stimulation of proline synthesis by abscisic acid and salt stress correlates with a striking activation of P5CS1 expression in *Arabidopsis* (Ábrahám *et al.*, 2003). By contrast, P5CS2 is weakly induced whereas PDH is inhibited to different extent by ABA and salt stress in shoots and roots of light-grown plants. Proline accumulation and light-dependent induction of P5CS1 by abscisic acid and salt stress is inhibited in dark-adapted plants. During dark adaptation P5CS2 is also down-regulated, whereas PDH expression is significantly enhanced in shoots. The inhibitory effect of dark

Table 7. Effects of brassinosteroids on saline stress

Species	Hormone and some concentrations applied	Organ which received BL	Mode of application	Physiological effects analyzed	Author
Oryza sativa	24-EBL (3 μM) 28-HBL (3 μM)	Seed	Soaking seeds (pre-treatment)	Photosynthetic pigments, nitrate reductase activity	Anuradha and Rao (2003)
Arabidopsis thaliana	24-EBL (0.1 μM)	Seedling	Medium constituent (Pre-treatment)	Proline content, proline metabolism enzymes	Ábrahám *et al.* (2003)
Oryza sativa	BB-16 (0.001, 0.01 mg/dm^3)	Seedling	Medium constituent (pre-treatment)	Antioxidant enzymes	Núñez *et al.* (2003)
Oryza sativa	24-EBL (3 μM)	Seed	Soaking seeds (pre-treatment)	Growth, antioxidative system, lipid peroxidation, proline and soluble protein content	Özdemir *et al.* (2004)
Brassica napus	24-EBL (1 and 2 μM)	Seed	Medium constituent Co-application (EBL and NaCl)	Germination and growth	Kagale *et al.* (2007)
Cicer arietinum	28-HBL (10^{-10} and 10^{-8} M)	Seed	Soaking seeds (applied before and after NaCl application)	nitrate reductase and carbonic anhydrase activities, nodule number	Ali *et al.* (2007)
Brassica junceae	28-HBL (10^{-10}, 10^{-8} and 10^{-6} M)	Seedling	HBL solution added to soil (after salt stress)	Nitrate reductase and carbonic anhydrase activities, chlorophyll content and photosynthetic rate, proline content	Hayat *et al.* (2007a)
Medicago sativa	BL (5 μM/L)	Seed	Seed soaking (pre-treatment to salt stress)	Germination and seedling growth, lipid peroxidation, antioxidant enzymes	Zhang *et al.* (2007)
Triticum aestivum	24-EBL (0.052, 0.104 and 0.156 μM)	Seed and seedling	Growth medium (co-application with salt stress)	Growth and grain yield	Ali *et al.* (2008a)
Triticum aestivum	24-EBL (0.0125, 0.025 and 0.0375 mg.l^{-1}/25 mL per pot)	Whole plant (~40-d-old)	Foliar spray	Growth and photosynthesis, enzymes	Shahbaz *et al.* (2008)
Zea mays	28-HBL (10^{-7}, 10^{-9} and 10^{-11} M	Seed	Seed soaking (Co-application with NaCl)	Growth, lipid peroxidation and antioxidative enzyme activities	Arora *et al.* (2008)
Spirulina platensis	24-EBL(0.5, 1.0 and 3.0 μM)	Culture	Constituent medium (Co-application with NaCl)	Growth and proline content	Saygideger and Deniz (2008)
Hordeum vulgare	24-EBL (3.0 μM)	Root meristem cells	Seed soaking	Mitotic activity	Tabur and Demir (2009)
Sorghum bicolor	24-EBL			Leaf development	Amzallag (2004)

adaptation on P5CS1 is mimicked by the application of 24-EBL (0.1 μM) for 3 days before the addition of either ABA or NaCl. The fact that both ABA and salt induction of P5CS1 transcription is inhibited by BL in light-grown plants suggests that steroid hormones may negatively regulate this common salt and ABA response pathway. Alternatively, BL may inhibit the light or sugar (or both) regulated maintenance of basal P5CS1 transcription which is essential for further induction by salt and ABA. Núñez *et al.*, (2003) studied the effects of a polyhydroxilated spirostanic analogue of brassinosteroid (BB-16) on the activities of antioxidant enzymes in rice seedlings (*Oryza sativa* L. cv. J-104), susceptible to saline stress, grown *in vitro* supplemented with NaCl. Seedlings exposed to 0.01 $mg.dm^{-3}$ for 16 days showed significant increase in the activities of catalase, superoxide dismutase and glutathione reductase and a slight increase in ascorbate peroxidase. On the other hand, 4 days exposure to BB-16 only increased superoxide dismutase and catalase activities at concentration 0.001 $mg.dm^{-3}$ BB-16. These results indicate that BB-16, which is structurally modified in the lateral chain in relation to natural brassinosteroids, changes the activity of key antioxidant enzymes, which might confer tolerance to saline stress. The effects of 24-EBL on seedling growth, antioxidative system, lipid peroxidation, proline and soluble protein content were investigated in seedlings of the salt-sensitive rice (*Oryza sativa* L.) cultivar IR-28 (Özdemir *et al.*, 2004). Seed application of 24-EBL (3 μM) improved seedling growth, alleviated the lipid damage and decreased proline accumulation caused by salt stress (120 mM) in a salt-sensitive rice variety. However, except for ascorbate peroxidase, it did not increase the activities of peroxidase, catalase and glutathione reductase under salinity stress. To determine the influence of 24-EBL salt-stress induced inhibition of *Brassica napus*, seed germination were allowed to germinate on a nutrient medium containing 1 or 2 μM of 24-EBL and different concentrations of NaCl (Kagale *et al.*, 2007). Presence of 24-EBL in the medium in particular at a concentration of 2 μM, considerably reduced the inhibitory effect of high salt on seed germination as evidenced by increase in germination and early seedling growth. Seeds of chickpea (*Cicer arietinum* L. cv. KPG-59) imbibed in aqueous solution of 10^{-10} or 10^{-8} M of 28-HBL and NaCl (1 or 10 mM) were evaluated (Ali *et al.*, 2007). The plants resulting from the seeds soaked in HBR (10^{-8} M) possessed higher leaf nitrate reductase and carbonic anhydrase activities, more dry mass, higher nodule number and more nodule fresh and dry mass, compared with water soaked, control. These values declined significantly in plants raised from the seeds soaked in NaCl. This effect was overcome, if NaCl treatment was given before or after HBR treatment. Other study verified the effect of 28-HBL (10^{-10}, 10^{-8}, 10^{-6} M) on salinity-induced changes in *Brassica juncea* Czern. and Coss cv.Varuna (Hayat *et al.*, 2007a). Plants that received only NaCl (50, 100 or 150 mM) treatment exhibited a decrease in nitrate reductase and carbonic anhydrase activities, chlorophyll content and photosynthetic rate 60 days after sowing. Subsequent treatment with HBR significantly increased all of these parameters. The 10^{-8} M concentration of HBR generated the best response and also overcame the detrimental effects when NaCl concentration was 50 mM. The HBR concentration of 10^{-8} M along with the NaCl concentration of 150 mM resulted in the increased concentration of tissue proline concentration compared to the other treatments.

Zhang *et al.* (2007) tested the seeds of three lucerne (*Medicago sativa* L.) varieties (cv. Victor, Victoria and Golden Empress) to investigate the effects of seed priming with 5 $\mu M.L^{-1}$ BL on germination and seedling growth under a high level of salt stress. Seed priming with BL improved the salt tolerance of lucerne seedlings. This was supported by increasing germination ability, root length, root vigour, root dry weight and shoot fresh and dry weight under a high level salt stress (13.6 dSm^{-1} NaCl solution). It was also demonstrated by the increase in peroxidase, catalase and superoxide dismutase activities and the lower malondialdehyde, reflecting the level of lipid peroxidation in lucerne seedlings.

Ali *et al.* (2008a) verified that root applied 24-EBL improved growth and yield of two wheat (*Triticum aestivum*) cultivars (S-24, salt tolerant and MH-97, moderately salt-sensitive). Plants were grown at 0 or 120 mM NaCl in continuously aerated Hoagland's nutrient solution. Different concentrations of 24-EBL (0.052, 0.104, 0.156 μM) were also maintained in the nutrient solution. Exogenous application of 24-EBL counteracted the salt-stress induced growth and grain yield inhibition of both wheat cultivars. The most effective concentrations for promoting growth were 0.104 and 0.052 μM under normal and saline conditions. However, root applied 0.052 μM 24-EBL enhanced the total grain yield and 100 grain weight of salt stressed plants of both cultivars and suggested that total grain yield was mainly increased by an increase in grain size which might have been due to 24-EBL induced increase in translocation of more photoassimilates towards grain. Growth improvement in both cultivars due to root-applied 24-EBL was found to be associated with improved photosynthetic capacity. Shahbaz *et al.* (2008) using the same wheat cultivars (S-24 and MH-97) under salinity stress (150 mM NaCl), verified that foliar spray of 24-EBL (0.0125, 0.025 and 0.0375 $mg.L^{-1}$) increased plant biomass and leaf area per plant of both cultivars under non-saline conditions. However, under saline conditions, improvement in growth due to exogenous EBR was observed only in S-24 (salt tolerant cultivar). Photosynthetic rate was reduced due to salt stress in both cultivars, but this inhibitory effect was ameliorated significantly by the exogenous application of EBR. The most effective dose in improving growth of both cultivars due to EBR spray under non-saline or saline conditions was found to be 0.025 mg.L^{-1}. EBR induced increase in growth was associated with improved photosynthetic capacity.

Arora *et al.* (2008) studied the effects of 28-HBL on seedling growth, lipid peroxidation and antioxidative enzyme activities in the seedlings of *Zea mays* L. var. Partap-1 under salt stress. The seeds were germinated in recipients containing different concentrations of NaCl (25, 50, 75 and 100 mM) only, 28-HBL (10^{-7}, 10^{-9} and 10^{-11} M) only and NaCl supplemented with 28-HBL for 7 days (Arora *et al.*, 2008). It was observed that 28-HBL treatments reduced the toxicity of salt on seedling growth considerably, 10^{-9} M concentration being the most effective. Lipid peroxidation level was significantly increased under saline stress, but lowered with HBR applications revealing less oxidative damage. Further HBR treatments to the seedlings showed an enhancement in activities of superoxide dismutase, guaiacol peroxidase, catalase and ascorbate peroxidase. The activities of all antioxidative enzymes were further increased in seedlings treated with solution containing HBR and salt together as compared to seedlings treated with different concentrations of salt solution only. The concentration of malondialdehyde (MDA) got increased by NaCl treatments but decreased with HBR supplementations. The MDA content of seedlings treated with different concentrations of salt in combination with various concentrations of HBR showed maximum decrease in 10^{-9} M concentration of HBR.

The biomass, growth and free proline concentration were investigated in *Spirulina platensis* treated with different concentrations of NaCl (50, 100, 150 and 200 mM) and 24-EBL (0.5, 1.0 and 3.0 μM) over 5 days (Saygideger and Deniz, 2008). Among the cultures supplied with different combinations of NaCl and EBR, growth rate was maximal for the culture containing 150 mM NaCl and 1.0 μM combination. Free proline concentration also increased in *S. platensis* under salinity stress, but EBR showed no notable effect on proline. Cytogenetic response of 24-EBL was evaluated under different NaCl conditions (0.3, 0.35 and 0.4M NaCl) on root meristem cells of barley (*Hordeum vulgare* L. cv. Bülbül 89) seeds (Tabur and Demir, 2009). EBR pretreatment in higher concentrations of salt (0.4 M NaCl) caused total inhibition of mitotic activity in root tip cells. However, comparison of all concentrations of salt and control revealed to have a successful performance in ameliorating the detrimental effects of salinity on chromosomal abnormalities. Leaf development of salt-treated *Sorghum bicolor* L. Moench plants are influenced by treatment with 24-EBL but only during a short period in development (Amzallag, 2004). The effects of EBR on leaf malformations during their unfolding in plants exposed to 150mM NaCl showed that treatments with 24-EBL enable modification of initiation, duration and intensity of this critical period of reorganization. It is suggested that BR, at specific concentrations and time in development may induce changes in cellular sensitivity to many growth regulators.

3.4 Thermal Stress

Plant chilling and plant heat injury inhibit growth by an effect on some essential metabolic enzymes from photosynthesis and respiration. BRs application induces the synthesis of heat-shock proteins, antioxidant enzymes and the expression of cold-related genes (Table 8).

Table 8. Effects of brassinosteroids on thermal stress (heat stress (HS) and cold stress (CS)

Species	**Type of thermal stress**	**BR and concentrations applied**	**Organ which received BR**	**Mode of application**	**Physiological effects analyzed**	**Author**
Lycopersicon esculentum	Heat stress	24-EBL (1, 10 and 20 µM) 24-EBL (1, 10 and 50 µM)	Leaves (4 weeks old) Pollen grain (in vitro)	Spray (before HS) Addition to medium (before HS)	Plant survival, CO_2 gas exchange Mitochondrial small heat shock proteins	Singh and Shono (2005)
Lycopersicon esculentum	Heat stress	24-EBL (2.12 and 10.6 nM) MH5 (2.12 and 10.6 nM)	Leaf discs	Incubation in Petri dishes (before HS)	Antioxidant enzymes activities Pollen viability	Mazorra *et al.* (2002)
Lycopersicon esculentum	Heat stress	24-EBL (0.01, 0.1 and 1.0 mg/L)	Whole plant	Spray (before HS)	CO_2 gas exchange	Ogweno *et al.* (2008)
Brassica napus	Heat stress	BL (10^{-6} M)	Seedling	Spray (before HS)	Endogenous ABA content	Kurepin *et al.* (2008)
Arabidopsis thaliana	Heat stress	24-EBL (1 µM)	Seed	Addition to medium (before HS)	Bleaching	Kagale *et al.* (2007)
Chorispora bungeana	Cold stress	24-EBL (0.05 mg/L)	Cultured cells	Medium constituent	Reactive oxygen species, lipid peroxidation, antioxidant enzymes	Liu *et al.* (2009)
Vigna radiata	Cold stress	EBL (3 mM)	Seedling (5-day-old)	Spray (applied after cold stress)	Proteins	Huang *et al.* (2006)
Brassica napus	Cold stress	24-EBL (0.05, 1 µM) 24-EBL (1 µM)	Seedling: Cotyledons Primary leaves	Injection into the apoplast by gentle pressure (applied before CS)	Membrane permeability, pigment content	Janeckzo *et al.* (2007)
Brassica napus	Cold stress	24-EBL (1 µM)	Seed	Addition to medium (before CS)	Transcripts of cold-related genes	Kagale *et al.* (2007)
Arabidopsis thaliana	Cold stress	24-EBL (1 µM)	Seed	Addition to medium (before CS)	Transcripts of cold-related genes	Kagale *et al.* (2007)
Cucumis sativus	Cold stress	24-EBL (0.1 µM)	Seedling	Spraying (before CS)	Electron transport rate	Xia *et al.* (2009)

3.1.1 Heat Stress

Tomato plants (*Lycopersicon esculentum* Mill.) treated with 24-EBL are more tolerant to high temperature stress than untreated plants (Singh and Shono, 2005). When it was analyzed the mitochondrial small heat shock proteins (Mt-sHSP), the authors verified that these proteins did not accumulate in EBR treated plants (1 μM) at 25°C, although treatment of plants at 38°C induced much more accumulation of Mt-sHSP proteins in EBR treated than in untreated plants. EBR possibly induced thermotolerance in tomato plants and these plants had better photosynthetic efficiency. Exposure to 45°C for 3 hours completely killed more than 90% untreated plants, while 1 μM EBR application was found to be most effective for survival of tomato plants at lethal temperature. Besides, *in vitro* pollen germination at high temperature showed a varied response to different EBR concentrations. About 50 μM EBR totally inhibited pollen germination; however, we observed a significant increase in *in vitro* pollen germination with the control at high temperature. Other effects of 1 μM EBR on pollen viability included enhanced pollen tube growth and reduced pollen bursting during heat stress. In other study with tomato discs (*Lycopersicon esculentum* Mill.) cv. Amalia, it was verified that the effects of 24-EBL (10.6 and 2.12 nM) and MH-5 (10.6 and 2.12 nM), a polyhydroxylated spirostanic analogue, in the activity of the enzymes catalase, peroxidase and superoxide dismutase at 25 and 40°C (Mazorra *et al.*, 2002). Both concentrations of EBR and MH-5 stimulated the activity of SOD at 25 and 40°C, the MH-5-stimulated increase of this enzyme was greater. Superoxide dismutase is a key enzyme in the detoxification of superoxide radicals. The increased superoxide dismutase activity after EBR treatment at 25°C suggests that EBR-promoted activation of SOD might decrease the possible toxic concentrations of O_2^- radicals. Peroxidase activity was unaffected at 25°C, while at 40°C the activity was enhanced by both compounds. The changes in catalase activity markedly depended on the structure of BRs, doses and temperature. The results suggest a possible role of EBR and MH-5 in the reduction of cell damage produced by heat stress due to induction of enzymatic antioxidants.

When BL (0.1% aqueous ethanol plus BL at 10^{-6} M) was applied to canola seedlings (*Brassica napus* L. cv Westar) at 20°C and 45°C (heat stress), Kurepin *et al.* (2008) verified that endogenous abscisic acid concentration was not affected in plants maintained at normal temperatures. However abscisic acid concentration was significantly elevated by heat stress alone and doubled by heat stress plus BL. These results suggest that the well-known enhancement of tolerance to high temperature stress that can be obtained by brassinosteroid applications may be caused by a brassinosteroid-induced elevation in endogenous abscisic acid concentration. When exogenously 24-EBL concentrations (0.01, 0.1 and 1.0 mg.L^{-1}) were applied in tomato (*Lycopersicon esculentum* Mill. cv. 9021) exposed to high temperature (40/30°C), the net photosynthetic rate, stomatal conductance and maximum carboxylation rate of Rubisco (ribulose 1,5-bisphosphate carboxylase oxygenase) were decreased (Ogweno *et al.*, 2008). The activities of antioxidant enzymes such as superoxide dismutase, ascorbate peroxidase, guaiacol peroxidase and catalase increased during heat treatments and these increases proved to be more significant in EBR-treated plants (0.1 mg.L^{-1} EBR). EBR application also reduced total hydrogen peroxide (H_2O_2) and malonaldehyde (MDA) contents, while significantly increased shoot weight following heat stress. It was concluded that EBR could alleviate the detrimental effects of high temperatures on plant growth by increasing carboxylation efficiency and enhancing antioxidant enzyme systems in leaves. EBR showed a concentration-dependent effect on net CO_2 assimilation at high temperatures (14.8%, 26.6% and 10.2% for 0.01, 0.1 and 1.0 mg.L^{-1} EBR treatments, respectively). The results indicated that the rate-limiting enzyme Rubisco in the Calvin Cycle and other enzymes involved in RuBP regeneration were protected by EBR pretreatment and functioned well under heat stress. It is clear that heat stress increased lipid peroxidation in plants and this was significantly alleviated by EBR pretreatment.

Arabidopsis thaliana seedlings were exposed to 43°C in the presence and absence of 1µM 24-EBL for 1, 2, 3 or 4 hours and then allowing them to recover at 22°C for 7 days (Kagale *et al.*, 2007). Untreated seedlings exposed to 2, 3 and 4 hours of heat stress exhibited increasing levels of bleaching, whereas in EBR-treated seedlings, mild to severe bleaching was observed only with 4 hours of heat stress.

3.1.2 Cold Stress

Suspension cultured cells of *Chorispora bungeana* with or without 24-EBL (EBR) application were exposed to 4 and 0°C for 5 days (Liu *et al.*, 2009). The 24-EBL (0.05 mg.L^{-1}) treated cells exhibited higher viability after exposure to low temperatures compared with the control. Under chilling stress, reactive oxygen species (ROS) levels and lipid peroxidation were increased in the cultured cells which were significantly inhibited by EBR treatment. The activities of antioxidant enzymes such as ascorbate peroxidase, catalase, peroxidase and superoxide dismutase were increased during chilling treatments and these increases were more significant in the EBR-applied suspension cells. The EBR treatment also greatly enhanced contents of ascorbic acid and reduced gluthatione under chilling stress. From these results, it can be concluded that EBR could play the positive roles in the alleviation of oxidative damage caused by ROS (reactive oxygen species) overproduction through enhancing antioxidant defense system, resulting in improving the tolerance of *C. bungeana* suspension cultures to chilling stress. BRs may reduce chilling injury of plant cell membranes due to lipid peroxidation, therefore protecting the structural integrity of the membranes and resulting in the enhancement of chilling tolerance.

Mung bean epycotils (*Vigna radiata* L.) whose growth was initially suppressed by chilling partly recovered their ability to elongate after treatment with 24-EBL (10 μM); 17 proteins down-regulated by this chilling were re-up-regulated and these up-regulated proteins are involved in methionine assimilation, ATP synthesis, cell wall construction and the stress response (Huang *et al.*, 2006). Other experiment using 24-EBL to investigate the effect on cold resistance of rape seedlings (*Brassica napus* L. cv. Lycosmos), verified that at 2°C, BR injection into cotyledons (0.05 and 1.00 μM) or primary leaves (1.0 μM) abolished the effect of cold on permeability as plants submitted to cold treatment without BR injection elevated the membrane permeability (Janeckzo *et al.*, 2007). BR solutions strongly elevated the membrane permeability at 20°C. In seedlings exposed to 2°C, pigments content was significantly higher in BR-treated leaves as compared to control. There were no differences between pigment contents of leaves injected with BR solutions or only water/ethanol at 20°C. The increase in membrane permeability at 20°C is probably due to the hormonal effect of 24-EBL on cell elongation. On the contrary, the elevated ion leakage induced by infiltration and cold stress at 2°C can be alleviated or abolished by BR treatments, which shows the stress protecting effect of BR.

Brassica napus and *Arabidopsis thaliana* grown on a nutrient solution containing 1 μM 24-EBL were exposed to cold stress by transferring seedlings to a growth chamber set at 2°C for a maximum of 3 days, the transcripts of cold-related genes analyzed clearly accumulated to higher levels in 24-EBL treated plants (Kagale *et al.*, 2007).

Xia *et al.* (2009) verified the effects of 24-EBL (0.1 μM) in cucumber (*Cucumis sativus* L.) cv.Jinyan No.4 seedlings submitted to chilling stress. Chilling stress (8°C) caused significant reduction in electron transport rate. EBR treatment alleviated chilling stress and enhanced the electron transport rate.

3.5 Pesticide stress

Pesticides induce destruction of pigments, hormonal imbalance and other harmful effects, recent studies have verified the potential to alleviate stress caused by pesticides (Table 9).

Table 9. Effects of brassinosteroids on pesticides stress

Species	Type of pesticide	BRs concentrations applied	Organ which received BR	Mode of application	Physiological effects analyzed	Author
Eucalyptus grandis	Imazapyr Glyphosate	BB-16 (0.08 and 0.16 mg/L)	Plants (3-month-old)	Spraying (before and after pesticide application)	Photosynthesis	Silva *et al.* (2009)
Cucumis sativus	Paraquat, fluazifop-ρ-butyl, haloxyfop, flusilazole, cuproxat, cyazofamid, imidacloprid, chlorpyrifos and abamectin	24-EBL (0.1 mg/L)	One half of the seedling	Spraying 1 day before pesticide treatment	gas exchange and chlorophyll fluorescence measurements	Xia *et al.* (2006)
Cucumis sativus	Paraquat	24-EBL (0.1 μM)	Seedling	Spraying before pesticide treatment	chlorophyll fluorescence measurements	Xia *et al.* (2009)

The effects of Imazapyr (0.750 $kg.ha^{-1}$) and Glyphosate (1.440 $kg.ha^{-1}$), two types of herbicides, and their interactions with a spirostanic analogue of castasterone (BB-16) (0.08 $mg.L^{-1}$ and 0.16 $mg.L^{-1}$) on the growth of seedlings clones of *Eucalyptus grandis* were evaluated (Silva *et al.*, 2009). A treatment verified BB-16 12 hours before herbicides application and other treatment verified the application of BB-16 after herbicide application. The seedlings that received the glyphosate application associated or not to BB-16, independently of concentration and time, exhibited necrosis before the seventh day, while the seedlings that received Imazapyr associated to BB-16 showed only necrotic lesion at the extremity of the lateral branches. The interaction of Glyphosate and BB-16 increased the potential to lead the plant to death while Imazapyr and BB-16 showed the potential to alleviate stress during a short time after the beginning of treatments when compared to the application of Imazapyr only. These finding suggests that the tolerance found in plants submitted to Imazapyr versus BB-16 may be related to the receptors of BB-16 and Imazapyr in plasma membrane.

The phytotoxicities of nine pesticides (paraquat, fluazifop-ρ-butyl, haloxyfop, flusilazole, cuproxat, cyazofamid, imidacloprid, chlorpyrifos and abamectin) at practical dosages on photosynthesis were investigated in cucumber (*Cucumis sativus* L. cv. Jynian No.4) at four leaf-stage by gas exchange and chlorophyll fluorescence measurements (Xia *et al.*, 2006). Inhibition of net photosynthetic rate (P_N) were alleviated by 24-EBL (0.1 $mg.L^{-1}$ 1 day before pesticide treatment) for the pesticides examined except paraquat and flusilazole. EBR pretreatment also increased quantum efficiency of photosystem II and photochemical quenching coefficient (qP). It is likely that EBR enhanced the resistance of cucumber seedlings to pesticides by increasing CO_2 assimilation capacity and activities of antioxidant enzymes. Ultrastructural studies showed that 22(S),23(S)-HBL (1.0 μM) applied to potato leaves (*Solanum tuberosum* L. cv. Désirée) significantly reduced H_2O_2 negative effects on cellular sub-structures, allowing better recovery of affected structures and reducing the macroscopic injury symptoms on leaves (Almeida *et al.*, 2005).

Xia *et al.* (2009) verified the effects of 24-EBL (0.1 μM) on cucumber (*Cucumis sativus* L.) plant sensitivity to paraquat which causes photo-oxidative stress. They compared the effects on photosynthetic efficiency by comparing the maximum photochemical efficiency of PSII in the dark adapted state (F_v/F_m). Fluorescence images showed that paraquat treatment resulted in a significant decrease in F_v/F_m in control plants and Paraquat-induced reduction in F_v/F_m was less in EBR-treated plants.

3.6 Heavy metals stress

BRs have ability to regulate the uptake of ions into the plant cells and they can be used to reduce the accumulation of heavy metals and radioactive elements in plants (Table 10).

Among pollutants of agricultural soils, Cu has become increasingly hazardous due to its involvement in fungicides, fertilizers and pesticides. However, Cu at high levels become strongly phytotoxic and cause inhibition of plant growth or even death. When *Brassica juncea* L. cv. PBR91 seeds were treated before germination with 24-EBL (10^{-7}, 10^{-9}, 10^{-11} M) and submitted to copper stress, there was an improvement in the shoot emergence and plant biomass production (Sharma and Bhardwaj, 2007). This compound at 10^{-7} M concentration was the most effective for lowering the Cu uptake and accumulation of ions. The fresh weight of the whole plant was increased in all the concentrations as compared to the control.

Although nickel is an essential element, required at low concentrations for urease metabolism, nickel at high concentration is toxic because it inhibits photosynthesis, respiration, enzymes activities and protein. Plants of *Brassica juncea* L. cv. T-59 were supplied with 50 or 100μM nickel at 10 days after sowing and sprayed with 28-HBL (10^{-8} M) at 20 days after sowing (Alam *et al.*, 2007). The plants treated with nickel alone exhibited reduced growth, net photosynthetic rate, content of chlorophyll and the activities of nitrate reductase and carbonic anhydrase, observed 40 days after sowing, whereas the contents of peroxidase, catalase and proline were increased. The spray of HBR partially neutralized the toxic effect of nickel on most of the parameters. Sharma *et al.* (2008) using *Brassica juncea* L. cv.

Table 10. Effects of brassinosteroids on heavy metals stress

Species	Heavy metal	BR and concentrations applied	Organ which received BL	Mode of application	Physiological effects analyzed	Author
Brassica junceae	Cu	24-EBL (10^{-7}, 10^{-9}, 10^{-11} M)	Seed	Seed soaking	Shoot emergence and plant biomass production	Sharma and Bhardwaj (2007)
Brassica junceae	Ni	28-HBL (10^{-8} M)	Leaf	Spraying	Growth, photosynthesis, enzymes activities	Alam *et al.* (2007)
Brassica junceae	Ni	28-HBL (0.01, 1.0 and 100 nM)	Seed	Seed soaking	Growth, protein content and antioxidative enzymes activities	Sharma *et al.* (2008)
Brassica napus	Cd	24-EBL (100 nM)	Seedling	Medium constituent *in vitro* culture	Chlorophyll fluorescence, photosynthetic pigments	Janeckzo *et al.* (2005)
Brassica junceae	Cd	28-HBL (0.01 μM)	Leaves	Spraying	Growth, chlorophyll pigments, enzymes, proline content	Hayat *et al.* (2007b)
Raphanus sativus	Cd	24-EBL (3 μM) 28-HBL (3 μM)	Seed	Seed soaking	Seed germination, seedling growth, proline content, enzymes, lipid peroxidation	Anuradha and Rao (2007a)
Cicer arietinum	Cd	28-HBL (0.01 μM)	Seedling	Spraying	number of nodules, leghemoglobin content, nitrogen and carbohydrate content, chlorophyll content, enzymes	Hasan *et al.* (2008b)
Vigna radiata	Al	24-EBL (10^{-8} M) 24-EBL (10^{-8} M)	Seedling	Foliar spray	Growth, photosynthesis, antioxidant enzymes	Ali *et al.* (2008b)
Phaseolus aureus	Al	BL (0.1, 10, 100 and 100,000 ng/L)	Seedling	Growth solution	Growth, chlorophyll content	Abdullahi (2003)
Chlorella vulgaris	Pb	20-hydroxyecdysone (10^{-10}–10^{-8} M)	Culture cells	Medium constituent	Growth, chlorophyll, sugar and protein content and phytochelatins synthesis	Bajguz and Godlewska-Zylkiewicz (2004)
Raphanus sativus	Pb	24-EBL			Antioxidant enzymes	Anuradha and Rao (2007b)

PBR91 seeds soaked for 8 hours in different concentrations of 28-HBL (0.01, 1.0 and 100nM) and submitted to nickel concentrations (25, 50 and 100 mg.dm^{-3}) verified, 7 days after germination, that the growth of seedlings was inhibited by Ni, however, less after HBL pre-treatment. The protein content and antioxidative enzymes activities (catalase, glutathione reductase, ascorbate peroxidase, superoxide dismutase, guaiacol peroxidase) were also increased by HBL treatment. The seed germination and seedling growth was significantly reduced by the Ni treatment but the HBL alone enhanced the germination percentage as well as shoot and root length (maximum germination observed with 100 $mg.dm^{-3}$ Ni and 1.0 nM HBL).

Cadmium is extremely toxic to plants. It retards biosynthesis of chlorophyll, alters water balance, decreases activity of various enzymes, favors stomatal closure, induces oxidative stresses in plants and slows down the rate of photosynthesis. Cadmium inhibits both the "light" and "dark" reactions of photosynthesis, but the Calvin cycle is more sensitive to its activity. The inhibition of photochemical processes by Cd may result from the limitation in the use of ATP and NADPH by the Calvin cycle and accompanying increase of pH gradient across the thylakoid membranes. Seedlings of winter rape (*Brassica napus* L.) cv. Górczánsky were cultured *in vitro* on media containing 24-EBL (100nM) and cadmium (300μM) (Janeckzo *et al.*, 2005). After 14 days of growth, fast fluorescence kinetics of chlorophyll a (Chl a) and contents of photosynthetic pigments and Cd in cotyledons were measured. Cd was strongly accumulated but its content in cotyledons was 14.7% smaller in presence of EBR. EBR reduces the toxic effect of Cd on photochemical processes by diminishing the damage of photochemical active reaction centers and the activity of O_2 evolving centers as well as maintaining efficient photosynthetic electron transport. The change in *Brassica juncea* plant growth and photosynthesis submitted to Cd (100 or 150 μM) and 28-HBL (0.01 μM) application was also verified by Hayat *et al.* (2007b). These authors observed that the plants fed with cadmium alone exhibited a decline in growth, in the levels of carbonic anhydrase and chlorophyll pigments and net photosynthetic rate. Nitrate content, the activity of nitrate reductase and the level of carbohydrate both in the leaves and roots decreased as the concentration of Cd increased. The toxic effect generated by Cd was overcome if the stressed plants were sprayed with HBL. The activities of antioxidant enzymes (catalase, peroxidase and superoxide dismutase) and the contents of proline increased over the control, irrespective of the treatments. Their level increased further, if the plants supplied with Cd were also supplemented with HBL. The effect of 24-EBL and 28-HBL on seed germination and seedling growth of radish (*Raphanus sativus* L.) was studied under Cd toxicity (Anuradha and Rao, 2007a). Both BRs at 3 μM concentration caused a considerable increase in seedling growth even under stress and restored the growth to the level of unstressed control seedlings. Besides, in response to Cd stress, radish seedlings accumulated proline (BR enhanced proline content), decreased catalase activity (BR increased) and peroxidase activity (BR reduced). However, Cd stress increased the activities of ascorbic peroxidase (BR enhanced), guaiacol peroxidase (BR enhanced), ascorbic acid oxidase (BR decreased), and superoxide dismutase

(BR enhanced). Lipid peroxidation induced by Cd was found reduced with the supplementation of BRs. Brassinosteroids strongly protect radish seedling from Cd induced oxidative stress by minimizing the impact of reactive oxygen species by increasing antioxidant enzyme activity, which may represent a secondary defensive mechanism against oxidative stress. The seedlings of *Cicer arietinum* L. cv. Uday were supplied with Cd (50, 100 and 150 μM) and sprayed with 0.01 μM of 28-HBL at 30-day stage (Hasan *et al.*, 2008). Plant fresh and dry mass, number of nodules, leghemoglobin content, nitrogen and carbohydrate content in the nodules, leaf chlorophyll content, nitrate reductase and carbonic anhydrase activities decreased proportionately with the increasing concentrations of Cd but the content of proline and the activities of catalase, peroxidase and superoxide dismutase increased. These effects were overcome if the stressed plants were sprayed with 28-HBL.

The aluminum toxicity is the major growth-limiting factor for crop cultivation on acidic soil. Seedlings of mung bean (*Vigna radiata* L. Wilczec) were subjected to aluminium (1 or 10 mM) stress at one week old stage and sprayed with 10^{-8} M of 24-EBL or 28-HBL at 14-day stage (Ali *et al.*, 2008b). The authors revealed that the level of antioxidant system (superoxide dismutase, catalase, peroxidase and proline) increased in response to Al stress that was further improved by brassinosteroid treatment (HBL and EBR). Therefore, it may be suggested that the ameliorated level of antioxidant system, at least in part, was responsible for the development of resistance against Al stress in mung bean seedlings. The increase in the degree of resistance due to the applications of BR was reflected in the improvement of plant growth, photosynthesis and related processes, in the presence of aluminium. It was also noticed that EBL was more effective than HBL. The difference between the effectiveness of these BR analogues is due to their structure and stability, because EBL is more stable than HBL under field conditions. Other study verified that BL (0.1, 10, 100, 100,000 ng.L^{-1}) promoted growth of mung bean (*Phaseolus aureus* Roxb.) seedlings under aluminium stress (2 and 5 mM) (Abdullahi, 2003). BL significantly increased fresh weights of shoots and roots and chlorophyll content under Al stress.

Lead is a heavy metal that accumulates in plant cell wall and causes growth inhibition. When *Chlorella vulgaris* cultures were inoculated with 10^{-6}–10^{-4} M lead, their growth and chemical composition decreased during the first 48 hours of cultivation. Application of 20-hydroxyecdysone (10^{-10}–10^{-8} M), considered a brassinosteroid-related compound, restored the decreased growth and composition of *Chlorella vulgaris* cells treated with lead (Bajguz and Godlewska-Zylkiewicz, 2004). This compound reduced the impact of lead stress on growth, prevented chlorophyll, sugar and protein loss and increased phytochelatins synthesis. Concentration-dependent stimulation was observed with increasing concentration of 20-hydroxyecdysone (20E) and decreasing concentration of lead. The supplementation of 24-EBL to radish seedlings (*Raphanus sativus* L.) reduced lead toxicity and enhanced the growth (Anuradha and Rao, 2007b). The activities of antioxidant enzymes (catalase, ascorbate peroxidase, guaiacol peroxidase, superoxide dismutase) showed an increase in brassinosteroid treated Pb-stressed seedlings when compared to control and a reduced peroxidase activity and an increase in the total glutathione content.

3.7 Hypoxia Stress

Kang *et al.* (2009) verified the effects of 24-EBL (1 $\mu g.L^{-1}$) added to nutrient solution on growth of cucumber (*Cucumis sativus* L.) under root-zone hypoxia, seedlings were hydroponically grown for 8 days in normoxic and hypoxic nutrient solutions with and without EBR. EBR added to hypoxic nutrient solution caused an increase in the concentration of fructose, sucrose and total soluble sugars in the roots but not in the leaves. EBR exerted little influence on plant performance in the nutrient solution, while EBR alleviated root-zone hypoxia-induced inhibition of root and shoot growth and net photosynthetic rate (P_N). EBR enhanced alcohol dehydrogenase activity but lowered lactate dehydrogenase activity in hypoxic roots. These results suggest that EBR may stimulate the photosynthate allocation down to roots and the shift from lactate fermentation to alcohol fermentation in hypoxic roots, resulting in the increase in ATP production through glycolysis and the avoidance of cytosolic acidosis and eventually enhanced tolerance of cucumber plants to root-zone hypoxia.

4. BRASSINOSTEROID POTENTIAL USE IN PLANT TISSUE CULTURE

As BR can affect plant elongation, cell division and vascular development influencing morphogenesis, the use of this plant regulator in plant biotechnology is promising as supplementation to medium (Table 11).

Spirostane analogues of brassinosteroids (BB-6 and MH-5) were tested for callus induction and plant regeneration in lettuce (*Lactuca sativa*) (Núñez *et al.*, 2004). The analogues enhanced both callus formation and shoot regeneration from cotyledons in lettuce when added at determined concentrations (0.001 or 0.01 $mg.L^{-1}$) with 0.1 $mg.L^{-1}$ 6-BA (benzyladenine, a cytokinin) in the culture medium. However, there was no callus induction when 6-BA was substituted by these analogues. These results showed that BB-6 and MH-5 stimulated callus formation in the presence of cytokinin.

Table 11. Effects of brassinosteroids on plant tissue culture as a constituent medium

Species	BR and concentrations applied	Physiological effects analyzed	Author
Lactuca sativa	BB-6 (0.001 mg/L or 0.01 mg/L) MH5 (0.001 mg/L or 0.01 mg/L)	Callus induction, shoot regeneration from cotyledons	Núñez *et al.* (2004)
Saccharum spp.	BB-6 (0.001 mg/L) MH5 (0.01 mg/L)	Protein metabolism	Nieves *et al.* (2007)
Nicotiana tabacum	BL (10^{-10}–10^{-6}M)	Cell division, cell-cycle related gene expression, organellar DNA content	Miyazawa *et al.* (2003)
Onosma paniculatum	BL (10, 10^3, 10^5 and 10^7 pg/L)	Growth and secondary metabolism (shikonin formation)	Yang *et al.* (2003)
Pinus taeda *Pseudotsuga menziesii* *Picea abies* *Oryza sativa*	BL (0.1 μM)	Somatic embryogenesis	Pullman *et al.* (2003, 2005, 2009)
Pinus wallichiana	24-EBL (2.0 μM)	Somatic embryogenesis	Malabadi and Nataraja (2007)
Spartina patens	BL (0.005–0.05 mg/L)	Callus growth and regeneration	Lu *et al.* (2003)
Arabidopsis thaliana	BL (1 μM)	Tracheary element differentiation	Oda *et al.* (2005)
Arabidopsis thaliana	28-HBL (1 mg/ml) 28-homocastasterone (1 mg/ml)	Cell expansion, membrane hyperpolarization	Zhang *et al.* (2005)
Gossypium hirsutum	BL (0.1, 0.5 and 1.0 μM)	Fiber development	Sun *et al.* (2005)
Gossypium hirsutum	BL (0.1, 0.5 and 1.0 μM)	Somatic embryogenesis	Aydin *et al.* (2006)
Brassica spp.	24-EBL (10^{-6}M) BL (10^{-7}M)	Microspore embryogenesis	Ferrie *et al.* (2005)
Hybrid (Eucalyptus grandis x *Eucalyptus urophylla)*	28-HCTS (4, 10, 25 and 62.5 mg/L) 5F-HCTS (4, 10, 25 and 62.5 mg/L)	Elongation and formation of new main shoots	Pereira-Netto *et al.* (2006a)
Malus prunifolia	5F-HCTS (0.5, 1.0, 5.0, and 10.0 μg/explants)	Branch elongation	Pereira-Netto *et al.* (2006b)
Nicotiana tabacum	BL (10^{-10}–10^{-8}M)	Shoot formation	Kim *et al.* (2008)
Cocos nucifera	22(S)23(S)-HBL (0.01, 0.1, 1.0, 2.0, 4 μM)	Initial callus formation, somatic embryogenesis	Azpeitia *et al.* (2003)

Taking into account the synergism reported between auxin and brassinosteroids, the presence of BB-6 or MH-5 in the culture medium may have increased the auxin/cytokinin ratio making it necessary to include 6-BA for callus formation. However, the brassinosteroid analogues did not show any effect on callus fresh weight nor shoot number per callus formation evaluated 25 days after culture initiation.

These two analogues, BB-6 and MH-5, were used in concentrations of 0.001–0.01 mg.L^{-1}, respectively, to evaluate the protein metabolism in sugarcane (*Saccharum* spp.) somatic embryogenesis (Nieves *et al.*, 2007). It was verified that BRs analogue treatments at high concentration did not differ from control (without analogues and one with NAA-naphthaleneacetic acid) regarding to somatic embryos production. Both BRs influenced total soluble proteins, storage proteins (albumins, globulins, prolamins and glutelins) and free-proline levels. Some storage proteins such as prolamins and glutelins showed decreases in their content in relation to control treatment. These results imply that the BRs studied were involved in differentiation and maturation of sugarcane somatic embryos caused by a decrease in proline synthesis. The authors suggest that BB-6 and MH-5 influenced proteins metabolism, in particular storage proteins, which are considered as an important nitrogen reserve in somatic and zygotic embryos.

To evaluate the BRs effects in cell division, the tobacco (*Nicotiana tabacum*) Bright Yellow 2 (BY-2) cell line, which is a widely-used model system in plant cell biology was used (Miyazawa *et al.*, 2003). BL (10^{-10}–10^{-6}M) promoted cell division only during the early phase of culture and in the absence of auxin (2,4D). At later stages in the culturing periods of BL-supplied and 2,4D-supplied BY-2 cells, differences in cell multiplication and cell-cycle related gene expression were observed. Moreover, the BL treated BY2 cells had morphological differences from the 2,4D treated cells. To determine whether suppressed organellar DNA replication limited this promotion of cell division during the early culture phase, this replication was examined and it was found that BL treatment had no effect on activating organellar (plastid and mitochondrial) DNA synthesis. These results suggest that the mechanism of the promotion of cell division by BL treatment is distinct from that regulated by the balance of auxin and cytokinin.

BL interact with IAA and 6-benzylaminopurine (BAP) to influence cell growth and secondary metabolism of cultured *Onosma paniculatum* cells (Yang *et al.*, 2003). In a BL and IAA interaction experiment, the optimal BL concentration for cell growth increased with IAA concentration. IAA concentrations of 0.05, 0.1, 1.0 and 10 mg.L^{-1} in the growth medium, the optimal BL concentrations for cell growth were 10, 10^3, 10^5 and 10^7 pg.L^{-1}, respectively. In a BL and BAP interaction experiment, cell growth decreased with increasing concentration of BL at any given concentration of BAP. The optimal concentrations of BL and IAA for cell growth were 10 pg.L^{-1} and 0.05 mg.L^{-1}, respectively. The optimal concentrations of BL and BAP for cell growth were 10 pg.L^{-1} and 0.5 mg.L^{-1}, respectively. The optimal concentrations of BL and IAA for enhanced shikonin (a compound resulted from the secondary plant metabolism with pharmaceutical potential) production were 10^7 pg.L^{-1} and 0.05 mg.L^{-1}, respectively and in BL and BAP combination the concentrations were 10^5 pg.L^{-1} (BL) and 0.5 mg.L^{-1} (BAP). BL increased phenylalanine ammonia-lyase (PAL) and p-hidroxybenzoic acid geranyltransferase (PHB-geraniltransferase) activities but decreased the activity of PHB-O-glucosyltransferase. These results suggest that enhanced shikonin formation induced by BL involves regulation of these key enzymes.

Somatic embryogenesis in rice and conifers can be improved by BL application (Pullman *et al.*, 2003). Using BL supplemented (0.1 μM) medium improved initiation percentages in loblolly (*Pinus taeda* L.) (15.0–30.1%), Douglas-fir (*Pseudotsuga menziesii* (Mirb.) Franco) (16.1–36.3%), Norway spruce (*Picea abies* L. Karst) (34.6–47.4%) and rice (*Oryza sativa* L.) (10%). BL (0.1 μM) increased the weight of loblolly pine embryogenic tissue by 66% and stimulated initiation in the more recalcitrant families of loblolly pine and Douglas-fir, thus compensating somewhat for genotypic differences in initiation. BL is part of the culture medium for embryogenic tissue initiation of Douglas fir (*Pseudotsuga menziesii* (Mirb.) Franco) (Pullman *et al.*, 2009) and is a constitutive of the culture medium for somatic embryogenesis of *Pinus taeda* and *Pseudotsuga menziesii* (Pullman *et al.*, 2005). *Pinus wallichiana* is one of the most recalcitrant species to *in vitro* propagation via somatic embryogenesis among all the Indian pines. The use of 24-EBL (2.0 μM) with 9.0 μM 2,4D in three genotypes of *Pinus wallichiana* enhanced the formation of embryogenic tissue from mature zygotic embryos on half-strength MSG basal medium. However the frequency of somatic embryogenesis was not similar in all the three genotypes tested (Malabadi and Nataraja, 2007).

The effect of BL on cultured calluses of *Spartina patens* (Ait.) Muhl., a halophyte monocot, was studied (Lu *et al.*, 2003). BL at 0.005 $m.L^{-1}$ (with benzyladenine at 0.2 $mg.L^{-1}$) and IAA (3 $mg.L^{-1}$) promoted regenerated shoot growth most significantly, increasing the shoot height increasing ratio by 395% after a 40-day culture. The authors suggested that BL at 0.03 $mg.L^{-1}$ is suitable for the callus growth and shoot regeneration, while BL at 0.005 $mg.L^{-1}$ effectively enhanced the regenerated shoot growth.

Brassinosteroids are thought to be related to tracheary element differentiation. In transgenic *Arabidopsis thaliana* cell suspension, BL applied exogenously at 1.0 μM promoted tracheary element differentiation in a dose dependent manner (Oda *et al.*, 2005). Zhang *et al.* (2005) verified that 28-HBL and its direct precursor 28-homocastasterone promoted cell expansion of *Arabidopsis thaliana* suspension cells and this cell expansion induced by HBL and HCS was correlated with the amplitude of the plasma membrane hyperpolarization they elicited. They observed that membrane hyperpolarization and cell expansion were partially inhibited by the proton pump inhibitor erythrosin B, suggesting that proton pumps were not the only ion transport system modulated by the two BRs. The authors also verified that anion currents were inhibited by HBL and HCS while outward rectifying K+ currents were increased by HBL but inhibited by HCS. The different electrophysiological behavior shown by these BRs indicates that small changes in the BR skeleton might be responsible for changes in bioactivity.

BRs regulate fiber development on cultured cotton (*Gossypium hirsutum* cv. Coker 312) ovules (Sun *et al.*, 2005). The application of BL (0.1 μM) stimulated fiber elongation while brassinazole (Brz2001) inhibited fiber development. Besides, treatment of cotton floral buds with brassinazole results in the complete absence of fiber differentiation, indicating that BR is required for fiber initiation as well as elongation. BL (0.1, 0.5 and 1 μM) was used to examine the potential effect on cotton somatic embryogenesis in cotton (*Gossypium hirsutum*) calli pieces (Aydin *et al.*, 2006). Somatic embryogenesis was stimulated especially for transition to cotiledonary phase at 0.5 $mg.L^{-1}$ BR. Histological preparations from embryogenic calli and somatic embryos at different stages of development revealed the spontaneous polyploidisation during early somatic embryogenesis on BR-treated calli. These results suggest that BR negatively affected calli growth, however, had a stimulating role in maturation of somatic embryos.

An increase in embryogenesis was observed in all *Brassica napus* lines evaluated including Topas 4079 and several recalcitrant cultivars. Garrisson, Westar and Allons treated with 24-EBL (10^{-6} M) and BL (10^{-7} M) (Ferrie *et al.*, 2005). The microspore embryogenesis, calculated as the number of embryos at 21 days of culture, was increased in the recalcitrant cultivars up to 12 times that of control. An increase in microspore embryogenesis was also observed for *Brassica juncea* when EBL or BL was added to the culture medium. In contrast, no significant increase in embryogenesis was observed for several other *Brassica* species evaluated (*Brassica nigra*, *Brassica carinata* and *Brassica rapa*). The addition of brassinosteroids to the induction media did not affect the subsequent conversion of the embryos to plantlets, but did appear to influence chromosome doubling.

28-Homocastasterone (28-HCTS) was used to treat *in vitro*-grown shoots of a hybrid between *Eucalyptus grandis* and *Eucalyptus urophylla* (Pereira-Netto *et al.*, 2006a). Treated shoots showed enhanced elongation and formation of new main shoots (the shoots originating directly from the initial explant) at low doses. Coincidently there was reduced elongation and formation of primary lateral shoots (shoots originating from the main shoot). However, a 5α-monofluoro derivative of 28-HCTS (5F-HCTS) was unable to either stimulate elongation or formation of new main shoots, although it did stimulate elongation of primary lateral shoots. The differential responses seen for these compounds on shoots of *Eucalyptus* suggest different BR bio-synthetic routes, differential chemical stability or perhaps different receptor sites for each compound. Although auxin and cytokinin were used in the culture medium, it was quite apparent that exogenously supplied brassinosteroids are able to change shoot patterns (apical dominance) in *Eucalyptus* and it seems likely that shooting in *Eucalyptus* might be influenced by the endogenous pool of bioactive brassinosteroids. Pereira-Netto *et al.* (2006b) demonstrated that 5F-HCTS (0.5, 5.0, 1.0 and 10.0 μg/explants) stimulated branch elongation in *in vitro*-grown shoots of *Malus prunifolia* (Wild.) Borkh, the marubakaido apple rootstock. The authors showed that this BR-stimulated branch elongation is paralleled by an increase in ethylene release.

When several concentrations of BL were added to a shoot induction medium that contained only benzyladenine, redifferentiation of adventious shoots from tobacco (*Nicotiana tabacum* L. cv. NT1) leaf discs was unaffected at low BL levels (10^{-10}–10^{-8} M), but was inhibited at higher concentrations (Kim *et al.*, 2008). When BL was applied without BA, only cell expansion occurred and no shoots formed. The determination time for shoot formation was shortened at low BL concentrations, but their formation was postponed at higher concentrations. In conclusion, at low concentrations, BL has no effect on shoot formation. However, it inhibits their formation at high concentrations when cytokinin is included in the media.

Explants from coconut (*Cocos nucifera* L.) were exposed to different concentrations of 22(S),23(S)-HBL (0.01, 0.1, 1.0, 2.0, 4 μM) and these explants responded favorably to the brassinosteroid, increasing their capacity to form initial callus, embryogenic callus and somatic embryos (Azpeitia *et al.*, 2003). The largest amount of somatic embryos formed was obtained exposing the explants for 3 days to the concentrations of 0.01 or 0.1 μM HBL.

5. CONCLUSION

Developing stage of the plant, the concentration and the time the brassinosteroid is applied as well as the types of brassinosteroid that are used are very important parameters to study the effects of exogenous BRs in plants. More studies should still be carried out in a great number of plants and other compounds with brassinosteroid-like responses should be tested. The great majority of works use brasinolides and EBLs but there are brassinosteroid analogues that are not commonly used and still have to be exploited. The beneficial effects of brassinosteroid application are incontestable mainly when plant stress conditions are evaluated. They enhance antioxidant enzymes activities and plant yield by improving the photosynthesis process. It seems that BRs effects are more detectable on seedlings and suspended cells than in older plants. The concentrations used vary a lot and the effects that were observed vary from plant to plant and from organ to organ.

Anyway, it is quite evident that BRs influence plant growth and development whether they are used alone or with other plant hormones such as auxins, gibberellins, ethylene or cytokinins or by interactions with other substances (phytochromes, salycilic acid, jasmonates and others). There really is a great expectation that BRs can elucidate other plant metabolic processes that have not been solved so far.

6. ACKNOWLEDGEMENTS

The author is grateful to CNPq and FAPERJ for financial support and Dr. Marco António Teixeira Zullo for precious information.

7. REFERENCES

Abdullahi, B.A., Gu, X.G., Gan, Q.L., and Yang, Y.H., 2003. Brassinolide amelioration of aluminum toxicity in mungbean seedling growth. *J. Plant Nutrition* **26**(9): 1725–1734.

Abraham, E., Rigo, G., Szekely, G., Nagy, R., Koncz, C., and Szabados, L., 2003. Light-dependent induction of proline biosynthesis by abscisic acid and salt stress is inhibited by brassinosteroid in *Arabidopsis*. *Plant Mol. Biol.* **51**(3): 363–372.

Alam, M.M., Hayat, S., Ali, B., and Ahmad, A., 2007. Effect of 28-homobrassinolide treatment on nickel toxicity in *Brassica juncea*. *Photosynthetica* **45**: 139–142.

Ali, B., Hayat, S., Hasan, S.A., and Ahmad, A., 2006. Effect of root applied 28-homobrassinolide on the performance of *Lycopersicon esculentum*. *Sci. Hort.* **110**: 267–273.

Ali, B., Hayat, S., and Ahmad, A., 2007. 28-Homobrassinolide ameliorates the saline stress in chickpea (*Cicer arietinum* L.). *Environ. Exp. Bot.* **59**: 217–223.

Ali, B., Hasan, S.A., Hayat, S., Hayat, Q., Yadav, S., Fariduddin, Q., and Ahmad, A., 2008b. A role for brassinosteroids in the amelioration of aluminium stress through antioxidant system in mung bean (*Vigna radiata* L. Wilczek). *Environ. Exp. Bot.* **62**: 153–159.

Ali, Q., Athar, H.U.R., and Ashraf, M., 2008a. Modulation of growth, photosynthetic capacity and water relations in salt stressed wheat plants by exogenously applied 24-epibrassinolide. *Plant Growth Regul.* **56**: 107–116.

Almeida, J.M., Fidalgo, F., Confraria, A., Santos, A., Pires, H., and Santos, I., 2005. Effect of hydrogen peroxide on catalase gene expression, isoform activities and levels in leaves of potato sprayed with homobrassinolide and ultrastructural changes in mesophyll cells. *Functional Plant Biol.* **32**: 707–720.

Altoé, J.A., Marinho, C.S., Muniz, R.A., Rodrigues, L.A., and Gomes, M.M.A., 2008. "Cleopatra" mandarin submitted to mycorrhization and to a brassinosteroid analogue. *Acta Sci. Agron.* **30**(1): 13–17.

Amzallag, G.N., 2004. Brassinosteroid: a modulator of the developmental window for salt-adaptation in *Sorghum bicolor*. *Israel J. Plant Sci.* **52**(1): 1–8.

Amzallag, G.N., and Vaisman, J., 2006. Influence of brassinosteroids on initiation of the root gravitropic response in *Pisum sativum* seedlings. *Biol. Plantarum* **50**: 283–286.

Anuradha, S., and Rao, S.S.R., 2003. Application of brassinosteroids to rice seeds (*Oryza sativa* L.) reduced the impact of salt stress on growth, prevented photosynthetic pigment loss and increased nitrate reductase activity. *Plant Growth Regul.* **40**(1): 29–32.

Anuradha, S., and Rao, S.S.R., 2007a. The effect of brassionosteroids on radish (*Raphanus sativus* L.) seedlings growing under cadmium stress. *Plant Soil Environ.* **53**: 465–472.

Anuradha, S., and Rao, S.S.R., 2007b. Effect of 24-epibrassinolide on the growth and antioxidant enzyme activities in radish seedlings under lead toxicity. *Indian J. Plant Physiol.* **12**: 396–400.

Arora, N., Bhardwaj, R., Sharma, P., and Arora, H.K., 2008. Effects of 28-homobrassinolide on growth, lipid peroxidation and antioxidative enzyme activities in seedlings of *Zea mays* L. under salinity stress. *Acta Physiol. Plantarum* **30**: 833–839.

Arteca, R.N., and Arteca, J.M., 2008. Effects of brassinosteroid, auxin, and cytokinin on ethylene production in *Arabidopsis thaliana* plants. *J. Exp. Bot.* **59**: 3019–3026.

Aval'baev, A.M., Bezrukova, M.V., and Shakirova, F.M., 2003. Effect of brassinosteroid on the hormonal balance in wheat seedlings. *Doklady Biol. Sci.* **391**(3): 337–339.

Aydin, Y., Talas-Ogras, T., Ipekci-Altas, Z., and Gozukirmizi, N., 2006. Effects of brassinosteroid on cotton regeneration via somatic embryogenesis, *Biologia* **61**: 289–293.

Azpeitia, A., Chan, J.L., Saenz, L., and Oropeza, C., 2003. Effect of 22(*S*), 23(*S*)-homobrassinolide on somatic embryogenesis in plumule explants of *Cocos nucifera* (L.) cultured in vitro. *J. Hort. Sci. Biotechnol.* **78**(5): 591–596.

Bajguz, A., and Asami, T., 2004. Effects of brassinazole, an inhibitor of brassinosteroid biosynthesis, on light- and dark-grown *Chlorella vulgaris*. *Planta* **218**(5): 869–877.

Bajguz, A., and Asami, T., 2005. Suppression of *Wolffia arrhiza* growth by brassinazole, an inhibitor of brassinosteroid biosynthesis and its restoration by endogenous 24-EBL. *Phytochem.* **66**: 1787–1796.

Bajguz, A., and Godlewska-Zykiewlu, B., 2004. Protective role of 20-hydroxyeedysone against lead stress in *Chlorella vulgaris* cultures. *Phytochem.* **65**(6): 711–720.

Bao, F., Shen, J., Brady, S.R., Muday, G.K., Asami, T., and Yang, Z., 2004. Brassinosteroids interact with auxin to promote lateral root development in *Arabidopsis*. *Plant Physiol.* **134**(4): 1624–1631.

Catunda, P.H.A., Marinho, C.S., Gomes, M.M.A., and Carvalho, A.J.C., 2008. Brassinosteroid and substrate in acclimatization of "Imperial" pineapple. *Acta Sci. Agron.* **30**(3): 345–352.

Chang, S.C., Kim, Y.S., Lee, J.Y., Kaufman, P.B., Kirakosyan, A., Yun, H.S., Kim, T.W., Kim, S.Y., Cho, M.H., Lee, J.S., and Kim, S.K., 2004. Brassinolide interacts with auxin and ethylene in the root gravitropic response of maize (*Zea mays*). *Physiol. Plantarum* **121**(4): 666–673.

Chon, N.M., Nishikawa-Koseki, N., Takeuchi, Y., and Abe, H., 2008. Role of ethylene in abnormal shoot growth induced by high concentration of brassinolide in rice seedlings. *J. Pest. Sci.* **33**: 67–72.

Cortes, P.A., Terrazas, T., Leon, T.C., and Larque-Saavedra, A., 2003. Brassinosteroid effects on the precocity and yield of cladodes of cactus pear (*Opuntia ficus-indica* (L) Mill.). *Sci. Hort.* **97**(1): 65–73.

Çag, S., Goren-Saglam, N., Cingil-Baris, C., and Kaplan, E., 2007. The effect of different concentration of epibrassinolide on chlorophyll, protein and anthocyanin content and peroxidase activity in excised red cabbage (*Brassica oleracea* L.) cotyledons. *Biotech. Biotechnol. Equipment* **21**: 422–425.

De Grauwe, L., Vandenbussche, F., Tietz, O., Palme, K., and Van Der Straeten, D., 2005. Auxin, ethylene and brassinosteroids: tripartite control of growth in the *Arabidopsis hypocotyls*. *Plant Cell Physiol.* **46**(6): 827–836.

Ding, J., Shi, K., Zhou, Y.-H., and Yu, J-Q, 2009a. Effects of root and foliar applications of 24-epibrassinolide on fusarium wilt and antioxidant metabolism in cucumber roots. *Hort. Sci.* **44**(5): 1340–1349.

Ding, J., Shi, K., Zhou, Y.-H., and Yu, J.-Q., 2009b. Microbial community responses associated with the development of *Fusarium oxysporum* f.sp.*cucumerinum* after 24-epibrassinolide applications to shoots and roots in cucumber. *Eur. J. Plant Pathol.* **124**: 141–150.

Fariduddin, Q., Ahmad, A., and Hayat, S., 2003. Photosynthetic response of *Vigna radiata* to pre-sowing seed treatment with 28-homobrassinolide. *Photosynthetica* **41**(2): 307–310.

Fariduddin, Q., Ahmad, A., and Hayat, S., 2004. Responses of *Vigna radiata* to foliar application of 28-homobrassinolide and kinetin, *Biologia Plantarum* **48**: 465–468.

Fariduddin, Q., Hasan, S.A., Ali, B., Hayat, S., and Ahmad, A., 2008. Effect of modes of application of 28-homobrassinolide on mung bean. *Turkish J. Biol.* **32**: 17–21.

Fedina, E.O., Karimova, F.G., Tarchevsky, I.A., Toropygin, I.Y., and Khripach, V.A., 2008. Effect of epibrassinolide on tyrosine phosphorylation of the calvin cycle enzymes. *Russian J. Plant Physiol.* **55**: 193–200.

Ferrie, A.M.R., Dirpaul, J., Krishna, P., Krochko, J., and Keller, W.A., 2005. Effects of brassinosteroids on microspore embryogenesis in *Brassica* species. *In Vitro Cell. Develop. Biol. Plant* **41**: 742–745.

Fu, F.Q., Mao, W.H., Shi, K., Zhou, Y.H., Asami, T., and Yu, J.Q., 2008. A role of brassinosteroids in early fruit development in cucumber. *J. Exp. Bot.* **59**: 2299–2308.

Golovatskaya, I.F., 2008. Interaction of gibberellic acid and 24-epibrassinolide in the regulation of *Arabidopsis thaliana* seedling scotomorphogenesis. *Russian J. Plant Physiol.* **55**: 663–669.

Gomes, M.M.A., Campostrini, E., Leal, N.R., Viana, A.P., Ferraz, T.M., Siqueira, L.N., Rosa, R.C.C., Núñez-Vázquez, M., and Zullo, M.A.T., 2006. Brassinosteroid analogue effects on the yield of yellow passion fruit plants. *Sci. Hort.* **110**: 235–240.

Hanano, S., Domagalska, M.A., Nagy, F., and Davis, S.J., 2006. Multiple phytohormones influence distinct parameters of the plant circadian clock. *Genes to Cells* **11**: 1381–1392.

Hasan, S.A., Hayat, S., Ali, B., and Ahmad, A., 2008. 28-homobrassinolide protects chickpea (*Cicer arietinum*) from cadmium toxicity by stimulating antioxidants. *Environ. Pollut.* **151**: 60–66.

Haubrick, L.L., Torsethaugen, G., and Assmann, S.M., 2006. Effect of brassinolide, alone and in concert with abscisic acid, on control of stomatal aperture and potassium currents of *Vicia faba* guard cell protoplasts. *Physiol. Plant.* **128**: 134–143.

Hayat, S., and Ahmad, A., 2003a. 28-Homobrassinolide induced changes favoured germinability of wheat seeds. *Bulg. J. Plant Physiol.* **29**(1–2): 55–62.

Hayat, S., and Ahmad, A., 2003b. Soaking seeds of *Lens culinaris* with 28-homobrassinolide increased nitrate reductase activity and grain yield in the field of India. *Annals Applied Biol.* **143**: 121–124.

Hayat, S., Ali, B., Hasan, S.A., and Ahmad, A., 2007b. Brassinosteroid enhanced the level of antioxidants under cadmium stress in *Brassica juncea*. *Environ. Exp. Bot.* **60**: 33–41.

Hayat, S., Ali, B., Hasan, S.A., and Ahmad, A., 2007a. Effect of 28-Homobrassinolide on salinity-induced changes in *Brassica juncea*. *Turk J. Biol.* **31**: 141–146.

Howell, W.M., Keller, G.E., Kirkpatrick, J.D., Jenkins, R.L., Hunsinger, R.N., and McLaughlin, E.W., 2007. Effects of the plant steroidal hormone, 24-epibrassinolide, on the mitotic index and growth of onion (*Allium cepa*) root tips. *Gen. Mol. Res.* **6**: 50–58.

Huang, B., Chu, C.H., Chen, S.L., Juan, H.F., and Chen, Y.M., 2006. A ptoteomics study of the mung bean epicotyl regulated by brassinosteroids under conditions of chilling stress. *Cellul. Mol. Biol. Lett.* **11**: 264–278.

Jager, C.E., Symons, G.M., Ross, J.J., Smith, J.J., and Reid, J.B., 2005. The brassinosteroid growth response in pea is not mediated by changes in gibberellin content. *Planta* **221**: 141–148.

Janeckzo, A., Koscielniak, J., Pilipowicz, M., Szarek-Lukaszewska, G., and Skoczowski, A., 2005. Protection of winter rape photosystem 2 by 24-epibrassinolide under cadmium stress. *Photosynthetica* **43**: 293–298.

Janeckzo, A., Gullner, G., Skoczowski, A., Dubert, F., and Barna, B., 2007. Effects of brassinosteroid infiltration prior to cold treatment on ion leakage and pigment contents in rape leaves. *Biol. Plant.* **51**: 355–358.

Jeong, D.-H., Lee, S., Kim, S.L., Hwang, I., and An, G., 2007. Regulation of brassinosteroid responses by Phytochrome B in rice. *Plant Cell Environ.* **30**: 590–599.

Kagale, S., Divi, U.K., Krochko, J.E., Keller, W.A., and Krishna, P., 2007. Brassinosteroid confers tolerance in *Arabidopsis thaliana* and *Brassica napus* to a range of abiotic stresses. *Planta* **225**: 353–364.

Kandenlinskaya, O.L., Topunov, A.F., and Grishchenko, E.R., 2007. Biochemical aspects of growth-stimulating effects of steroid phytohormones on lupine plants. *Appl. Biochem. Microbiol.* **43**: 324–331.

Kang, Y.-Y., Guo, S.-R., Li, J., and Duan, J.-J., 2009. Effect of root-applied 24-epibrassinolide on carbohydrate status and fermentative enzyme activities in cucumber (*Cucumis sativus* L.) under hypoxia. *Plant Growth Regul.* **57**: 259–269.

Kartal, G., Temei, A., Arican, E., and Gozukirmizi, N., 2009. Effects of brassinosteroids on barley root growth, antioxidant system and cell division. *Plant Growth Regul.* **58**: 261–267.

Kesy, J., Trzaskalska, A., Galoch, E., and Kopcewicz, J., 2003. Inhibitory effect of brassinosteroids on the flowering of the short-day plant *Pharbitis nil*. *Biol. Plant.* **47**(4): 597–600.

Kim, S.-K., Chang, S.C., Lee, E.J., Chung, W.-S., Kim, Y.-S., Hwang, S., and Lee, J.S., 2000. Involvement of brassinosteroids in the gravitropic response of primary root of maize. *Plant Physiol.* **123**: 997–1004.

Kim, T.W., Lee, S.M., Joo, S.H., Yun, H.S., Lee, Y., Kaufman, P.B., Kirakosyan, A., Kim, S.H., Nam, K.H., Lee, J.S., Chang, S.C., and Kim, S.K., 2007. Elongation and gravitropic responses of *Arabidopsis* roots are regulated by brassinolide and IAA. *Plant Cell Environ.* **30**: 679–689.

Kim, S.L., Lee, Y., Lee, S.H., Kim, S.H., Han, T.J., and Kim, S.K., 2008. Brassinolide influences the regeneration of adventitious shoots from cultured leaf discs of tobacco. *J. Plant Biol.* **51**: 221–226.

Kitanaga, Y., Jian, C., Hasegawa, M., Yazaki, J., Kishimoto, N., Kikuchi, S., Nakamura, H., Ichikawa, H., Asami, T., Yoshida, S., Yamaguchi, I., and Suzuki, Y., 2006. Sequential regulation of gibberellin, brassinosteroid, and jasmonic acid biosynthesis occurs in rice coleoptiles to control the transcript levels of anti-microbial Thionin genes. *Biosci. Biotechnol. Biochem.* **70**(10): 2410–2419.

Kurepin, L.V., Qaderi, M.M., Back, T.G., Reid, D.M., and Pharis, R.P., 2008. A rapid effect of applied brassinolide on abscisic acid concentrations in *Brassica napus* leaf tissue subjected to short-term heat stress. *Plant Growth Regul.* **55**: 165–167.

Leubner-Metzger, G., 2001. Brassinosteroids and gibberellins promote tobacco seed germination by distinct pathways. *Planta* **213**: 758–763.

Li, L., Xu, J., Xu, Z.H., and Xue, H.W., 2005. Brassinosteroids stimulate plant tropisms through modulation of polar auxin transport in *Brassica* and *Arabidopsis*. *Plant Cell* **17**: 2738–2753.

Li, K.R., Wang, H.H., Han, G., Wang, Q.J., and Fan, J., 2008. Effects of brassinolide on the survival, growth and drought resistance of *Robinia pseudoacacia* seedlings under water-stress. *New Forests* **35**: 255–266.

Liu, Y., Zhao, Z., Si, J., Di, C., Han, J., and An, L., 2009. Brassinosteroids alleviate chilling oxidative damage by enhancing antioxidant defense system in suspension cultured cells of *Chorispora bungeana*. *Plant Growth Regul.* **59**(3): 207–214.

Lu, Z., Huang, M., Ge, D.P., Yang, Y.H., Cao, X.N., Qin, P., and She, J.M., 2003. Effect of brassinolide on callus growth and regeneration in *Spartina patens* (Poaceae). *Plant Cell Tissue Organ Cult.* **73**(1): 87–89.

Malabadi, R.B., and Nataraja, K., 2007. 24-Epibrassinolide induces somatic embryogenesis in *Pinus wallichiana* A. B. Jacks. *Journal Plant Sci.* **2**: 171–178.

Mazorra, L.M., Núñez, M., Hechavarria, M., Coll, F., and Sánchez-Blanco, M.J., 2002. Influence of brassinosteroids on antioxidant enzymes activity in tomato under different temperatures. *Biol. Plant.* **45**(4): 593–596.

Mazorra, L.M., Núñez, M., Nápoles, M.C., Yoshida, S., Robaina, C., Coll, F., and Asami, T., 2004. Effects of analogs of brassinosteroids on the recovery of growth inhibition by a specific brassinosteroid biosynthesis inhibitor. *Plant Growth Regul.* **44**: 183–185.

Miyazawa, Y., Nakajima, N., Abe, T., Sakai, A., Fujioka, S., Kawano, S., Kuroiwa, T., and Yoshida, S., 2003. Activation of cell proliferation by brassinolide application in tobacco BY-2 cells: effects of brassinolide on cell multiplication, cell-cycle-related gene expression, and organellar DNA contents. *J. Exp. Bot.* **54**: 2669–2678.

Müssig, C., Shin, G.H., and Altmann, T., 2003. Brassinosteroids promote root growth in *Arabidopsis*. *Plant Physiol.* **133**(3): 1261–1271.

Nakashita, H., Yasuda, M., Nitta, T., Asami, T., Fujioka, S., Arai, Y., Sekimata, K., Takatsuto, S., Yamaguchi, I., and Yoshida, S., 2003. Brassinosteroid functions in a broad range of disease resistance in tobacco and rice. *Plant J.* **33**(5): 887–898.

Nieves, N., Rodríguez, K., Cid, M., Castillo, R., González, J.L., and Núnez, M., 2007. Effect of the brassinosteroid analogs BB-6 and MH-5 on proteins metabolism in sugarcane somatic embryogenesis. *Agron. Costarricence* **31**(2): 71–77.

Núñez, M., Mazzafera, P., Mazorra, L.M., Siqueira, W.J., and Zullo, M.A.T., 2003. Influence of a brassinosteroid analogue on antioxidant enzymes in rice grown in culture medium with NaCl. *Biol. Plant.* **47**(1): 67–70.

Núñez, M., Siqueira, W.J., Hernandez, M., Zullo, M.A.T., Robaina, C., and Coll, F., 2004. Effect of spirostane analogues of brassinosteroids on callus formation and plant regeneration in lettuce (*Lactuca sativa*). *Plant Cell Tissue Organ Cult.* **78**(1): 97–99.

Oda, Y., Mimura, T., and Hasezawa, S., 2005. Regulation of secondary cell wall development by cortical microtubules during tracheary element differentiation in *Arabidopsis* cell suspensions. *Plant Physiol.* **137**: 1027–1036.

Ogweno, J.O., Song, X.S., Shi, K., Hu, W.H., Mao, W.H., Zhou, Y.H., Yu, J.Q., and Nogues, S., 2008. Brassinosteroids alleviate heat-induced inhibition of photosynthesis by increasing carboxylation efficiency and enhancing antioxidant systems in *Lycopersicon esculentum*. *J. Plant Growth Regul.* **27**: 49–57.

Ono, E.O., Nakamura, T., Machado, S.R., and Rodrigues, J.D., 2000. Application of brassinosteroid to *Tabebuia alba* (Bignoniaceae) plants. *Braz. J. Plant Physiol.* **12**(3): 187–194.

Özdemir, F., Bor, M., Demiral, T., Turkan, I., 2004. Effects of 24-epibrassinolide on seed germination, seedling growth, lipid peroxidation, proline content and antioxidative system of rice (*Oryza sativa* L.) under salinity stress. *Plant Growth Regul.* **42**(3): 203–211.

Papadopoulou, E., and Grumet, R., 2005. Brassinosteriod-induced femaleness in cucumber and relationship to ethylene production. *Hortsci.* **40**: 1763–1767.

Peng, J., Tang, X., and Feng, H., 2004. Effects of brassinolide on the physiological properties of litchi pericarp (*Litchi chinensis* cv. Nuomoci). *Sci. Hort.* **101**: 407–416.

Pereira-Netto, A.B., Cruz-Silva, C.T.A., Schaefer, S., Ramirez, J.A., and Galagovsky, L.R., 2006a. Brassinosteroid-stimulated branch elongation in the marubakaido apple rootstock. *Trees-Struct. Funct.* **20**: 286–291.

Pereira-Netto, A.B., Cruz-Silva, C.T.A., Schaefer, S., Ramirez, J.A., and Galagovsky, L.R., 2006b. Brassinosteroid-stimulated branch elongation in the marubakaido apple rootstock. *Trees-Struct. Funct.* **20**: 286–291.

Pullman, G.S., Zhang, Y., and Phan, B.H., 2003. Brassinolide improves embryogenic tissue initiation in conifers and rice. *Plant Cell Rep.* **22**(2): 96–104.

Pullman, G.S., Johnson, S., Van Tassel, S., and Zhang, Y., 2005. Somatic embryogenesis in loblolly pine (*Pinus taeda*) and Douglas fir (*Pseudotsuga menziesii*): improving culture initiation and growth with MES pH buffer, biotin, and folic acid. *Plant Cell Tissue Organ Cult.* **80**: 91–103.

Pullman, G., Johnson, S., and Bucalo, K., 2009. Douglas fir embryogenic tissue initiation. *Plant Cell Tissue Organ Cult.* **96**: 75–84.

Saglam-Çag, S., 2007. The effects of epibrassinolide on senescence in wheat leaves. *Biotecnol. Biotechnol. Equipment* **21**: 63–65.

Saka, H., Fujii, S., Imakawa, A.M., Kato, N., Watanabe, S., Nishizawa, T., and Yonekawa, S., 2003. Effect of brassinolide applied at the meiosis and flowering stages on the levels of endogenous plant hormones during grain-filling in rice plant (*Oryza sativa* L.). *Plant Prod. Sci.* **6**(1): 36–42.

Saygideger, S., and Deniz, F., 2008. Effect of 24-epibrassinolide on biomass, growth and free proline concentration in *Spirulina platensis* (*Cyanophyta*) under NaCl stress. *Plant Growth Regul.* **56**: 219–223.

Shahbaz, M., Ashraf, M., and Athar, H.U.R., 2008. Does exogenous application of 24-epibrassinolide ameliorate salt induced growth inhibition in wheat (*Triticum aestivum* L.)? *Plant Growth Regul.* **55**: 51–64.

Sharma, P., and Bhardwaj, R., 2007. Effects of 24-epibrassinolide on growth and metal uptake in *Brassica juncea* L. under copper metal stress. *Acta Physiol. Plant.* **29**: 259–263.

Sharma, P., Bhardwaj, R., Arora, N., Arora, H.K., and Kumar, A., 2008. Effects of 28-homobrassinolide on nickel uptake, protein content and antioxidative defence system in *Brassica juncea*. *Biol. Plant.* **52**: 767–770.

Silva, C.M.M., Gomes, M.M.A., and Freitas, S.P., 2009. Effects of herbicides associated to a brassinosteroid analogue on the photosynthetic apparatus of *Eucalyptus grandis* seedlings. *Planta Daninha* **27**(4): 789–797.

Singh, I., and Shono, M., 2005. Physiological and molecular effects of 24-epibrassinolide, a brassinosteroid on thermotolerance of tomato. *Plant Growth Regul.* **47**: 111–119.

Song, W.J., Zhou, W.J., Jin, Z.L, Cao, D.D., Joel, D.M., Takeuchi, Y., and Yoneyama, K., 2005. Germination response of *Orobanche* seeds subjected to conditioning temperature, water potential and growth regulator treatments. *Weed Res.* **45**: 467–476.

Song, W.J., Zhou, W.J., Jin, Z.L., Zhang, D., Yoneyama, K., Takeuchi, Y., and Joel, D.M., 2006. Growth regulators restore germination of *Orobanche* seeds that are conditioned under water stress and suboptimal temperature. *Aust. J. Agric. Res.* **57**: 1195–1201.

Sun, Y., Veerabomma, S., Abdel-Mageed, H.A., Fokar, M., Asami, T., Yoshida, S., and Allen, R.D., 2005. Brassinosteroid regulates fiber development on cultured cotton ovules. *Plant Cell Physiol.* **46**(8): 1384–1391.

Symons, G.M., Davies, C., Shavrukov, Y., Dry, I.B., Reid, J.B., and Thomas, M.R., 2006. Grapes on steroids. Brassinosteroids are involved in grape berry ripening. *Plant Physiol.* **140**: 150–158.

Tabur, S., and Demir, K., 2009. Cytogenetic response of 24-epibrassinolide on the root meristem cells of barley seeds under salinity. *Plant Growth Regul.* **58**: 119–123.

Tanaka, K., Nakamura, Y., Asami, T., Yoshida, S., Matsuo, T., and Okamoto, S., 2003. Physiological roles of brassinosteroids in early growth of *Arabidopsis*: brassinosteroids have a synergistic relationship with gibberellin as well as auxin in light-grown hypocotyl elongation. *J. Plant Growth Regul.* **22**(3): 259–271.

Upreti, K.K., and Murti, G.S.R., 2004. Effects of brassinosteroids on growth, nodulation, phytohormone content and nitrogenase activity in French bean under water stress. *Biol. Plant.* **48**: 407–411.

Vardhini, B.V., and Rao, S.S.R., 2002. Acceleration of ripening of tomato pericarp discs by brassinosteroids. *Phytochem.* **16**: 843–847.

Vardhini, B.V., and Rao, S.S.R., 2003. Amelioration of osmotic stress by brassinosteroids on seed germination and seedling growth of three varieties of sorghum. *Plant Growth Regul.* **41**(1): 25–31.

Whippo, C.W., and Hangarter, R.P., 2005. A brassinosteroid-hypersensitive mutant of BAK1 indicates that a convergence of photomorphogenic and hormonal signalling modulates phototropism. *Plant Physiol.* **139**: 448–457.

Xia, X.J., Huang, Y.Y., Wang, L., Huang, L.F., Yu, Y.L., Zhou, Y.H., and Yu, J.Q., 2006. Pesticides-induced depression of photosynthesis was alleviated by 24-epibrassinolide pretreatment in *Cucumis sativus* L. *Pest. Biochem. Physiol.* **86**: 42–48.

Xia, X.-J., Wang, Y.-J., Zhou, Y.-H., Tao, Y., Mao, W.-H., Shi, K., Asami, T., Chen, Z., and Yu, J.-Q., 2009. Reactive oxygen species are involved in brassinosteroid-induced stress tolerance in cucumber. *Plant Physiol.* **150**: 801–814.

Yang, Y.H., Huang, J., and Ding, J., 2003. Interaction between exogenous brassinolide, IAA and BAP in secondary metabolism of cultured *Onosma paniculatum* cells. *Plant Growth Regul.* **39**(3): 253–261.

Yu, J.Q., Huang, L.F., Hu, W.H., Zhou, Y.H., Mao, W.H., Ye, S.F., and Nogues, S., 2004. A role for brassinosteroids in the regulation of photosynthesis in *Cucumis sativus*. *J. Exp. Bot.* **55**(399): 1135–1143.

Yun, H.R., Joo, S.-H., Park, C.H., Kim, S-K, Chang, S.C., and Kim, S.Y., 2009. Effects of brassinolide and IAA on ethylene production and elongation in maize primary roots. *J. Plant Biol.* **52**: 268–274.

Zhang, Z.S., Ramirez, J., Reboutier, D., Brault, M., Trouverie, J., Pennarun, A.M., Amiar, Z., Biligui, B., Galagovsky, L., and Rona, J.P., 2005. Brassinosteroids regulate plasma membrane anion channels in addition to proton pumps during expansion of *Arabidopsis thaliana* cells. *Plant Cell Physiol.* **46**: 1494–1504.

Zhang, S., Hu, J., Zhang, Y., Xie, X.J., and Knapp, A., 2007. Seed priming with brassinolide improves lucerne (*Medicago sativa* L.) seed germination and seedling growth in relation to physiological changes under salinity stress. *Aust. J. Agric. Res.* **58**: 811–815.

Zhang, M.C., Zhai, Z.X., Tian, X.L., Duan, L.S., and Li, Z.H., 2008. Brassinolide alleviated the adverse effect of water deficits on photosynthesis and the antioxidant of soybean (*Glycine max* L.). *Plant Growth Regul.* **56**: 257–264.

Chapter 8

GENOMIC AND NON-GENOMIC EVENTS INVOLVED IN THE BRASSINOSTEROID-PROMOTED PLANT CELL GROWTH

A.B. PEREIRA-NETTO
Department of Botany-SCB, Centro Politecnico-UFPR, C.P. 19031 Curitiba, PR, 81531-970, Brazil

Abstract: In all multicellular organisms growth and morphogenesis must be coordinated. In plants, coordinate control of growth is regulated by both external stimuli and internal mechanisms. In most multicellular organisms, steroids act as internal mediators for physiological and developmental regulation. Brassinosteroids (BRs) are steroids known to induce a broad spectrum of responses in plants; however, promotion of cell growth is a major biological effect of BRs. In this chapter, an insight into the genomic and non-genomic events involved in the brassinosteroids-promoted plant cell growth is provided.

Key words: cell expansion, cell division, cell weight, brassinazole, uniconazole

1. INTRODUCTION

Because plants are sessile, the proper functioning of developmental programs that control growth patterns is critical for their survival. Coordinated plant growth is modulated through networked actions of plant growth regulators which results in orderly cell division and tightly regulated cell expansion (Azpiroz *et al.*, 1998; Kim *et al.*, 2006).

Brassinosteroids (BRs) comprise a specific class of low-abundance plant steroids of ubiquitous occurrence in plants (Fujioka 1999; Belkhadir and Chory, 2006). BRs induce a broad spectrum of responses (Clouse and Sasse,

S. Hayat and A. Ahmad (eds), *Brassinosteroids: A Class of Plant Hormone*,
DOI 10.1007/978-94-007-0189-2_8,

1998), however, stimulation of growth via cell elongation and cell division is a major biological effect of BRs (Zurek *et al.*, 1994; Hu *et al.*, 2000). Genetic and biochemical approaches have contributed to an impressive progress in our understanding of the BRs metabolism and its regulation (Fujioka and Sakurai, 1997; Choe *et al.*, 1999; Noguchi *et al.*, 2000; Yokota, 1997), as well as of the BR-induced signaling, including the identification of BR receptors, key signaling elements, and BR-induced gene expression (Choe *et al.*, 2002; Clouse *et al.*, 1996; Friedrichsen *et al.*, 2000; Geldner *et al.*, 2007; Hu *et al.*, 2000; Li and Nam, 2002; Li *et al.*, 2002; Mora-García *et al.*, 2004; Pérez-Pérez *et al.*, 2002; Schumacher *et al.*, 1999; Wang *et al.*, 2001, 2002; Yin *et al.*, 2002). In contrast with the fast progress in understanding how BRs are transduced to affect gene expression, little is known about the downstream events that result on the establishment of plant cell growth patterns. This chapter is focused on the genomic and non-genomic events closely involved in the brassinosteroids-induced plant cell growth. Aspects of the involvement of BRs on the regulation of expression of cell wall enzymes, D-type cyclins, and regulation of activity of proteins such as vacuolar H^+-ATPase and aquaporins, among others important for the control of cell growth are discussed. Although BRs have been shown to interact with several other plant growth regulators on the determination of plant cell growth patterns, possible cross talk between BRs and other plant growth regulators, important for cell growth, will not be discussed in this chapter due to space limitations.

2. CONTROL OF CELL EXPANSION

A distinguishing feature of plant cells is the presence of a cell wall outside the plasmalemma. It has been accepted for quite some time that the cell wall is not only a complex network of polysaccharides and other high molecular weight molecules but also a dynamic structure with key roles in plant growth (Wu *et al.*, 1996; Jauneau *et al.*, 1998). Plant growth is largely accomplished by cell expansion, which depends on the *de novo* biosynthesis and modification of cell wall components (activated mainly by gene expression), the deposit orientation of nascent cellulose microfibrils (determined by cortical microtubules), and the internal turgor pressure (generated by water intake). Cell expansion, a developmental process regulated by environmental cues such as light and internal growth regulators, including brassinosteroids (BRs), is of much greater importance for plants than for most of the other organisms and the final size reached by all plant organs depends upon a period of significant cell expansion. This period of cell expansion usually follows an active period of cell division (Azpiroz *et al.*, 1998). Before reaching their final size, plant cells usually enlarge 10- to 1000-fold in volume by a process that entails massive vacuolation and irreversible expansion of the cell wall (Cosgrove, 1997).

Cellulose comprises the main structural component of cell walls (Aspinall, 1980). Cellulose is essentially a linear homopolymer of β-4-linked D-glucosyl residues. These polymers bond together to form crystalline structures, and amorphous regions, which are more prone to enzymatic degradation (Malberg *et al.*, 1992; Beguin and Aubert, 1994). Except for grasses and cereals, all flowering plants have type I wall, in which the principal cellulose cross-linking glycan is xyloglucan (Carpita and Gibeaut, 1993; Cosgrove, 1997). All xyloglucans, consist of a cellulose-like backbone carrying single α-D-xylopyranosyl units attached to *O*-6. And, some xylosyl residues are further substituted at *O*-2 by β-D-galactopyranosyl units (Fry, 1989a; Carpita and Gibeaut, 1993). In the type II cell wall of the grasses and cereals, the major glycans cross linking the cellulose microfibrils are glucuronoarabinoxylan and (1,3)(1,4)-β-D-glucan (Buckeridge *et al.*, 2004). Type II cell walls contain a relatively low amount of xyloglucan, which could, nevertheless, be very important for the establishment of the expansion rate (Yokoyama *et al.*, 2004). It is considered that structural changes in these networks are regulated by enzymatic modification, and therefore wall-modifying enzymes would be expected to play an important role in regulating the plasticity of the cell walls.

Cell expansion, critical for growth in all plant organs, is controlled by coordinated alterations in wall mechanical properties, cell hydraulics, biochemical processes and gene expression (Cosgrove, 1997). The water intake is the driving force for cell expansion. In order for turgor-driven cell expansion to proceed, the cell wall must transiently go through relaxation or loosening, by slippage or breakage of the hemicellulose tethers, and proper incorporation of new polymers into the expanding wall to maintain cell wall thickness and integrity (Roberts, 1994; Clouse and Sasse, 1998; Clouse, 2002; Rose *et al.*, 2002). Xyloglucan and other wall polymers are constantly being secreted and incorporated into the wall and the precise balance between xyloglucan incorporation and breakdown determine the expansion rate (Takeda *et al.*, 2002).

Various proteins with possible roles in cell wall modification processes have been reported, including glucanases, expansins and xyloglucan endotransglycosylase/hydrolase (XTHs), the later, a well-defined class of enzymes that exhibit xyloglucan endohydrolase (XEH) and xyloglucan endo-transglucosylase activities (XET) (Campbell and Braam, 1999; Cosgrove, 1997; Rose *et al.*, 2002). Korrigan glucanase, for example, is considered to be critical for wall assembly, cell expansion, and cytokinesis, besides acting as a cellulase whose activity is required for cellulose synthesis (reviewed in He *et al.*, 2003), while expansin loosens the cell wall by disrupting hydrogen bonding (Cosgrove, 2000; Lee *et al*., 2001). It has been pointed out that expansins might be primarily responsible for wall relaxation, but glucanases and xyloglucan endotransglucosylase (XETs) would affect the extent of expansin activity by changing the viscosity of the hemi-cellulose matrix (Cosgrove, 1997). Thus, by altering the viscosity of the matrix, glucanases, other wall hydrolases, and XET affect the amount of wall enlargement that results from expansin activity (Cosgrove, 1997).

In dicotyledonous plants, the major cell wall load-bearing network consisted of cellulose microfibrils coated and cross-linked by hydrogen-bonded xyloglucan is considered by many to be the main component endowing the wall with its mechanical properties (Fry, 1989b; Carpita and Gibeaut, 1993; Carpita and McCann, 2000). Xyloglucan endohydrolase (XEH) catalyzes an endolytic cleavage of a cross-linking xyloglucan polymer at any of the non-xylosylated Glc residues (Fanutti *et al*., 1993) allowing cellulose microfibrils to separate and the cell to expand while xyloglucan endotransglucosylase activities (XET) cleaves a xyloglucan chain (=donor substrate) endolytically and then transfers the newly generated terminus on to the non-reducing end of an acceptor substrate (reviewed in Johansson *et al*., 2004 and Takeda *et al*., 2008). Cell wall loosening is a temporary requirement for cell expansion that must be followed by rapid reinforcement of the wall structure (Johansson *et al*., 2004; Knox, 2008). XET activity is largely responsible for cutting and rejoining xyloglucan chains within the cell wall matrix, thereby controlling wall plasticity (Fry *et al*., 1992). In fact, a purified XTH displaying exclusively XET activity, was shown to stimulate cell wall expansion, being the enzyme also considered to act as a primary cell wall-loosening agent (Van Sandt *et al*., 2007). Using split pea (*Pisum sativum*) stem segments, Takeda and co-workers (2002) demonstrated that the integration of whole exogenous xyloglucan in stem cell walls, mediated by the action of wall-bound XET, resulted in increased rigidity and suppression of elongation of pea stems. In addition, those authors found that the whole xyloglucan was incorporated into the cell wall and induced the rearrangement of cortical microtubules from transverse to longitudinal orientation. As expected, suppression of elongation of pea stems was related to integrations of whole xyloglucan. Conversely, integration of a xyloglucan-derived oligosaccharide in cell walls of pea stems resulted in wall loosening. This wall loosening oligosaccharide was shown to solubilize xyloglucan from the cell wall and to keep the microtubules in a transverse orientation. Not surprisingly, enhanced elongation of pea stems was related to integrations of the xyloglucan oligosaccharide. More recently, Cavalier and co-workers (2008) demonstrated that in mutants of *Arabidopsis thaliana* impaired in xyloglucan biosynthesis, the cell wall mechanical strength is reduced.

Light and electron microscopy analysis of cell in wild-type and various *Arabidopsis* BR mutants have provided direct physical evidence that cell expansion is considerably reduced in BR mutants (Azpiroz *et al*., 1998; Kauschmann *et al*., 1996; Szekeres *et al*., 1996; Takahashi *et al*., 1995). In addition, analysis of the mechanical features of cell walls demonstrated that BRs promote wall loosening in soybean epicotyls (Zurek *et al*., 1994) and hypocotyls of *Brassica chinensis* (Wang *et al*., 1993) and *Cucurbita maxima* (Tominaga *et al*., 1994).

3. THE ROLE OF CELL WALL ENZYMES IN THE BRASSIONOSTEROIDS-PROMOTED CELL EXPANSION

It is now well established that BRs promote cell elongation through various events, including changes in transcript levels of genes encoding cell wall remodeling enzymes required for cell expansion such as xyloglucan endotransglycosylase/hydrolase (XTH), Pectin Lyase-like (PLL), glucanases (KOR) and expansins (EXP). And, mounting reports are clarifying the role of these cell wall enzymes in the BR-induced cell expansion. For example, the expression of the Korrigan (kor) gene, which encodes a membrane-bond endo-β-1,4-glucanase required for the correct assembly of the walls of elongating cells, has been shown to be unaffected by auxin, gibberellin, or ethylene (Nicol *et al*., 1998). However, KOR expression is reduced in the *Arabidopsis* BR-deficient mutants *det2* (He *et al*., 2003; Nicol *et al*., 1998), *dwf4* and *dim1* (He *et al*., 2003), while it is up-regulated by BL (He *et al*., 2003).

XETs are encoded by a large multi-gene family in *Arabidopsis* and only a subset of which is regulated by BR (Xu *et al*., 1996). In elongating soybean (*Glycine max*) epicotyls, BR application resulted in increased plastic extensibility of the walls with a concomitant enhancement in the mRNA level of bru1 (Zurek *et al*., 1994). The regulation of bru1 expression under these conditions is specific to BR and occurs at the post-transcriptional level (Zurek and Clouse, 1994). Recombinant BRU1 protein exhibits xyloglucan-specific transglycosylation *in vitro*, and shares significant sequence identity with numerous XETs from several plant species. In addition, increasing concentrations of applied BR during early stages of elongation resulted in a linear enhancement of extractable XET activity in the soybean epicotyls (Oh *et al*., 1998). *In situ* hybridization in cross-sections of apical epicotyl revealed that bru1 was more highly expressed in inner vs. outer stem tissue, particularly in phloem cells, in parenchyma cells surrounding the xylem elements and in the starch sheath layer (Oh *et al*., 1998). Besides soybean,

BR-regulated XETs have also been identified in various plants such as *Arabidopsis* (Xu *et al.*, 1995; Goda *et al.*, 2002; He *et al.*, 2003), tobacco (Vissenberg *et al.*, 2000, 2005) and tomato (*Lycopersicon lycopersicum*) (Catala *et al.*, 1997; Koka *et al.*, 2000).

The *Arabidopsis* tch4 gene, which encodes an XET with sequence similarity to bru1 (Xu *et al.*, 1995), was shown to be up-regulated by BR (Xu *et al.*, 1995; He *et al.*, 2003). In addition, tch4 was also shown to be strongly expressed in expanding tissues, particularly in dark-grown hypocotyls, and in organs that undergo cell wall modification such as vascular elements. In rice (*Oryza sativa*), internodal elongation accompanies panicle formation, and, in the internode, analysis of the expression pattern demonstrated that two BR-upregulated XTH genes, osxtr1 and osxtr3, were preferentially expressed in the elongating zone (Uozo *et al.*, 2000), which suggests a role for these genes in rice cell expansion.

Various reports on genome-wide microarray analysis of BR-regulated genes (Goda *et al.*, 2002, 2004; Mussig *et al.*, 2002; Nemhauser *et al.*, 2004) showed or confirmed that several genes involved in cell expansion or cell wall organization such as genes encoding XETs, pectin lyase-like and expansins were up-regulated by BRs. And, especially in the case of expansin genes, the kinetics of induction of these genes was similar to those of BR-induced cell expansion (Goda *et al.*, 2002).

Creep measurements of internode segments of the *lkb* mutant, a pea (*Pisum sativum* L.) BR biosynthesis dwarf mutant (Nomura *et al.*, 1997, 1999), and wild type plants showed that the yield threshold was distinctly lower in the wild type than in *lkb* which suggests that some load-bearing bonds in the cell wall of *LKB* are more easily broken than in *lkb*. Although no statistical difference was found for cell wall extensibility for wild type, *lkb* and BR-treated *lkb*, the data from creep measurements suggested that BR is required for acid-induced cell wall loosening and that BR modulates the activity of wall loosening enzymes which are limiting the yield threshold (Wadaa and Katsumi, 2005).

Accumulating evidences indicate that cell wall enzymes, especially XETs and expansins, play important role(s) in the BR promoted cell expansion. However, further studies, especially at the post-transcription level, are necessary to fully elucidate the role of these enzymes in the BR promoted cell expansion.

4. THE ROLE OF VACUOLAR H^+-ATPASE IN THE CONTROL OF CELL EXPANSION BY BRASSIONOSTEROIDS

Although gene expression is important for many BR-elicited cellular processes, similar to animal steroids, BRs can exert non-genomic effects on cell physiology that are independent of DNA transcription and protein synthesis. Vacuolar H^+-ATPase (V-ATPase) together with the H^+-pyrophosphatase, a second proton pump specific to plants and phototrophic bacteria (Rea and Poole, 1993), is known to drive solute uptake into the vacuole. Because of that, V-ATPase activity is considered to be important for cell expansion. Indeed, in transgenic carrot (*Daucus carota*) lines, cell expansion has been shown to be reduced by antisense inhibition of subunit A of V-ATPase (Gogarten *et al*., 1992). During the early 1990s, dicyclohexyl carbodiimide (DCCD), an inhibitor of the plasma membrane ATPase was shown to prevent BR-stimulated cell elongation (Katsumi, 1991). Later, treatment of a vacuolar ATPase with BL was shown to be able to activate the enzyme, indicating that the BR-induced cell expansion might be due to a reduction on the water potential of the vacuole in response to increased ion and/or sugar uptake (Tominaga and Sakurai, 1996) (Figure 1). Afterwards, Schumacher and colleagues (1999) demonstrated that the *det3 Arabidopsis* mutant was defective in a gene encoding the ubiquitous subunit C of the V-ATPase, and a close

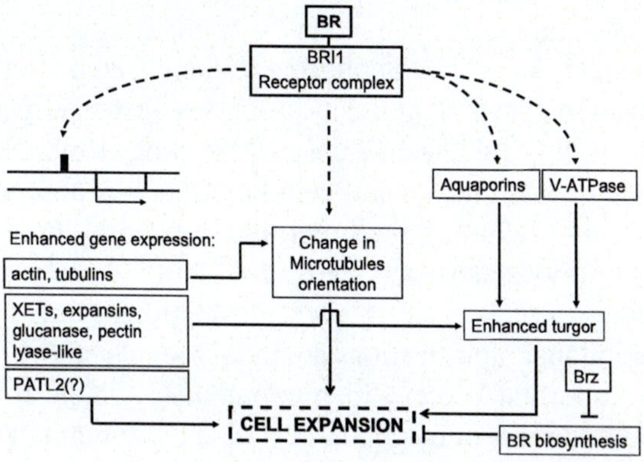

Figure 1. Schematic representation of the regulation of cell expansion by BRs. BR binding to BRI1, the BR receptor, signals through a phosphorylation cascade (not shown) that includes rapid changes in the rate of cell expansion through regulated assembly of the V-ATPase and enhanced activity of aquaporins, changes in the orientation of microtubules and enhanced gene expression which leads to cell wall loosening or re-organization.

relationship between V-ATPase activity and cell expansion was also found. In addition, those authors proposed a model in which the V-ATPase activity would be regulated differentially by various phytohormones, among them BRs, through different modes of assembly of the enzyme. That hypothesis found support in the finding that the regulatory subunit H of the V-ATPase interacts with and is phosphorylated *in vitro* by BRI1, the BR receptor. It has also been hypothesized that the activated BRI1 could regulate the proton-pumping activity of a plasma membrane or vacuolar ATPase (Peng and Li, 2003). However, further studies are required to reveal how an activated BRI1, could control the activities of the plasma membrane and/or vacuolar ATPase.

Somewhat surprisingly, in systems such as the plasma membrane vesicles of potato (*Solanum tuberosum* L.) dormant tubers, epi-brassinolide (epi-BL) has been shown to inhibit ATP-dependent accumulation of H^+ (Ladyzhenskaya and Korableva, 2001). Thus, the effect of BRs on the activity of ATPases seems to depend upon the localization of the enzymes in the different subcellular compartments and the consequences of that for the BR-induced cell expansion remain to be elucidated.

5. THE ROLE OF MICROTUBES IN THE CONTROL OF CELL EXPANSION BY BRASSIONOSTEROIDS

Microtubules are the basic components of the eukaryotic cell cytoskeleton and are involved in several cellular processes, especially cell expansion (Goddard *et al.*, 1994; Takahashi *et al.*, 1995; Nick, 1998; He *et al.*, 2003). Microtubules also constitute several cellular structures observed specifically during cell division, including the preprophase band, mitotic spindle fibers, and phragmoplast (Wasteneys and Yang, 2004).

Tubulins play a crucial role in plant development and their expression and post-transcriptional modifications are tightly regulated (Yang *et al.*, 2009). Tubulin is an important protein in microtubules, which are composed of polymerized heterodimers of α-tubulin (TUA) and β-tubulin (TUB) (Sunohara *et al.*, 2009; Yang *et al.*, 2009). It has been predicted that microtubules assembly as well as stability is regulated through the transcription of different isotypes of tubulin, the folding of tubulin monomers, formation of functional dimers and also through post-translational modifications (Nogales, 2000). Control of the orientation of microtubules is crucial for the determination of the orientation of cell expansion (Williamson, 1991; Cyr and Palevitz, 1995). In fact, in *tid1-1*, a rice mutant defective in the synthesis of an α-tubulin, abnormalities in cell expansion and division were associated with the aberrant orientation of cortical microtubules (Sunohara *et al.*, 2009).

Reorganization of cortical microtubules is known to be important for BR-induced plant growth and it is thought that microtubule-membrane interaction is involved in this process (Mayumi and Shibaoka, 1995). BRs are known for many years to change re-configuration of microtubules to transverse orientation, in order to allow longitudinal growth. For example, in epicotyls of azuki beans (*Vigna angularis*), it has been demonstrated that inhibitors of cellulose biosynthesis or microtubules re-orientation inhibit BR promoted stem elongation. In addition, BL was shown to enhance the percentage of cortical microtubules transversally oriented (Mayumi and Shibaoka, 1995), which contributes to the establishment of the growth direction once the orientation of cortical microtubules usually correlates with the orientation of microfibrils (Clouse and Sasse, 1998). In another example, in the *bul1/dwf7-3* mutant, a BR-deficient mutant, BR treatment induces cortical microtubule orientation and restores cell expansion (Mussig and Altmann, 2003).

Takahashi *et al.* (1995) and Szekeres *et al.* (1996) demonstrated that the dwarf phenotype of the *Arabidopsis* BR-deficient (Klahre *et al.*, 1998) mutant *dim* could be restored to that of the wild type by treatment with exogenous BL. Interestingly, in this mutant, expression of 6α-tubulins and 5β-tubulins were similar to that of the wild type, while a specific β-tubulin gene, tub1, was reduced (Takahashi *et al.*, 1995). In addition, expression of a putative β-tubulin gene in chick pea (*Cicer arietinum*) correlated with BL-induced growth (Munoz *et al.*, 1998). Studies using a rice BR-insensitive (Yamamuro *et al.*, 2000) and an *Arabidopsis* BR-deficient mutant, *bul1-1* (Catterou *et al.*, 2001), confirmed the effect of BR on the organization of cortical microtubules during cell elongation. Indirect immunofluorescence of α-tubulin demonstrated that very few microtubules were present in the *bul1-1* mutant, and also that the parallel microtubules organization that is typical of elongating cells in the wild type was missing in the mutant. Following BR treatment, microtubules reorganized and became correctly oriented, which suggested the involvement of BRs in microtubules organization. Furthermore, molecular analyses in the study with the *bul1-1* mutant showed that total tubulin was slightly lower in the mutant, compared to the wild type, although BR treatment induced no significant change in total tubulin for the mutant or for the wild type plants. At the transcriptional level, Northern blot analysis using independent probes showed a considerable reduction in β1-tubulin mRNA (TUB1) for the mutant, however, BR treatment increased α and β-tubulin transcript levels only in the wild type plants. Similar results were also found for another *Arabidopsis* BR-deficient mutant *dim* (Catterou *et al.*, 2001). Thus, the microtubules reorganization observed in BR-treated *bul1-1* plants was shown not to result either from an up regulation of tubulin gene expression, or from an enhancement in tubulin content. Instead, the BR promoted microtubules nucleation/organization was proposed to result from the effect of BRs on other factors such as microtubule-associated proteins (MAPs), proteins known to be important for the polymerization (Desai and Mitchison, 1997) and stabilization (Yang *et al.*, 2009) of microtubules. In fact, plant microtubules are highly dynamic and their stability depends on the activity of various MAPs (Yang *et al.*, 2009). Thus, BRs could, indeed, change microtubules dynamics and stability through changes in the activity of MAPs, however, that remains to be demonstrated. The expression of a β-tubulin also shown to be up-regulated by BL in wild-type *Arabidopsis*, is reduced in the *Arabidopsis* insensitive mutant *bri1*, which indicates that β- tubulin is not directly controlled by BR signaling (He *et al.*, 2003). Although studies involving the identification of BR-regulated genes/proteins have been demonstrating that BRs promote the expression of both, α-tubulin and β-tubulins, accumulating evidence indicate that the BR promoted cell wall expansion might not require increased tubulin expression.

The actin cytoskeleton in plants is essential for plant growth (Muday *et al.*, 2000). Actin is associated with microtubules to determine their orientation (Bishop and Yokota, 2001). BR-upregulated actin genes are known for quite some time (Goda *et al.*, 2004). And, more recently, a protemics study on plasma membrane proteins using Two-dimensional Electrophoresis and Image Scanning (2-D DIGE), revealed a BR-upregulated actin, 24 hours after BR treatment (Tang *et al.*, 2008). bru2, a BR-upregulated gene that may encode an actin effector protein that controls polymerization of actin molecules was isolated from BL-treated rice seedlings (Sasuga *et al.*, 2000). Thus, the BR-induced cell expansion might be, at least partially, dependent on the regulation of the orientation of microtubules by promoting the synthesis of an actin effector protein. Also, the formation of cortical F-actin has been demonstrated to be under the control ROP (for Rho-of-plant) GTPases (Fu *et al.*, 2002). And, Li and co-workers (2005) demonstrated that BR increases expression of ROP GTPases, which implicates BR in the establishment of F-actin assembly/reorganization patterns.

A proteomics study of BR-regulated proteins in *Arabidopsis* identified 42 BR-regulated proteins, among them three BR-induced cytoskeleton proteins, including actin 2 (At3g18780), tubulin α-6 chain (At4g14960), and tubulin β-4 chain (At5g44340) (Deng *et al.*, 2007). Although the importance of the tubulin genes for the BR-induced microtubules organization is still under debate (Bishop and Yokota, 2001; Catterou *et al.*, 2001; Munoz *et al.*, 1998; Takahashi *et al.*, 1995), the finding that various tubulins and especially an actin are BR-regulated proteins, provide further evidence that BR promoted cell expansion depends on changes in the pattern of microtubules organization. In addition to the cytoskeleton proteins, a member of the Sec14-like proteins,

PATL2, was also shown to be BR-regulated (Deng *et al.*, 2007). Although the precise function of the PATLs in *Arabidopsis* remains to be elucidated, there is evidence that supports their role in cell wall formation. More recently, another protemics study in *Arabidopsis* revealed PATL1 and confirmed PATL2 as BR-upregulated proteins (Tang *et al.*, 2008). PATL2 had been previously identified in microarray studies and had also shown to be affected in *bri1-116* and *bzr1-1D* mutants (Deng *et al.*, 2007), which indicates that the BR promoted cell expansion might rely, partially, in changes in PATLs's activity.

Finally, it has also been hypothesized that the activated BR receptor, BRI1, could affect the reorganization of cortical microtubules (Peng and Li, 2003). However, further studies are required to reveal how an activated BRI1 could control the nucleation/organization of the microtubules.

6. THE ROLE OF AQUAPORINS IN THE CONTROL OF CELL EXPANSION BY BRASSINOSTEROIDS

Water uptake and flow across membranes is a critical requirement for plant cell expansion. Aquaporins are proteinaceous channels formed by a superfamily of proteins containing over 450 members (Sasaki, 2008). Aquaporins in the plasma and vacuolar membranes effectively facilitate the intercellular and intracellular water transport in plants (Suga *et al.*, 2002). Therefore, aquaporins are considered to be involved in the cell expansion process in plants. Analysis of the aquaporin activities of *Arabidopsis* wild type and BR biosynthetic and insensitive mutants, *cpd* and *bri1*, respectively, demonstrated that protoplasts from both mutants presented significantly lower water permeability when compared to the wild-type. In addition, BL treatment enhanced osmotic permeability of hypocotyl protoplasts from the *cpd* mutant significantly, though BL had no effect on the *bri1* mutant (Morillon *et al.*, 2001). In radish (*Raphanus sativus*), mRNA and protein levels of aquaporin isoforms in root and shoot were unaffected by BL treatment (Suga *et al.*, 2002). This finding is consistent with various reports on microarray analysis of BR-regulated genes in which no BR-regulated aquaporin gene was found (Goda *et al.*, 2002, 2004; Mussig *et al.*, 2002), except for a single putatively identified aquaporin gene (Goda *et al.*, 2004). In addition, although a proteomics study of BR-regulated proteins in *Arabidopsis* identified 42 BR-regulated proteins, none of them was an aquaporin gene (Deng *et al.*, 2007). Quantitative regulation of aquaporins is assumed to be a critical step on the control of the water flow and its pathway in plant tissues. This assumption, along with data discussed in this chapter, suggests that more likely the BR promoted cell expansion might rely, at least partially, in changes in aquaporins activity rather than changes in aquaporins at the transcription level.

7. CONTROLL OF CELL DIVISION BY BRASSIONOSTEROIDS

The cell cycle is regulated at multiple points, but major controls operate at the G1/S and G2/M phase boundary (Oakenfull *et al.*, 2002). At the molecular level, cell-cycle transitions in all eukaryotes are controlled by serine/threonine protein kinases known as CDKs. These are subunits of a catalytic domain, the CDK itself, and a regulatory subunit or cyclin, which activates the CDK complex and determines its substrate specificity (Oakenfull *et al.*, 2002; Inzé and De Veylder, 2006). Five different classes of CDKs, CDKA to CDKF, have been identified. The patterns of expression and translation of CDKAs are constitutive during the cell cycle, being CDKA activity essential for both the G1/S and G2/M transitions of the cell cycle, while CDKBs specifically regulates the G_2/M check point. Only CDKDs phosphorylate the C-terminal domain (CTD) of the largest subunit of RNA polymerase II (Inzé and De Veylder, 2006). Moreover, in contrast to the CDKDs, activation of CDKF does not require the association with H-type cyclin (CYCH) (Yamaguchi *et al.*, 2003). Down-regulation of CDKF activity results in a gradual reduction in CDK activity (Umeda *et al.*, 2000), whereas increased expression of the rice CDKD;1 accelerates S-phase progression. CDKC and CDKE have no clear role on the control of the cell cycle (Inzé and De Veylder, 2006).

A nomenclature system for plant cyclins, based on their functional similarity to mammalian cyclins, designates them as CycA, CycB, CycC, CycD, CycH, CycL, CycP and CycT (Inzé and De Veylder, 2006). Sequence data was used to further subdivide them into five groups: CycAl, A2, A3, and CycBl and B2. D-type cyclins (CycD), also known as G_1 proteins, are thought to regulate the G1/S transition (reviewed in Inzé and De Veylder, 2006) and, therefore, represent key regulators of cell commitment to division (Fuerst *et al.*, 1996; Gutierrez *et al.*, 2002; Shen, 2002; Trimarchi and Lees, 2002; Fu *et al.*, 2008). CycDs play key roles in linking the *Arabidopsis* cell cycle to extracellular and developmental signals such as phytohormones and carbohydrate levels that are important in influencing decisions by plant cells to divide (Healy *et al.*, 2001; Oakenfull *et al.*, 2002). Among the groups of cyclins identified so far, A-type cyclins regulate the S/M phase control and B-type cyclins regulates both the G2/M transition and intra-M-phase control.

Early work using the bean second internode bioassay suggested that besides affecting cell expansion, BRs also affected cell division (Steffens,

1991). However, whether BR played a role in cell division was an open question for quite some time because of the contradictory results reported. For example, treatment of parenchyma cell cultures of Jerusalem artichoke (*Helianthus tuberosus*) with nanomolar concentrations of BR, in the presence of cytokinin and auxin, led to an enhancement of at least 50% in the total number of cells (Clouse and Zurek, 1991). In addition, in protoplasts of Chinese cabbage (*Brassica rapa*), BR promoted cell division in a dose-dependent manner and enhanced formation of cell clusters when applied with 2,4-dichlorophenoxyacetic acid (2,4-D) and kinetin (Nakajima *et al.*, 1996). Similar results were reported later for *Petunia* protoplast cultures (Oh and Clouse, 1998). However, studies carried out on carrot cell cultures (Sala and Sala, 1985), and with hormone autonomous callus or suspension cultures of *Agrobacterium*-transformed tobacco (Roth *et al.*, 1989), demonstrated that BR had no stimulatory effect on cell division. Furthermore, microscopy analysis of BR-deficient and BR-insensitive *Arabidopsis* mutants showed that the dwarfism was mainly due to reduction in cell size rather than in cell number (Kauschmann *et al.*, 1996). Because several of the reports on the effects of BRs on cell division originated from *in vitro*-grown tissue or protoplasts, and, because the effects of mitogenic factors on cell division and their interaction are very complex, the contradictory reports on the effects of BRs on cell division were suggested to be due to unbalanced concentration or combination of phytohormones in the culture media (Nakajima *et al.*, 1996). Indeed, it became apparent that a high BR concentration in the culture medium was less effective, or even inhibitory rather than stimulatory for cell division, and that interaction with auxin(s) and cytokinin(s) was also critical in determining BR effect (Oh and Clouse, 1998). In fact, at the biochemical level, BRs are known for a long time to change endogenous cytokinin concentrations in various plant species. For example, when provided via a culture medium containing growth-limiting level of auxin, epi-BL increased the endogenous predominant cytokinins N-6-(δ-2-isopentenyl) adenine and trans-zeatin in tobacco callus tissue (Gaudinova *et al.*, 1995). Thus, BRs might stimulate cell division by itself but also through a BR-driven enhancement of the endogenous levels of cytokinins. In a later study, Bajguz demonstrated that in synchronously dividing cultures of the alga *Chlorella vulgaris*, accelerated increases in cell number following BR treatment were related to increased nucleic acid and protein content (Bajguz, 2000). Furthermore, in roots of wheat (*Triticum aestivum*), similarly to what had been found after cytokinin treatments, increased mitotic rate and nucleoli volumes were related to treatment with epi-BL (Fatkhutdinova *et al.*, 2002). Howell *et al.* (2007) observed great increase in the number of cells in prophase and telophase within 48 h of exposure to epiBL. More recently, a 2.8-fold increase in the mitotic index was found for *Hordeum vulgare* cv. Zafer-160 root cells originated from seeds treated with 0.5 μM homobrassinolide (Kartal *et al.*, 2009).

In leaves of *Arabidopsis*, overexpression of cycd3;1 leads to a considerably enhanced number of smaller cells, and cell division replaces cell expansion as the driven mechanism for leaf growth (Dewitte *et al.*, 2003). The identification of BR-regulated genes involved in cell division such as cycd3 (Hu *et al.*, 2000) pointed to cell division as a further mechanism, besides cell expansion, contributing to BR promoted growth, being the coordination of these diverse processes a result, at least partially, from interactions between BRs and other phytohormones (Lisso *et al.*, 2005). During the late 90's, evidence on the effect of BRs on cell division started to accumulate at the molecular level. In plants, CDKs are often given the nonspecific designation cdc2 because they share high amino acid sequence identity with cdc2 of fission yeast. In *Arabidopsis*, a cdc2b was found to be BR-induced in darkness; however, cdc2b was demonstrated to play a role in various physiological events but not in cell-cycle control (Yoshizumi *et al.*, 1999). After that, Hu and co-workers (2000) identified BR-responsive genes in cell suspension cultures of *det2*, an *Arabidopsis* mutant deficient on BRs biosynthesis, and found that epi-BL upregulated transcription of the cycd3, a cyclin gene involved in the regulation of the G1/S transition (Figures 2 and 3). Cytokinin activates cell division through cycd3 and it is quite interesting that epi-BL was also shown to be able to effectively replace zeatin in culturing of *Arabidopsis* callus and suspension cells. Those findings led Hu and co-workers to conclude that epi-BL stimulated cell division through the induction of cycd3. The epi-BL-driven induction of cycd3 was further demonstrated to involve *de novo* protein synthesis, but no protein phosphorylation or dephosphorylation. Since this finding was apparently inconsistent with the BRI1 signal transduction pathway, Hu and co-workers performed an RNA gel-blot analysis in cell suspension culture of *bri1*, a BR-mutant in which the BRI1 pathway is blocked (Li and Chory, 1997). When treated with BL, *bri1* cells accumulated cycd3 mRNA in a dose-dependent way similar to that of a wild-type. That finding suggested that BR-driven induction of cycd3 involves a BR signalling pathway that differs from the BRI1 pathway.

Evidence on the role of BRs in the regulation of cell cycle-related genes was, for quite some time, essentially restricted to the enhancement of cdc2 and cycd3 genes by BRs in *Arabidopsis* (Yoshizumi *et al.*, 1999; Hu *et al.*, 2000). However, RT-PCR analysis carried out in Jinchun No. 4, a non-parthenocarpic cultivar of cucumber (*Cucumis sativus*) in which flow cytometric analysis showed that epi-BL treatment triggers cell division, demonstrated that epi-BL induced expression of two cyclin genes (cyca and cycb) besides cycd3;1, cycd3;2 and cdkb after anthesis, although the expression of another cdk gene, cdka, was slightly down-regulated after treatment with epi-BL (Fu *et al.*, 2008) (Figure 3).

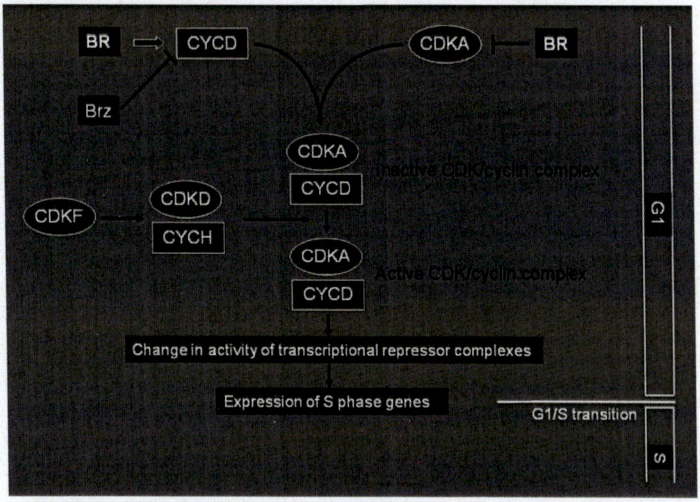

Figure 2. Schematic representation of the regulation of the G1/S transition in the cell cycle by BRs. In the presence of BRs, cyclins associate with CDKA forming an inactive CDKA/CYCLIN complex. This complex is likely activated by phosphorylation through the CDK-activating kinase pathway, which involves CDKF and CDKD associated with an H-type cyclin. CDKA/CYCLIN complex trigger the G1/S transition through changes in the activity of transcriptional repressor complexes. Dynamics of the transcriptional factors is not shown.

Brassinazole (Brz) is a triazole derivative known to selectively and directly binds to the DWF4 protein, a cytochrome P450 monooxygenase that catalyzes the hydroxylation of the 22-position of the side chains of BRs, an BR activation step (Choe *et al.*, 1998; Asami *et al.*, 2000, 2001, 2003; Yamamoto *et al.*, 2001). Northern blot analysis showed that Brz treatment considerably repressed the expression of cscycd3;1 and cscycd3;2 in cucumber, while accumulation of cscycd3;1 transcripts after Brz treatment could be rescued by epi-BL treatment. Accumulation of the cscycd3;2 transcripts followed a similar trend (Fu *et al.*, 2008). In a study aimed to elucidate the involvement of BRs in the progression of tracheary element differentiation in cultured *Zinnia elegans* cells, uniconazole was shown to suppress the accumulation of transcripts for genes that are involved in secondary wall formation. Somewhat surprisingly, the inhibitor of BRs biosynthesis did not inhibit cell division in *Zinnia* cells. The suppression of transcript accumulation was recoverable upon BL application (Yamamoto *et al.*, 1997).

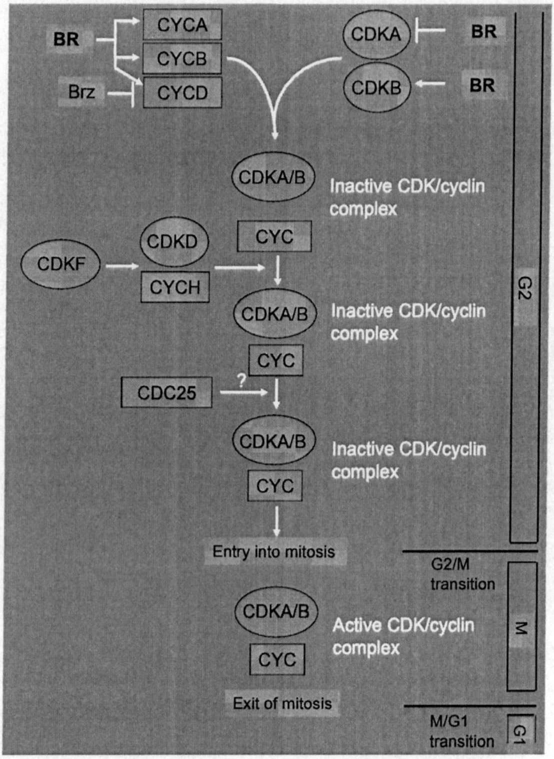

Figure 3. Schematic representation of the regulation of the G2/M transition in the cell cycle by BRs. In the presence of BRs, cyclins associate with CDKA forming an inactive CDKA/CYCLIN complex. This complex is likely activated by phosphorylation through the CDK-activating kinase pathway, which involves CDKF, CDKD associated with an H-type cyclin, and possibly a CDC25. Activated CDK/CYC complexes trigger the G2-to-M transition through the phosphorylation of a plethora of different substrates. Exit from mitosis requires the proteolytic destruction of the cyclin subunits. This destruction is initiated by the activation of the anaphase-promoting complex (APC) through its association with the CCS52 protein (not shown).

During the past 10 years, the identification of BR-responsive genes contributed significantly to our understanding about the mechanisms behind the BR promoted cell division, however, similarly to what occur with the BR-promoted cell expansion, further studies, especially at the post-transcription level, are necessary to fully elucidate the role BRs in the promotion of plant cell division.

8. CONTROL OF CELL WEIGHT ACCUMULATION BY BRASSINOSTEROIDS

In *Wolffia arrhiza*, an aquatic monocotyledon of the duckweed-family, treatment with 10^{-9} M epi-BL results in 33% enhancement in fresh weight, relative to the control (Bajguz and Asami, 2005). This result indicates that limited BR supply constrain cell fresh weight accumulation in *W. arrhiza*.

uzu is known as a major semi-dwarf, single gene, a recessive natural mutation on the BR receptor kinase gene *hvbri1* in barley (*Hordeum vulgare* L.). Somewhat surprisingly, in the "Ryohu" isogenic line of barley, fresh weight accumulation of 1 μM 2,4dichlorophenoxy acetic (2,4-D) acid-grown calli were higher in the *uzu* line, compared to the non-mutant line (Rikiishi *et al*., 2008). However, in the "Hoshimasari" isogenic lines, no significant difference in fresh weight accumulation was found for 1 μM 2,4-D-grown calli of either *uzu* or non-mutant line (Rikiishi *et al*., 2008). The reason(s) for the differential responses, towards fresh weight accumulation, found for the *uzu* lines and non-mutant lines of the "Ryohu" and "Hoshimasari" isogenic lines remain to be elucidated. However, the results reported for *uzu* line and non-mutant lines of the "Ryohu" and "Hoshimasari" isogenic lines of barley indicate that an impaired ability to sense BRs is not critical for fresh weight accumulation in barley calli.

9. CONTROL OF CELL EXPANSION BY INHIBITORS OF BR BIOSYNTHESIS

Mutants deficient in BRs biosynthesis or response have significantly contributed to increase the knowledge about BRs and their actions (Clouse and Sasse, 1998; Li and Chory, 1999; Bishop and Yokota, 2001; Clouse, 2002). However, the use of specific biosynthesis inhibitors is an alternative way for the determination of physiological functions of BRs. From the morphological standpoint, brassinazole (Brz)-treated plants, such as *Arabidopsis* and *Lepidium sativum*, display features similar to those presented by BR-deficient mutants, including dwarfism in the light (Asami *et al*., 2000, 2001). In contrast to non-treated *Arabidopsis*, hypocotyl cell expansion is reduced in the brassinazole-treated plants, whereas no differences can be detected in the number of cells between the brassinazole-treated and non-treated plants (Asami *et al*., 2000). Furthermore, morphological changes induced by brassinazole are rescued by the addition of exogenous BL but not by gibberellin.

Uniconazole, a triazole-type plant-growth retardant (Iwasaki and Shibaoka, 1991), is known to inhibit BRs biosynthesis, besides inhibiting gibberellins biosynthesis. More specifically, and similarly to Brz, uniconazole blocks the step catalyzed by DWF4 in BR biosynthesis (Asami *et al.*, 2001). In isolated *Zinnia* mesophyll cells, exogenously supplied uniconazole prevents uncommitted cells from trans-differentiating into tracheary elements without inhibiting cell division, being BL, but not gibberellins, able to overcome the effect (Iwasaki and Shibaoka, 1991). When the synthesis of BRs is blocked by uniconazole, not only hypocotyl elongation (Asami *et al.*, 2000) but also preceding actin filament aggregation and microtubule bundling are inhibited (Iwasaki and Shibaoka, 1991), indicating that not proper assembly or orientation of components of the plant cell cytoskeleton might be involved in the uniconazole-induced inhibition of hypocotyl elongation.

10. CONTROL OF CELL WEIGHT BY INHIBITORS OF BR BIOSYNTHESIS

Brz, applied as a diastereomeric mixture of Brz2001 has been shown to inhibit fresh weight accumulation in *Chlorella vulgaris*, an alga, while, co-treatment with Brz2001 and BL results in normal growth in light-grown cells (Bajguz and Asami, 2004). Similarly to what had been found for *C. vulgaris* cells, Brz2001 was shown to reduce fresh weight accumulation, in a concentration-dependent way, in cultures of *Wolffia arrhiza* (Bajguz and Asami, 2005). However, co-treatment with 10^{-6}–5.10^{-6} M Brz2001 and 10^{-9} M epi-BL showed a weaker promotive effect on fresh weight accumulation, compared to treatment with epiBL alone. It seems to be straightforward that Brz is able to reduce fresh weight accumulation in plant cells, and this inhibitory effect is possibly due, at least partially, to an eventual Brz-induced inhibition of cell expansion. However, it remains to be elucidated.

11. CONCLUSION

Data presented in this chapter demonstrate that BR control of cell growth rely in a broad variety of mechanisms. These mechanisms are starting to be understood at the transcriptional level. However, further studies, especially at post-transcriptional level, are necessary in order to fully elucidate the mechanism(s) behind the BR-promoted cell growth.

Since plant architecture is largely determined by the pattern of cell growth, increased knowledge about the networked pathways for the BRs biosynthesis and BR signaling, along with the use of selected or structurally modified BRs is expected to make it possible for us to effectively: (1) Control excessive vegetative growth in young orchards, or conversely, to promote shooting in producing orchards; (2) Inhibit lateral growth in forestry; (3) Produce plants more resistant to damage by wind and rain; 3. Improve clonal propagation techniques not only for conventional commercial applications, but also for the fastly increasing number of desirable genetically transformed plants; among many other practical applications.

12. ACKNOWLEDGEMENTS

The author thanks CNPq-Brazil for financial support.

13. REFERENCES

Asami, T., Min, Y.K., Nagata, N., Yamagishi, K., Takatsuto, S., Fujioka, S., Murofushi, N., Yamaguchi, I., and Yoshida, S. 2000. Characterization of brassinazole, a triazole-type brassinosteroid biosynthesis inhibitor. *Plant Physiol.*, **123**: 93–100.

Asami, T., Mizutani, M., Fujioka, S., Goda, H., Min, Y.K., Shimada, Y., Nakano, T., Takatsuto, S., Matsuyama, T., Nagata, N., Sakata, K., and Yoshida, S. 2001. Selective interaction of triazole derivatives with dwf4, a cytochrome p450 monooxygenase of the brassinosteroid biosynthetic pathway, correlates with brassinosteroid deficiency *in planta*. *J. Biol. Chem.*, **276**: 25687–25691.

Asami, T., Nakano, T., Nakashita, H., Sekimata, K., Shimada, Y., and Yoshida, S. 2003. The influence of chemical genetics on plant science: shedding light on functions and mechanism of action of brassinosteroids using biosynthesis inhibitors. *J. Plant Growth Reg.*, **22**: 336–349.

Aspinall, G.O. 1980. Chemistry of cell wall polysaccharides. *In* J Preiss J (Ed), Carbohydrates: Structure and Function. The Biochemistry of Plants. vol. 3. Academic Press, New York, USA: pp. 473–500.

Azpiroz, R., Wu, Y., LoCascio, J.C., and Feldmann, K.A., 1998. An *Arabidopsis* brassinosteroid-dependent mutant is blocked in cell elongation. *Plant Cell*, **10**: 219–230.

Bajguz, A. 2000. Effect of brassinosteroids on nucleic acids and protein content in cultured cells of *Chlorella vulgaris*. *Plant Physiol. Biochem.*, **38**: 209–215.

Bajguz, A., and Asami, T. 2004. Effects of brassinazole an inhibitor of brassinosteroid biosynthesis on light- and dark-grown *Chlorella vulgaris*. *Planta*, **218**: 869–877.

Bajguz, A., and Asami, T. 2005. Suppression of *Wolffia arrhiza* growth by brassinazole, an inhibitor of brassinosteroid biosynthesis and its restoration by endogenous 24-epibrassinolide. *Phytochem.*, **66**: 1787–1796.

Beguin, P., and Aubert, J.P. 1994. The biological degradation of cellulose. *FEMS Microbiol. Rev.*, **13**: 25–28.

Belkhadir, Y., and Chory, J. 2006. Brassinosteroid Signaling: a paradigm for steroid hormone signaling from the cell surface. *Science*, **314**: 1410–1411.

Bishop, G.J., and Yokota, T. 2001. Plants steroid hormones, brassinosteroids: current highlights of molecular aspects on their synthesis/metabolism, transport, perception and response. *Plant, Cell Physiol.*, **42**: 114–120.

Buckeridge, M.S., Rayon, C., Urbanowics, B., Tine, A.M.A.S., and Carpita, N.C. 2004. Mixed linkage (1→3), (1→4) β-D-glucans of grasses. *Cereal Chem.*, **18**: 115–127.

Campbell, P., and Braam, J. 1999. Xyloglucan endotransglycosylases: diversity of genes, enzymes and potential wall-modifying functions. *Trends Plant Sci.*, **4**: 361–366.

Carpita, N.C., and Gibeaut, D.M. 1993. Structural models of primary cell walls in flowering plants: consistency of molecular structure with the physical properties of the walls during growth. *Plant J.*, **3**: 1–30.

Carpita, N., and McCann, M. 2000. The cell wall. *In* BB Buchanan, W Gruissem, RL Jones (Eds), Biochemistry and Molecular Biology of Plants. *American Society of Plant Physiologists*, Rockville, USA: pp. 52–108.

Catala, C., Rose, J.K.C., and Bennett, A. 1997. Auxin-regulation and spatial localization of an endo-1,4-β-D-glucanase and a xyloglucan endotransglycosylase in expanding tomato hypocotyls. *Plant J.*, **12**: 417–426.

Catterou, M., Dubois, F., Schaller, H., Aubanelle, L., Vilcot, B., Sangwan-Norreel, B.S., and Sangwan, R.S. 2001. Brassinosteroids, microtubules and cell elongation in *Arabidopsis thaliana*. II. Effects of brassinosteroids on microtubules and cell elongation in the *bul*1 mutant. *Planta*, **212**: 673–683.

Cavalier, D.M., Lerouxel, O., Neumetzler, L., Yamauchi, K., Reinecke, A., Freshour, G., Zabotina, O.A., Hahn, M.G., Burgert, I., Pauly, M., Raikhel, N.V., and Keegstra, K. 2008. Disrupting two *Arabidopsis thaliana* xylosyltransferase genes results in plants deficient in xyloglucan, a major primary cell wall component. *Plant Cell*, **20**: 1519–1537.

Choe, S., Dilkes, B.P., Fujioka, S., Takatsuto, S., Sakurai, A., and Feldmann, K.A. 1998. The DWF4 gene of *Arabidopsis* encodes a cytochrome P450 that mediates multiple 22α-hydroxylation steps in brassinosteroid biosynthesis. *Plant Cell*, **10**: 231–243.

Choe, S., Dilkes, B.P., Gregory, B.D., Ross, A.S., Yuan, H., Noguchi, T., Fujioka, S., Takatsuto, S., Tanaka, A., Yoshida, S., Tax, F.E., and Feldmann, K.A. 1999. The *Arabidopsis* dwarfl mutant is defective in the conversion of 24-methylenecholesterol to campesterol in brassinosteroid biosynthesis. *Plant Physiol.*, **119**: 897–907.

Choe, S., Schmitz, R.J., Fujioka, S., Takatsuto, S., Lee, M.O., and Yoshida, S. 2002. *Arabidopsis* brassinosteroid insensitive *dwarf12* mutants are semidominant and defective in a glycogen synthase kinase 3β-like kinase. *Plant Physiol.*, **130**: 1506–1515.

Chory, J. 2001. Light, brassinosteroids, and *Arabidopsis* development. Proceedings of the Symposium: Plant Physiology 2000 and Beyond: Breaking the Mold, Plant Biology 2001-ASPP, Providence, Rhode Island, USA: Abstract 30005.

Clouse, S.D. 1997. Molecular genetic analysis of brassinosteroid action. *Physiol. Plant.*, **100**: 702–709.

Clouse, S.D. 2002. Brassinosteroids: plant counterparts to animal steroid hormones? *Vitam Horm*, **65**: 195–223.

Clouse, S.D., and Sasse, J.M. 1998. Brassinosteroids: essential regulators of plant growth and development. *Ann. Rev. Plant Phys. Plant Mol. Biol.*, **49**: 427–451.

Clouse, S.D, and Zurek, D. 1991. Molecular analysis of brassinolide action in plant growth and development. *In* HG Cutler, T Yokota, G Adam (Eds), Brassinosteroids: Chemistry, Bioactivity and Applications. American Chemical Society, Washington, D.C., USA: pp. 122–140.

Clouse, S.D., Langford, M., and McMorris, T.C. 1996. A brassinosteroid-insensitive mutant in *Arabidopsis thaliana* exhibits multiple defects in growth and development. *Plant Physiol.*, **111**: 671–678.

Cosgrove, D. 1997. Relaxation in a high-stress environment: the molecular basis of extensible cell walls and enlargement. *Plant Cell*, **9**: 1031–1041.

Cosgrove, D.J. 2000. Loosening of plant cell walls by expansins. *Nature*, **407**: 321–326.

Cyr, R.J., and Palevitz, B.A. 1995. Organization of cortical microtubules in plant cells. *Curr. Opin. Cell Biol.*, **7**: 65–71.

Deng, Z., Zhang, X., Tang, W., Oses-Prieto, J.A., Suzuki, N., Gendron, J.M., Chen, H., Guan, S., Chalkley, R.J., Peterman, T.K., Burlingame, A.L., and Wang, Z.-Y. 2007. A proteomic study of brassinosteroid response in *Arabidopsis*. *Mol. Cell. Proteomics*, **6**: 2058–2071.

Desai, A., and Mitchison, T.J. 1997. Microtubule polymerization dynamics. *Ann. Rev. Cell Dev. Biol.*, **13**: 83–117.

Dewitte, W., Riou-Khamlichi, C., Scofield, S., Healy, J.M.S., Jacqmard, A., Kilby, N.J., and Murray, J.A.H. 2003. Altered cell cycle distribution, hyperplasia, and inhibited differentiation in *Arabidopsis* caused by the D-type cyclin CYCD3. *Plant Cell*, **15**: 79–92.

Fanutti, C., Gidley, M.J., and Reid, J.S.G. 1993. Action of a pure xyloglucan endo-transglycosylase (formerly called xyloglucan-specific endo-(1→4)-β-D-glucanase) from the cotyledons of germinated nasturtium seeds. *Plant J.*, **3**: 691–700.

Fatkhutdinova, R.A., Shakirova, F.M., Chemeris, A.V., Sabirzhanov, B.E., and Vakhitov, V.A. 2002. NOR activity in wheat species with different ploidy levels treated with phytohormones. *Russ. J. Genet.*, **38**: 1335–1338.

Friedrichsen, D.M., Joazeiro, C.A.P., Li, J., Hunter, T., and Chory, J. 2000. Brassinosteroid-insensitive-1 is a ubiquitously expressed leucine-rich repeat receptor serine/threonine kinase. *Plant Physiol.*, **123**: 1247–1256.

Fry, S.C. 1989a. The structure and functions of xyloglucan. *J. Exp. Bot.*, **40**: 1–11.

Fry, S.C. 1989b. Cellulases, hemicelluloses and auxin-stimulated growth: a possible relationship. *Physiol. Plant.*, **75**: 532–536.

Fry, S.C., Smith, R.C., Renwick, K.F., Martin, D.J., Hodge, S.K., and Matthews, K.J. 1992. Xyloglucan endotransglycosylase, a new wall-loosening enzyme activity from plants. *Biochem. J.*, **282**: 821–828.

Fu, F.Q., Mao, W.H., Shi, K., Zhou, Y.H., Asami, T., and Yu, J.Q. 2008. A role of brassinosteroids in early fruit development in cucumber. *J. Exp. Bot.*, **59**: 2299–2308.

Fu, Y., Li, H., and Yang, Z. 2002. The ROP2 GTPase controls the formation of cortical fine F-actin and the early phase of directional cell expansion during *Arabidopsis* organogenesis. *Plant Cell*, **14**: 777–794.

Fuerst, R.A.U.A., Soni, R., Murray, J.A.H., and Lindsey, K., 1996. Modulation of cyclin transcript levels in cultured cells of *Arabidopsis thaliana*. *Plant Physiol.*, **112**: 1023–1033.

Fujioka, S. 1999. Natural occurrence of brassinosteroids in the plant kingdom. *In* A Sakurai, T Yokota, SD Clouse (Eds), Brassinosteroids: Steroidal Plant Hormones. Springer Verlag, Tokyo: pp. 21–45.

Fujioka, S., and Sakurai, A. 1997. Biosynthesis and metabolism of brassinosteroids. *Physiol. Plant.*, **100**: 710–715.

Gaudinova, A., Sussenbekova, H., Vojtechova, M., Kaminek, M., Eder, J., and Kohout, L. 1995. Different effects of 2 brassinosteroids on growth, auxin and cytokinin content in tobacco callus-tissue. *Plant Growth Reg.*, **17**: 121–126.

Geldner, N., Hyman, D.L., Wang, X., Schumacher, K., and Chory, J. 2007. Endosomal signaling of plant steroid receptor kinase BRI1. *Genes & Develop.*, **21**: 1598–1602.

Goda, H., Sawa, S., Asami, T., Fujioka, S., Shimada, Y., and Yoshida, S. 2004. Comprehensive comparison of auxin-regulated and brassinosteroid-regulated genes in *Arabidopsis*. *Plant Physiol.*, **134**: 1555–1573.

Goda, H., Shimada, Y., Asami, T., Fujioka, S., and Yoshida, S. 2002. Microarray analysis of brassinosteroid-regulated genes in *Arabidopsis*. *Plant Physiol.*, **130**: 1319–1334.

Goddard, R.H., Wick, S.M., Silflow, C.D., and Snustad, D.P. 1994. Microtubule components of the plant cell cytoskeleton. *Plant Physiol.*, **104**: 1–6.

Gogarten, J.P., Fichmann, J., Braun, Y., Morgan, L., Styles, P., and Taiz, S.L., 1992. The use of antisense mRNA to inhibit the tonoplast H^+ ATPase in carrot. *Plant Cell.*, **4**: 851–864.

Gutierrez, C., Ramirez-Parra, E., Castellano, M.M., and del Pozo, J.C. 2002. G(1) to S transition: more than a cell cycle engine switch. *Curr. Opin. Plant Biol.*, **5**: 480–486.

He, J.-X., Fujioka, S., Li, T.-C., Kang, S.G., Seto, H., Takatsuto, S., Yoshida, S., and Jang, J.-C. 2003. Sterols regulate development and gene expression in *Arabidopsis*. *Plant Physiol.*, **131**: 1258–1269.

Healy, J.M.S., Menges, M., Doonan, J.H., and Murray, J.A.H. 2001. The *Arabidopsis* D-type cyclins CycD2 and CycD3 both interact *in vivo* with the PSTAIRE cyclin-dependent Kinase Cdc2a but are differentially controlled. *J. Biol. Chem.*, **276**: 7041–7047.

Howell, W.M., Keller, G.E. III, Kirkpatrick, J.D., Jenkins, R.L., Hunsinger, R.N., and McLaughlin, E.W. 2007. Effects of the plant steroidal hormone, 24-epibrassinolide, on the mitotic index and growth of onion (*Allium cepa*) root tips. *Genet. Mol. Res.*, **6**: 50–58.

Hu, Y., Bao, F., and Li, J., 2000. Promotive effect of brassinosteroids on cell division involves a distinct CycD3-induction pathway in *Arabidopsis*. *Plant J.*, **24**: 693–701.

Inzé, D., and De Veylder, L. 2006. Cell cycle regulation in plant development. *Ann. Rev. Genet.*, **40**: 77–105.

Iwasaki, T., and Shibaoka, H. 1991. Brassinosteroids act as regulators of tracheary-element differentiation in isolated *Zinnia* mesophyll cells. *Plant Cell Physiol.*, **32**: 1007–1014.

Jauneau, A., Roy, S., Reis, D., and Vian, B. 1998. Probes and microscopical methods for the localization of pectins in plant cells. *Int. J. Plant Sci.*, **159**: 1–13.

Johansson, P., Brumer, III H., Baumann, M.J., Kallas, A.M., Henriksson, H., Denman, S.E., Teeri, T.T., and Jonesa, T.A. 2004. Crystal structures of a poplar xyloglucan endotransglycosylase reveal details of transglycosylation acceptor binding. *Plant Cell.*, **16**: 874–886.

Kartal, G., Temel, A., Arican, E., and Gozukirmizi, N. 2009. Effects of brassinosteroids on barley root growth, antioxidant system and cell division. *Plant Growth Regul.*, **58**: 261–267.

Katsumi, M. 1991. Physiological modes of brassinolide action in cucumber hypocotyls growth. *In* HG Cutler, T Yokota, G Adam (Eds), Brassinosteroids: Chemistry, Bioactivity, and Application. ACS Symp Ser 474. *American Chemical Society*, Washington, DC, USA: pp. 246–254.

Kauschmann, A., Jessop, A., Knocz, C., Szekeres, M., Willmitzer, L., and Altmann, T. 1996. Genetic evidence for an essential role of brassinosteroids in plant development. *Plant J.*, **9**: 701–713.

Kim, H.B., Kwon, M., Ryu, H., Fujioka, S., Takatsuto, S., Yoshida, S., An, C.S., Lee, I., Hwang, I., and Choe, S. 2006. The regulation of DWARF4 expression is likely a critical mechanism in maintaining the homeostasis of bioactive brassinosteroids in *Arabidopsis*. *Plant Physiol.*, **140**: 548–557.

Klahre, U., Noguchi, T., Fujioka, S., Takatsuto, S., Yokota, T., Nomura, T., Yoshida, S., and Chua N.-H. 1998. The *Arabidopsis DIMINUTO/DWARF1* gene encodes a protein involved in steroid synthesis. *Plant Cell*, **10**: 1677–1690.

Knox, J.P. 2008. Revealing the structural and functional diversity of plant cell walls. *Curr. Opin. Plant Biol.*, **11**: 308–313.

Koka, C.V., Cerny, R.E., Gardner, R.G., Noguchi, T., Fujioka, S., Takatsuto, S., Yoshida, S., and Clouse, S.D. 2000. A putative role for the tomato genes *DUMPY* and *CURL*-3 in brassinosteroid biosynthesis and response. *Plant Physiol.*, **122**: 85–98.

Ladyzhenskaya, E.P., and Korableva, N.P. 2001. Effects of growth regulators on H^+ translocation across the membranes of plasma membrane vesicles from potato tuber cells. *Appl. Biochem. Microbiol.*, **37**: 521–523.

Lee, Y., Choi, D., and Kende, H. 2001. Expansins: ever-expanding numbers and functions. *Curr. Opin. Plant Biol.*, **4**: 527–532.

Li, J., and Chory, J. 1997. A putative leucine-rich receptor kinase involved in brassinosteroid signal transduction. *Cell*, **90**: 929–938.

Li, J., and Chory, J. 1999. Brassinosteroid actions in plants. *J. Exp. Bot.*, **50**: 275–282.

Li, J., and Nam, K.H. 2002. Regulation of brassinosteroid singling by a GSK3/SHAGGY-like kinase. *Science*, **295**: 1299–1301.

Li, J., Wen, J., Lease, K.A., Doke, J.T., Tax, F.E., and Walker, J.C. 2002. BAK1, an *Arabidopsis* LRR receptor-like protein kinase, interacts with BRI1 and modulates brassinosteroid signaling. *Cell*, **110**: 213–222.

Li, L., Xu, J., Xu, Z.-H., and Xue, H.-W. 2005. Brassinosteroids stimulate plant tropisms through modulation of polar auxin transport in *Brassica* and *Arabidopsis*. *Plant Cell*, **17**: 2738–2753.

Lisso, J., Steinhauser, D., Altmann, T., Kopka, J., and Mussig, C. 2005. Identification of brassinosteroid-related genes by means of transcript co-response analyses. *Nucleic Acids Res.*, **33**: 2685–2696.

Malberg, L.M., Tamblyn Lee J.M., and Forsberg, C.W. 1992. Degradation of cellulose and hemicelluloses by rumen microorganisms. *In* G Winkelmann (Ed), Microbial Degradation of Natural Products. Weinheim, VCH Verlagsgesellschaft mbH: pp. 1127–1159.

Mayumi, K., and Shibaoka, H. 1995. A possible double role for brassinolide in the reorientation of cortical microtubules in the epidermal cells of Azuki bean epicotyls. *Plant Cell Physiol.*, **36**: 173–181.

Mora-García, S., Vert, G., Yin, Y., Caño-Delgado, A., Cheong, H., and Chory, J. 2004. Nuclear protein phosphatases with Kelch-repeat domains modulate the response to brassinosteroids in *Arabidopsis*. *Genes & Develop.*, **18**: 448–460.

Morillon, R., Catterou, M., Sangwan, R.S., Sangwan, B.S., and Lassalles, J.P. 2001. Brassinolide may control aquaporin activities in *Arabidopsis thaliana*. *Planta*, **212**: 199–204.

Muday, G.K., Hu, S., and Brady, R. 2000. The actin cytoskeleton may control the polar distribution of an auxin transport protein. *Gravit. Space Biol. Bull.*, **13**: 75–83.

Munoz, F.J., Labrador, E., and Dopico, B. 1998. Brassinolides promote the expression of a new *Cicer arietinum* tubulin gene involved in the epicotyl elongation. *Plant Mol. Biol.*, **37**: 807–817.

Mussig, C., and Altmann, T. 2003. Genomic brassinosteroid effects. *J. Plant Growth Regul.*, **22**: 313–324.

Mussig, C., Fischer, S., and Altmann, T. 2002. Brassinosteroid-regulated gene expression. *Plant Physiol.*, **129**: 1241–1251.

Nakajima, N., Shida, A., and Toyama, S. 1996. Effects of brassinosteroid on cell division and colony formation of Chinese cabbage mesophyll protoplasts. *Japan J. Crop Sci.*, **65**: 114–118.

Nemhauser, J.L., Mockler, T.C., and Chory, J. 2004. Interdependency of brassinosteroid and auxin signaling in *Arabidopsis*. *PLoS Biol.* **2**: e258.

Nick, P. 1998. Signaling to the microtubular cytoskelton in plants. *Int. Rev. Cytol.*, **184**: 33–80.

Nicol, F., His, I., Jauneau, A., Vernhettes, S., Canut, H., and Hofte, H. 1998. A plasma membrane-bound putative endo-1,4-β-Dglucanase is required for normal wall assembly and cell elongation in *Arabidopsis*. *EMBO J.*, **17**: 5563–5576.

Nogales, E. 2000. Structural insights into microtubule function. *Ann. Rev. Biochem.*, **69**: 277–302.

Noguchi, T., Fujioka, S., Choe, S., Takatsuto, S., Tax, F.E., Yoshida, S., and Feldmann, K.A. 2000. Biosynthetic pathways of brassinolide in *Arabidopsis*. *Plant Physiol.*, **124**: 201–209.

Nomura, T., Nakayama, M., Reid, J.B., Takeuchi, Y., and Yokota, T. 1997. Blockage of brassinosteroid biosynthesis and sensitivity causes dwarfism in garden pea. *Plant Physiol.*, **113**: 31–37.

Nomura, T., Kitasaka, Y., Takatsuto, S., Reid, J.B., Fukami, M., and Yokota, T. 1999. Brassinosteroid/sterol synthesis and plant growth as affected by *lka* and *lkb* mutations of pea. *Plant Physiol.*, **119**: 1517–1526.

Oakenfull, E.A., Riou-Khamlichi, C., and Murray, J.A.H. 2002. Plant D-type cyclins and the control of G1 progression. *Philos. Transac. Royal Soc. Biol. Series*, **357**: 749–760.

Oh, M.H., and Clouse, S.D. 1998. Brassinolide affects the rate of cell division in isolated leaf protoplasts of *Petunia hybrida*. *Plant Cell Rep.*, **17**: 921–924.

Oh, M.H., Romanow, W., Smith, R., Zamski, E., Sasse, J., and Clouse, S. 1998. Soybean BRU1 encodes a functional xyloglucan endo-transglycosylase that is highly expressed in inner epicotyl tissues during brassinosteroid-promoted elongation. *Plant Cell Physiol.*, **39**: 124–130.

Peng, P., and Li, J. 2003. Brassinosteroid signal transduction: a mix of conservation and novelty. *J. Plant Growth Regul.*, **22**: 298–312.

Pérez-Pérez, J.M., Ponce, M., and Micol, J.L. 2002. The *UCU1Arabidopsis* gene encodes a SHAGGY/GSK3-like kinase required for cell expansion along the proximodistal axis. *Develop Biol.*, **242**: 161–173.

Rea, P.A., and Poole, R.J. 1993. Vacuolar H^+-translocating pyrophosphatase. *Ann. Rev. Plant Physiol. Plant Mol. Biol.*, **44**: 157–180.

Rikiishi, K., Saisho, D., and Takeda, K. 2008. *Uzu*, a barley semi-dwarf gene, suppresses plant regeneration in calli derived from immature embryos breeding. *Science*, **58**: 149–155.

Roberts, K. 1994. The plant extracellular matrix: in a new expansive mood. *Curr. Opin. Cell Biol.*, **6**: 688–694.

Rose, J.K., Braam, J., Fry, S.C., and Nishitani, K. 2002. The XTH family of enzymes involved in xyloglucan endotransglucosylation and endohydrolysis: current perspectives and a new unifying nomenclature. *Plant Cell Physiol.*, **43**: 1421–1435.

Roth, P.S., Bach, T.J., and Thompson, M.J. 1989. Brassinosteroids: potent inhibitors of transformed tobacco callus cultures. *Plant Sci.*, **59**: 63–70.

Sala, C., and Sala, F. 1985. Effect of brassinosteroid on cell division and enlargement in cultured carrot (*Daucus carota* L.) cells. *Plant Cell Rep.*, **4**: 144–147.

Sasaki, S. 2008. Introduction for special issue for aquaporin expanding the world of aquaporins: new members and new functions. *Pflugers Arch.- Euro. J. Physiol.*, **456**: 647–649.

Sasuga, Y., Tanaka, S., and Takenaga, H. 2000. Possible mechanism of the cell elongation induced by brassinolide in rice seedlings. *Plant Cell Physiol.*, **41**(supplement): s200.

Schumacher, K., Vafeados, D., McCarthy, M., Sze, H., Wilkins, T., and Chory, J. 1999. The *Arabidopsis* det3 mutant reveals a central role for the vacuolar H^+-ATPase in plant growth and development. *Genes & Develop.*, **13**: 3259–3270.

Shen, W.H. 2002. The plant E2F-Rb pathway and epigenetic control. *Trends Plant Sci.*, **7**: 505–511.

Steffens, G.L. 1991. U.S. Department of Agriculture Brassins Project: 1970–1980. *In* HG Cutler, T Yokota, G Adam (Eds), Brassinosteroids: Chemistry, Bioactivity, & Applications. American Chemical Society, Washington, D.C., USA: pp. 2–17.

Suga, S., Komatsu, S., and Maeshima, M. 2002. Aquaporin isoforms responsive to salt and water stresses and phytohormones in radish seedlings. *Plant Cell Physiol.*, **43**: 1229–1237.

Sunohara, H., Kawai, T., Shimizu-Sato, S., Sato, Y., Sato, K., and Kitano, H. 2009. A dominant mutation of *TWISTED DWARF 1* encoding an α-tubulin protein causes severe dwarfism and right helical growth in rice. *Genes Genet. Syst.*, **84**: 209–218.

Szekeres, M., Nemeth, K., Koncz-Kalman, Z., Mathur, J., Kauschmann, A., Altmann, T., Redei, G.P., Nagy, F., Schell, J., and Koncz, C. 1996. Brassinosteroids rescue the deficiency of CYP90, a cytochrome P450, controlling cell elongation and de-etiolation in *Arabidopsis*. *Cell*, **85**: 171–182.

Takahashi, T., Gasch, A., Nishizawa, N., and Chua, N.H. 1995. The DIMINUTO gene of *Arabidopsis* is involved in regulating cell elongation. *Genes & Develop.*, **9**: 97–107.

Takeda, T., Furuta, Y., Awano, T., Mizuno, K., Mitsuishi, Y., and Hayashi, T. 2002. Suppression and acceleration of cell elongation by integration of xyloglucans in pea stem segments. *Proc. Nat. Acad. Sci. USA*, **99**: 9055–9060.

Takeda, T., Miller, J.G., and Fry, S.C. 2008. Anionic derivatives of xyloglucan function as acceptor but not donor substrates for xyloglucan endotransglucosylase activity. *Planta*, **227**: 893–905.

Tang, W., Deng, Z., Oses-Prieto, J.A., Suzuki, N., Zhu, S., Zhang, X., Burlingame, A.L., and Wang, Z.-Y. 2008. Proteomics studies of brassinosteroid signal transduction using prefractionation and two-dimensional DIGE. *Mol. Cell. Proteomics*, **7**: 728–738.

Tominaga, R., and Sakurai, N. 1996. Brassinolide induces vacuolar H^+-ATPase activation and stem elongation. *Plant Cell Physiol.*, **37**(Suppl.): S152.

Tominaga, R., Sakurai, N., and Kuraishi, S. 1994. Brassinolide-induced elongation of inner tissues of segments of squash (*Cucurbita maxima* Duch.) hypocotyls. *Plant Cell Physiol.*, **35**: 1103–1106.

Trimarchi, J.M., and Lees, J.A., 2002. Sibling rivalry in the E2F family. *Nature. Rev. Mol. Cell Biol.*, **3**: 11–20.

Umeda, M., Umeda-Hara, C., and Uchimiya, H. 2000. A cyclin-dependent kinaseactivating kinase regulates differentiation of root initial cells in *Arabidopsis*. *Proc. Nat. Acad. Sci. USA*, **97**: 13396–13400.

Uozo, S., Tanaka-Ueguchi, M., Kitano, H., Hattori, K., and Matsuoka, M. 2000. Characterization of XET-related genes of rice. *Plant Physiol.*, **122**: 853–860.

Van Sandt, V.S.T., Suslov, D., Verbelen, J.-P., and Vissenberg, K. 2007. Xyloglucan endotransglucosylase activity loosens a plant cell wall. *Ann. Bot.*, **100**: 1467–1473.

Vissenberg, K., Martinez-Vilchez, I.M., Verbelen, J.P., Miller, J.G., and Fry, S.C. 2000. *In vivo* colocalization of xyloglucan endotransglycosylase activity and its donor substrate in the elongation zone of *Arabidopsis* roots. *Plant Cell*, **12**: 1229–1237.

Vissenberg, K., Fry, S.C., Pauly, M., Hofte, H., and Verbelen, J.P. 2005. XTH acts at the microfibril-matrix interface during cell elongation. *J. Exp. Bot.*, **56**: 673–683.

Xu, W., Campbell, P., Vargheese, A.K., and Braam, J. 1996. The *Arabidopsis* XET-related gene family: environmental and hormonal regulation of expression. *Plant J.*, **9**: 879–889.

Xu, W., Prugganan, M.M., Polisensky, D.H., Antosiewicz, D.M., Fry, S.C., and Braam, J. 1995. *Arabidopsis* TCH4, regulated by hormones and the environment, encodes a xyloglucan endotransglycosylase. *Plant Cell*, **7**: 1555–1567.

Wadaa, Y., and Katsumi, M. 2005. Brassinolide as a modulator of the activities of cell wall loosening proteins. *Plant Biotech.*, **22**: 33–38.

Wang, T.-W., Cosgrove, D.J., and Arteca, R.N. 1993. Brassinosteroid stimulation of hypocotyl elongation and wall relaxation in pakchoi (*Brassica chinensis* cv *Lei-choi*). *Plant Physiol.*, **101**: 965–968.

Wang, Z.Y., Seto, H., Fujioka, S., Yoshida, S., and Chory, J. 2001. BRI1 is a critical component of a plasma-membrane receptor for plant steroids. *Nature*, **410**: 380–383.

Wang, Z.Y., Nakano, T., Gendron, J., He, J.X., Chen, M., Vafeados, D., Yang, Y.L., Fujioka, S., Yoshida, S., Asami, T., and Chory, J. 2002. Nuclear-localized BZR1 mediates brassinosteroid-induced growth and feedback suppression of brassinosteroid biosynthesis. *Develop. Cell*, **2**: 505–513.

Wasteneys, G.O., and Yang, Z. 2004. New views on the plant cytoskeleton. *Plant Physiol.* **136**: 3884–3891.

Williamson, R.E. 1991. Orientation of cortical microtubules in interphase plant cells. *Int. Rev. Cytol.*, **129**: 135–206.

Wu, S.C., Blumer, J.M., Darvill, A.G., and Albersheim, P. 1996. Characterization of an endo-β-1,4-glucanase gene induced by auxin in elongating pea epicotyls. *Plant Physiol.*, **110**: 163–170.

Yamaguchi, M., Kato, H., Yoshida, S., Yamamura, S., Uchimiya, H., and Umeda, M. 2003. Control of *in vitro* organogenesis by cyclin-dependent kinase activities in plants. *Proc. Nat. Acad. Sci. USA*, **100**: 8019–8023.

Yamamoto, R., Demura, T., and Fukuda, H. 1997. Brassinosteroids induce entry into the final stage of tracheary element differentiation in cultured *Zinnia* cells. *Plant Cell Physiol.*, **38**: 980–983.

Yamamoto, R., Fujioka, S., Demura, T., Takatsuto, S., Yoshida, S., and Fukuda, H. 2001. Brassinosteroid levels increase drastically prior to morphogenesis of tracheary elements. *Plant Physiol.*, **125**: 556–563.

Yamamuro, C., Ihara, Y., Wu, X., Noguchi, T., Fujioka, S., Takatsuto, S., Ashikari, M., Kitano, H., and Matsuoka, M. 2000. Loss of function of a rice brassinosteroid insensitive1 homolog prevents internode elongation and bending of the lamina joint. *Plant Cell*, **12**: 1591–1606.

Yang, G., Jan, A., and Komatsu, S. 2009. Characterization of β-tubulin 4 regulated by gibberellins in rice leaf sheath. *Biol. Plant.*, **53**: 422–428.

Yin, Y., Wang, Z.Y., Mora-Garcia, S., Li, J., Yoshida, S., Asami, T., and Chory, J. 2002. BES1 accumulates in the nucleus in response to brassinosteroids to regulate gene expression and promote stem elongation. *Cell*, **109**: 181–191.

Yokota, T. 1997. The structure, biosynthesis and function of brassinosteroids. *Trends Plant Sci.*, **2**: 137–143.

Yokoyama, R., Rose, J.K., and Nishitani, K. 2004. A surprising diversity and abundance of xyloglucan endotransglucosylase/hydrolases in rice. Classification and expression analysis. *Plant Physiol.*, **134**: 1088–1099.

Yoshizumi, T., Nagata, N., Shimada, H., and Matsui, M. 1999. An *Arabidopsis* cell cycle-dependent kinase-related gene CDC2b, plays a role in regulating seedling growth in darkness. *Plant Cell*, **11**: 1883–1896.

Zurek, D.M., and Clouse, S.D. 1994. Molecular cloning and characterization of a brassinosteroid-regulated gene from elongating soybean (*Glycine max* L.) epicotyls. *Plant Physiol.*, **104**: 161–170.

Zurek, D.M., Rayle, D.L., McMorris, T.C., and Clouse, S.D. 1994. Investigation of gene expression, growth kinetics, and wall extensibility during brassinosteroid-regulated stem elongation. *Plant Physiol.*, **104**: 505–513.

Chapter 9

ROLE OF BRASSINOSTEROIDS ON HORTICULTURAL CROPS

Y.Y. KANG[1] AND S.R. GUO[2]
[1]*College of Horticulture, South China Agricultural University, Guangzhou, 510642,* China;
[2]*College of Horticulture, Nanjing Agricultural University, Nanjing, 210095,* China

Abstract: With the progress of chemical synthesis technology, structurally modified brassinosteroids (BRs) with greater stability, under field conditions have been synthesized on a commercial scale and registered as plant growth regulators for specific horticultural crops. In both fundamental and application-oriented research, BRs and their analogues play prominent roles in various physiological processes including, seed development and germination, flower sex expression, fruit development, improvement of quantity and quality of crops, and resistance to various biotic and abiotic stresses. It is worthy to note here that the involvement of BRs in plant protection from adverse environmental stress and pesticides seems to have good prospects, since BRs appear nontoxic and environmentally friendly. It is well known that horticultural crops have a great variety of produce organs as well as high yield and output values. Moreover, their production is susceptible to sub-optimum environmental conditions, especially in facilities cultivation. Thus, practical application of BRs to horticultural crops for enhancing crops production and protection may have a promising prospect in the near future.

Key words: Brassinosteroids, horticultural crop, seed germination, growth, environmental stress

1. INTRODUCTION

Plant hormones are a group of naturally occurring, organic substances. They regulate essential processes in a plant's life cycle at low concentrations. Some plant hormones have been well studied, including abscisic acid (ABA),

S. Hayat and A. Ahmad (eds.), *Brassinosteroids: A Class of Plant Hormone*,
DOI 10.1007/978-94-007-0189-2_9, © Springer Science+Business Media B.V. 2011

auxins, cytokinins, ethylene, gibberellins (GA_S) and brassinosteroids (BRs). They have been in use in horticultural applications since 1940s. Well studied expressions include, the promotion of fruit ripening by ethylene, regulation of the cell cycle by auxin and cytokinin, induction of seed germination and stem elongation by GA, and the maintenance of seed dormancy by ABA (Gray, 2004).

BRs are a group of naturally occurring plant steroids and are important for a broad spectrum of cellular and physiological processes, including stem elongation, pollen tube growth, leaf bending and epinasty, root inhibition, fruit development, ethylene biosynthesis, proton pump activity, xylem differentiation, photosynthesis, and gene expression. Moreover, BRs can induce plant tolerance to a variety of biotic and abiotic stresses (Xia *et al.*, 2009a). The work with mutant plants of rice, pea and tomato involved in BR synthesis, metabolism, signaling, and response provides further strong evidence that BRs are essential for crop growth and development (Bishop, 2003). Experiments to investigate the potential economic benefit of BRs in agriculture and horticulture have been studied as early as the 1980s. Afterwards, the chemical synthesis of BRs analogs confirms structure-activity relationships and provides a method for the synthesis of active BRs on a commercial scale for greenhouse and field evaluations. 24-epibrassinolide (EBR) and 28-homobrassinolide (HBR) have been registered as plant growth regulators for some kinds of horticultural crops. Extensive testings of EBR and HBR in China, Japan and Russia show that exogenously applied BRs have the ability to substantially increase yield and quality in a variety of plant species, and the results can be variable depending on the mode of application, growth stage of application and the environmental conditions (Divi and Krishna, 2009).

Horticultural plants, include fresh fruits and vegetables, herbaceous annual and perennial plants, which exhibit wide variation and diversity in their cultivated varieties with differences in flower or fruit color and plant shape, form, size, color, or flavor and aroma adding to that diversity and to the plants' value. Horticulture and the associated green industries are rapidly developing professional fields with increasing importance to society. However, it is well known that the production of horticultural plants is also typified by intense management, high management cost, environmental control, significant technology use, and high risk. Thus, in order to meet the development of horticulture industry, the focus of this review is to detail the role of BRs, a kind of nontoxic and environmentally friendly hormone, on horticultural crops. And with the progress of chemical synthesis technology, practical application of BRs to horticultural crops for enhancing yield, quality and stress tolerance may have a promising prospect in near future.

2. SEED DEVELOPMENT AND GERMINATION

As early as 1949, it was reported that several plant growth regulators can act as germination regulators. It was well established that BRs not only promoted seed germination, but also reversed the inhibitory effects of ABA (Rao *et al.*, 2002). Recent study demonstrated that BRs may be required for normal seed development and germination. It was reported that the *lk*, a severely BR deficient mutant of pea, produced irregularly shaped seeds (Nomura *et al.*, 2004). Furthermore, when pea seeds were rapidly growing, the level of biologically active brassinolide (BL) and castasterone (CS), and the transcript levels of two BR C-6 oxidases (*CYP85A1* and *CYP85A6*) reached a maximum. In the early stages of germination, the level of CS, but not that of BL, increased in the growing tissues, accompanied by high transcript level of *CYP85A1* and *CYP85A6*, however, the level of 6-Deoxocathasterone (6-deoxoCT) decreased with an increase in the levels of downstream intermediates. These results suggest that during germination and at early growth stage, CS played significant role, but 6-deoxoCT is the major BR for storage and utilization. In addition, investigation of the level of mRNAs for BR biosynthesis and perception genes showed that in mature seeds, most of mRNAs were present, but the level was generally lower, compared with immature seeds, except that *CYP90A9* mRNA rapidly increased during seed development and reached the maximum in mature seeds. These mRNAs were stored in mature pea seeds and seemed to be utilized on their germination. However, it was found that *de novo* transcription of mRNAs started as early as during the seed imbibition (Nomura *et al.*, 2007).

The average pod length and seed size of the BR-deficient mutant faba bean 'Rinrei' were reduced by about 80% of those in wild-type 'Mild Green', and the reduction in seed size was mainly due to physical restriction within pods. However, using exogenous application of 0.2 nmol of BL to the surface of the pod partially restored the length of pods in the mutant containing three seeds. However, seed length was not reduced for those that developed individually in pods containing only single seed. The authors concluded that here the seed size is regulated by BR, but the role is somehow indirect (Fukuta *et al.*, 2006).

3. BRs AND PLANT GROWTH

BRs are a new group of plant hormones with significant growth-promoting activity. BR-deficient or BR-insensitive mutants display dwarfism. Exogenous BR_S have specific effects on cell elongation, division, and differentiation. Underlying physiological pathways include modification of cell wall properties, effects on carbohydrate assimilation and allocation, and the control of aquaporin activities. BR apparently coordinates and integrates diverse processes required for growth, partly via interactions with other phytohormones setting the frame for BR responses. Ultimately, BR-promoted growth is mediated through genomic pathways. Positive regulators of BR response (such as BZR1 and BES1) and putative downstream components (such as EXO) are involved in the regulation of BR-responsive genes and growth promotion (Müssig, 2005). The following paragraph details the effects of BRs in different tissues, the dependence on experimental conditions, the interaction with other phytohormones, and underlying mechanisms.

3.1 Vegetative growth

3.1.1 Shoot

Many studies have demonstrated the role of BRs, alone and in interaction with other plant hormones, in cell elongation. As is well known, BL has an additive effect with GA or a synergistic effect with auxin on stem segment elongation. Sequential treatment of cucumber hypocotyl sections, first with HBR and then with IAA, resulted in synergistic enhancement of auxin-induced elongation. However, when the order of treatment was reversed, HBR was inactive. The findings suggest modulation of the response capacity to IAA by BRs (Katsumi, 1985). Moreover, elongation of *diageotropica* (*dgt*) hypocotyls, a mutant of tomato insensitive to auxin, was promoted by IAA only in the presence of EBR, suggesting that EBR restores the auxin responsiveness of *dgt* hypocotyls in respect to elongation (Park, 1998). Combined treatment of excised sweet pepper shoots with EBR, zeatin and GA demonstrated that EBR did not always act directly on stem elongation but may be an elicitor and/or an enhancer of elongation in association with endogenous and/or other exogenously added growth regulators (Franck-Duchenne *et al.*, 1998). However, studies of the BR and GA mutants of pea established that BRs-mediated growth response was not mediated by changes in bioactive GA levels, providing the evidence that BRs are important regulators of stem elongation (Jager *et al.*, 2005).

Growth is associated with increase in carbohydrate demand for biosynthetic metabolism. Root immersed with EBR markedly promoted the elongation of the epicotyls in cucumber seedlings, accompanied by the transport of sugar from the primary leaf to the epicotyl, suggesting that the sugar accumulation is closely related to the promotion of elongation (Nakajima and Toyama, 1995, 1999). Goetz *et al.* (2000) provided further evidence that BR-induced stimulation of hypocotyl growth is also linked to an elevated carbohydrate supply via the up-regulation of an extracellular invertase. The expression of the tomato *Lin6* gene (encoding an extracellular invertase) was specifically BR-induced in the first section of 10 mm below the hypocotyl hook (comprising the primary elongation zone). Increased Lin6 mRNA levels were also detected following BR treatment in tomato suspension cell culture. This induction is correlated with increased invertase activity and sucrose uptake rate into the suspension cell culture (Goetz *et al.*, 2000).

Mutant analysis has provided further evidence that the ability to synthesize, perceive and respond to BRs is essential for normal plant growth and development. Similar to *Arabidopsis* mutants, BR-deficient mutants in tomato plants showed reduced leaf growth and altered leaf morphology (Müssig, 2005). Tomato plants homozygous for the *extreme dwarf* allele (d^x) had both reduced cell size and number (Bishop *et al.*, 1996, 1999). The dumpy (*dpy*) mutant of tomato (*Lycopersicon esculentum* Mill.), a BR-deficient mutant, exhibited condensed, dark-green rugose and very hairy leaves that were downward curling and well spaced along the stem. Upon treatment with BL, *dpy* leaves regained a wild-type appearance. The *curl-3* (*cu-3*) mutant, a BRs insensitive mutant of *Lycopersicon pimpinellifolium* Mill., showed dark-green curled leaves, and genetic restoration of the *cu3* mutant phenotype using the CaMV 35S promoter to express the tomato BRI1 exhibited less serration of the leaf margins, indicating a role for BRs in leaf margin development (Koka *et al.*, 2000; Holton *et al.*, 2007; Campos *et al.*, 2009).

3.1.2 Root

It is widely accepted that BRs are necessary for the growth of not only shoot, but also that of root. In support of this, Yokota *et al.* (2001) demonstrated the existence of several BRs in tomato roots, and thick and less lateral roots of the *lk,* a BR deficient mutant of pea. Earlier studies predominantly reported inhibitory effects of BRs on root growth, which conflicted with the growth-promoting effect in shoots and with the reduced root system of BR-deficient or BR-insensitive mutants. Recent studies showed that the effects of BRs on root growth were strongly dependent on the BR concentration used. In general, low concentrations (nM range) of BRs can stimulate primary root growth, while higher concentrations (μM range) of BRs inhibited the growth (Haubrick and Assmann, 2006).

Guan and Roddick (1988) reported inhibitory effects of EBR on the growth of both aseptically-cultured excised tomato roots and the roots in intact tomato seedlings. Further studies found that the inhibitory effects were observed only at 0.1 μM of EBR or greater. At 10 μM basally- and apically-applied EBR inhibited growth in apical regions of excised tomato roots, but not in basal regions. Lower concentrations (1 and 0.1 μM) also inhibited growth, only in apical regions and predominantly when applied directly to those regions. The reduced responsiveness to EBR observed in roots grown by this method is thought to be due to the larger inoculum used rather than the physiological age of the roots (Roddick *et al.*, 1993). Likewise, seed treatment with 2 mM of BL also inhibited root growth of *Lepidium sativum*, but had no effect on ethylene levels, suggesting that at this concentration BL is acting independently of ethylene to inhibit cress root elongation. DNA synthesis remained unchanged within 24 hours after seed treatment, but decreased after 48 hours, illustrating a significant specific effect on very early cress root growth by seed treatment with BL (Jonesheld *et al.*, 1996). In contrast, Howell *et al.* (2007) found that 0.005 ppm of EBR nearly doubled the mean root length and the number of mitoses in onion (*Allium cepa*) root tips over that of controls, providing the evidence that exogenously applied BRs can increase the number of mitoses in plants, but failed to show cytogenetic data.

Studies of the BR synthesis mutants *lk* and *lkb* and the BR responsive mutant *lka* of pea showed fewer and shorter lateral roots and reduced nodule numbers in the mutants compared with wild-type plants, and grafting studies also revealed that grafted plants having an *lkb* shoot developed fewer nodules than those having an LKB shoot on a perplant, as well as per-milligram root on DW basis. These results suggest that BRs have an important role in lateral root development and may be influencing nodulation mechanism of the shoot that is involved in regulating the nodule numbers of the root (Ferguson *et al.*, 2005).

3.2 Reproductive growth

Very high BRs levels are found in pollen and seed, and extensive studies in *Arabidopsis* have demonstrated the significance of BRs in the regulation of reproductive growth. Müssig (2005) reviewed that BRs could determine branching and flower formation through modulating metabolic pathways and nutrient allocation or interacting with other signalling pathways. BRs may also affect fertilization via the stimulation of filament and pollen growth, and modify pollen properties. However, the function of BRs in the generative phase is largely unclear.

3.2.1 Flowers

Korableva *et al.* (2002) found that immersing potato tubers in EBR solution resulted in prolonged deep dormancy and inhibited sprouting, accompanied by increased production of ethylene and ABA. Moreover, spray of two BR analogues, BB-6 and BB-16 to the cladodes of cactus pear initiated the development of vegetative buds of cactus pear earlier than the controls and increased the rate of growth of cladodes under both greenhouse and field conditions. The BRs also promoted precocity, accelerating growth during the first stages of vegetative bud development (Aristeo Cortes *et al.*, 2003).

Short-day plant *Pharbitis nil* treated with BL and CS formed less number of flowers than control plants, but the degree of inhibition depended on the concentration and the method of BRs application as well as the length of the inductive dark period. In plants regenerated from sub-induced apices, treated with BL at concentrations of 1 and 10 μM, the flower formation was inhibited completely (Kesy *et al.*, 2003). Moreover, Papadopoulou and Grumet (2005) reported that application of EBR to cucumber plants caused earlier and increased female flower production, and that the effects may be mediated, at least in part, by BR-induced production of ethylene.

3.2.2 Fruits

Recently, extensive research has revealed the importance of BRs in the ripening of tomato fruit. A study on the ripening of pericarp discs from tomato showed lycopene and carbohydrate levels and ethylene production increased, whereas the level of chlorophyll and ascorbic acid decreased, after immersing in EBR or HBR, which was consistent with accelerated ripening (Vardhini and Rao, 2002). Similar results were reported in ripe fruits of tomato after root applied HBR (Ali *et al.*, 2006). Montoya *et al.* (2005) also found that biosynthesis of BRs was enhanced in the developing fruits of tomato. Using d^x mutant of tomato, a BR-deficient mutant (*extreme dwarf*), as plant material, Lisso *et al.* (2006) found that d^x produced bioactive BRs in fruits but not in the shoot. By surveying metabolic processes in d^x fruits, reduced starch and sugar levels, and elevated amino acid levels were found. EBR application to leaves partly normalized metabolic changes in d^x fruits, suggesting that shoot-derived BR-dependent factors are required for proper fruit metabolism. Studies in two cucumber cultivars with different parthenocarpic capacity showed that parthenocarpic growth was induced by exogenous BRs but inhibited by the loss of BR synthesis. BRs triggered active cell division associated with increased transcripts of cell cycle-related genes, especially that of cyclin *D3* genes. These results strongly suggest that BRs play an import role during early fruit development in cucumber (Fu *et al.*, 2008). Expression analysis of the genes encoding BR biosynthesis enzymes and receptors during grape berry development revealed transcript accumulation patterns that were consistent with a dramatic increase in endogenous BR levels observed at the onset of fruit ripening. These results provide evidence that increase in endogenous BR levels are associated with ripening in grapes, and application of EBR to grape berries significantly promoted ripening, while brassinazole (Brz), an inhibitor of BR biosynthesis, significantly delayed fruit ripening (Symons *et al.*, 2006).

Besides the effects on the fruit ripening, BRs are also reported to improve fruit yield and quality. Spray of BL solution on litchi leaves before blossom increased the activities of pectin methyesterase and polygalacturonase and the content of water-soluble pectin, protopectin and calcium in the fruit pericarp, and reduced fruit cracking rate, suggesting an important role in increasing the commercial value of litchi fruits (Peng *et al.*, 2004). Likewise, BR analogue BB-16 sprayed during several periods of reproductive development can increase the number of fruits per plant in yellow passion (*Passiflora edulis f.* flavicarpa). BR application in three consecutive weeks after the appearance of the first flowers was the most promising treatment because it produced the highest number of fruits plant^{-1} (Gomes *et al.*, 2006).

3.2.3 Seeds

Very high BR levels are also found in seeds, suggesting the significance of BRs in the seed development. Foliar treatment with HBR enhanced the growth and seed yield of mustard, together with increased carbonic anhydrase (CA) activity and Pn (Hayat *et al.*, 2000). Moreover, soaking with different concentrations of HBR decreased root length and nodule number per plant in *Lens culinaris*, but increased nitrate reductase (NR) activity and grain yield at the final harvest, 140 DAS (Hayat and Ahmad, 2003).

3.3 Micropropagation

Besides whole plants, excised tissues also have been used to analyse the mechanisms involved in BR-promoted growth. Exogenous BR significantly stimulated adventitious bud formation from hypocotyl segments of cauliflower cultured *in vitro* in the light. However, regeneration was much lower in the dark, which was not due to a possible increase in ethylene synthesis, in the dark because regeneration was not improved by added AVG (an ethylene biosynthesis inhibitor) and AOA (an inhibitor of ethylene action) to the culture medium (Sasaki, 2002). It was also reported that exogenous application of different concentrations of BR significantly increased the capacity of coconut plumule explants to form initial callus, embryogenic callus and somatic embryos. A larger number of somatic embryos formed were obtained on exposing the explants to BR, compared to untreated explants. This is a very promising result, considering the very slow progress of micropropagation research for this very recalcitrant species that has taken now three decades since it started (Azpeitia *et al.*, 2003).

Treatment of shoot apices of marubakaido apple rootstock with a fluoro derivative of 28-homoethylcastasterone (5F-HCTS) in the range of 100-10,000 ng per shoot increased the apple rootstock's multiplication rate, mainly due to the increase in the number of primary and secondary lateral branches (Schaefer *et al*., 2002). Further study found that 5F-HCTS stimulated branch elongation in *in vitro*-grown shoots of *Malus prunifolia*, accompanied by increased ethylene release. However, either the presence of 1-amino-cyclopropane-1-carboxylic acid (ACC) in the culture medium or an ethylene-enriched atmosphere resulted in the inhibition of branch elongation, indicating that the stimulation of branch elongation observed for 5F-HCTS-treated shoots was not, at least directly, related to BR-induced enhancement in ethylene level (Pereira-Netto *et al*., 2006). These results show that shoot proliferation induced by 5F-HCTS is an effective method to enhance the *in vitro* multiplication rate in *Malus prunifolia*. Using exogenous applications of BL and Brz220 to shoot apices. Pereira-Netto *et al*. (2009) further clarified that 5F-HCTS-induced stimulation of shoot elongation and formation of new shoots in the Marubakaido shoots was under the control of changes in the endogenous BR pool and its likely involvement in a variety of mechanisms and consequently did not result from changes in the endogenous levels of any single metabolite.

4. BRs AND STRESS RESPONSES

Horticultural crops production and maintenance requires extensive use of soil manipulation, irrigation, fertilization, plant growth regulation, pruning/pinching/trimming, and environmental control. Besides, plants can be grown in natural environments, or in very confined environments, such as in nurseries, greenhouses, growth rooms, or in pots. It has been shown that similar to agricultural crops, horticultural crops are also constantly exposed to a variety of abiotic and biotic stresses, and especially susceptible to sub-optimum environmental conditions in facilities cultivation.

BRs are a group of naturally occurring plant steroidal compounds with wideranging biological activity that offer the unique possibility of increasing crop yields through both changing plant metabolism and protecting plants from environmental stresses. Recently, Bajguz and Hayat (2009) have reviewed the role of BRs in response to various kinds of stresses via activation of different mechanisms. The following paragraph mainly brings together evidences for BRs on horticultural crops stress tolerance.

4.1 Oxidative stress

It has been shown that plant responses to different types of stresses are associated with the generation of reactive oxygen species (ROS), suggesting that ROS may function as a common signal in pathways of plant stress responses (Xia *et al.*, 2009a). Until recently, the underlying mechanisms for BR-mediated oxidative stress responses have not been well understood.

Almeida *et al.* (2005) reported that HBR spraying of potato increased the basal level of CAT2 and did not change the isozyme pattern of catalase (CAT) in response to 100mM H_2O_2, but affected the magnitude of increase of total activity. Even so, ultrastructural studies showed that HBR significantly reduced H_2O_2 negative effects on cellular sub-structures, allowing better recovery of affected structures and reduced the macroscopic injury symptoms on leaves, thus data point to a HBR protective role. Recent research further demonstrated that H_2O_2 mediated the transcriptional induction of defense or antioxidant genes by BRs. BRs treatment enhanced NADPH oxidase activity and elevated H_2O_2 level in apoplast. H_2O_2 levels were elevated as early as 3 hours and returned to basal level 3 days after BR treatment. BR-induced H_2O_2 accumulation was accompanied by increased tolerance to oxidative stress. Inhibition of NADPH oxidase and chemical scavenging of H_2O_2 reduced BRs-induced oxidative and cold tolerance and defense gene expression. BRs treatment induced expression of both regulatory genes, such as RBOH, MAPK1, and MAPK3, and genes involved in defense and antioxidant responses. These results strongly suggest that elevated H_2O_2 level resulting from enhanced NADPH oxidase activity are involved in BRs-induced stress tolerance (Xia *et al.*, 2009a).

4.2 Heavy metals stress

The toxic effects of heavy metals on plants and BRs-mediated detoxification mechanisms have been elaborated by Bajguz and Hayat (2009). On the basis of previous studies in horticultural crops, it is concluded that the elevated level of antioxidant system and activities of CA, NR are one of the important mechanisms responsible for BRs-alleviated injury of heavy metals to plants (Hayat *et al.*, 2007; Ali *et al.*, 2008). The recent results of Fariduddin *et al.* (2009b) also supported above conclusion. Treatment of seeds with HBR improved the growth, photosynthetic parameters and antioxidant enzymes in the plants grown under copper stress. The elevated antioxidant enzymes and proline might be responsible to overcome the toxic effects of copper in *Brassica juncea*. Anuradha and Rao (2009) also found that seed application of EBR reduced the toxic effects of cadmium (Cd) on plant growth, pigment content, photosynthesis and activities of CA, NR in *Raphanus sativus* L. The studies clearly demonstrated the ameliorating effect of EBR in mitigating the toxicity of Cd in plants.

It was also reported that there was an improvement in the shoot emergence and biomass production of mustard plants under the influence of pre-germination seed treatment with EBR. Moreover, EBR blocked copper metal uptake and its accumulation in the plants (Sharma and Bhardwaj, 2007). Similar inhibitory role of BRs on the uptake of Ni was also reported by Sharma *et al.* (2008).

4.3 Thermal stress

High temperature stress caused the appearance of granules in the nucleus, nucleolus and cytoplasm of tomato leaf disks. In chloroplasts and in mitochondria the internal membrane system was disorganized and in chloroplasts some starch granules were detected. These symptoms were more marked in the cells treated with BB6, a Cuban synthetic BR. The influence of BB6 on the ultrastructure of leaf cells was apparent also before being subject to heat stress (Sam *et al.*, 2001). Furthermore, several studies have shown that BR-induced thermotolerance in tomato plants was associated with higher synthesis of heat shock proteins (HSPs) and higher expression of mitochondrial small hsps in the leaves. Significantly higher *in vitro* pollen germination, enhanced pollen tube growth and low pollen bursting have been observed in the presence of EBR at 35°C, a temperature high enough to induce heat-stress symptoms in tomato, indicating a positive role for BRs, during plant growth and reproduction (Dhaubhadel *et al.*, 1999; 2002; Singh and Shono, 2005).

The ROS scavenging system was also reported to play an important role in protecting cells from oxidative damage induced by high-temperature stress. Mazorra *et al.* (2002) demonstrated a positive role of EBR and MH5 (a polyhydroxylated spirostanic analogue of BRs) in the reduction of cell damage in tomato plants produced by heat stress due to the induced synthesis of enzymatic antioxidants. Likewise, foliar application of EBR could alleviate the detrimental effects of high temperatures on the growth of tomato plants by increasing carboxylation efficiency and enhancing antioxidant enzyme systems in leaves (Ogweno *et al.*, 2008).

The underlying mechanisms for BR-mediated chilling stress responses in horticultural crops have not been understood. Studies in cucumber seedlings showed that improvement of photosynthesis was partly responsible for EBR-alleviated chilling injury (Yu *et al.*, 2002).

4.4 Water stress

Water stress is one of the most common environmental stresses that affect plant growth and development. A deficit of water leads to various alterations in plants, including stomatal closure, leaf abscission and changes in the composition of cell wall or plasma membrane, and can result in a decline in growth as photosynthesis and turgor are decreased (Jager *et al*., 2008).

The elevated antioxidant system, at least in part, was responsible for the amelioration of water stress. Soaking *Robinia pseudoacacia* L. roots in BL prior to planting significantly increased the survival and growth of seedlings. BL treatment decreased the transpiration rate, stomatal conductance and malondialdehyde (MDA) content of seedlings growing under moderate or severe water stress, compared to untreated seedlings. Leaf water content, predawn water potential, soluble sugar content, free proline content, and superoxide dismutase (SOD), peroxidase (POD) and CAT activities were all greater in water-stressed seedlings treated with BL, compared to the control. The results indicate that treatment of seedlings with BL may be a useful management tool for afforestation projects in arid and semiarid areas (Li *et al*., 2008). A BR-induced increase in tolerance to water stress has also been reported in Indian mustard plants. Foliar spray of HBR elevated the level of the antioxidant system (CAT, POD, SOD, proline) both in the plants treated with HBR and/or exposed to drought stress. The elevation in antioxidant system detoxified the ROS generated by drought stress and thereby improved the altered physiological performance of stressed plants. Therefore, it maybe suggested that HBR ameliorated the drought stress in plants, which was mediated through the modification of their antioxidant system (Fariduddin *et al*., 2009a).

It is well established that ABA is involved in many of the processes related to water stress. Water loss or reduced water uptake results in an accumulation of ABA. However, it is not known whether changes in endogenous BR levels are normally involved in mediating the plant's response to water stress. Recently, there has been some evidence for the interaction between ABA and BRs in many physiological processes, such as stomatal closure in response to water stress (Jager *et al*., 2008). Upretik and Murtig (2004) found that the ABA contents in the nodules of *Phaseolus vulgaris* L. were not involved in mediating the BRs-enhanced water stress tolerance of plants. In wild-type (WT) pea plants, water stress caused a significant increase in ABA level, but it did not result in altered BR levels in either shoot apex, internode or leaf tissue. Furthermore, the plant's ability to increase ABA levels in response to water stress was not affected by BR deficiency, as there was no significant difference in ABA levels between WT, *lkb* (a BR-deficient mutant) and *lka* (a BR-perception mutant) plants before or 14 days after the cessation of watering. In addition, the effect of water stress on traits such as height, leaf size and water potential in *lkb* and *lka* was similar to that observed in WT plants. Therefore, it appears that, at least in pea, changes in endogenous BR levels are not normally the part of plant's response to water stress (Jager *et al*., 2008).

4.5 Hypoxia stress

Oxygen deficiency in the root-zone is one of the major causes of yield reductions in crop plants (Drew, 1997). Previous studies show that oxygen is necessary for the biosynthesis of BRs, and oxygen levels influence their concentrations and physiological functions in plants (Cervantes, 2001; Ramonell *et al*., 2001). Ershova and Khripach (1996) demonstrated that in oxygen-deprived conditions, EBR treatment decreased the conjugated dienoic acids (by 13%) and MDA (by 38%) in pea seedlings, compared to aerated control. It was also reported that root applied EBR, a highly active and stable steroidal hormone, alleviated oxidative damage induced by root-zone hypoxia in cucumber seedlings and rendered the plants more tolerant to it (Kang *et al*., 2007). Further studies demonstrated that EBR confers hypoxia tolerance of cucumber plants probably by enhancing supply of sugars to hypoxic roots and shifting hypoxic metabolism from lactate fermentation to alcohol fermentation which favors ATP production through glycolysis and alleviation of cytosolic acidification (Kang *et al*., 2009).

4.6 Pathogen stress

Earlier studies in potato tubers infected by *Phytophthora infenstans* showed entirely contradictory results (Vasyukova *et al*., 1994; Korableva *et al*., 2002; Bajguz and Hayat, 2009), indicating that whether BRs protect the plants against pathogen infection or not but depended on the method and time of BRs application and was connected with different stimulating points of either the plant or the pathogen.

On the basis of previous studies, the induction of systemic acquired resistance (SAR) was reported to be involved in mediating the BRs-enhanced pathogen resistance of plants. Exogenous application of 24-epicastasterone and 24-episecasterone and the extract of *Lychnis viscaria* seeds, enhanced the resistance of cucumber and tomato to viral or fungal pathogens, accompanied by a stimulation of three pathogenesis-related (PR) proteins, such as chitinase, β-1,3-glucanase and POD, which were the molecular markers of SAR (Roth *et al*., 2000). A protective effect of EBR against fungi was also established in field trials with cucumber. Seed treatment with EBR, then foliar spray of EBR at the flowering stage, significantly suppressed the growth of mildew, mainly through an increase in the activities of POD and polyphenoloxidase (PPL) (Khripach *et al*., 2000).

In contrast, Ding *et al*. (2009a) found that both root and foliar applications of EBR significantly reduced disease severity caused by fusarium wilt, together with improved plant growth and reduced losses in biomass, regardless of application method. EBR treatments significantly reduced pathogen-induced accumulation of ROS, flavonoids, and phenolic compounds, and also the activities of defense-related and ROS-scavenging enzymes. The enzymes included SOD, ascorbate peroxidase, guaiacol peroxidase, CAT as we as phenylalanine ammonia-lyase and PPL. EBR applications triggered a slight increase in H_2O_2 concentration followed by increases in the transcript levels of WRKY transcription factor and defense-related genes. This study demonstrated that EBR enhanced resistance to fusarium wilt by a novel mechanism that was not related to its active transport or increase in antioxidant system. Further studies of EBR-induced disease resistance in cucumber plants showed that root and foliar-applied EBR decreased the *Fusarium* population on root surfaces and in nutrient solution, but increased the population of fungi and *actinobacteria* on root surfaces. PCR-DGGE analysis showed that EBR treatment significantly alleviated the decrease of diversity index and evenness index of bacterial community on root surfaces caused by FO-inoculation. Moreover, several kinds of decomposing bacteria and growth-promoting bacteria were identified from root surfaces of FO-inoculated plants and EBR-pre-treated plants, respectively. Overall, these results show that the microbial community on root surfaces was affected by a complex interaction between phytohormone-induced resistance and plant pathogens (Ding *et al*., 2009b).

4.7 Pesticides application

BRs are known to protect crops from the toxicity of herbicides, fungicides and insecticides. It was reported that treatment with EBR, which was isolated from *Aegle marmelos,* Correa, significantly declined the percentage of chromosomal aberrations of *Allium cepa* induced by maleic hydrazide (Sondhi *et al*., 2008). In addition, based on earlier studies, it was concluded that increased capacity of CO_2 assimilation was partly responsible for BRs-enhanced pesticides resistance. Xia *et al*. (2006) found that EBR pretreatment can increase the resistance of cucumber plants to pesticides, which might be mediated by enhanced activities of CO_2 assimilation and antioxidant enzymes. Likewise, the application of EBR to *Vicia faba* seeds before sowing strongly diminished the inhibitory effect of herbicide Terbutryn on fluorescence parameters and CO_2 assimilation, which recovered 13 days after plant emergence and showed values similar to those of control plants, and thus counteracted the decrease in plant growth caused by Terbutryn, and plants registered the same growth rate as that of the controls (Piñol and Simón, 2009).

Recent studies showed that application of EBR accelerated metabolism of various pesticides and consequently reduced their residual levels in cucumber (*Cucumis sativus* L). EBR pretreatment alleviated the decline of Pn and Φ_{PSII} caused by chlorpyrifos application, and this effect of EBR was associated with reductions of chlorpyrifos residues. Moreover, EBR had a positive effect on the activation of glutathione *S*-transferase (GST), POD, and glutathione reductase (GR) after treatment with chlorpyrifos, although the effect on GR was attenuated at later points when plants were treated with 1 mM chlorpyrifos. In addition, EBR enhanced the expression of *P450* and *MRP*, which encoded P450 monooxygenase and ABC-type transporters, respectively. However, the expression of *GST* was consistently lower than that of plants treated with only chlorpyrifos. Importantly, the stimulatory effect of EBR on pesticide metabolism was also observed for cypermethrin, chlorothalonil, and carbendazim, which was attributed to the enhanced activity and genes involved in pesticide metabolism. The results suggest that BRs may be promising, environmentally friendly and natural substances suitable for wide application to reduce the risks of human and environment exposure to pesticides (Xia *et al.*, 2009b).

5. CONCLUSION

From the current review, it is clear that BRs have the ability to improve quantity and quality of horticultural crops and protect plants against several kinds of stresses. Horticulture is characterized by high benefit and quick return of investment. Thus, with more recent advances in the synthesis of synthetic analogues, it seems more likely that new formulations will be possible, enabling better crop performance and high economic output. Besides, genetic manipulation of BR activity can also be a practical and promising strategy for crop improvement.

6. ACKNOWLEDGEMENT

This work was supported by the National Natural Science Foundation of China (No. 30871736).

7. REFERENCES

Ali, B., Hayat, S., Aiman Hasan, S., and Ahmad, A., 2006. Effect of root applied 28-homobrassinolide on the performance of *Lycopersicon esculentum*. *Sci. Hortic.* **110**: 267–273.

Ali, B., Hayat, S., Fariduddin, Q., and Ahmad, A., 2008. 24-Epibrassinolide protects against the stress generated by salinity and nickel in *Brassica juncea*. *Plant Physiol. Biochem.* **72**: 1387–1392.

Almeida, J., Fidalgo, F., Confraria A., Santos, A., Pires, H., and Santos, I., 2005. Effect of hydrogen peroxide on catalase gene expression, isoform activities and levels in leaves of potato sprayed with homobrassinolide and ultrastructural changes in mesophyll cells. *Funct. Plant Biol.* **32**: 707–720.

Anuradha, S., and Rao, S., 2009. Effect of 24-epibrassinolide on the photosynthetic activity of radish plants under cadmium stress. *Photosynthetica* **47**: 317–320.

Aristeo Cortes, P., Terrazas, T., Colinas León, T., and Larqué-Saavedra, A., 2003. Brassinosteroid effects on the precocity and yield of cladodes of cactus pear (*Opuntia ficus-indica* (L) Mill.). *Sci. Hortic.* **97**: 65–73.

Azpeitia, A., Chan, J., Saenz, L., and Oropeza, C., 2003. Effect of 22(S), 23(S)-homobrassinolide on somatic embryogenesis in plumule explants of *Cocos nucifera* (L.) cultured *in vitro*. *J. Hortic. Sci. Biotech.* **78**: 591–596.

Bajguz, A., and Hayat, S., 2009. Effects of brassinosteroids on the plant responses to environmental stresses. *Plant Physiol. Biochem.* **47**: 1–8.

Bishop, G., 2003. Brassinosteroid mutants of crops. *J. Plant Growth Regul.* **22**: 325–335.

Bishop, G., Harrison, K., and Jones, J., 1996. The tomato *Dwarf* gene isolated by heterologous transposon tagging encodes the first member of a new cytochrome P450 family. *Plant Cell* **8**: 959–969.

Bishop, G., Nomura, T., Yokota, T., Harrison, K., Noguchi, T., Fujioka, S., Takatsuto, S., Jones, J., and Kamiya, Y., 1999. The tomato DWARF enzyme catalyses C-6 oxidation in brassinosteroid biosynthesis. *Proc. Natl. Acad. Sci. USA.* **96**: 1761–1766.

Campos, M., Almeida, M., Rossi, M., Martinelli, A., Martinelli, A., Junior, C., Figueira, A., Rampelotti-Ferreira, F., Vendramim, J., Benedito, V., and Peres, L., 2009. Brassinosteroids interact negatively with jasmonates in the formation of anti-herbivory traits in tomato. *J. Exp. Bot.* **60**: 4347–4361.

Cervantes, E., 2001. Brassinolide plays in the hypoxia band. *Trends Plant Sci.* **6**: 240.

Dhaubhadel, S., Browning, K., Gallie, D., and Krishna, P., 2002. Brassinosteroid functions to protect the translational machinery and heat-shock protein synthesis following thermal stress. *Plant J.* **29**: 681–691.

Dhaubhadel, S., Chaudhary, S., Dobinson, K., and Krishna, P., 1999. Treatment of 24-epibrassinolide, a brassinosteroid, increases the basic thermotolerance of *Brassica napus* and tomato seedlings. *Plant Mol. Biol.* **40**: 333–342.

Ding, J., Shi, K., Zhou, Y., and Yu, J., 2009a. Effects of root and foliar applications of 24-epibrassinolide on Fusarium wilt and antioxidant metabolism in cucumber Roots. *Hortsci.* **44**: 1340–1345.

Ding, J., Shi, K., Zhou, Y., and Yu, J., 2009b. Microbial community responses associated with the development of *Fusarium oxysporum f.sp.cucumerinum* after 24-epibrassinolide applications to shoots and roots in cucumber. *Eur. J. Plant Pathol.* **124**: 141–150.

Divi, U., Krishna, P., 2009. Brassinosteroids: a biotechnological target for enhancing crop yield and stress tolerance. *New Biotechnol.* **26**: 131–136.

Drew, M., 1997. Oxygen deficiency and root metabolism: Injury and acclimation under hypoxia and anoxia. *Annu. Rev. Plant Phys. Plant Mol. Biol*. **48**: 223–250.

Ershova, A., and Khripach, V., 1996. Effect of epibrassinolide on lipid peroxidation in *Pisum sativum* at normal aeration and under oxygen deficiency. *Russ. J. Plant Physiol*. **43**: 750–752.

Fariduddin, Q., Khanam, S., Hasan, S., Ali, B., Hayat, S., and Ahmad, A., 2009a. Effect of 28-homobrassinolide on the drought stress-induced changes in photosynthesis and antioxidant system of *Brassica juncea* L. *Acta Physiol. Plant.* **31**: 889–897.

Fariduddin, Q., Yusuf, M., Hayat, S., and Ahmad, A., 2009b. Effect of 28-homobrassinolide on antioxidant capacity and photosynthesis in *Brassica juncea* plants exposed to different levels of copper. *Environ. Exp. Bot*. **66**: 418–424.

Ferguson, B., Ross, J., and Reid, J., 2005. Nodulation phenotypes of gibberellin and brassinosteroid mutants of pea. *Plant Physiol*. **138**: 2396–2405.

Franck-Duchenne, M., Wang, Y., Tahar, S., and Beachy, R., 1998. *In vitro* stem elongation of sweet pepper in media containing 24-epi-brassinolide. *Plant Cell, Tissue & Org. Cult*. **53**: 79–84.

Fu, F., Mao, W., Shi, K., Zhou, Y., Asami, T., and Yu, J., 2008. A role of brasssinosteroids in early fruit development in cucumber. *J. Exp. Bot*. **59**: 2299–2308.

Fukuta, N., Fukuzono, K., Kawaide, H., Abe, H., and Nakayama, M., 2006. Physical restriction of pods causes seed size reduction of a brassinosteroid-deficient Faba bean (*Vicia faba*). *Ann. Bot*. **97**: 65–69.

Goetz, M., Godt, D., and Roitsch, T., 2000. Tissue-specific induction of the mRNA for an extracellular invertase isoenzyme of tomato by brassinosteroids suggests a role for steroid hormones in assimilate partitioning. *Plant J*. **22**: 515–522.

Gomes, M., Campostrini, E., Leal, N., Viana, A., Ferraz, T., Siqueira, L., Rosa, R., Netto, A., Nuñez-Vázquez, M., and Zullo, M., 2006. Brassinosteroid analogue effects on the yield of yellow passion fruit plants (*Passiflora edulis* f. *flavicarpa*). *Sci. Hortic*. **110**: 235–240.

Gray, W., 2004. Hormonal regulation of plant growth and development. *PLoS Biol*. **2**: 1270–1273.

Guan, M., and Roddick, J., 1988. Epibrassinolide inhibition of development of excised, adventitious and intact roots of tomato (*Lycopersicon esculentum*): comparison with the effects of steroidal estrogens. *Physiol. Plant*. **74**: 720–726.

Haubrick, L., and Assmann, S., 2006. Brassinosteroids and plant function: some clues, more puzzles. *Plant Cell Env*. **29**: 446–457.

Hayat, S., and Ahmad, A., 2003. Soaking seeds of *Lens culinaris* with 28-homobrassinolide increased nitrate reductase activity and grain yield in the field in India. *Ann. Appl. Biol*. **143**: 121-124.

Hayat, S., Ahmad, A., Mobin, M., Hussain, A., and Fariduddin Q., 2000. Photosynthetic rate, growth, and yield of mustard plants sprayed with 28-homobrassinolide. *Photosynthetica* **38**: 469–471.

Hayat, S., Ali, B., Aiman Hasan, S., and Ahmad, A., 2007. Brassinosteroid enhanced the level of antioxidants under cadmium stress in *Brassica juncea*. *Environ. Exp. Bot*. **60**: 33–41.

Holton, N., Caño-Delgado, A., Harrison, K., Montoya, T., Chory, J., and Bishop, G., 2007. Tomato *BRASSINOSTEROID INSENSITIVE1* is required for systemin-induced root elongation in *Solanum pimpinellifolium* but is not essential for wound signaling. *Plant Cell* **19**: 1709–1717.

Howell, W., Keller III, G., Kirkpatrick, J., Jenkins, R., Hunsinger, R., and McLaughlin, E., 2007. Effects of the plant steroidal hormone, 24-epibrassinolide, on the mitotic index and growth of onion (*Allium cepa*) root tips. *Genet. Mol. Res*. **6**: 50–58.

Jager, C., Symons, G., Ross, J., Smith, J., and Reid, J., 2005. The brassinosteroid growth response in pea is not mediated by changes in gibberellin content. *Planta* **221**: 141–148.

Jager, C., Symons, G., Ross, J., and Reid, J., 2008. Do brassinosteroids mediate the water stress response? *Physiol. Plant.* **133**: 417–425.

Jonesheld, S., Vandoren, M., and Lockwood, T., 1996. Brassinolide application to *Lepidium sativum* seeds and the effects on seedling growth. *J. Plant Growth Regul.* **15**: 63–67.

Kang, Y., Guo, S., Li, J., and Duan, J., 2007. Effects of 24-epibrassinolide on antioxidant system in cucumber seedling roots under hypoxia stress. *Agri. Sci. China* **6**: 281–289.

Kang, Y., Guo, S., Li, J., and Duan, J., 2009. Effect of root applied 24-epibrassinolide on carbohydrate status and fermentative enzyme activities in cucumber (*Cucumis sativus* L.) seedlings under hypoxia. *Plant Growth Regul.* **57**: 259–269.

Katsumi, M., 1985. Interaction of a brassinosteroid with IAA and GA_3 in the elongation of cucumber hypocotyl sections. *Plant Cell Physiol.* **26**: 615–626.

Kesy, J., Trzaskalsky, A., Galoch, E., and Kopcewicz, J., 2003. Inhibitory effect of brassinosteroids on the flowering of the short-day plant *Pharbitis nil*. *Biol. Plant.* **47**: 597–600.

Khripach, V., Ahabinskii, V., and de Groot A., 2000. Twenty years of brassinosteroids: steroidal plant hormones warrant better crops for the XXI century. *Ann. Bot.* **86**: 441–447.

Koka, C., Eric Cerny, R., Gardner, R., Noguchi, T., Fujioka, S., Takatsuto, S., Yoshida, S., and Clouse, S., 2000. A Putative role for the tomato genes *DUMPY* and *CURL-3* in brassinosteroid biosynthesis and response. *Plant Physiol.* **122**: 85–98.

Korableva, N., Platonova, T., Dogonadze, M., and Evsunina, A., 2002. Brassinolide effect on growth of apical meristems, ethylene production, and abscisic acid content in potato tubers. *Biol. Plant.* **45**: 39–43.

Li, K., Wang, H., Han, G., Wang, Q., and Fan, J., 2008. Effects of brassinolide on the survival, growth and drought resistance of *Robinia pseudoacacia* seedlings under water-stress. *New Forest*, **35**: 255–266.

Lisso, J., Altmann, T., and Müssig, C., 2006. Metabolic changes in fruits of the tomato d^x mutant. *Phytochem.* **67**: 2232–2238.

Mazorra, L., Núñez, M., Hechavarria, M., Coll, F., and Sánchez-Blanco, M., 2002. Influence of brassinosteroids on antioxidant enzymes activity in tomato under different temperatures. *Biol. Plant.* **45**: 593–596.

Montoya, T., Nomura, T., Yokota, T., Farrar, K., Harrison, K., Jones, J., Kaneta, T., Kamiya, Y., Szekeres, M., and Bishop, G., 2005. Patterns of Dwarf expression and brassinosteroid accumulation in tomato reveal the importance of brassinosteroid synthesis during fruit development. *Plant J.* **42**: 262–269.

Müssig, C., 2005. Brassinosteroid-promoted growth. *Plant Biol.* **7**: 110–117.

Nakajima, N., and Toyama, S., 1999. Effects of epibrassinolide on sugar transport and allocation to the epicotyl in cucumber seedlings. *Plant Prod. Sci.* **2**: 165–171.

Nakajima, N., and Toyama, S., 1995. Study on brassinosteroid-enhanced sugar accumulation in cucumber epicotyls. *Jpn. J. Crop Sci.* **64**: 616–621.

Nomura, T., Jager, C., Kitasaka, Y., Takeuchi, K., Fukami, M., Yoneyama, K., Matsushita, Y., Nyunoya, H., Takatsuto, S., Fujioka, S., Smith, J., Kerckhoffs, H., Reid, J., and Yokota, T., 2004. Brassinosteroid deficiency due to truncated steroid 5α-reductase causes dwarfism in the *lk* mutant of pea. *Plant Physiol.* **135**: 2220–2229.

Nomura, T., Ueno, M., Yamada Y., Takatsuto, S., Takeuchi, Y., and Yokota, T., 2007. Roles of brassinosteroids and related mRNAs in pea seed growth and germination. *Plant Physiol.* **143**: 1680–1688.

Ogweno, J., Song, X., Shi, K., Hu, W., Mao, W., Zhou, Y., Yu, J., and Nogués, S., 2008. Brassinosteroids alleviate heat-induced inhibition of photosynthesis by increasing carboxylation efficiency and enhancing antioxidant systems in *Lycopersicon esculentum. J. Plant Growth Regul.* **27**: 49–57.

Papadopoulou, E., and Grumet, R., 2005. Brassinosteriod-induced femaleness in cucumber and relationship to ethylene production. *Hortsci.* **40**: 1763–1767.

Park, W., 1998. Effect of epibrassinolide on hypocotyl growth of the tomato mutant *diageotropica*. *Planta* **207**: 120–124.

Peng, J., Tang, X., and Feng, H., 2004. Effects of brassinolide on the physiological properties of litchi pericarp (*Litchi chinensis* cv. nuomoci). *Sci. Hortic.* **101**: 407–416.

Pereira-Netto, A., Cruz-Silva, C., Schaefer, S., Ramirez, J., and Galagovsky, L., 2006. Brassinosteroid-stimulated branch elongation in the marubakaido apple rootstock. *Trees* **20**: 286–291.

Pereira-Netto, A., Roessner, U., Fujioka, S., Bacic, A., Asami, T., Yoshida, S., and Clouse, S., 2009. Shooting control by brassinosteroids: metabolomic analysis and effect of brassinazole on Malus prunifolia, the Marubakaido apple rootstock. *Tree Physiol.* **29**: 607–620.

Piñol, R., and Simón, E., 2009. Effect of 24-epibrassinolide on chlorophyll fluorescence and photosynthetic CO_2 assimilation in *Vicia faba* plants treated with the photosynthesis-inhibiting herbicide Terbutryn. *J. Plant Growth Regul.* **28**: 97–105.

Ramonell, K., Kuang, A., Porterfield, D., Crispi, M., Xiao, Y., McClure, G., and Musgrave, M., 2001. Influence of atmospheric oxygen on leaf structure and starch deposition in *Arabidopsis thaliana*. *Plant Cell Env.* **24**: 419–428.

Rao, S., Vardhini, B., Sujatha, E., and Anuradha, S., 2002. Brassinosteroids-A new class of phytohormones. *Curr. Sci.* **82**: 1239–1245.

Roddick, J., Rijnenberg, A., and Ikekawa, N., 1993. Developmental effects of 24-epibrassinolide in excised roots of tomato grown *in vitro*. *Physiol. Plant.* **87**: 453–458.

Roth, U., Friebe, A., and Schnabl, H., 2000. Resistance induction in plants by a brassinosteroid-containing extract of *Lychnis viscaria* L. *Zeitschrift Naturforschung* **55**: 552–559.

Sam, O., Núñez, M., Ruiz-Sánchez, M., Dell'amico, J., Falcón, V., Dela Rosa, M., and Seoane, J., 2001. Effect of a brassinosteroid analogue and high temperature stress on leaf ultrastructure of *Lycopersicon esculentum*. *Biol. Plant.* **44**: 213–218.

Sasaki, H., 2002. Brassinolide promotes adventitious shoot regeneration from cauliflower hypocotyl segments. *Plant Cell Tiss. Org.* **71**: 111–116.

Schaefer, S., Medeiro, S., Ramirez, J., Galagovsky, L., and Pereira-Netto, A., 2002. Brassinosteroid-driven enhancement of the *in vitro* multiplication rate for the marubakaido apple rootstock [*Malus prunifolia* (Willd.) Borkh]. *Plant Cell Rep.* **20**: 1093–1097.

Sharma, P., and Bhardwaj, R., 2007. Effects of 24-epibrassinolide on growth and metal uptake in *Brassica juncea* L. under copper metal stress. *Acta Physiol. Plant.* **29**: 259–263.

Sharma, P., Bhardwaj, R., Arora, N., Arora, H., and Kumar, A., 2008. Effects of 28-homobrassinolide on nickel uptake, protein content and antioxidative defence system in *Brassica juncea*. *Biol. Plant.* **52**: 767–770.

Singh, I., and Shono, M., 2005. Physiological and molecular effects of 24-epibrassinolide, a brassinosteroid on thermotolerance of tomato. *Plant Growth Regul.* **47**: 111–119.

Sondhi, N., Bhardwaj, R., Kaur, S., Kumar, N., and Singh, B., 2008. Isolation of 24-epibrassinolide from leaves of *Aegle marmelos* and evaluation of its antigenotoxicity employing *Allium cepa* chromosomal aberration assay. *Plant Growth Regul.* **54**: 217–224.

Symons, G., Davies, C., Shavrukov, Y., Dry, I., Reid, J., and Thomas, M., 2006. Grapes on steroids. Brassinosteroids are involved in grape berry ripening. *Plant Physiol.* **140**: 150–158.

Upretik, K., and Murtig, S., 2004. Effects of brassinosteroids on growth, nodulation, phytohormone content and nitrogenase activity in French bean under water stress. *Biol. Plant.* **48**: 407–411.

Vardhini, B., and Rao, S., 2002. Acceleration of ripening of tomato pericarp discs by brassinosteroids. *Phytochem.* **61**: 843–847.

Vasyukova, N., Chalenko, G., Kaneva, I., Khripach, V., and Ozeretskovskaya, O., 1994. Brassinosteroids and potato late blight. *Appl. Biochem. Microbiol.* **30**: 464–470.

Xia, X., Huang, Y., Wang, L., Huang, L., Yu, Y., Zhou, Y., and Yu, J., 2006. Pesticides-induced depression of photosynthesis was alleviated by 24-epibrassinolide pretreatment in *Cucumis sativus* L. *Pestic. Biochem. Phys.* **86**: 42–48.

Xia, X., Wang, Y., Zhou, Y., Tao, Y., Mao, W., Shi, K., Asami, T., Chen, Z., and Yu, J., 2009a. Reactive oxygen species are involved in brassinosteroid-induced stress tolerance in cucumber. *Plant Physiol.* **150**: 801–814.

Xia, X., Zhang, Y., Wu, J., Wang, J., Zhou, Y., Shi, K., Yu, Y., and Yu J., 2009b. Brassinosteroids promote metabolism of pesticides in cucumber. *J. Agric. Food Chem.* **57**: 8406–8413.

Yokota, T., Sato, T., Takeuchi, Y., Nomura, T., Uno, K., Watanabe, T., and Takatsuto, S., 2001. Roots and shoots of tomato produce 6-deoxo-28-norcathasterone, 6-deoxo-28-nortyphasterol and 6-deoxo-28-norcastasterone, possible precursors of 28-norcastasterone. *Phytochem.* **58**: 233–238.

Yu, J., Zhou, Y., Ye, S., and Huang, L., 2002. 24-epibrassinolide and abscisic acid protect cucumber seedlings from chilling injury. *J. Hortic. Sci. Biotech.* **77**: 470–473.

Chapter 10

BRASSINOSTEROID ACTION AND ITS RELATION WITH HEAT STRESS MECHANISMS IN PLANTS

L.M. MAZORRA

Department of Plant Physiology and Biochemistry, National Institute of Agricultural Sciences, Carretera a Tapaste Km 3½, San José de las Lajas, Havana, Cuba

Abstract: Brassinosteroids (BRs) are plant steroid hormones which can regulate several physiological effects in plants, including promotion of cell growth and induction of heat stress tolerance. BR-biosynthesis and signaling mutants show a striking dwarf phenotype and these mutants may therefore lead to intrinsic cellular stress. However, how stress mechanisms could function during BR action is poorly understood. This review will focus on stress gene/protein and metabolites that have showed to be altered during BR-induced heat tolerance and as a result of changed BR content and perception. Recent studies on proteomics and micro-array of BR-treated tissues or BR-related mutants have revealed up/down regulation of specific enzymes/genes related to reactive oxygen scavenging, redox control, protein folding, and others. We will specifically discuss potential roles for Heat Shock Proteins, Antioxidant metabolites and enzymes in BR-induced thermal tolerance. In addition, as stress mechanisms are not exclusive of plants under stressful situations, the review will also discuss on how these protective factors may be implicated in classical BR effects during normal growth stimulation. Putative models to explain the role of Antioxidants, Oxidants and Heat Shock Proteins in BR action will be presented.

Key words: Brassinosteroids, heat shock proteins, oxidative stress, antioxidants, abiotic stress, thermotolerance

S. Hayat and A. Ahmad (eds.), *Brassinosteroids: A Class of Plant Hormone*,
DOI 10.1007/978-94-007-0189-2_10, © Springer Science+Business Media B.V. 2011

1. INTRODUCTION

Steroidal hormones called brassinosteroids (BRs) can act as modulators of plant response to a diversity of abiotic stresses (Clouse and Sasse, 1998; Krishna, 2003). BRs have the ability of inducing tolerance in plants to salinity, drought, high and low temperature, heavy metals and others (Bajguz and Hayat, 2009). However, the mechanisms by which these hormones regulate stress tolerance are largely unknown.

Many metabolites and proteins have been implicated in stress mechanisms, including Heat Shock Proteins (HSPs) and Antioxidants. HSPs are molecular chaperones with basic functions in normal plant growth and development; however, they are better known due to accumulation in response to heat stress (Liberek *et al.*, 2008). HSPs constitute a large family of five members (HSP100, HSP90, HSP70, HSP60) and small Heat Shock Proteins (sHSPs) (Kotak *et al.*, 2007).

Protection against stress includes mechanisms beyond the HSP expression. Stress also induces reactive oxygen species (ROS), which can lead to oxidative damage if not efficiently scavenged. Antioxidants function to protect plants against damaging effect of ROS and include ROS-scavenging metabolites and enzymes. Although the established role of Antioxidants in plant responses to stress, these mechanisms also are present during normal growth and development (Mittler, 2002; Apel and Hirt, 2004; Foyer and Noctor, 2005).

This review critically discusses up-to-date research on the induction of heat shock responses, including thermotolerance by BRs and seeks to do an integral analysis of Heat Shock Proteins and Antioxidants as important protective stress mechanisms in this process. It pretends to put together data from heat shock (HS) experiments, BR microarray and proteomic studies using BR-deficient/signaling mutants to provide a comprehensive view about the putative role of HSPs and Oxidative stress in BR-mediated effects of growth and heat tolerance as well as potential practical implications for the use of BRs as anti-stress hormones in agriculture.

2. BR INDUCE HEAT STRESS TOLERANCE IN PLANTS

In 1995, the first report demonstrating that application of 24-epibrassinolide (EBL) markedly enhances viability of bromegrass cells following exposure of cell suspension cultures to high-temperature stress was published (Wilen *et al.*, 1995). Also, Zhu *et al.* (1996) demonstrated the increase of heat tolerance in cucumber seedlings. Three years later, Dhaubhadel and his colleagues at Dr Krishna's lab extended the studies to include other plant species and proved that 24-epibrassinolide (EBL) also has the ability to induce heat tolerance in *Brassica napus* and tomato (*Solanum lycopersicum L*) seedlings. They noted that these seedlings, when grown on MS medium supplemented with 1 μM of EBL, are significantly and consistently more tolerant (measured as survival) to a lethal heat treatment than are untreated seedlings (Dhaubhadel *et al*., 1999). An increased survival of seedlings grown on presence of EBL was also confirmed in *Arabidopsis thaliana* (Kagale *et al*., 2007), suggesting that EBL-induced heat tolerance can not be specific for species.

Using other way of BR application, an increased plant performance under heat stress also has been achieved through spraying of leaves with EBL (Table 1). Thus, the application of EBL on leaf tissue of tomato seedlings increases accumulation of dry biomass by 7% and survival by about 80% (Singh and Shono, 2005; Ogweno *et al*., 2008).

In stress tolerance experiments, the way of hormone application can determine how is it taken and therefore enhance the efficiency of these compounds in practical applications. It is known that the transport and metabolic transformation of BRs are related to their uptake and distribution occurring into different plant tissues (Clouse and Sasse, 1998). BRs directly penetrate plants through root when added into growth medium or directly applied to leaf tissues as a foliar spray. The application of BR into medium has the advantage of ensuring constant exposure of plants to these compounds. However, the fact of that induction of heat tolerance by BR foliar spraying can also be observed, even several days after BR application (Singh and Shono, 2005; Ogweno *et al*., 2008) that indicates long-lasting effects for these compounds.

Heat tolerance can be effectively induced at 1 μM of EBL, although, some studies reported induction at 0.01 μM (Mazorra and Núñez, 2006) or at concentrations between 1 and 50 μM (Dhaubhadel *et al*., 1999; Ogweno *et al*., 2008; Singh and Shono, 2005). It suggests that induction of heat tolerance can be obtained at relatively high BR amount. Nevertheless, concentrations lower than 1 μM have generally shown to be less effective in heat tolerance induction, whereas higher concentrations are sometimes inhibitory for normal plant growth. Classical BR effect of growth stimulation can be obtained at very low concentrations (<0.01 μM). Between years 2006 and 2009, we also demonstrated induction of thermotolerance to heat in tomato seedlings exposed constantly to EBL in the growth medium (Mazorra and Núñez, 2006; Mazorra, 2009). However, we detected BR-increased survival lower than those reported in previous studies. Results suggest that BR-induced heat tolerance might be a phenomenon that highly depends on BR concentration and assay conditions. This has practical implication with optimal BR does, need to be empirically determined in order to achieve both tolerance to heat and growth stimulation.

Table 1. Overview of application strategies and major effects of BRs on heat stress tolerance of seedlings and pollen

BR treatment		**HS treatment and recovery**		**BR-induced effects**		**Crop**
Way of BR application	**Concentration**	**HS treatment**	**HS recovery**	**Survival**	**Growth and yield**	
EBL uptake by root	1µM	45°C–4 hours (in dark)	7 days–20°C	~6-fold increase of seedling survival	–	*Brassica napus* (Seedling stage)[a]
EBL uptake by root	1µM	45°C–4 hours (in dark)	7 days–23°C	~2-fold increase of seedling survival	–	*Solamum Lycopersicon* (Seedling stage)[a]
EBL uptake by root	1µM	45°C–3 hours (in dark)	7 days–22°C	~3-fold increase of seedling survival	–	*Arabidopsis thaliana* (Seedling stage)[b]
EBL uptake by root	0.01µM	45°C–45 hours (in dark)	7 days–25°C	~0.2-fold increase of seedling survival	–	*Solamum Lycopersicon* (Seedling stage)[c]
EBL uptake by root	1µM	45°C–7 hours (in dark)	7 days–25°C	~1-fold increase of seedling survival	~40–50% increase of stem length	*Solamum Lycopersicon* (Seedling stage)[d]
EBL foliar spraying	0.1mg.L^{-1} (~50 mL/plant)	45°C light/30°C dark–8 days)	25°C light/ 18°C dark–4 days	–	~7% of increase (seedling dry weight)	*Solamum Lycopersicon* (Seedling stage)[e]
EBL foliar spraying	1µM (~12mL/plant)	45°C–3 hours (in hight)	25°C light/ 20°C dark–7 days	>80% increase of seedling survival	–	*Solamum Lycopersicon* (Seedling stage)[f]
EBL foliar spraying	1µM (~25 mL/plant)	35°C hight/27°C dark–2 months	–	–	>100% of increase (Fruit weight and number)	*Solamum Lycopersicon* (Seedling stage)[f]
EBL uptake by pollen	1µM	35°C–4 hours 40°C–4 hours	–	~13.31% increase of pollen germination ~8–26% reduction of pollen bursting	~8–50% increase of pollen tube growth	Solamum Lycopersicon (Seedling stage)[f]

Sources: Data from Dhaubhadel *et al.*, 1999[a]; Kagale *et al.*, 2007[b]; Mazorra and Núñez, 2006[c]; Mazorra, 2009[d]; Ogweno *et al.*, 2008[e]; Singh and Shono, 2003, 2005[f].

Although, data from several published experiments should be analysed carefully because of different BR concentrations require the use of several control treatments that should contain the same amount of ethanol (solvent for EBL) added in each BR concentration. Sometimes, a unique control treatment (0.01% ethanol) or distilled water has been included. It has been suggested that the effectiveness of BR-induced heat tolerance is affected by the concentration of ethanol into growth medium (Confraria *et al.*, 2007). Besides, specifically in spraying experiments, additional care should be taken to reproduce the same amount of BR per plant between experiences because

Table 2. General overview of BR-induced HSP responses in plants, highlighting key differences of HSP expression at protein and gene level between *Arabidopsis* and *Brassica napus*

Class of HSP		Non-stress	Heat stress tolerance		Analysis
			Arabidopsis	*Brassica napus*	
HSP101	Protein	No detected	+(20h recovery)	+	Western blot
	Transcripts	No detected	–	–	Northern blot
HSP90	Protein	+	+(20h recovery)	+	Western blot; Proteomic
	Transcripts	+	–	–	Northern blot
HSP70	Protein	+	–	+	Proteomic; western blot
	Transcripts	+	–	–	Microarray
sHSP	Protein	No detected	–	+	Western blot
	Transcripts	No detected	–	–	Northern blot

Sources: Dhaubhadel *et al*., 1999; Goda *et al*., 2004; Kagale *et al*., 2007; Deng *et al*., 2007.

it can vary significantly with plant biomass, volume of BR solution applied, etc, even when equivalent BR concentrations are used.

Published data also show that BRs can improve plant performance under a variety of high temperature regimes (Table 1). Thus, this can be observed following short-term HS treatments (a few minutes to hours) or after several days/months of plant growth under elevated temperature (Dhaubhadel *et al*., 1999; Singh and Shono, 2005; Ogweno *et al*., 2008). It indicates that BRs could increase the ability of plants to acclimate rapidly, within hours or days, to elevated temperatures that would be damaging or lethal without such acclimation. Also, BR-improved heat tolerance can be independent of whether or not high temperature treatment was applied in dark or light conditions (Dhaubhadel *et al*., 1999; Singh and Shono, 2005; Ogweno *et al*., 2008).

2.1 BR-induced heat tolerance in the reproductive development

Induction of heat tolerance by BRs seems to be a phenomenon not only of vegetative growth but also of reproductive development (Table 1). Singh and Shono (2003) observed significantly higher *in vitro* tomato pollen germination, enhanced pollen tube growth and low pollen bursting in the presence of EBL at 35°C, a temperature high enough to induce heat-stress symptoms in tomato. A beneficial effect of EBL foliar spraying on fruit yield was also observed under heat stress, which was mainly due to the increase in fruit number (Singh and Shono, 2005). Maybe, the high abundance of naturally occurring BRs into pollen could reflect a role of steroidal hormones in other reproductive processes other than the well-known stimulation of pollen tube growth (Clouse and Sasse, 1998).

Practical implications of protective effects of BRs in reproduction under high temperature stress can be enormous because of flowering and fruit set are highly heat sensitive processes of tomato plant (Weaver and Timm 1989; Dane *et al*., 1991) and relate directly to yield. Nevertheless, up-to-date results about the effect of BRs in reproductive process are limited; their interpretation is complex so that more tests should be made to confirm the hypothesis.

2.2 Induction of heat tolerance by spirostanic analogues of BRs

Interest for studies of BR analogues as inducers of heat tolerance has gained momentum because the synthesis of these compounds is cheaper and presents long-lasting effects. A variety of spirostanic analogues of BRs were obtained, many of them have been tested in field experiments with good results (Núñez *et al*., 2003).

We have tested the ability of two spirostanic analogues of castasterone (Figure 1) to increase survival of tomato seedlings. Exposure of tomato seedlings grown on MS medium within plastic pots to a sub-lethal heat stress of 45°C during 4 hours left about 10% of seedlings alive, whereas, ~90% of seedlings survived with 1μM of EBL (Figure 2).

However, the BR analogues A and B at the concentration of 1μM were slightly less effective than EBL to increase survival (Figure 2). It is important to note that these analogues still keep biological activity despite they have drastic changes in their chemical structure (Mazorra and Núñez, 2006).

Although BR analogues can be biologically less active than natural BRs, their long-lasting effects and lower prices make them good candidates for practical applications.

2.3 Heat tolerance of mutants with altered BR signaling

Cross-links between heat stress tolerance and BR-signaling have begun to be revealed. Koh *et al*. (2007) found that, in rice, the disruption of the gene BIN2 that encodes for a negative regulator of the BR receptor (BRI1) leads to a phenotype of increased tolerance to a variety of abiotic stress, including heat shock. They observed that this mutant has normal growth, hypersensitivity to BL and concluded that elevated stress tolerance and up-regulated stress genes are likely due to enhanced activity in BR signal transduction pathways. Consistent with this hypothesis, it was observed that a semi-dwarf tomato mutant with altered BR sensitivity (*abs1*) is responsive to EBL and shows increased heat tolerance (Mazorra, 2009). These findings indicate that future research should focus on elucidating the mechanism involved with these increased BR-signaling activities, which leads to the induction of stress signal-transduction pathways. This should help us devise approaches that use molecular engineering to confer increased stress tolerance in plants.

2.4 Interactions of BRs with other hormones in plant heat tolerance

It is believed that BRs likely induce heat tolerance indirectly through other plant hormones such as abscisic acid (ABA), salicylic acid (SA) and ethylene, which have also been linked to HS signaling in different plant species and might interact through complex networks to regulate heat stress responses (Larkindale and Huang, 2004; Larkindale *et al.*, 2005a, 2005b). A transient peak in ABA level was reported in response to HS in pea plants (Liu *et al.*, 2006) and during recovery from HS treatments in creeping bentgrass (Larkindale and Huang, 2004). BRs have demonstrated to interact with auxins, ethylene, jasmonic acid (JA) and salicylic acid in specific bioassays (Clouse and Sasse, 1998; Khripach *et al.*, 1999). However, direct evidences of BR-hormone interactions in heat tolerance are greatly lacking. Although, it has been demonstrated that BR-increased heat tolerance of *Brassica napus* seedlings subjected to short-term heat shocks was associated with marked increases of endogenous ABA concentration (Kurepin *et al.*, 2008). Increased ABA level was also confirmed in cellular culture of *Chlorella vulgaris* (Bajguz, 2009). Practical exploitation of possible interactions of BRs and other plant regulators could be useful to design novel protective products based on mixtures of these compounds.

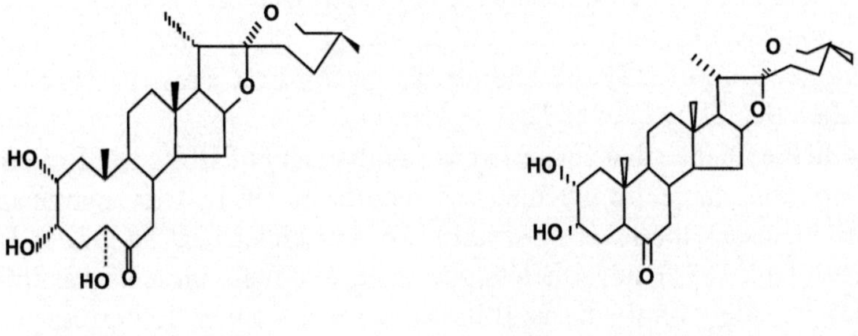

A: 25(R) 2*a*,3*a*,5*a* trihydroxi-spirostan-6-one

B: 25(R) 2*a*,3*a*, dihydroxi 5α spirostan-6-one

Figure 1. Chemical structure of two spirostanic analogues of castasterone. They have a spiroketalic ring instead of classic lateral chain of natural BRs.

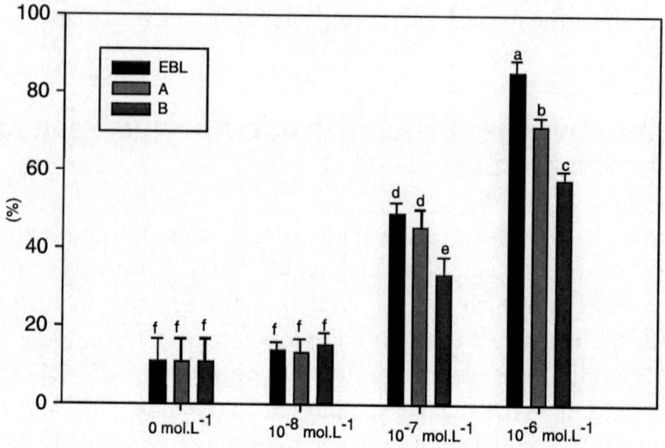

Figure 2. Survival percentage of heat tolerant tomato seedlings treated with different BRs.

3. INVOLVEMENT OF HEAT SHOCK PROTEINS IN BR-MEDIATED EFFECTS

HSPs are proposed to be chaperones, a large and diverse group of unrelated proteins that assist in correcting the noncovalent assembly and/or disassembly of other polypeptide-containing structures. Chaperones include major classes of proteins: HSP100, HSP90, HSP70, HSP60 and small HSP (sHSP). They are essential for proper functioning of the cell under normal growth; however, HSPs are one of the molecular factors involved in heat shock response in plants (Liberet *et al.*, 2008). There are no direct evidences about how HSP70, HSP90 and HSP60 contribute in the survival of plants heat stress in plants, whereas, the role of HSP100 and sHSPs during heat stress has been better established (Kotak *et al.*, 2007).

The first evidence that suggested an involvement of HSPs in BR-mediated heat responses came from Kulaeva's reports in 1991. They demonstrated that BR increases the level of several proteins under heat stress conditions (Kulaeva *et al.*, 1991); nevertheless, the identity of these proteins was inferred to be HSPs, solely on the basis of molecular mass. Since 1999, research has provided direct demonstrations that EBL can regulate HSPs under normal growth of seedlings or during the development of heat tolerance. Effects of BRs on major classes of HSPs are described below.

3.1 HSP101

It is known that HSP101 is expressed at low levels at normal temperatures and is strongly induced by heat (Vierling, 1991) and considered as an important determinant of thermotolerance in plants (Hong *et al.*, 2003; Lee *et al.*, 2005). EBL treatment does not modify HSP101 accumulation in *Brassica napus* under normal conditions; however, this compound stimulated HS-induced accumulation of HSP101 during stress exposure and recovery in *Brassica napus*. By contrast with other HSPs, this protein starts enhancing its level during HS treatment. In *Arabidopsis*, HSP101 was not altered, except at 20 hour recovery time point (Kagale *et al.*, 2007). Results suggest that HSP101 may be involved in EBL-induced heat tolerance, likely playing a role against protein disaggregation.

3.2 HSP90

In the absence of heat strees, EBL has the ability of increasing levels of hsp90 mRNA in bromegrass cells (Wilen *et al.*, 1995) and the amount of HSP90 in *Brassica napus* (Dhaubhadel *et al.*, 1999), but it does not affect *hsp90* gene expression in *Arabidopsis* (Kagale *et al.*, 2007). The protein Shepherd (SHD), an ortholog of GRP94, an endoplasmatic reticulum (ER)-resident HSP90-like protein was induced by BR (Deng *et al.*, 2007). HSP90 is essential for the proper functioning of signaling pathways involving steroid receptors and a variety of protein kinases in animal cells (Young *et al.*, 2001). It is therefore possible that HSP90 participates in brassinosteroid-mediated signal transduction during normal growth and development. The EBL-mediated accumulations of HSP90 and *hsp90* transcripts do not occur during HS treatment. However, HSP90 up-regulation appears later at the recovery from HS in *Brassica napus* and *Arabidopsis thaliana* and it paralleled with an EBL-induced heat tolerance (Dhaubhadel *et al.*, 1999). Maybe, this enhanced expression of HSP90 indicates that the machinery of BR-signaling might be more important after returning heated seedlings to normal temperatures.

3.3 HSP70

Studies indicated that the HSP70 does not accumulate in response to BRs in seedlings without heat stress (Dhaubhadel *et al.*, 1999; Kagale *et al.*, 2007). However, proteomic analyses have detected an HSP70 in response to BL (Deng *et al.*, 2007), indicating that this HSP could participate during BR-induced growth. As molecular chaperones, HSP70 could act together with HSP90, BiP2 and other ER-localised chaperones during folding of a variety of BR-induced cell wall proteins (Goda *et al.*, 2002, 2004; Deng *et al.*, 2007).

Similar to HSP90, HSP70 also accumulates during the period of recovery in thermotolerant seedlings of *Brassica napus* but not of *Arabidopsis thaliana* (Dhaubhadel *et al*., 1999; Kagale *et al*., 2007).

3.4 sHSPs

sHSPs are found at low levels during normal growth, are strongly heat stress induced and have similarly to HSP101, could be key for thermotolerance (Kotak *et al*., 2007). EBL does not affect the expression of these proteins and their transcript levels before and during HS treatment; nevertheless, increases HS-enhanced accumulation of class II sHSPs and a mitochondrial sHSP, preferentially during recovery from HS (Dhaubhadel *et al*., 1999; Singh and Shono, 2005). sHSPs might play a role in protecting proteins against HS not only in cytosol but also in organelles and it could be essential during BR-induced heat tolerance. Overexpression of the chloroplast-localized sHSP in tomato and tobacco provides evidence that this sHSP protects photosystem II under some stress conditions (Neta-Sharir *et al*., 2005). Overexpression of a mitochondrial sHSP enhanced thermotolerance in tobacco (Sanmiya *et al*., 2004).

It has been seen that EBL promotes higher accumulation of *hsp* transcripts without changes at protein level during heat stress; however, induces higher accumulation of HSPs accompanied with a decrease of *hsp* transcripts during recovery (Dhaubhadel *et al*., 1999). Clearly, HSPs accumulation does not correlate with *hsp* gene expression, which could be due to a higher HSP synthesis. Subsequent studies proved that EBL treatment limited the loss of some of the components of the translational apparatus during prolonged heat stress and increased the level of expression of some of the components of the translational machinery during recovery, which was correlated with a more rapid resumption of cellular protein synthesis following heat-stress and a higher survival rate (Dhaubhadel *et al*., 2002).

Microarrays and proteomics of plants treated with BRs have revealed a weak correlation between protein and gene expression (Goda *et al*., 2002; Mussig *et al*., 2002; Konishi and Komatsu, 2003; Goda *et al*., 2004; Huang *et al*., 2006; Deng *et al*., 2007). It has been shown that most BR-induced RNAs respond slowly to BR treatment with mild fold changes (Goda *et al*., 2004; Nemhauser *et al*., 2004). As the change in protein abundance is expected to lag behind that of RNA, the proteins regulated at the transcript level would mostly show slow response to BR treatment. The observation of

HSP expression at later time points than transcript increases is coherent with these findings.

Another important observation is that all major classes of HSPs accumulate during recovery of *Brassica napus* seedlings, whereas, HSP101 and HSP90 do in *Arabidopsis thaliana*. It could likely reflect collaborations among different HSPs to induce heat tolerance.

Genetic analyses in *Arabidopsis* indicates that HSP101 interacts with sHSP chaperone system to re-solubilize protein aggregates after heat stress, a process that requires complex interactions of HSP101 protein domains (Lee *et al.*, 2005). Substrates that are denatured can be refolded and reactivated by Hsp70 and HSP90 with the participation, in some cases, of HSP101 (Friedrich *et al.*, 2004).

In heat tolerance experiments, it has been noted that seedlings can die several days after HS treatment, even if they were induced to producing HSPs. Dhaubhadel *et al.* (1999) observed that HSP accumulation is present in *Brassica napus* seedlings that grow without EBL and later mostly resulted dead (90%). Likely, accumulation of all major classes of HSPs in response to heat would not be enough to proper thermotolerance and other mechanisms could have a critical function. Recent research has demonstrated that pathways beyond the induction of HSPs are involved in the acquisition of thermotolerance in plants (Larkindale *et al.*, 2005a, 2005b).

Therefore, despite of BR-induced heat tolerance which is common for *Brassica napus*, *Solanum lycopersicum L* and *Arabidopsis thaliana*, the pattern of BR-induced expression of HSP during heat tolerance seems to be specie-specific (Dhaubhadel *et al.*, 2002; Kagale *et al.*, 2007). In *Arabidopsis thaliana*, the accumulation of HSP101 and HSP90, later during recovery was necessary for BR-enhanced heat survival (Kagale *et al.*, 2007). It should be noted that EBL-induced thermotolerance was lower in *Arabidopsis* than in *Brassica napus* (Dhaubhadel *et al.*, 1999; Kagale *et al.*, 2007), which could also explain these differences.

Based on the analysing of all these HS experiments, it should be highlighted that they are made under artificial conditions; it means that the seedlings are grown in constant presence of external BR at relatively high concentrations, controlled growth conditions without the natural fluctuations of light/dark temperature and sunlight illumination and it could likely lead to changes in HS response that does not reflect the process of thermotolerance in nature. Future studies about HSP expression should focus this limitation and further BR-induced heat tolerance experiments, in field conditions will be welcomed.

4. OXIDANTS AND ANTIOXIDANTS IN BR-MEDIATED HEAT STRESS EFFECTS

In addition to the involvement of HSP in BRs-mediated heat response, oxidative metabolism and signaling have been implicated. There is evidence that high-temperature stress is one of the conditions that disrupts the cellular metabolic homeostasis and promotes the production of reactive oxygen species (ROS) (Mittler, 2002). Oxidative stress occurs when there is an imbalance in any cell compartment in the production of ROS and antioxidant defence, thereby causing damage (Mittler, 2002; Apel and Hirt 2004).

Antioxidant defences include low-molecular weight metabolites such as ascorbic acid, glutathione, α-tocopherol and enzymatic antioxidants that include Ascorbate peroxidase (APX), Peroxidase (POX), Glutathione reductase (GR), Dehydroascorbate reductase (DHAR), Catalase (CAT), Superoxide dismutase (SOD), and others. Increased level of SOD, CAT, POX and CAT has been observed in response to heat stress (Rivero *et al.*, 2004; Ogweno *et al.*, 2008) and antioxidant genes, for example an isoform of APX enzyme (APX2), are required for heat acclimation (Larkindale and Vierling, 2008).

Mazorra *et al.* (2002) found that BRs can enhance the activity of SOD, POX and CAT in heated tomato leaves. In addition, foliar spray of tomato seedlings with EBL increased the activity of SOD, APX, POX, CAT at normal temperature, during HS treatment and recovery (Ogweno *et al.*, 2008). Incubation of cucumber seedlings with this BR increased antioxidant activity during heat tolerance (Zhu *et al.*, 1996). Data indicate that BRs could improve plant performance under heat conditions through the activation of the antioxidant system.

Generally, plant responses to environmental stress such as high-temperature stress have been associated with the accumulation of ROS such as hydrogen peroxide (H_2O_2), singlet oxygen ($^1O_2$), superoxide ($O^{2\bullet-}$), and the hydrogen radical (•OH), as a result of oxidative stress (Anderson 2002; Apel and Hirt, 2004). EBL is able of decreasing the HS-induced H_2O_2 accumulation in tomato plants during HS treatment and recovery and it is correlated with lower oxidative stress (assessed as lipid peroxidation) and dry biomass (Ogweno *et al.*, 2008). It suggests that BRs could also act to decrease HS-induced oxidative damage. Several Calvin-cycle enzymes within chloroplasts are extremely sensitive to high levels of H_2O_2, causing reductions in CO_2 fixation and foliar biomass (Willenkens *et al.*, 1997; Zhou *et al.*, 2004, 2006). In correspondence with this, EBL has demonstrated to increase photosynthesis under high temperature stress (Singh and Shono, 2005; Ogweno *et al.*, 2008).

It should be said that the increased Antioxidant capacity by treatments with BRs has been detected in other types of abiotic stress, including low temperature, heavy metals (Liu *et al.*, 2009; Wang *et al.*, 2009). So, oxidative stress could function to regulate, directly or indirectly, the BR action for a wide range of stressful situations.

4.1 Relationship between HSP/oxidative responses and BR signaling

A recently studied aspect is how the alterations of BR content and signaling can influence HSP response. The data suggest that HSP expression can be, in part, independent of endogenous BR level (Kagale *et al.*, 2007). Transcripts of all four classes of HSPs were present at high levels in BR-deficient mutants (*det2-1* and *dwf4*). Notably, *det2-1* seedlings accumulated *hsp* transcripts, including the heat-induced hsp101 and *shsp*, even in the absence of any HS. HSP101 protein was also present in *det2-1* seedlings in the absence of HS. Hsp transcripts are enhanced in BR-deficient mutant *cpd* (Szekeres *et al.*, 1996). These observations coincided with studies of HSP regulation by other hormones, which show that HSP expression neither seems to entirely depend on hormone-signaling. Thus, the accumulation of HSPs was present in the *Arabidopsis* ABA signaling mutants, abscisic acid insensitive 1 (*abi1*) and *abi2* (Larkindale *et al.*, 2005a). Like ABA, salicylic acid (SA) and ethylene do not appear to be required for HSP synthesis during HS (Larkindale *et al.*, 2005a; Clarke *et al.*, 2004).

In addition to altered HSP expression, disrupted redox balance and oxidative stress can be detected in BR-defective/perception dwarf mutants. Dwarfism in most BR-related mutants can lead to intrinsic cellular stress, as has been suggested by Szekeres *et al.* (1996). BR-biosynthesis/signaling mutants have shown altered oxidative stress. Seedlings of the BR-deficient mutant (*det2-9*) show tolerance to oxidative stress in response to treatment with oxidant agents (paraquat, menadione and H_2O_2).

These seedlings had higher SOD and CAT activity and increased expression of *cat* genes (Cao *et al.*, 2005). Also, we have seen increased oxidative stress (measured as lipid and protein oxidation), altered antioxidant activity and redox state in tomato BR-deficient/perception mutants (unpublished data).

Links between HSP response and oxidative stress in BR-mediated heat effects can be suggested. It is possible that BR-induced *hsp* transcripts can be regulated by H_2O_2 in BR-treated and heated tissues. A burst of H_2O_2 was reported to occur after very short periods at high temperature, apparently as a result of NADPH oxidase activity (Vacca *et al.*, 2004). This burst has been correlated with the induction of HS-responsive genes, a process assumed to be mediated through direct sensing of H_2O_2 by heat shock factors (HSFs) (Volkov *et al.*, 2006). In addition, hormones such as ethylene, ABA and salicylic acid regulate ROS levels and antioxidant capacity (Larkindale and Knight, 2002). Interesting, in two independent studies, H_2O_2-regulating antioxidant enzymes and *hsp* transcripts are up-regulated during HS treatment in EBL-treated thermotolerant plants, before visible damage can be observed (Dhaubhadel *et al.*, 1999; Ogweno *et al.*, 2008). Therefore, a hypothetical model that considers ROS signaling in regulating HSP accumulation and heat tolerance can be considered (Figure 3). A potential inter-link H_2O_2 – BR could generate practical interest in the combined use of both compounds as anti-heat stress product in field and tissue culture applications.

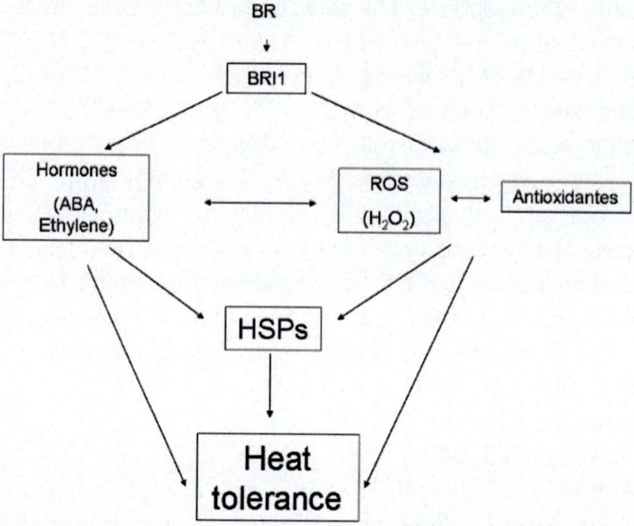

Figure 3. Hypothetical model for BR action in heat tolerance: BR is perceived at membrane by BRI1 receptor, leading to signal-transduction cascade. Signals such as ABA, ethylene and ROS can be second messengers to induce, directly or indirectly synthesis of HSPs and tolerance to heat.

Similar to HSPs, Antioxidants are not exclusive of plants under stressful situations and therefore they may be implicated in classical BR effects during normal growth stimulation. There are a few evidences that link oxidative stress and BR signaling during normal growth. Some cross-talks could be suggested from microarray and proteomic studies in BR perception/ biosynthesis mutants. BR-microarrays have revealed up/down regulation of oxidative stress-related genes in BR-defective mutants and in response to BR application. Several oxidative stress-related genes such as Superoxide dismutase 3[Fe] are down-regulated by brassinolide (Goda *et al.*, 2004).

In addition, a diversity of *pox* genes has demonstrated to be up-regulated or down-regulated by brassinolide or BR-deficiency (Goda *et al.*, 2002, 2004). POX enzymes could likely function to regulate ROS levels in BR-enhanced cell wall loosening during BR-induced growth. BR-proteomic analysis also has indicated a role for steroidal hormones in regulating proteins related to oxidative stress and cellular redox state (Konishi and Komatsu, 2003; Deng *et al.*, 2007).

Up-regulation of Glutathione-S-transferases protein and genes by BRs has been found in BR signaling mutants (Konishi and Komatsu, 2003; Goda *et al.*, 2004; Deng *et al.*, 2007); it suggests a role for these hormones in the control of detoxification processes. Monodehydroascorbate reductase and Superoxide dismutase [Cu–Zn], two enzymes that control, directly or indirectly, the levels of H_2O_2 and superoxide ($O^{2\bullet-}$) were induced by brassinolide (Konishi and Komatsu, 2003; Deng *et al.*, 2007), which could suggest the participation of these antioxidant enzymes in regulating redox balance and antioxidant capacity in BR-treated plants.

Some antioxidant genes such as *pox*, whose levels are markedly regulated by BRs in microarrays (Goda *et al.*, 2002, 2004), have not been detected in proteomic studies (Konishi and Komatsu, 2003; Huang *et al.*, 2006; Deng *et al.*, 2007).

Nevertheless, several Glutathione S-transferases were detected in both proteomic and genomic analyses related to BR effects on growth and stress, suggesting that this group of enzyme could be implicated in BR-regulated processes such as growth stimulation, detoxification and stress protection (Konishi and Komatsu, 2003; Huang *et al.*, 2006; Deng *et al.*, 2007). Likely, BRs have the ability of regulating, at post-transcriptional level, specific proteins related to redox state and signaling during growth promotion.

Taken together, evidences indicate that steroidal hormones might control, directly or indirectly, oxidative stress mechanisms during normal plant growth. Oxidative stress could be a central pathway in the complex network of signals by which BRs regulate plant growth. Nevertheless, these data should be carefully analysed so that dwarfism and oxidative stress components can have complex interactions.

In particular, results related to the use of BR-related mutants for testing the influence of BR content and signaling on HS response could reflect, in part, the effect of the striking dwarf phenotype of most BR-mutants leading to intrinsic cellular stress. BR-mediated response could arise, in part, from differences in stature and developmental stage between mutant and wild type plants. An aspect that is worth to be studied is how alterations of BR signaling influence oxidative mechanisms, under heat.

5. CONCLUSION

BRs have proved to modify a variety of plant stress responses to heat; however, this does not mean they are essential hormones for heat tolerance. Two key factors linked with heat tolerance (accumulation of HSPs, Antioxidant activity) have been postulated to regulate BR-mediated effects. In addition, altered level of HSPs and Antioxidants in BR-biosynthetic and perception mutants also suggests cross-talks between BR signaling and stress responses. Nevertheless, further research is necessary to reveal direct links among HS response, Oxidative stress, Plant regulators and other stress mechanisms with BR and heat signaling pathways. Physiological/biochemical studies of mutants with altered BR and heat signaling will allow elucidating important questions. In particular, micro-arrays and proteomics of BR-treated plants or mutants with increased heat tolerance could be very informative to reveal key connexions between steroidal hormones and stress protection.

BRs and their analogues as anti-heat stress substances have also generated considerable practical interest for agricultural uses (i.e, increments of crop yields in hot environments). However, most evidences about their potentialities have come from assays, done in controlled plant growth cabinets or in plant tissue culture conditions. It of course does not necessary reflect what happens under crop production systems subjected to stressful conditions so that field tests should be implemented to help confirm lab findings. In addition, most HS works have been conducted in dicotyledonous plants (*Arabidopsis thaliana, Solanum lycopersicum* and *Brassica napus*) and using 24-epibrassinolide as active BR. Experiments with other agronomically-important species and types of BRs, including BR analogues should be made.

6. ACKNOWLEDGEMENT

We are grateful to International Foundation for Science (IFS), Grant Agreement No. C/4162-1, and the Cuban Ministry of Science, Technology and Environment. Special thanks for Dr. Miriam Núñez and Dr. José Roberto Martín Triana.

7. REFERENCES

Anderson, J.A. 2002. Catalase activity, hydrogen peroxide content and thermotolerance of pepper leaves. *Sci. Hort.*, **95**(4): 277–284.

Apel, K., and Hirt, H. 2004. Reactive oxygen species: metabolism, oxidative stress, and signal transduction. *Ann. Review Plant Biol.*, **55**: 373–399.

Bajguz, A. 2009. Brassinosteroid enhanced the level of abscisic acid in *Chlorella vulgaris* subjected to short-term heat stress. *J. Plant Physiol.*, **166**(8): 882–886.

Bajguz, A., and Hayat, S. 2009. Effects of brassinosteroids on the plant responses to environmental stresses. *Plant Physiol. Biochem.* **47**(1): 1–8.

Cao, S., Xu, Q., Cao, Y., Qian, K., An, K., Zhu, Y., Binzen, H., Zhao, and Kuai, B. 2005. Loss-of-function mutations in DET2 gene lead to an enhanced resistance to oxidative stress in *Arabidopsis*. *Physiol. Plant.*, **123**: 57–64.

Clarke, S.M., Mur, L.A., Wood, J.E., Scott, I.M. 2004. Salicylic acid dependent signalling promotes basal thermotolerance but is not essential for acquired thermotolerance in *Arabidopsis thaliana*., *Plant J.*, **38**: 432–447.

Clouse, S.D., and Sasse, J.M. 1998. Brassinosteroids: essential regulators of plant growth and development. *Ann. Rev. Plant. Physiol. Plant Mol. Biol.*, **49**: 427–451.

Confraria, A., Desikan, R., Santos, I., and Neill, S. 2007. Brassinosteroids protect plants against heat stress. *Comparative Biochem. Physiol.*, **146**(4): S279.

Dane, F., Hunter, A.G., and Chambliss, O.L. 1991. Fruit set, pollen fertility, and combining ability of selected tomato genotypes under high temperature field conditions. *J. Am. Hort. Sci.*, **116**: 906–910.

Deng, Z., Zhang, X., Tang, W., Oses-Prieto, J.A., Suzuki, N., Gendron, J.M., Chen, H., Guan, S., Chalkley, R.J., Peterman, T.K., Burlingame, A.L., and Wang, Z.Y. 2007. A proteomic study of brassinosteroid response in *Arabidopsis*. *Mol. Cell. Proteomics.*, **11**: 451–467.

Dhaubhadel, S., Browning, K.S., Gallie, D.R., and Krishna P. 2002. Brassinosteroid functions to protect the translational machinery and heat-shock protein synthesis following thermal stress. *Plant J.*, **29**(6): 681–691.

Dhaubhadel, S., Chaudhary, S., Dobinson, K., and Krishna, P. 1999. Treatment with 24-epibrassinolide, a brassinosteroid increases the basic thermotolerance of *Brassica napus* and tomato seedlings. *Plant Mol. Biol.*, **40**: 333–342.

Foyer, C.H., and Noctor, G. 2005. Redox Homeostasis and antioxidant signaling: A metabolic interface between stress perception and physiological responses. *Plant Cell.*, **17**: 1866–1875.

Friedrich, K.L., Giese, K.C., Buan, N.R., and Vierling, E. 2004. Interactions between small heat shock protein subunits and substrate in small heat shock protein–substrate complexes. *J. Biol. Chem.*, **279**: 1080–1089.

Goda, H., Sawa, S., Asami, T., Fujioka, S., Shimada, Y., and Yoshida, S. 2004. Comprehensive comparison of auxin-regulated and brassinosteroid-regulated genes in *Arabidopsis*. *Plant Physiol.*, **134**: 1555–1573.

Goda, H., Shimada, Y., Asami, T., Fujioka, S., and Yoshida, S. 2002. Microarray analysis of brassinosteroid-regulated genes in *Arabidopsis*. *Plant Physiol.*, **130**: 1319–1334.

Hong, S.W., Lee, U., and Vierling, E. 2003. *Arabidopsis* hot Mutants Define Multiple Functions Required for Acclimation to High Temperatures. *Plant Physiol.*, **132**: 757–767.

Huang, B., Chu, C.-H., Chen, S.-L., Juan, H.-F., and Chen, Y.-M. 2006. A proteomics study of the mung bean epicotyl regulated by brassinosteroids under conditions of chilling stress. *Cell. Mol. Biol. Lett.*, **11**: 264–278.

Kagale, S., Divi, U.K., Krochko, J.E., Séller, W.A., and Krishna, P. 2007. Brassinosteroid confers tolerance in *Arabidopsis thaliana* and *Brassica napus* to a range of abiotic stresses. *Planta.*, **225**: 353–364.

Khripach, V., Zhabinskii, V., and De Groot, A. 1999. Brassinosteroids. A New Class of Plant Hormones. Academic Press. New York: 456 p.

Koh, S., Lee, S. Ch., Kim, M.K., Koh, J.H., Lee, S., An, G., Choe, S., and Kim, S.R. 2007. T-DNA tagged knockout mutation of rice OsGSK1, an orthologue of *Arabidopsis* BIN2, with enhanced tolerance to various abiotic stress. *Plant Mol. Biol.*, **65**: 453–466.

Konishi, H., and Komatsu, S. 2003. A proteomics approach to investigating promotive effects of brassinolide on lamina inclination and root growth in rice seedlings. *Biol. Pharm. Bull.*, **26**: 401–408.

Kotak, S., Larkindale, J., Lee, U., Koskull-Doring, P.V., Vierling, E., and Scharf, K.D. 2007. Complexity of the heat stress response in plants. *Curr. Opin. Plant Biol.*, **10**: 310–316.

Krishna, P. 2003. Brassinosteroid-mediated stress responses. *J. Plant Growth Regul.*, **22**(4): 289–297.

Kulaeva, O.N., Burkhanova, E.A., Fedina, A.B., Kaokhlova, V.A., and Bokebayeva, G.A. 1991. Effect of brassinosteroids on protein synthesis and plant-cell-ultrastructure under stress conditions. *In* HG, Cutler, T Yokota, G Adam (Eds), Brassinosteroids: Chemistry, Bioactivity and Applications: American Chemical Society, Washington DC: pp. 141–155.

Kurepin, L.V., Qaderi, M.M., Back, T.G., Reid, D.M., and Pharis, R.P. 2008. A rapid effect of applied brassinolide on abscisic acid concentrations in *Brassica napus* leaf tissue subjected to short-therm heat stress. *Plant Growth Regul.*, **55**: 165–167.

Larkindale, J., Mishkind, M., and Vierling, E. 2005a. Plant responses to high temperature. *In* MA Jenks, PM, Hasegawa (Eds), Plant Abiotic Stress. Blackwell Publishing. Oxford. UK: pp. 100–144.

Larkindale, J., Hall, J.D., Knight, M.R., and Vierling, E. 2005b. Heat stress phenotypes of *Arabidopsis* mutants implicate multiple signaling pathways in the acquisition of thermotolerance. *Plant Physiol.*, **138**: 882–897.

Larkindale, J., and Huang, B. 2004. Thermotolerance and antioxidant systems in *Agrostis stolonifera*: involvement of salicylic acid, abscisic acid, calcium, hydrogen peroxide, and ethylene. *J. Plant Physiol.*, **161**: 405–413.

Larkindale, J., and Knight, M.R. 2002. Protection against heat stress-induced oxidative damage in *Arabidopsis* involves calcium, abscisic acid, ethylene, and salicylic acid. *Plant Physiol.*, **128**: 682–695.

Larkindale, J., and Vierling, E. 2008. Core genome responses involved in acclimation to high temperature. *Plant Physiol.*, **146**: 748–761.

Lee, U., Wie, C., Escobar, M., Williams, B., Hong, S.W., and Vierling, E. 2005. Genetic analysis reveals domain interactions of *Arabidopsis* Hsp100/ClpB and cooperation with the small heat shock protein chaperone system. *Plant Cell.*, **17**: 559–571.

Liberek, K., Lewndowska, A., and Zietkiewicz, S. 2008. Chaperones in control of protein disaggregation. *EMBO J.*, **27**: 328–335.

Liu, H.T., Liu, Y.Y., Pan, Q.H., Yang, H.R., Zhan, J.C., and Huang, W.D. 2006. Novel interrelationship between salicylic acid, abscisic acid, and PIP2-specific phospholipase C in heat acclimation-induced thermotolerance in pea leaves. *J. Exp. Bot.*, **57**: 3337–3347.

Liu, Y., Zhao, Z., Si, P., Di, C., Han, J., and An, L. 2009. Brassinosteroids alleviate chilling-induced oxidative damage by enhancing antioxidant defense system in suspension cultured cells of *Chorispora bungeana*. *Plant Growth Regul.*, **59**: 207–214.

Mazorra, L.M. 2009. Los brasinoesteroides y su relación con la tolerancia de plántulas de tomate al calor. Doctoral Thesis. Faculty of Biology. University of Havana. Cuba, 100p.

Mazorra, L.M., and Núñez, M. 2006. Inducción de termo-tolerancia por la 24-epibrasinólida y dos análogos espirostánicos de brasinoesteroides en plantas de tomate. *Cult. Trop.*, **27**(3): 89–93.

Mazorra, L.M., Núñez, M., Hechavarría, M., Coll, F., and Sánchez-Blanco, M.J. 2002. Influence of brassinosteroids on antioxidant enzymes activity in tomato under different temperatures. *Biologia Plantarum.*, **45**(4): 593–596.

Mittler, R. 2002. Oxidative stress, antioxidants and stress tolerance. *Trends Plant Sci.*, **7**: 405–410.

Mussig, C., Fischer, S., and Altmann, T. 2002. Brassinosteroid-regulated gene expression. *Plant Physiol.*, **129**: 1241–1251.

Nemhauser, J.L., Mockler, T.C., and Chory, J. 2004. Interdependency of brassinosteroid and auxin signaling in *Arabidopsis*. *PLoS Biol.*, **2**: E258.

Neta-Sharir, I., Isaacson, T., Lurie, S., and Weiss, D. 2005. Dual role for tomato heat shock protein 21: protecting photosystem II from oxidative stress and promoting color changes during fruit maturation. *Plant Cell.* **17**: 1829–1838.

Núñez, M., Robaina, C., and Coll, F. 2003. Synthesis and practical applications of brassinosteroid analogs. *In* S. Hayat, A. Ahmad, Brassinosteroids. Bioactivity and Crop Productivity. Kluwer Academic Publishers. The Netherlands: pp. 87–117.

Ogweno, J.O., Son, X.S., Shi, K., Hu, W.H., Mao, W.H., Zhou, Y.H., Yu, J.Q., and Nogués, S. 2008. Brassinosteroid alleviate heat-induced inhibition of photosynthesis by increasing carboxylation efficiency and enhancing antioxidant systems in *Lycopersicon esculentum*. *J. Plant Growth Regul.*, **27**: 49–57.

Rivero, R.M., Ruiz, J.M., and Romero, L. 2004. Oxidative metabolism in tomato plants subjected to heat stress. *J. Hort. Sci. Biotechnol.*, **79**: 560–564.

Sanmiya, K., Suzuki, K., Egawa, Y., and Shono, M. 2004. Mitochondrial small heat-shock protein enhances thermotolerance in tobacco plants. *FEBS Lett.,* **557**: 265–268.

Singh, I., and Shono, M. 2003. Effect of 24-epibrassinolide on pollen viability during heat-stress in tomato. *Indian J. Exp. Biol.*, **41**: 174–176.

Singh, I., and Shono, M. 2005. Physiological and molecular effects of 24-epibrassinolide, a brassinosteroid, on thermotolerance of tomato. *Plant Growth Regul.*, **47**: 111–119.

Szekeres, M., Nemeth, K., Konez-Kalman, Z., Mathur, J., Kauschann, A., Altmann, T., Redei, G.P., Nagy, F., Schell, J., and Koncz, C. 1996. Brassinosteroids rescue the deficiency of CYP90, a cytochrome P_{450} controlling cell elongation and de-etiolation in *Arabidopsis*. *Cell.*, **85**: 171–182.

Vacca, R.A., de Pinto M.C., Valenti, D., Passarella, S., Marra, E., and De Gara, L. 2004. Production of reactive oxygen species, alteration of cytosolic ascorbate peroxidase, and impairment of mitochondrial metabolism are early events in heat shock induced programmed cell death in tobacco bright-yellow 2 cells. *Plant Physiol.*, **134**: 1100–1112.

Vierling, E. 1991. The roles of heat proteins in plants. *Ann. Rev. Plant Mol. Biol.*, 42: 579–620.

Volkov, R.A., Panchuk, I.I., Mullineaux, P.M., and Schoffl, F. 2006. Heat stress-induced H_2O_2 is required for effective expression of heat shock genes in *Arabidopsis*. *Plant Mol. Biol.*, **61**: 733–746.

Wang, H., Feng, T., Peng, G.X., Yan, M., Zhou, P., and Tang, X. 2009. Ameliorative effects of brassinosteroid on excess manganese-induced oxidative stress in *Zea mays* L. leaves. *Agric. Sci. China.*, **8**: 1063–1074.

Weaver, M.L., and Timm, H. 1989. Screening tomato for high temperature tolerance through pollen viability tests. *Hort. Sci.*, **24**: 493–495.

Wilen, R.W., Sacco, M., Gusta, L.V., and Krishna, P. 1995. Effects of 24-epibrassinolide on freezing and thermotolerance of bromegrass (*Bromus inermis*) cell cultures. *Physiol. Plant.*, **95**: 195–202.

Willenkens, H., Chamnongpol, S., Davey, M., Schraudner, M., Langebartels, C., Van Montagu, M., Inze, D., and Van Camp, W. 1997. Catalase is a sink for H_2O_2 and is indispensable for stress defence in C_3 plants. *EMBO J.*, **16**: 4806–4816.

Young, J.C., Moarefi, I., and Hartl, F.U. 2001. Hsp90: a specialized but essential protein-folding tool. *J. Cell Biol.*, **154**: 267–273.

Zhou, Y.H., Yu, J.Q., Huang, L.F., and Nogués, S. 2004. The relationship between CO_2 assimilation, photosynthetic electron transport and water-water cycle in chill-exposed cucumber leaves under low light and subsequent recovery. *Plant Cell Environ.*, **27**: 1503–1514.

Zhou, Y.H., Yu, J.Q., Mao, W.H., Huang, L.F., Song, S., and Nogués, S. 2006. Genotypic variation on Rubisco expression, photosynthetic electron flow and antioxidant metabolism in the chloroplasts of chill-exposed cucumber plants. *Plant Cell Physiol.*, **47**: 192–199.

Zhu, C., Zeng, G.W., and Liu, F. 1996. Effect of epi-brassinolide on the heat shock tolerance and antioxidant metabolism in cucumber seedling. *J. Zhejiang Agric. Univ.*, **22**(3): 284–288.

Chapter 11

PROTECTIVE EFFECTS OF BRASSINOSTEROIDS AGAINST HERBICIDES

R. PIÑOL AND E. SIMÓN
Cátedra de Fisiología Vegetal, Facultad de Biología, Universidad de Barcelona, Diagonal 645, E-08028 Barcelona, Spain

Abstract: The role of brassinosteroids in protecting plants from herbicide damage has increasing interest owing to the projected effects of herbicides on modern agriculture. Earlier data suggested that brassinosteroids can alleviate the decreases in plant growth caused by herbicides, and brassinosteroids have been applied to palliate this undesirable effect. However, rigorous and systematic studies on this topic have scarcely been reported. At present, the action of brassinosteroids in protecting plants from environmental stresses is clearly known, and herbicide application is considered a man-made environmental stress. Herbicides that inhibit photosynthetic electron transport at PSII level compete with plastoquinone bound at the Q_B site and thus inhibit the electron transfer from Q_A to Q_B. Chlorophyll fluorescence measurements are used as a non invasive accurate method to analyze damage caused to the photosynthetic apparatus. Such measurements detect the effect of environmental stresses on photosynthesis. Analyses of chlorophyll fluorescence together with measures of photosynthetic CO_2 assimilation and plant growth indicate that the harmful effects caused by s-triazine herbicides can be alleviated by brassinosteroids. The protective effect of brassinosteroids persists over time and leads to a more rapid recovery of plants. Brassinosteroids afford no protection against the damaging effect of other herbicides that inhibit photosynthetic electron transport but do not belong to the s-triazine group. The present chapter reports data on the protective effect of brassinosteroids on plants treated with herbicides that inhibit photosynthetic electron transport at the PSII level. It also discusses the biochemical and physiological causes of their protective effect.

Key words: Brassinosteroids, cyanazine, epibrassinolide, herbicides, metribuzin, methabenzthiazuron, simazine, triazine, terbutryn

DOI 10.1007/978-94-007-0189-2_11, © Springer Science+Business Media B.V. 2011

1. INTRODUCTION

Brassinosteroids (BRs) have a broad spectrum of stimulatory and protective activities that cause a positive effect on the quantity and quality of crops (Takematsu and Takeuchi, 1989; Vardhini *et al.*, 2006). Under field conditions, 24-epibrassinolide (EBR), one of the most widely used BR, increases not only crop yield but also crop quality (Ikekawa and Zhao, 1991; Khripach *et al.*, 1996, 2000). Thus, BRs have a potential application in agriculture to increase yield and to stimulate crop growth under unfavourable conditions, such as high salinity, low and high temperatures, drought or nutrient deficiency (Khripach *et al.*, 2000; Sasse, 2003; Divi and Krishna, 2009). The benefits of treatment with BRs have been described in plants subjected to chill stress, mild drought and salt stress (Clouse and Sasse, 1998; Krishna 2003; Kagale *et al.*, 2007). EBR enhances both cold and heat tolerance (Wilen *et al.*, 1995; Singh and Shono, 2005), and also reduces the impact of salt stress on growth, restores pigment levels and increases nitrate reductase activity in rice (Anuradha and Rao, 2003). EBR protects barley and cucumber plants against fungi (Pshenichnaya *et al.*, 1997; Khripach *et al.*, 2000) and other BRs also stimulate resistance to viral infection (Bobrik *et al.*, 1998). In summary, BRs have a protective effect on plants growing under biotic and abiotic stress, as is reviewed in detail in the various chapters of present issue.

Herbicides represent a special stress imposed by humans in agricultural, horticultural and other plant cultures. BRs seem able to protect plants from herbicide damage but detailed data on this topic are scarce. It is known that BRs protect plants against pesticides and herbicides; however, only a few preliminary studies have addressed this topic (Krishna, 2003). Besides, the mechanism of BR action on herbicide-treated plants has not been studied in depth.

Here we summarize the information reported on the effects of BRs in plants treated with photosynthesis inhibitor herbicides, and we report new experimental data to clarify this topic.

2. PHOTOSYNTHESIS-INHIBITOR HERBICIDES

Herbicides are chemicals used to manage unwanted plants in agriculture and horticulture which are usually referred to as “weeds”. This term, which admits several definitions, is taken here to mean “a plant that forms populations that are able to enter habitats cultivated, markedly disturbed or occupied by man and potentially depress or displace the resident populations that are deliberately cultivated or are of ecological or aesthetic interest”

(Navas, 1991). Weeds seriously deplete the quality, quantity and economic value of crops by competing for space, light, water and nutrients, by acting as a potential host to plant pathogens, and by contaminating harvests (Percival and Baker, 1991).

The more than 300 herbicides used today, empirically developed, correspond to more than 40 chemical families and target many metabolic pathways in plants. Thus, they cause many different physiological stresses on plants, and plant responses to herbicides are also highly variable. They range from no response or relatively minor, short-term responses, from which the plant easily recovers, to severe lethal damage. Herbicides also trigger a multitude of plant defence responses, and plant species vary widely in their abilities to show such responses (Dyer and Weller, 2005).

Herbicides affect not only weeds, but also crops plants. To prevent damage to the crops, herbicide selectivity has been widely exploited. Selectivity refers to the capacity of a particular herbicide to eliminate weeds without affecting crop yield or quality. The main mechanism of herbicide selectivity derives from the differential metabolism of weed and crop plant species. Differential interception and uptake of herbicides is also involved and it is important to strictly follow the recommendations of herbicide application to minimize crop yield losses (Pinto de Carvalho *et al.*, 2009).

However, even selective herbicides can induce stress in crop plants, usually transient stress, since the herbicide is metabolized or the metabolic damage is overcome when the enzymes inhibited by herbicides are replaced by *de novo* synthesis (Dyer and Weller, 2005). Both plant crops and weeds can develop resistance to herbicides (Holt *et al.*, 1993; Oettmeier, 1999; Szigeti and Lehoczki, 2003). Herbicide resistance can be conferred by several mechanisms, among which the loss of sensitivity of the target site seems to be the most important. Thus, some biotypes of weeds have developed resistance to triazine herbicides by a gene mutation of D1 protein (Gronwall, 1994; Shukla and Devine, 2008). Resistance can also be due to rapid degradation or conjugation of herbicides. Glutathione-S-transferase and cytochrome P450 monooxygenase are the main enzymes involved in these processes (Janjic *et al.*, 2007).

Among the long list of herbicides used in agriculture, we will focus on photosynthesis-inhibitory herbicides, since BR protective effect has been shown in plant crops treated with them (Piñol and Simón, 2009).

Two herbicide groups inhibit photosynthesis: those that inhibit photosynthesis at the level of PSII, and those that do so at the level of PSI.

The first group includes triazines, urea and uracils, which are soil-applied compounds widely used in cropping systems. Atrazine, simazine, cyanazine and terbutryn, among others, belong to the triazine group. Other compounds in this group include foliar-applied herbicides such as bentazon, bromoxynil and ioxynil, desmedipham and phenmedipham, and propanil. Triazines are used for selective pre-emergence and early-postemergence control of annual weeds in crops. They contain a heterocyclic nitrogen structure that is either symmetrical (*s*-triazines) or asymmetrical (*as*-triazinones). Substituted ureas, including diuron and chlorotoluron, are produced by substitution of the hydrogen atoms of urea with other chemical groups, such as phenyl, methyl or methoxy (Fedtke, 1982; Bowyer *et al.*, 1991; Percival and Baker, 1991; Wakabayashi and Böger, 2004; Dyer and Weller, 2005).

About half of all known herbicides inhibit photosynthetic electron transport at the level of the PSII acceptor site. They compete with plastoquinone to bind at the Q_B site and thus inhibit the electron transfer from Q_A to Q_B. Their specific binding at the Q_B site of PSII reduces electron transfer to the plastoquinone pool (Xiong *et al.*, 1997), inhibits photosynthetic O_2 evolution and affects flash-induced chlorophyll (Chl) fluorescence (Zimmermann *et al.*, 2006). Thus, these herbicides inhibit PS II function by displacing Q_B from its binding site on the D1 protein of this system (Edelman and Mattoo, 2008).

The second group includes PSI inhibitors such as the bipyrilidium herbicides, of which paraquat is the best known example. Paraquat, which is used extensively throughout the world, is applied to foliage and rapidly causes wilting and desiccation. These herbicides bind close to ferredoxin in the acceptor site of PSI (Dyer and Weller, 2005).

Since all these herbicides inhibited photosynthesis, it is reasonable to conclude that the damage they inflict would be due to the loss of carbon fixation. However, a more rapid toxicity caused by blocking the photosynthetic electron transfer is related with the photo-oxidative stress caused by herbicides The damage observed in intact plants after treatment with photosynthetic-inhibiting herbicides may be attributable to the production of high-energy reactive oxygen species (ROS) such as superoxide and singlet oxygen. Singlet oxygen and other ROS are highly toxic and cause acute membrane damage (Halliwell, 1991). Although these molecules have short half-lives and low mobility but they can accumulate locally with serious consequences (Blokhina *et al.*, 2003).

A quite important number of weed species have developed resistance to PSII and PSI inhibitors. The resistance to PSII inhibitors appears to be attributable to a serine-to-glycine substitution at the binding site of D1 protein. The resistance is associated with reduced photosynthesis efficiency. Enhanced degradative pathways could also cause crop resistance to triazines (Dyer and Weller, 2005, Janjic *et al.*, 2007; Shukla and Devine, 2008). A limited absorption and translocation, and rapid degradation or oxidation could cause crop resistance to substituted ureas (Holt *et al.*, 1993).

The foliar-applied bipyridilium herbicides have no soil activity and their selection intensity causing resistance should be reduced. However, repeated application in orchard and nurseries may force the evolution of resistant biotypes. Resistance could be explained either by herbicide sequestration in cell walls or by enhanced detoxification of ROS (Dyer and Weller, 2005).

3. CHLOROPHYLL FLUORESCENCE AS AN INDICATOR OF THE HERBICIDES EFFECT IN THE PHOTOSYNTHETIC APPARATUS

Chlorophyll (Chl) fluorescence-based methods have been used for many years to study the effect of environmental factors, including herbicides, on plants. Chl fluorescence *in vivo* and its kinetics of relaxation are sensitive early indicators of damage to photosynthetic apparatus and are thus suitable parameters to study stress damage (Maxwell and Johnson, 2000; Rohácek, 2002). Quenching phenomena are strongly influenced by several stress factors, such as drought or high irradiance, and also herbicides (Krause and Weis, 1991). Changes in the quantum yield of non-cyclic electron transport *in vivo* can be evaluated from measurements of the Chl fluorescence yield in the steady-state and maximal levels (Genty *et al.*, 1989). Thus, Chl fluorescence parameters mirror the effects of stress on the photosynthetic apparatus and other physiological effects which feed back on photosynthesis (Bolhar-Nordenkampf *et al.*, 1989). Thus, Chl fluorescence measurements in combination with simultaneous analysis of leaf gas exchange provide information about the partitioning of excitation energy between photochemical processes, which are responsible for CO_2 reduction, and non-photochemical processes, of which radiationless dissipation is the most relevant (Sivak and Walker, 1985).

Inhibition of photosynthesis or associated biological processes by herbicides may affect the physiological state of a plant. Besides, plant biochemical parameters such as ATP-formation, CO_2-fixation and O_2-evolution have been used as reliable indicators of the herbicide effect. However, due to the complexity of the methods involved and the time required to obtain results with these parameters, Chl fluorescence became a simple and rapid method to study herbicide effects. It offers higher sensitivity than assays based on growth inhibition. Thus, Chl fluorescence parameters have been used to evaluate damage caused by many herbicides, even those affecting metabolic processes that are not directly linked to photosynthesis (Juneau *et al.*, 2007). Chl fluorescence parameters in the dark- or light-adapted state are easily obtained by using Pulse-Amplitude-Modulated (PAM)-fluorometry. The effective quantum yield ($\Delta F/F'_M$) and the non-photochemical quenching (NPQ) appear to be the most sensitive indicators of herbicide damage although the maximum quantum yield in dark-adapted state (F_V/F_M) is the parameter most frequently measured. The effect of herbicides on $\Delta F/F'_M$ and F_V/F_M depends on the specific herbicide, the exposure time and the dose applied. Thus, simazine treatment reduced the $\Delta F/F'_M$ of *Pontederia cordata* leaves by up to 96%, while paraquat application reduced the $\Delta F/F'_M$ of *Arabidopsis* seedlings by 62% (Juneau *et al.*, 2007).

4. EFFECT OF BRASSINOSTEROIDS ON CROP PLANTS TREATED WITH PHOTOSYNTHESIS-INHIBITING HERBICIDES

4.1 Effect on crop plants treated with photosynthesis-inhibiting herbicides at the level of PS II

Evidence that BR treatment can ameliorate biotic and abiotic stress is discussed by Khripach *el al.* (2000), Bajguz and Hazat (2009) and also in some of the chapters of present book. However, only a few preliminary studies have addressed the protection afforded by BRs against pesticides and herbicides (Cutler, 1991; Krishna, 2003). Thus, BRs diminish herbicidal injury to rice caused by symetrin, (an s-triazine herbicide), butachlor and pretilachlor (both chloroacetamide herbicides that inhibit germination) and damage to wheat caused by simazine (*s*-triazine). This protection may be a result of reducing transpiration and herbicidal absorption and counteracting the herbicide-induced inhibition of photosynthesis (Hamada, 1986; Mandava, 1988).

Simultaneous measurements of Chl fluorescence and photosynthetic CO_2 assimilation (*A*) in *Vicia faba* leaves were studied during the early weeks of growth in order to evaluate the protective effect of EBR (1.47 or 1.96 kg h^{-1}) against damage caused by terbutryn (Terb) at pre-emergence. *V. faba* seeds were incubated for 24 h in EBR solutions (2.10^{-6} or 2.10^{-5} mM) and immediately sown. The highest dose of Terb strongly decreased *A*, F_V/F_M, NPQ and $\Delta F/F'_M$ during the first 3–4 weeks after plant emergence (Piñol and Simón, 2009). Terb also increased the basal quantum yield of non-photochemical processes (F_0/F_M), the degree of reaction centre closure (1- q_P) and the fraction of light absorbed in PSII antennae that was lost via thermal energy dissipation in the antennae ($1 - F'_V/F'_M$). The herbicide had significantly reduced plant growth at the end of the experiment. The application of EBR to *V. faba* seeds before sowing strongly diminished the effect of Terb on Chl fluorescence parameters and *A* values, which recovered 13 days after plant emergence and showed similar values to control plants. The protective effect of EBR on *A* was detected at photosynthetic photon flux density (PFD) of 650 μmol m^{-2} s^{-1} and the effect on $\Delta F/F'_M$ and photosynthetic electron transport (*J*) was detected under actinic lighting up to 1750 μmol m^{-2} s^{-1}. The highest dose of EBR also counteracted the decrease in plant growth caused by Terb, and plants registered the same growth values as controls (Piñol and Simón, 2009).

4.2 Effect on crop plants treated with photosynthesis-inhibiting herbicides at the level of PS I

Xia *et al.* (2006) studied the protective effect of EBR on the damage caused by paraquat, fluazifop-p-butyl and haloxyfop on *Cucumis sativus*. Fluazifop-p-butyl and haloxyfop inhibit the biosynthesis of fatty acid, paraquat is a photosynthesis-inhibiting herbicide that acts at the level of PS I. These authors reported a pronounced effect of paraquat on suppressing photosynthetic CO_2 assimilation and stomata conductance. However, EBR pre-treatment (0.1 mg L^{-1} sprayed on seedlings) alone had no effect on any of the gas exchange parameters. Chl fluorescence parameters were also affected by paraquat, since F_V/F_M was almost undetectable in treated cucumber leaves. EBR inhibited the strong decline of F_V/F_M but it was not restored to that of the control. Quantum efficiency of PSII, photochemical quenching coefficient q_P and NPQ decreased strongly in paraquat-treated plants. EBR pre-treatment only had a slight ameliorative effect on q_P but had no effect on the quantum efficiency of PSII or NPQ after paraquat treatment. EBR pre-treatment alleviated the adverse effects of fluazifop-p-butyl and haloxyfop herbicides on *A* and the decline of q_P caused by fluazifop-p-butyl. Haloxyfop treatment did not affect fluorescence parameters. EBR pre-treatment alone increased the capacity of photosynthetic CO_2 assimilation in cucumber control seedlings but did not significantly affect Chl fluorescence parameters. However, Piñol and Simón (2009) did not find a significant effect of EBR on photosynthetic CO_2 assimilation in *V. faba* plants. This discrepancy could be explained by the fact that EBR (10^{-2} mg L^{-1}) was applied to *V. faba* seeds before sowing but EBR (0.1 mg L^{-1}) was sprayed on cucumber seedlings.

5. NEW DATA ON THE EFFECT OF 24-EPIBRASSINOLIDE ON PLANTS TREATED WITH HERBICIDES THAT INHIBIT ELECTRON TRANSPORT AT THE PS II LEVEL

5.1 Effect of EBR on *Vicia faba* plants treated with cyanazine, an *s*-triazine herbicide

Cyanazine (Cyan) is an important selective triazine herbicide used to control several annual grassy and broad-leafed weeds in the fields of corn, soybean, cotton and other crops. Since it has a short soil half life (approximately 30 days), in cold climates Cyan is preferred over atrazine, one the most popular triazine herbicides (Ferrer *et al.*, 2000; Gu *et al.*, 2003). The combination of atrazine plus Cyan is commonly used in forestry situations, with most weed species appearing to be susceptible or moderately susceptible (Dixon *et al.*, 2006). The effect of Cyan on barley plants has been studied using Chl fluorescence induction for three weeks after herbicide application. An immediate response of plants and a regeneration of plants from Cyan action were detected. The reduction in yield was dose dependent (Matouškova *et al.*, 1999).

The protective effect of EBR on Cyan-induced damage has been studied in *V. faba* plants cultivated in growth chamber (16 h light/8 h darkness, day/night temperatures of 20/15°C and relative humidity between 65 and 60%), with a PFD of 700 µmol m^{-2} s^{-1}. Seeds of *V. faba* cv. Reina Blanca (Semillas Fitó, Barcelona, Spain) were imbibed for 24 h in 2.10^{-6} mM or 2.10^{-5} mM EBR solutions (EBR_1 and EBR_2, respectively) and then immediately sown in plastic pots with garden earth, perlite, and peat (2:7:1). Cyan treatment was applied at pre-emergence at the recommended dose of 0.75 kg ha^{-1} to pots containing EBR_1- and EBR_2-treated seeds 5 days after sowing. The Cyan doses applied followed Mohammadzamani *et al.* (2009) recommendations. A higher dose of Cyan was also tested but the effects on *V. faba* were drastic. There were five replicates per treatment. One set of plants was left untreated as a control. The other sets were treated only with one of the following: EBR_1, EBR_2 and Cyan as herbicide or EBR controls. Sampling began on the 6–7th day after plant emergence and was repeated every 4–5 days for the next 45 days.

Chl fluorescence was monitored using a pulse modulation fluorometer (PAM 101-103, Walz, Effeltrich, Germany), which provides a low-intensity, pulsed measuring beam (peak wavelength = 650 nm) from a light-emitting diode at frequencies of 1.6 or 100 kHz.

Detached leaves were dark-adapted for 30 min at room temperature before receiving a saturating pulse (PFD of 4500 µmol m^{-2} s^{-1}) of 700 ms duration from a halogen lamp pulse source (FL 103, Walz), in order to determine the maximum level of Chl fluorescence (F_M) in dark-adapted state. The minimal fluorescence, F_0, was measured using a modulated red radiation beam of 1 µmol m^{-2} s^{-1}. An actinic radiation source (Schott KL1500 halogen lamp, PFD 650 µmol m^{-2} s^{-1}) was then applied and a train of saturating pulses of white light (700 ms, PFD of 4500 µmol m^{-2} s^{-1}) was applied repetitively at 20 s intervals. The non-photochemical quenching coefficient (NPQ), was calculated as $(F_M - F'_M)/ F'_M$; the degree of reaction centre closure in light-adapted state, 1-q_P, was calculated as $1 - ((F'_M - F_S)/(F'_M - F'_0))$; the fraction of light absorbed in PSII antennae that is lost via thermal energy dissipation in the antennae, $1 - F'_V/F'_M$, was calculated as $1 - ((F'_M - F'_0)/F'_M)$ The maximal quantum yield of PS II (F_V/F_M) and the effective quantum yield ($\Delta F/F'_M$) were also calculated following Schreiber and Bilger (1993). F'_0 was determined immediately after turning off the photosynthetic radiation, following Oxborough and Baker (1997).

Measurements of *A* were performed on intact plants using an infrared gas analyzer portable photosynthesis system LICOR-6200, (LI-COR Inc, Lincoln, Nebraska, USA), working in a closed circuit. *A* was measured at 20ºC, at a CO_2 concentration of 350–370 ppm, air flux of 500 µmol s^{-1} and relative humidity 40–50%. Conditions of measurements were close to environmental conditions in the growth chamber. Chl fluorescence and *A* measurements were performed on the most recent fully expanded leaf.

Statistical analysis: Mean values of fluorescence parameters, *A*, plant growth and plant biomass in shoots and roots were compared by a one way ANOVA; differences between treatments were compared with the LSD test.

Table 1. Effect of cyanazine and 24-epibrassinolide on growth parameters of *Vicia faba* plants at the end of the experiment

Treatment	**Plant length (cm)**	**Dry weight of roots (g)**	**Dry weight of shoots (g)**	**Number of flowers**
Control	29.00 ± 0.79^a	3.02 ± 0.18^a	3.69 ± 0.28^a	8.25 ± 3.64^a
Cyan 0.75 kg. ha^{-1}	22.80 ± 2.25^b	1.11 ± 0.20^c	1.40 ± 0.20^c	2.60 ± 0.96^a
Cyan 0.75 kg. ha^{-1} + EBR_1 2.10^{-6} mM	23.83 ± 1.77^{ab}	1.64 ± 0.14^b	2.04 ± 0.15^b	5.67 ± 1.73^a
Cyan 0.75 kg. ha^{-1} + EBR_2 2.10^{-5} mM	24.20 ± 0.62^{ab}	1.75 ± 0.12^b	2.09 ± 0.14^b	6.00 ± 1.85^a

Data represent mean values ± standard errors of five replicates. Different letters in the same data column represent significant differences at the 5% level.

Table 1 shows that Cyan treatment reduced plant growth. Thus, plant length, and the dry weight of roots and shoots decreased significantly in Cyan-treated plants, but the number of flowers were the same as in the control group.

In Cyan plus EBR_2 and in Cyan plus EBR_1 treatments, the decrease in plant growth parameters was less marked than in plants treated with herbicide alone. The strongest protective effect of EBR on *V. faba* growth corresponded to the highest doses of EBR, which improved plant length and the dry weight of shoots and roots.

Photosynthetic CO_2 assimilation was also affected by Cyan treatment. Thus, *A* decreased by 53% in Cyan plants with respect to controls at the first sampling (Figure 1). *A* values in Cyan-treated plants remained lower than untreated plants during the growing period, and at the last sampling *A* values were still 38% lower than controls.

As reported previously (Piñol and Simón, 2009), *A* values of EBR-treated plants showed the same values as control plants and they were omitted from Figure 1.

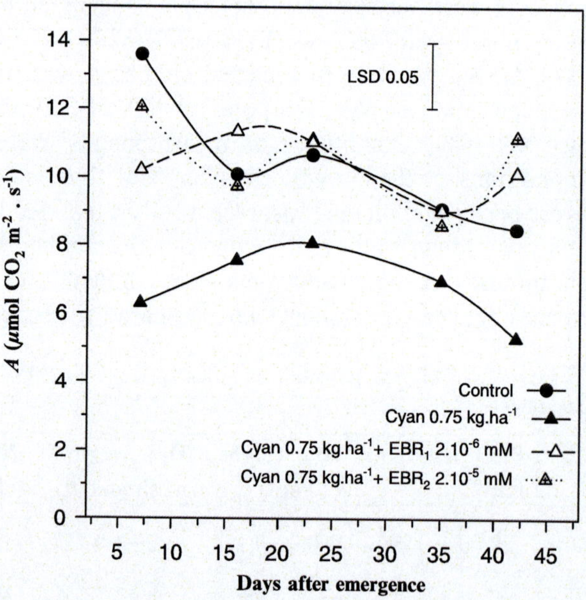

Figure 1. Effect of cyanazine 0.75 kg ha^{-1} (Cyan) and Cyan plus 24-epibrassinolide at doses of 2.10^{-6} and 2.10^{-5} mM (EBR_1 and EBR_2, respectively) on photosynthetic CO_2 assimilation (*A*) of *Vicia faba* leaves during 43 days after plant emergence. Each value represents the mean values of five replicates.

In plants treated with Cyan plus EBR_1 *A* decreased by only 24% and plants treated with Cyan plus EBR_2 showed no significant differences compared to controls at the first sampling (Figure 1). Thus, EBR_1 and EBR_2 treatments maintained the rate of *A* 90% and 61%, respectively, higher than plants treated with Cyan only, at the first sampling. At the end of the experiment, EBR_1 maintained the rate of *A* 96% higher than plants treated with Cyan alone.

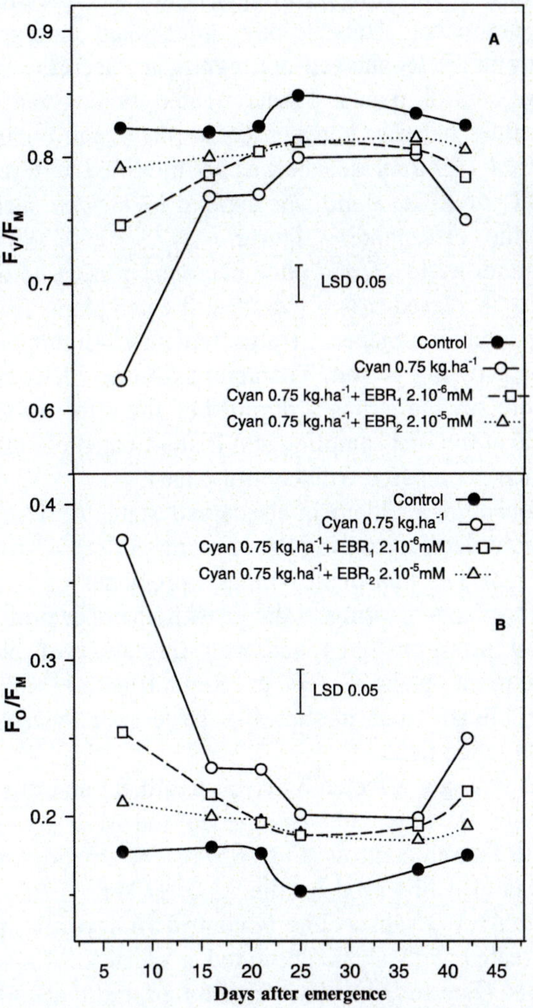

Figure 2. Effect of cyanazine 0.75 kg ha^{-1} (Cyan) and Cyan plus 24-epibrassinolide at doses of 2.10^{-6} and 2.10^{-5} mM (EBR_1 and EBR_2, respectively) on the maximum quantum yield F_V/F_M (**A**) and on the basal quantum yield of non-photochemical processes F_0/F_M (**B**) of *Vicia faba* leaves during 43 days after plant emergence. Each value represents the mean values of five replicates.

Control plants maintained an almost constant F_V/F_M ratio throughout the experiment (Figure 2A). Cyan treatment caused a decrease in this parameter of 26%, 7 days after the emergence of the plants. However, this initial decrease diminished rapidly and 16 days after plant emergence Cyan plants showed F_V/F_M values 7% lower than control plants. This small difference was maintained during the growing period. The application of EBR_1 and EBR_2 to *V. faba* seeds before sowing strongly diminished the effect of Cyan on fluorescence parameters. Thus, 7 days after plant emergence, plants treated with Cyan plus EBR_2 showed not significant decrease in the F_V/F_M ratio compared to control plants. Plants treated with Cyan plus EBR_1 presented F_V/F_M values halfway between Cyan plants and controls (Figure 2A). As occurred in *A*, Chl fluorescence was not modified with respect to the controls when EBR_1 or EBR_2 alone was applied to *V. faba* seeds and data were omitted from the corresponding figures.

The basal quantum yield of non-photochemical processes in the dark-adapted state, F_0/F_M, increased strongly in Cyan-treated plants (Figure 2B) at the first sampling after emergence. It also remained slightly higher than controls during the growing period. The application of EBR_1 and EBR_2 to *V. faba* seeds, before sowing strongly diminished the effect of Cyan on the basal quantum yield at the first sampling and EBR_2-treated plants showed no significant differences, compared with control plants.

The effective quantum yield in light-adapted state, $\Delta F/F'_M$, of control plants showed a slight decrease during the growing period (43 days) (Figure 3A). This finding indicates that photosynthesis processes of *V. faba* plants were not optimal for plants growing in the growth chamber probably due to the PDF rate (700 μmol m^{-2} s^{-1}) and/or a limitation of pots to root development and nutrient uptake. *V. faba* cv. Reina Blanca is well adapted to the Mediterranean climate and it shows a large root system and high photosynthesis rates in the field.

The $\Delta F/F'_M$ was strongly affected by Cyan treatment and this effect was maintained for longer: 21 days after emergence, the $\Delta F/F'_M$ was still 23% lower in Cyan plants than in controls (Figure 3A). The ameliorative effect of EBR treatments was also observed on the $\Delta F/F'_M$. Thus, EBR_1 and EBR_2 plants showed mean $\Delta F/F'_M$ values close to control 16 days after emergence, while the plants treated with Cyan presented a mean $\Delta F/F'_M$ value 40% lower than controls (Figure 3A). The fraction of light absorbed in the PSII antennae that is lost via thermal energy dissipation in the antennae, $1 - F'_V/F'_M$, was increased by 50% in Cyan-treated plants at the first sampling, but both EBR_1 and EBR_2 treatments counteracted the enhanced caused by Cyan, since plants showed no significant differences with respect to control (data not shown).

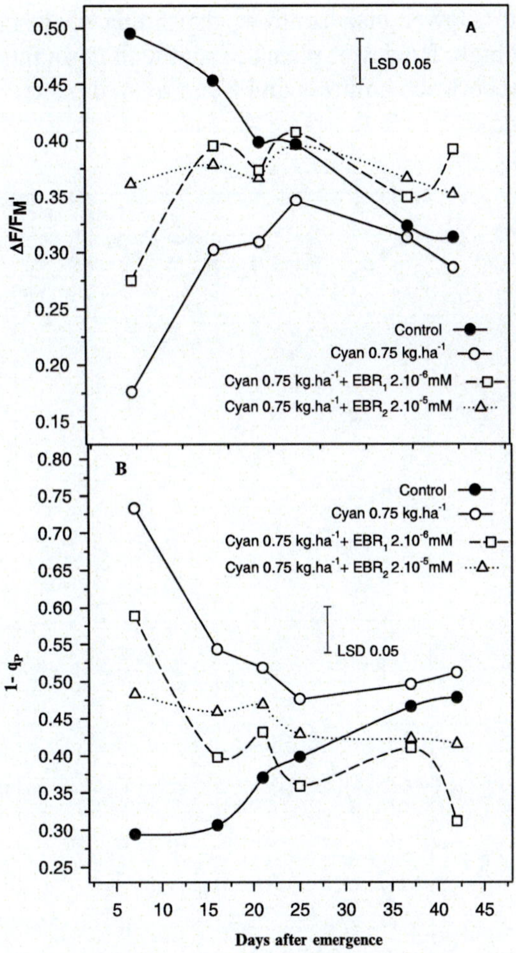

Figure 3. Effect of cyanazine 0.75 kg ha^{-1} (Cyan) and Cyan plus 24-epibrassinolide at doses of 2.10^{-6} and 2.10^{-5} mM (EBR_1 and EBR_2, respectively) on the effective quantum yield, $\Delta F/F'_M$ (**A**) and on the $1 - q_P$ (**B**) of *Vicia faba* leaves during 43 days after plant emergence. Each value represents the mean values of five replicates.

The palliative effect of EBR_1 and EBR_2 on Cyan damage on *V. faba* plants was also observed in the proportion of reaction center closure ($1 - q_P$) (Figure 3B) since EBR_2 reduced by a half the high level of this parameter caused by Cyan at the first sampling. NPQ also revealed the palliative effect of EBR in *V. faba* leaves (data not shown).

The protective effect of EBR was also reflected in the measurements of $\Delta F/F'_M$ in plants subjected to an increasing range of actinic light (up to 1750 μmol m^{-2} s^{-1}) (Figure 4A). Thus, the $\Delta F/F'_M$ of Cyan plants showed lower values than control plants in this range of actinic light. These values were from 59% to 86% lower than controls, depending on whether the actinic light was low or high. However, plants treated with Cyan plus EBR_2 showed $\Delta F/F'_M$ halfway between controls and Cyan treated plants. The protective

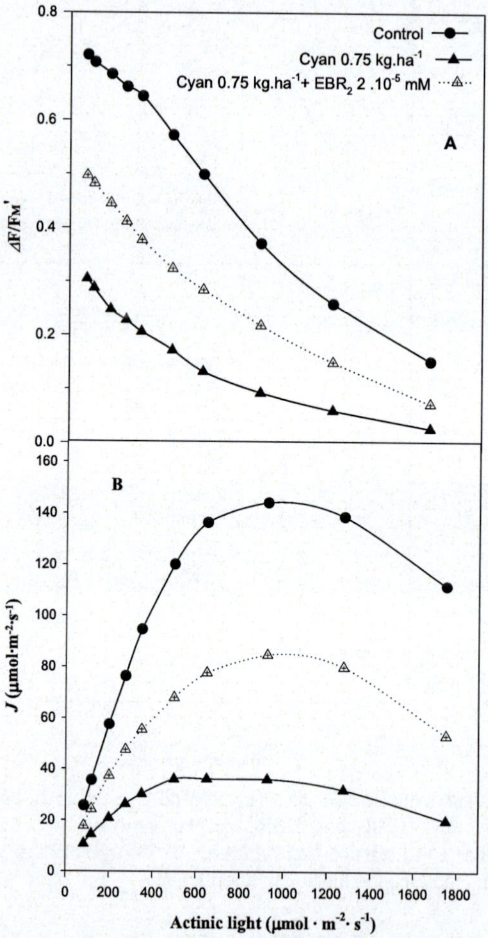

Figure 4. Effect of cyanazine 0.75 kg ha^{-1} (Cyan) and Cyan plus 24-epibrassinolide at dose of 2.10^{-5} mM on (**A**) effective quantum yield, $\Delta F/F'_M$ and on (**B**) photosynthetic electron transport, *J*, of *Vicia faba* leaves 22 days after plant emergence. Data correspond to one representative experiment.

effect of EBR_2 on the photosynthetic electron transport (*J*) inhibited by Cyan in *V. faba* leaves was strongest at PFD of 500–1800 µmol m^{-2} s^{-1} (Figure 4B). This finding indicates that EBR may protect plants from Cyan damage in the Mediterranean field conditions, since the experimental fields at the University of Barcelona receive PFD ranging from about 750 to 1750 μmol m^{-2} s^{-1} for most of the year (Munné-Bosch and Alegre, 2000). In summer PFD may exceed these values. In the present study the photosynthetic electron transport decreased slightly even in control plants at high PFD, possibly due to other environmental stresses (Figure 4B). Thus, the higher EBR concentration did not totally protect herbicide-treated plants, although it improved the response of $\Delta F/F'_M$. This indicates that the damage to the photosynthetic apparatus had been at least partially prevented or repaired. Thus *A* values and the final growth of plants were increased.

5.2 Effect of EBR on *Vicia faba* plants treated with simazine, an *s*-triazine herbicide

The herbicide simazine (Sim) is a widely used selective systemic herbicide. Sim is phytotoxic to many non-target species at concentrations below the recommended dose. In some environmental conditions Sim can remain for a long time in the soil and still be toxic to plants. Despite its phytotoxicity many plant species respond to repeated use by developing tolerance and, in some cases, resistance (Strandberg and Scott-Fordsmand, 2002).

The protective effect of EBR on the damage caused by Sim has been tested in the experimental fields at the Faculty of Biology of the University of Barcelona. *V. faba* seeds were imbibed for 24 h in 2.10^{-6} mM or 2.10^{-5} mM EBR solutions (EBR_1 and EBR_2, respectively) and then immediately sown in plastic pots with garden earth and sand (2:1). Pots were placed outdoor under a removable tunnel of plastic which protected them from rain and low temperatures. Plants were regularly irrigated during the growing period. Sim was applied at pre-emergence at the recommended dose of 1.5 kg ha^{-1} to pots containing EBR_1- and EBR_2-treated seeds 5 days after sowing. There were five replicates per treatment. One set of plants was left untreated as controls. The other sets were treated only with one of the following: EBR_1, EBR_2, or Sim, as EBR or herbicide controls. Sampling was begun on the 4th day after plant emergence. Chl fluorescence parameters were measured every 4–5 days for the next 38 days. *A* was measured every 2 weeks for the next 2 months. *A* measurements, fluorescence parameters and statistical analysis were as described above (see section 5.1 of this chapter).

Table 2 showed that Sim treatment reduced plant growth. Thus, plant length, dry weight of roots, dry weight of shoots and number of flowers significantly decreased in Sim-treated plants, compared to controls.

In Sim plus EBR_2 and in Sim plus EBR_1 treatments, the decrease in plant growth parameters was less marked than in plants treated with herbicide alone. However, none of the plant growth parameters studied reached the values of control plants. The strongest protective effect of EBR on *V. faba* growth corresponded to the highest doses of EBR, in which the length of plants and the dry weigh of shoots improved.

The effect of Sim on plant growth was also detected in the photosynthetic parameters studied. Thus, *A* decreased by 46% in Sim-treated plants with respect to controls at the first sampling (Figure 5). *A* values in Sim-treated plants maintained lower values than control plants throughout the growing period and at the last sampling (58 days after plant emergence) *A* values were still 19% lower than controls. The effect of Sim on *A* was persistent, since 44 days after emergence Sim-treated plants showed *A* values close to the first sampling and 50% lower than control plants. The lower dose of EBR clearly protected *A* from herbicide damage. Thus, in the Sim plus EBR_1 treatments, *A* also decreased by 45% at the first sampling compared to controls but after the second sampling *A* values showed no significant differences with respect to the controls. At the last samplings Sim plus EBR_1 plants showed the same *A* values as control plants (Figure 5).

Table 2. Effect of simazine and 24-epibrassinolide on growth parameters of *Vicia faba* plants at the end of the experiment

Treatment	Plant length (cm)	Dry weight of roots (g)	Dry weight of shoots (g)	Number of flowers
Control	70.00 ± 2.00^{a}	1.82 ± 0.25^{a}	7.14 ± 0.38^{a}	12.37 ± 0.91^{a}
Sim 1.50 kg. ha^{-1}	48.36 ± 1.54^{c}	1.24 ± 0.15^{b}	3.68 ± 0.22^{c}	7.50 ± 0.91^{b}
Sim 1.50 kg. ha^{-1} + EBR_1 2.10^{-6} mM	49.78 ± 1.92^{bc}	1.08 ± 0.14^{b}	3.52 ± 0.18^{c}	8.59 ± 0.81^{b}
Sim 1.50 kg. ha^{-1} + EBR_2 2.10^{-5} mM	53.93 ± 2.21^{b}	1.25 ± 0.16^{b}	4.50 ± 0.25^{b}	8.30 ± 0.87^{b}

Data represent mean values ± standard errors of five replicates. Different letters in the same data column represent significant differences at the 5% level.

Figure 6 shows the distribution of classes of *V. faba* plants according to the rate of *A* at the last sampling. All control plants exhibited *A* values above 8 μmol CO_2 m^{-2} s^{-1}, while 20% of Sim-treated plants showed *A* values between 3 and 7.99 μmol CO_2 m^{-2} s^{-1}, and 4% of Sim-treated plants presented *A* values lower than 3 μmol CO_2 m^{-2} s^{-1} (Figure 6A and 6B). Even the lower doses of BR (EBR_1) induced a strong protective effect, since only 12% of plants presented *A* values between 3 and 7.99 μmol CO_2 m^{-2} s^{-1} and the percentage of plants that showed the highest values of *A* was greater than in the control groups (Figure 6C).

When the same experimental approach was assayed with another cv of *V. faba* (Luz de Otoño, Semillas Fitó, Barcelona, Spain) the results were quite similar (Figure 7). The ameliorative effect of EBR_1 and EBR_2 on *A* values in Sim-treated plants was clear both at 35 days and at 57 days after plant emergence.

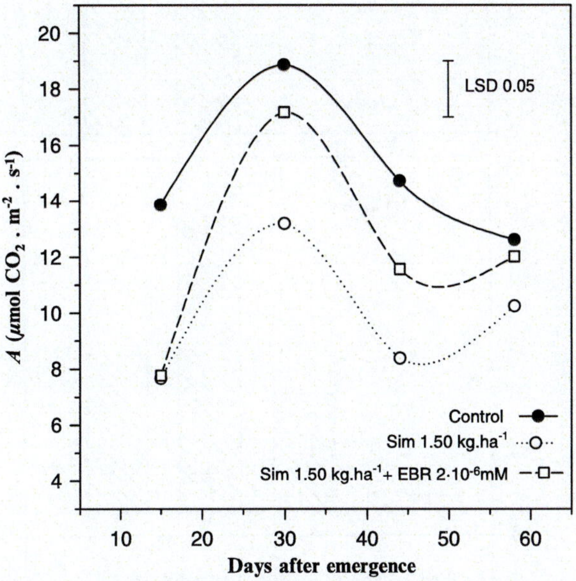

Figure 5. Effect of simazine 1.5 kg ha^{-1} (Sim) and Sim plus 24-epibrassinolide at doses of 2.10^{-6} mM (EBR) on photosynthetic CO_2 assimilation (*A*) of *Vicia faba* leaves during 58 days after plant emergence. Each value represents the mean values of five replicates.

Fluorescence parameters in dark-adapted state were less affected than *A* or Chl fluorescence parameters in light-adapted state. Thus, F_V/F_M showed a strong decrease at the first sampling in Sim-treated plants but at the second sampling the data did not differ from the controls. However, the persistent effect of Sim was also observed in F_V/F_M since 32 and 38 days after emergence a strong decrease in F_V/F_M was observed (Figure 8A). EBR_1 did not reduce the initial decrease of F_V/F_M caused by the herbicide but it totally counteracted the damage effect caused by Sim detected 32 and 38 days after emergence. As can be seen in Figure 8B, Sim strongly increased F_0/F_M, which recovered at the second sampling. A delayed effect of the herbicide was also detected at 32 and 38 days after emergence. EBR_1 did not counteract the Sim early effects but it cancelled the delayed effects observed at the end of period of Chl fluorescence measurements.

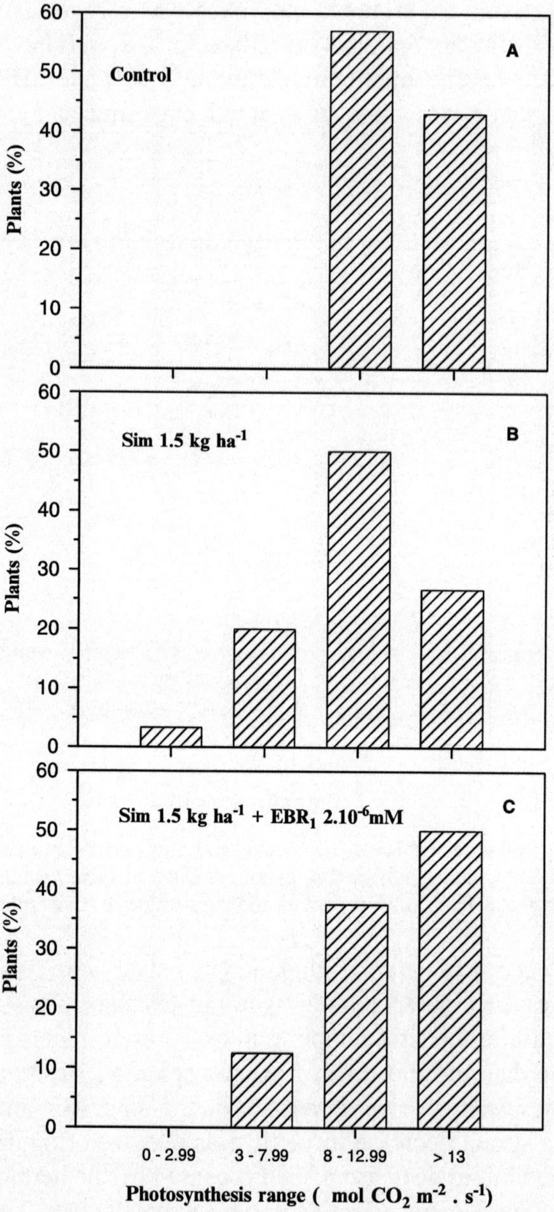

Figure 6. Effect of simazine 1.5 kg ha^{-1} (Sim) and Sim plus 24-epibrassinolide at doses of 2.10^{-6} mM (EBR_1) on the distribution of classes according the photosynthetic CO_2 assimilation (*A*) of *Vicia faba* cv. Reina Blanca leaves 58 days after plant emergence. Each treatment included 15 replicates.

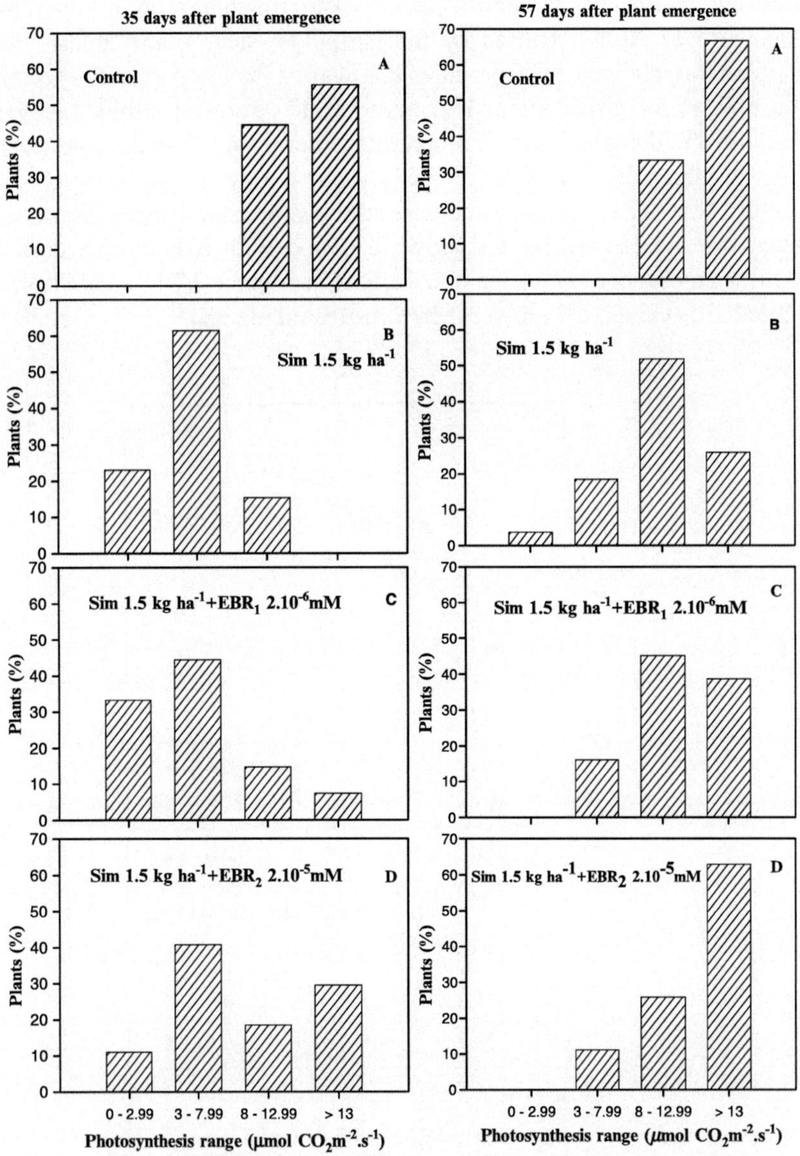

Figure 7. Effect of simazine 1.5 kg ha^{-1} (Sim) and Sim plus 24-epibrassinolide at doses of 2.10^{-6} mM and 2.10^{-5} mM (EBR_1 and EBR_2, respectively) on the distribution of classes according to the photosynthetic CO_2 assimilation (*A*) of *Vicia faba* cv. Luz de Otoño leaves 35 and 57 days after plant emergence. Each treatment included 10 replicates.

Chl fluorescence in light-adapted state was more sensitive to Sim. Thus, $\Delta F/F'_M$ showed values always lower than controls throughout the growing period and 38 days after plant emergence Sim-treated plants showed values that were 43% lower than control plants. EBR_1 treatment did not overcome the decrease of $\Delta F/F'_M$ caused by the herbicide, since plants treated with EBR_1 plus Sim showed $\Delta F/F'_M$ values that were 25% lower than the controls at the end of the experiment. However, $\Delta F/F'_M$ showed strong recovery, since it was 32% higher than Sim controls (Figure 9A). The damage caused by Sim and the palliative effect of EBR_1 application was also observed in 1-q_P (Figure 9B). The important increase in the proportion of reaction center closure was showed during the growing period with maximum values 32 days after emergence. EBR_1 palliated this effect since EBR_1-treated plants presented 1-q_P values 23% lower than Sim-treated plants.

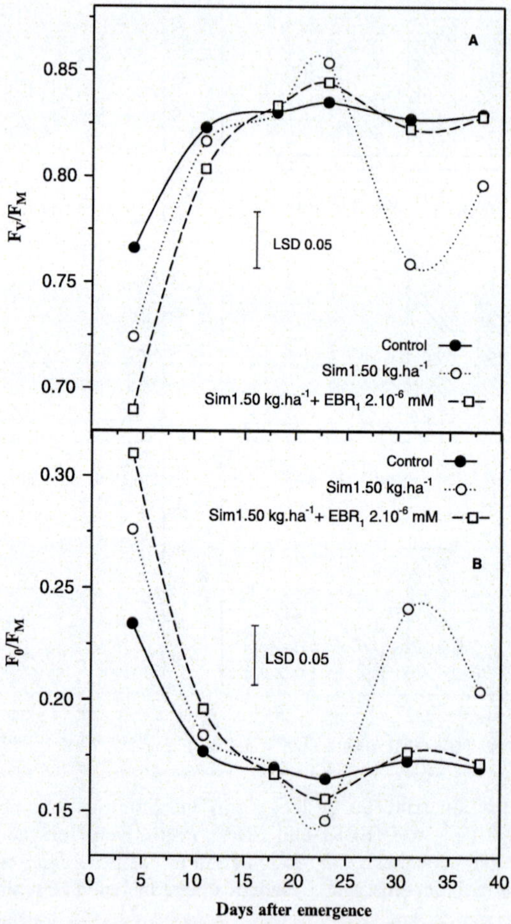

Figure 8. Effect of simazine 1.50 kg ha^{-1} (Sim) and Sim plus 24-epibrassinolide at dose of 2.10^{-6} mM (EBR_1) on the variable fluorescence ratio F_V/F_M (**A**) and on the basal quantum yield of non-photochemical processes F_0/F_M (**B**) of *Vicia faba* leaves during 38 days after plant emergence. Each value represents the mean values of five replicates.

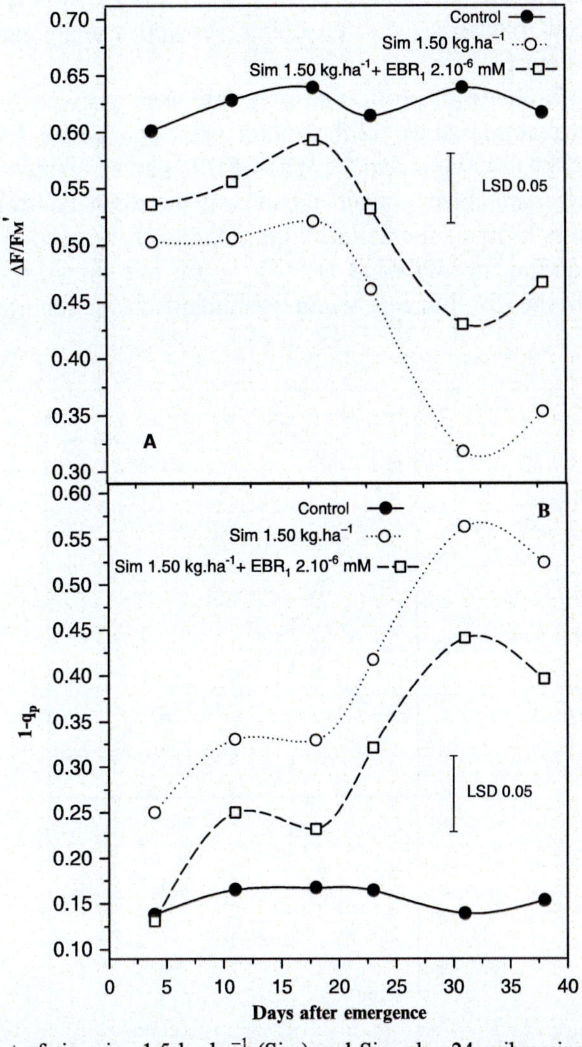

Figure 9. Effect of simazine 1.5 kg ha^{-1} (Sim) and Sim plus 24-epibrassinolide at doses of 2.10^{-6} mM (EBR_1 and EBR_2) on the effective quantum yield, $\Delta F/F'_M$ (**A**) and on the $1 - q_P$ (**B**) of *Vicia faba* leaves during 38 days after plant emergence. Each value represents the mean values of five replicates.

The effect of Sim on $\Delta F/F'_M$ of *V. faba* leaves depended on the actinic light applied (Figure 10A). Thus, when the actinic light was lower than 500 μmol m^{-2} s^{-1} Sim reduced the $\Delta F/F'_M$ by about 35% compared to the controls. This reduction slowly decreased when the actinic light was between 500 and 1000 μmol m^{-2} s^{-1}, and it was strongly diminished at actinic light above 1000 μmol m^{-2} s^{-1}. The protective effect of EBR_2 was observed throughout the range of PFD tested, since the BR maintained $\Delta F/F'_M$ values close to the controls. Only when PFD was below 250 µmol m^{-2} s^{-1} did Sim plus EBR_2-treated plants showed a slight decrease (8%) respect to the controls.

The photosynthetic electron transport rate also showed the persistent effect of Sim treatment since 62 days after plant emergence *J* was strongly diminished (75% at 950 µmol m^{-2} s^{-1}) and EBR_2 had an positive effect since *J* values were maintained almost a half way between controls and Sim-treated plants in most of the actinic light conditions assayed (Figure 10B). Data corresponding to NPQ and $1-F'_V/F'_M$, are not shown but they were consistent with the Chl fluorescence in light-adapted state reported.

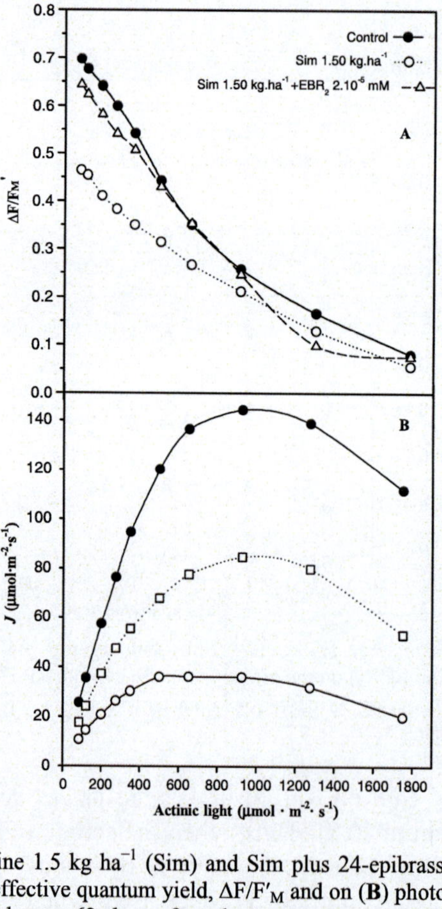

Figure 10. Effect of simazine 1.5 kg ha^{-1} (Sim) and Sim plus 24-epibrassinolide at dose of 2.10^{-5} mM (EBR_2) on (**A**) effective quantum yield, $\Delta F/F'_M$ and on (**B**) photosynthetic electron transport, *J*, of *Vicia faba* leaves 62 days after plant emergence. Data correspond to one representative experiment.

5.3 Effect of EBR on *Vicia faba* plants treated with methabenzthiazuron, a urea derivative herbicide

Methabenzthiazuron (MBT) is one of the most common herbicide used in Mediterranean areas for the control of weeds in cereal and bean crops. It is an urea derivative that, when applied at the recommended doses, strongly inhibits photosynthetic electron transport at the PSII level (Fedtke, 1982). MBT (0.25–0.40 g m^{-2}) reduced photosynthetic O_2-evolution, Chl fluorescence parameters such as F_V/F_M, $\Delta F/F'_M$, q_P, NPQ and aerial biomass of *V. faba* plants, under Mediterranean field conditions. However, photosynthesis and Chl fluorescence recovered 1 month after plant emergence (Vidal *et al.*, 1995).

The protective effect of EBR on the damage caused by MBT, in *V. faba* plants was tested in the experimental fields of the Faculty of Biology of the University of Barcelona, as described in 5.2. above. MBT was applied at doses of 1.4, 2.8 and 5.6 kg h^{-1} and EBR was applied at doses of 2.10^{-6} mM and 2.10^{-5} mM EBR (EBR_1 and EBR_2, respectively).

At the doses tested EBR did not alleviate the damage caused by MBT in *V. faba* (Table 3): MBT plus EBR-treated plants showed no significant differences from MBT-treated plants in length or dry weight of roots and shoots, which were the parameters most strongly affected by MBT.

Table 3. Effect of methabenzthiazuron and 24-epibrassinolide on growth parameters of *Vicia faba* plants at the end of the experiment

Treatment	Plant length (cm)	Dry weight of roots (g)	Dry weight of shoots (g)	Number of flowers
Control	72.00 ± 2.00^a	1.82 ± 0.25^a	7.14 ± 0.38^a	12.37 ± 0.91^a
MBT 5.6 kg. ha^{-1}	55.20 ± 3.46^b	1.78 ± 0.38^a	5.64 ± 0.64^b	10.25 ± 1.15^a
MBT 5.6 kg. ha^{-1} + EBR_1 2.10^{-6} mM	50.22 ± 3.95^b	1.40 ± 0.20^a	5.22 ± 0.53^b	9.93 ± 1.95^a
MBT 5.6 kg. ha^{-1} + EBR_2 2.10^{-5} mM	58.18 ± 4.20^b	1.62 ± 0.36^a	5.70 ± 0.47^b	10.76 ± 0.83^a

Data represent mean values ± standard errors of six replicates. Different letters in the same data column represent significant differences at the 5% level.

All the doses of MBT applied also decreased the *A* values (by 26% when applied at 5.6 kg ha^{-1}) and this decrease was maintained almost constant during the first month after plant emergence. However, EBR did not significantly modify the effect of MBT. Chl fluorescence parameters were also affected by MBT but the two doses of EBR assayed did not have a significant palliative effect on the herbicide-induced damage. The experiment was repeated using another cv of *V. faba* but the results were similar (data not shown).

5.4 Effect of EBR on *Vicia faba* plants treated with metribuzin, an *as*-triazinone herbicide

Metribuzin (MET) belongs to the group of *as*-triazinone herbicides which also inhibit photosynthetic electron transport at the PSII level. It is one the most active herbicides, only requiring an application of 0.3–0.7 kg h^{-1} (Fedtke, 1982). The protection afforded by EBR against the damage caused by MET in *V. faba* plants was tested as described in 5.2 above. MET was applied at the recommended dose of 0.49 kg h^{-1} and EBR was applied at doses of 2.10^{-6} mM and 2.10^{-5} mM EBR (EBR_1 and EBR_2, respectively).

MET significantly decreased plant growth of *V. faba* plants, measured as the length of plants and the dry weigh of roots and shoots (Table 4). The application of EBR did not modify the growth inhibition caused by MET, as measured at the end of the experiment (75 days after plant emergence).

Table 4. Effect of metribuzin and 24-epibrassinolide on growth parameters of *Vicia faba* plants at the end of the experiment

Treatment	**Plant length (cm)**	**Dry weight of roots (g)**	**Dry weight of shoots (g)**	**Number of flowers**
Control	65.00 ± 3.25^a	2.02 ± 0.42^a	6.18 ± 0.52^a	16.20 ± 2.87^a
MET 0.49 kg. ha^{-1}	54.74 ± 4.86^b	1.34 ± 0.25^b	4.11 ± 0.74^b	14.82 ± 3.25^a
MET 0.49 kg. ha^{-1} + EBR_1 2.10^{-6} mM	53.88 ± 4.91^b	1.28 ± 0.28^b	4.14 ± 1.01^b	15.91 ± 2.78^a
MET 0.49 kg. ha^{-1} + EBR_2 2.10^{-5} mM	56.27 ± 4.08^b	1.33 ± 0.32^b	4.48 ± 0.47^b	13.75 ± 3.83^a

Data represent mean values ± standard errors of five replicates. Different letters in the same data column represent significant differences at the 5% level.

Photosynthetic CO_2 assimilation was strongly inhibited by MET: *A* values decreased by 50% at the first sampling. 15 and 30 days after plant emergence, *A* values were still 13% lower than controls. However, when MET was applied to EBR-treated plants *A* values showed the same values as MET-treated plants (data not shown).

Chl Fluorescence measurements during the growth of *V. faba* plants also indicated the damage caused by MET. This was not modified by EBR treatments. $\Delta F/F'_M$, was diminished by MET until 40 days after the plant emergence, and the control values were only reached 2 months after plant emergence. Neither EBR_1 nor EBR_2 afforded protection against the effect of the herbicide (data not shown).

6. COMPARATIVE EFFECTS OF 24-EPIBRASSINOLIDE AND PHOTOSYNTHESIS-INHIBITOR HERBICIDES AT THE PS II LEVEL

The application of s-triazine (Cyan, Sim), urea derivative (MBT) and *as*-triazinone (MET) herbicides at the recommended doses before plant emergence clearly reduced photosynthetic CO_2 assimilation and plant growth parameters. The treatments also affected Chl fluorescence parameters in both dark- and light-adapted states of *V. faba* leaves during the early weeks of growth (Figures 1–10, Tables 1–4). The $\Delta F/F'_M$ was the fluorescence parameter most strongly reduced: it decreased by between 65% and 94%, depending on the age of plants and the photosynthetic actinic lighting applied. As $\Delta F/F'_M$ computes the proportion of light absorbed by Chl associated with PSII, which is used in photochemistry (Maxwell and Johnson 2000), the data indicated that these herbicides impaired PSII photochemistry in a light-adapted state. This fluorescence parameter is considered the best measure of herbicide damage to *V. faba* plants (Vidal *et al.*, 1995). The decrease in $\Delta F/F'_M$ was also associated with a drastic increase in $1 - F'v/F'_M$, the fraction of light absorbed in PSII that is lost via thermal energy dissipation in the antennae, and in $1-q_P$ which gives an indication of the proportion of PSII reaction centres that are closed. These changes indicate inhibition of the electron transport rate. This inhibition is the predicted effect of herbicides that bind strongly at the Q_B side (Zimmerman *et al.*, 2006) and it blocks PS II. This prevents the re-oxidation of the primary acceptor Q_A and, consequently, electron transport to PSI. Thus, herbicides inhibit PSII function by displacing Q_B from its binding site on the D1 protein of this system (Xiong *et al.*, 1997).

Chl fluorescence parameters relating to the dark-adapted state were also affected. Thus, F_V/F_M, showed a considerable decrease, albeit weaker than that indicated by fluorescence parameters related to the light-adapted state. The basal quantum yield of non-photochemical processes in PSII, F_0/F_M, also increased strongly at the first sampling, which was attributed mainly to an increase in F_0 (data not shown) since F_M values did not change significantly in these plants. The increase in F_0 may be due to the initially damaged reaction centres of PSII (Lazár 1999).

EBR_1 and EBR_2 applied to seeds before sowing protected *V. faba* plants from the damage caused by Cyan and Sim, since they both improved plant growth (Table 1 and 2), photosynthetic CO_2 assimilation (Figures 1, 5–7) and reduced the herbicide effects on all fluorescence parameters tested (Figures 2–4 and 8–10). The protective effect of EBR was clearly shown even in the case of Sim, whose effects on the crop were delayed (Figures 8–10). Thus, 1 month after plant emergence, $\Delta F/F'_M$, of *V. faba* leaves decreased by half in EBR_1-treated plants with respect to Sim-treated plants (Figure 9A).

In addition, EBR induced the recovery from damage caused by s-triazine herbicides in *V. faba* plants, as detected by Chl fluorescence and photosynthetic CO_2 assimilation. This pattern was previously found in *V. faba* plants treated with Terb (Piñol and Simón, 2009). Cyan and Terb reduced *A* and Chl fluorescence parameters during the first 3 weeks after emergence, but Cyan seemed to cause stronger damage since it was still detected 43 after emergence. The effects of Sim on *V. faba* plants were more persistent, since at the last sampling (58 days after emergence) plants maintained lower levels of *A* and $\Delta F/F'_M$. Nevertheless, the protective effect of EBR was apparent in plants treated with each of the three herbicides.

As occurs with Terb, in Cyan- and Sim-treated plants also the Chl fluorescence parameters related to the light-adapted state were most affected by EBR treatments. The protective effect of EBR on $\Delta F/F'_M$, was even more apparent in response to a range of actinic lights (Figures 4A and 10A). Table 5 summarizes the variations with respect to the control of some of the more significant parameters of photosynthesis in response to herbicides and EBR application.

However, at the doses applied EBR did not counteract the damage caused by MBT, a urea derivative herbicide, or MET, an *as*-triazinone herbicide. MBT and MET also inhibited photosynthetic electron transport at Q_B site. MBT was tested at three concentrations but at the highest dose, when it induced a moderate decrease of *A* (23%) and a strong decrease of $\Delta F/F'_M$ (73%), EBR did not palliate the damage on *V. faba* (Table 5).

Table 5. Comparative efffects of cyanazine (Cyan), simazine (Sim), methabenzthiazuron (MBT) and metribuzin (MET) at doses of 0.75, 1.5, 5.6 and 0.49 kg ha^{-1} respectively on photosynthetic CO_2 assimilation (*A*), fluorescence parameters and days of recovery of *Vicia faba* plants and the palliative effects caused by the applications of 24-epibrassinolide (EBR) to seeds.

Measures at the first sampling % of the control						Recovery of $\Delta F/F'_M$ in days		Measures at the last sampling % of the control	
A		F_V/F_M		$\Delta F/F'_M$				$\Delta F/F'_M$	
Cyan	*Cyan + EBR₂*	*Cyan*	*Cyan + EBR₂*	*Cyan*	*Cyan + EBR₂*	*Cyan*	*Cyan + EBR₂*	*Cyan*	*Cyan + EBR₂*
↓ 53%	↓ 11%	↓ 26%	↓ 3%	↓ 66%	↓ 29%	21 days	16 days	–	–
Sim	*Sim + EBR₁*	*Sim*	*Sim + EBR₁*	*Sim*	*Sim + EBR₁*	*Sim*	*Sim + EBR₁*	*Sim*	*Sim + EBR₁*
↓ 46%	↓ 45%	↓ 5%	↓ 5%	↓ 16%	↓ 10%	No	No	↓ 43%	↓ 25%
MBT	*MBT + EBR₂*	*MBT*	*MBT + EBR₂*	*MBT*	*MBT + EBR₂*	*MBT*	*MBT + EBR₂*	*MBT*	*MBT + EBR₂*
↓ 23%	↓ 23%	↓ 12%	↓ 12%	↓ 73%	↓ 73%	21 days	21 days	–	–
MET	*MET + EBR₂*	*MET*	*MET + EBR₂*	*MET*	*MET + EBR₂*	*MET*	*MET + EBR₂*	*MET*	*MET + EBR₂*
↓ 50%	↓ 38%	↓ 10%	↓ 11%	↓ 31%	↓ 30%	58 days	58 days	–	–

EBR_1: 2.10^{-6} mM; EBR_2: 2.10^{-5} mM

7. HOW DO BRASSINOSTEROIDS ALLEVIATE THE DAMAGE EFFECT CAUSED BY HERBICIDES?

The mechanism whereby the EBR counteracts the herbicide treatment is not clear. However, we can suggest possible explanations. EBR could affect the herbicide inhibition of PSII by displacement of Q_B from its binding site on D1 protein of PSII. This protein is degraded when the photosynthetic system cannot process the energy of photons accumulated, but little is known about the PSII repair process, either at the level of protein synthesis, insertion, and concomitant assembly of the D1 protein or later functional post-translational assembly steps (Zang *et al.*, 2000). Thus, D1 protein of PSII must be degraded, re-synthesized *de novo* and reinserted into the PSII reaction centre to repair the damage and re-establish PSII function (Asada, 1999). It is known that BR could affect gene expression and protein synthesis (Vert *et al.*, 2005; Wang *et al.*, 2006; Lisso *et al.*, 2005). Thus, Deng *et al.* (2007) describe two 29-kDa chloroplast ribonucleoproteins depending on BRs, and several mutants of BR with proteomic changes which could directly or indirectly affect the D1 turnover in the thylakoid membranes (Ye and Sugiura, 1992). Thus, ERB could be implicated in the control of D1 damage and repair.

The fact that only plants treated with *s*-triazine herbicides showed the protective effect of EBR is difficult to explain. However, urea herbicides and *s*-triazine herbicides have different binding sites on D1 protein. According to van Rensen (1989) both kinds of herbicides have two binding sites but only one of these sites is shared by urea and *s*-triazine herbicides. This suggests that the strength of binding to D1 and the turnover rate of D1 could be different in plants treated with urea herbicides or *s*-triazine herbicides.

However, other effects on the photosynthetic system cannot be ruled out. Thus, Chen *et al.* (1995) reported a 29 kDa ribonucleoprotein involved in the mRNA stability of subunit IV of the cytochrome b_6/f complex. Besides, seed application of BR restored the Chl level, as well as nucleic acids and soluble protein, in rice plants grown in saline medium (Anuradha and Rao, 2003).

EBR may also cause the palliative effect on *s*-triazine herbicide-treated plants by enhancing the antioxidant enzyme system, as described by Ogweno *et al.* (2008) in tomato plants subjected to high temperatures. The Chl fluorescence parameters in dark-adapted state indicate damage to the photosynthetic apparatus which could be alleviated by EBR protection from oxidative stress. Little is known about the role of BRs in the plant response to oxidative stress. However, it was shown that exogenous application of BRs modified antioxidant enzymes in rice, maize and sorghum (see Bajguz and Hayat, 2009). It has also been reported that BR protects from cadmium and aluminium toxicity or from salt stress by stimulating the antioxidative enzymatic defence system (Hasan *et al.*, 2008; Ali *et al.*, 2008; Nuñez *et al.*, 2003). EBR application also reduced lipid peroxidation (expressed as malonaldehyde content) in tomato plants subjected to high temperature stress (Ogweno *et al.*, 2008). This could explain the effect of BR on herbicides that inhibit elongation reactions in fatty acid synthesis, as described by Hamada (1986). Thus, an enhancement of the antioxidant system could explain the palliative effect of BR on damage caused by triazine herbicides.

Respect to the fact that BRs cannot alleviate the damage caused by MBT and MET, it should be borne in mind that different herbicides could cause a variable degree of oxidative stress, as they elicit different physiological responses. Thus, in spite of the fact that MET caused similar effects on *A* to Cyan, the effect of MET on Chl fluorescence parameters was much weaker, since the decrease in $\Delta F/F'_M$ was half that caused by Cyan (Table 5). However, EBR did not palliate the damage caused by MET. It has been suggested that oxidative stress is differentially induced in maize by different herbicides and that MET produced stronger oxidative stress than other herbicides (Nemet Alla *et al.*, 2008a). Besides, MET caused a significant accumulation of ammonium and a decrease in protein formation which could contribute to an increase in MET damage compared with other herbicides (Nemet Alla *et al.*, 2008b). This would explain why EBR did not afford protection from this additional damage.

On the other hand, EBR could increase herbicide detoxification, by the reaction mechanisms of herbicide degradation

Plants transform herbicides by means of a phased detoxification system (Coleman *et al.*, 1997; Cherian and Oliveira, 2005). Herbicides are first metabolized by Phase I (transformation), which usually involves hydrolysis or oxidation. The major reactions of phase I enzymes include oxidation catalysed by cytochrome P450 monooxygenases (Cyt P450) (Siminszky, 2006). The multiplicity of isoforms of Cyt P450 accounts for the wide range of substrates accommodated and the variety of reactions catalysed. Phase II (conjugation) involves conjugation to glutathione and glucose catalysed by GST and UDP-glycosyltransferase, respectively. Phase III involves modified herbicides becoming compartmentalized in vacuoles or being bound to cell wall components (Frear *et al.*, 1985; Coleman *et al.*, 1997; Komives and Gullner, 2005). The differential toxicity of herbicides has long been known and initial detoxification differences can bring an important contribution to this differential toxicity (Eshel *et al.*, 1975).

Triazine herbicides are metabolised through hydroxylation, and N-dealkylation (phase I) and glutathione conjugation (phase II). The enzymes responsible have not been well studied but N-dealkylation appears to be catalysed by Cyt P450 in several plants (Shimabukuro *et al.*, 1973; Lamoureux *et al.*, 1998; see Simoneaux and Gould, 2008). Transgenic rice plants expressing a human cytochrome P450 *CYP1A1* gene metabolized simazine and atrazine more rapidly than control plants (Kawahigashi *et al.*, 2005). Detoxification of triazine herbicides has also been related with the activity of GST in several crops and weeds (Gray *et al.*, 1995; Kreuz *et al.*, 1996; Cole *et al.*, 1997; Hatzios and Burgos, 2004; Cho and Kong, 2007). As in the case of Cyt P450, differences in the properties of GST isoenzymes account for species-specific differences in the susceptibility of different plant species to certain herbicides (Kreuz *et al.*, 1996; van Eerd *et al.*, 2003). It has also been suggested that certain GSTs function through peroxidase activity, to protect plant from oxidative stress (Sommer and Böger, 2001). Besides, triazine herbicides induce detoxification enzymes. Thus, Terb induces detoxification enzymes in *Stodoptera frugiperda*, such as microsomal oxidases, GSTs and hydrolases, apparently by synthesis *de novo* (Yu, 2004).

Xia *et al.* (2009b) suggested that BRs reduced the phytotoxic effect of herbicides and pesticides, by accelerating their metabolism. According to these authors, BRs can be considered as "safeners", which induce the activity of numerous plant Cyt P450s and enhance glutathione conjugation involved in the biodegradation of herbicides (Hatzios and Burgos, 2004). Thus, transcriptional analyses of BR-deficient or BR-treated *Arabidopsis* and cucumber plants have shown that BR-regulated genes include those detoxification genes encoding Cyt P450, GST and UDP-glycosyltransferase (Müssig *et al.*, 2002; Xia *et al.*, 2009a). A proteomic study of BR responses in *Arabidopsis* indicated at least six GSTs, two of them showing proteomic changes in BR mutants (Deng *et al.*, 2007). Proteins that showed homology to GST were also found to be increased with BR application in rice (Konishi and Komatsu, 2003). However, the list of BR-regulated genes has significant differences dependent on the authors, possibly due to differences in their respective experimental conditions. Thus, Goda *et al.* (2002) found five Cyt P450 genes down-regulated by BR. More work will be necessary to clarify this point.

8. CONCLUSION

As discussed in this chapter, exogenous BRs cause a palliative effect on the damage produced by *s*-triazine herbicides. The protection is reflected in plant growth, photosynthetic CO_2 assimilation, and Chl fluorescence parameters measured in both dark- and light-adapted state. This protective effect is not found for photosynthesis-inhibiting herbicides acting at the PSI level, or for other such herbicides that act at PSII level but belong to the *as*-triazinone or urea derivative groups. BR effects on plants treated with *s*-triazine herbicides are persistent since exogenous application of BRs to seeds before sowing induces ameliorative effects in the crop that are detected at the end of growth period. The protective action of BRs in plants treated with *s*-triazine herbicides could be explained by reference to various hypotheses, but new experimental work will be necessary to clarify this.

9. ACKNOWLEDGEMENT

The authors are grateful to the *Ministerio de Ciencia e Innovación* for financial support (Financial project ref. CGL2008-01443), Dr. Carmen Brosa from I.Q.S., University Ramon Llull, Barcelona, for the gift of 24-epibrassinolide, the *Campos Experimentales de la Universidad de Barcelona* for technical support, and Robin Rycroft for correcting the English text.

10. REFERENCES

Ali, B., Hasan, S.A., Hayat, S., Hayat, Q., Yadav, S., Fariduddin, Q., and Ahmad, A. 2008. A role for brassinosteroids in the amelioration of aluminium stress through antioxidant system in mung bean (*Vigna radiata* L. Wilczek). *Environ. Exp. Bot.*, **62**: 153–159.

Anuradha, S., and Rao, S.S.R. 2003. Application of brassinosteroids to rice seeds (*Oryza sativa* L.) reduced the impact of salt stress on growth, prevented photosynthetic pigment loss and increased nitrate reductase activity. *Plant Growth Regul.*, **40**: 29–32.

Asada, K. 1999. The water-water cycle in chloroplasts: scavenging of active oxygens and dissipation of excess photons. *Ann. Rev. Plant Physiol. Plant Mol. Biol.*, **50**: 601–639.

Bajguz, A., and Hayat, S. 2009. Effects of brassinosteroids on the plant responses to environmental stress. *Plant Physiol. Biochem.*, **47**: 1–8.

Blokhina, O., Virolainen, E., and Fagerstedt, K.V. 2003. Antioxidants, oxidative damage and oxygen deprivation stress: a review. *Ann. Botany*, **91**: 179–194.

Bobrik, A.O., Khripach, V.A., Zhabinskii, V.N., Zavadskaya, M.I., and Litvinovska, R.P. 1998. A method of production of sanitated seed potato. *Pat. Appl. BY*, **19**: 981,189.

Bolhar-Nordenkampf, H.R., Long, S.P., Baker, N.R., Öquist, G., Schreiber, U., and Lechner, E.G. 1989. Chlorophyll fluorescence as a probe of the photosynthetic competence of leaves in the field: a review of current instrumentation. *Func. Ecol.*, **3**: 497–514.

Bowyer, J.R., Camilleri, P., and Vermaas, W.F.J. 1991. Photosystem II and its interaction with herbicides. *In* NR Baker, MP Percival (Eds.) Herbicides. Elsevier Sci. Pub. B.V., Amsterdam: pp. 27–85.

Chen, Q., Adams, C.C., Usack, L., Yang, J., Monde, R.A., and Stern, D.B. 1995. An AU-rich element in the 3′ untranslated region of the spinach chloroplast *petD* gene participates in sequence-specific RNA-protein complex formation. *Mol. Cell. Biol.*, **15**: 2010–2018.

Cherian, S., and Oliveira, M.M. 2005. Transgenic plants in phytoremediation: recent advances and new possibilities. *Environ. Sci. Technol.*, **39**: 9377–9390.

Cho, H.Y., and Kong, K.H. 2007. Study on the biochemical characterization of herbicide detoxification enzyme, glutathione S-transferase. *BioFactor*, **30**: 281–287.

Clouse, S.D., and Sasse, J.M. 1998. Brassinosteroids: essential regulators of plant growth and development. *Ann.Rev.Plant Physiol. Plant Mol. Biol.*, **49**: 427–451.

Cole, D.J., Cummins, I., Hatton, P.J., Dixon, D., and Edwars, E. 1997. Glutathione transferases in crops and major weeds. *In* KK Hatzios (Ed.) Regulation of enzymatic systems detoxificatying xenobiotics in plants. Kluwer Aca. Publ., Dordretch: pp. 139–154.

Coleman, J.O.D., Blake-Kalff, M.M.A., and Davies, T.G.E. 1997. Detoxification of xenobiotics by plants: chemical modification and vacuolar compartmentation. *Trends in Plant. Sci.*, **2**: 144–151.

Cutler, H.G. 1991. Brassinosteroids through the looking glass: an appraisal. *In* HG Cutler, T Yokota, G Adam G (Eds.) Brassinosteroids: chemistry, bioactivity and applications. ACS Symp Ser 474, American Chemical Society. Washington, DC: pp. 334–345.

Deng, Z., Zhang, X., Tang, W., Oses-Prieto, J.A., Suzuki, N., Gendron, J.M.., Chen, H., Guan, S., Chalkley, R.J., Peterman, T.K., Burlingame, A.L., and Wang, Z. Y., 2007. A proteomics study of brassinosteroid response in *Arabidopsis*. *Mol. Cell. Proteomics*, **6**: 2058–2071.

Divi, U.K., and Krishna, P. 2009. Brassinosteroids confer stress tolerance. *In* H Hirt (Ed.) Plant stress biology. Wiley-VCH, Weinheim: pp. 119–135.

Dixon, F.L., Clay, D.V., and Willoughby, I. 2006. The efficacy of pre-emergence herbicides on problem weeds in woodland regeneration. *Crop Protection*, **25**: 259–268.

Dyer, W.E., and Weller, S.C. 2005. Plant response to herbicide. *In* MA Jenks, PM Hasegawa (Eds.) Plant abiotic stress. Blackwell Pub. Ltd., Oxford: pp. 171–214.

Edelman, M., and Mattoo, A.K. 2008. D1-protein dynamics in photosystem II: the lingering enigma. *Photosynth. Res.*, **98**: 609–620.

Eshel, Y., Kovacs, M., and Rubin, B. 1975. Differential tolerance of peas to prometryne and terbutryn. *Pestic. Biochem. Physiol.,* **5**: 295–302

Fedtke, C. 1982. Biochemistry and Physiology of Herbicide Action. Springer-Verlag, Berlin: 202 pp.

Ferrer, I., Thurman, E.M., and Barceló, D. 2000. First LC/MS determination of cyanazine amide, cyanazine acid and cyanazine in groundwater samples. *Environ. Sci. Technol.*, **34**: 714–718.

Frear, D.S., Swanson, H.R., and Mansager, E.R. 1985. Alternate pathways of metribuzin metabolism in soybean: formation of *N*-glucoside and homoglutathione conjugates. *Pest. Biochem. Physiol.*, **23**: 56–65.

Genty, B., Briantais, J.M., and Baker, N.R. 1989. The relationships between the quantum yield of photosynthetic electron transport and quenching of chlorophyll fluorescence. *Biochim. Biophys. Acta*, **900**: 87–92.

Goda, H., Shimada, Y., Asami, T., Fujioka, S., and Yoshida, S. 2002. Microarray analysis of brassinosteroid-regulated genes in *Arabidopsis*. *Plant Physiol.*, **130**: 1319–1334.

Gray, J.A., Stoltenberg, D.E., and Balke, N.E. 1995. Absence of herbicide cross-resistance in two atrazine-resistant velvetleaf (*Abutillon theophrasti*) biotopes. *Weed Sci.*, **43**: 352–357.

Gronwall, J.W. 1994. Resistance to photosystem II inhibiting herbicides. *In* SB Powles, JAM Holtum (Eds.) Herbicide resistance in plants: biology and biochemistry. CRC Press, Boca Raton: pp. 83–139.

Gu, J.G., Quiao, C., and Gu, J.D. 2003. Biodegradation of the herbicides atrazine, cyanazine, and dicamba by methanogenic enrichment cultures from selective soils in China. *Bull. Environ. Contam. Toxicol.*, **71**: 924–932.

Halliwell, B. 1991. Reactive oxygen species in living systems: source, biochemistry, and role in human disease. *Am. J. Med.*, **91**: S14–S22.

Hamada, K. 1986. Brassinolide in crop cultivation. Plant growth regulators in agriculture, *FFTC Book Ser.*, **34**: 188–196.

Hasan, S.A., Hayat, S., Ali, B., and Ahmad, A. 2008. 28-Homobrassinolide protects chickpea (*Cicer arietinum*) from cadmium toxicity by stimulating antioxidants. *Environ. Pollution*, **151**: 60–66.

Hatzios, K.K and Burgos, N. 2004. Metabolism-based herbicide resistance: regulation by safeners. *Weed Science*, **52**: 454–467.

Holt, J.S., Powles, S.B., and Holtum, J.A.M. 1993. Mechanisms and agronomic aspects of herbicide resistance. *Ann. Rev. Plant Physiol. Plant Mol. Biol.*, **44**: 203–229.

Ikekawa, N., and Zhao, Y.J. 1991. Application of 24-epibrassinolide in Agriculture. *In* HG Cutler, T Yokota, G Adam (Eds.) Brassinosteroids: chemistry, bioactivity and applications. ACS Symp Ser 474, American Chemical Society. Washington, DC: pp. 280–291.

Janjic, V., Milosevic, D., Djalovic, I., and Týr, S. 2007. Weed resistance to herbicides - Mechanisms and molecular basis. *Acta Agric. Serbica*, **XII**(24): 19–40.

Juneau, P., Qiu, B., and Deblois, C.P. 2007. Use of chlorophyll fluorescence as a tool for determination of herbicide toxic effect: review. *Toxicol. Environ. Chem.*, **89**: 609–625.

Kagale, S., Divi, U.K., Krochko, J.E., Keller, W.A., and Krishna, P. 2007. Brassinosteroid confers tolerance in *Arabidopsis thaliana* and *Brassica napus* to a range of abiotic stresses. *Planta*, **225**: 353–364.

Kawahigashi, H., Hirose, S., Ohkawa, H., and Ohkawa, Y. 2005. Transgenic rice plants expressing human *CYP1A1* remediate the triazine herbicides atrazine and simazine. *J. Agric. Food Chem.*, **53**: 8557–8564.

Khripach, V.A., Zhabinskii, V.N., Litvinovskaya, M.I., Savel'eva, E.A., Karas, I.I., and Vakulenko, V.V. 1996. A method of increasing of potato food value. Pat. Appl. BY 960,345.

Khripach, V., Zhabinskii, V., and de Groot, A. 2000. Twenty years of Brassinosteroids: steroidal plant hormones waeeant better crops for the XXI century. *Annals of Botany*, **86**: 441–447.

Komives, T., and Gullner, G. 2005. *Phase I* xenobiotic metabolism system in plants. *Z. Naturforsch.*, **60c**: 179–185.

Konishi, H., and Komatsu, S. 2003. A proteomics approach to investigating promotive effects of brassinolide on lamina inclination and root growth in rice seedlings. *Biol. Pharm. Bull.*, **26**: 401–408.

Krause, G.H., and Weis E. 1991. Chlorophyll fluorescence and photosynthesis: the basics. *Ann. Rev. Plant Physiol. Plant Mol. Biol.*, **42**: 313–349.

Kreuz, K., Tommasini, R., and Martinoia, E. 1996. Old enzymes for a new job. Herbicide detoxification in plants. *Plant Physiol.*, **111**: 349–353.

Krishna, P. 2003. Brassinosteroid-mediated stress responses. *J. Plant Growth Regul.*, **22**: 289–297.

Lamoureux, G.L., Simoneaux, B., and Larson, J. 1998. The metabolism of atrazine and related 2-chloro-4,6-bis (alkylamino)-*s*-triazines in plants. *In* LG Ballentine, JE McFarland, DS Hackett (Eds.) Triazine Herbicides: risk Assessment. American Chemical Society, Washington DC: pp. 60–81.

Lazár, D. 1999. Chlorophyll a fluorescence induction. *Biochim. Biophys. Acta*, **1412**: 1–28.

Lisso, J., Steinhauser, D., Altmann, T., Kopka, J., and Müsig, C. 2005. Identification of brassinosteroid-related genes by means of transcript co-response analyses. *Nucleic Acids Res.*, **33**: 2685–2696.

Mandava, N.B. 1988. Plant growth-promoting brassinosteroids. *Ann. Rev. Plant Physiol. Plant Mol. Biol.*, **39**: 23–52.

Matouškova, M., Nauš, J., and Flašarová, M. 1999. A long-term response of chlorophyll fluorescence induction to one-shot application of cyanazine on barley plants and its relation to crop yield. *Photosynthetica*, **37**: 281–294.

Maxwell, K., and Johnson, GN. 2000. Chlorophyll fluorescence – a practical guide. *J. Exp. Bot.*, **51**: 659–668.

Mohammadzamani, D., Minaei, S., Alimardani, R., Almassi, M., Rashidi, M., and Norouzpour, H. 2009. Variable rate herbicide application using the global positioning system for generating a digital management map. *Int. J. Agric. Biol.*, **11**: 178–182.

Munné-Bosch, S., and Alegre, L. 2000. Changes in carotenoids, tocopherols and diterpenes during drought and recovery, and the biological significance of chlorophyll loss in *Rosmarinus officinalis* plants. *Planta*, **210**: 925–931.

Müssig, C., Fischer, S., and Altmann, T. 2002. Brassinosteroid-regulated gene expression. *Plant Physiol.*, **129**: 1241–1251.

Navas, M.L. 1991. Using plant populations biology in weed research: a strategy to improve seed management. *Weed Res.*, **13**: 171–179.

Nemet Alla, M.M., Badawi, A.M., Hassan, N.M., El-Bastawisy, Z.M., and Badran, E.G. 2008a. Effect of metribuzin, butachlor and chlorimuron-ethyl on amino acid and protein formation in wheat and maize seedlings. *Pest. Biochem. Physiol.*, **90**: 8–18.

Nemet Alla, M.M., Badawi, A.M., Hassan, N.M., El-Bastawisy, Z.M., and Badran, E.G. 2008b. Herbicide tolerance in maize is related to increased levels of glutathione and glutathione-associated enzymes. *Acta Physiol. Plant.*, **30**: 371–379.

Nuñez, M., Mazzafera, P., Mazorra, L.M., Siqueria, W.A., and Zullo, M.A.T. 2003. Influence of brassinosteroid analogue on antioxidant enzymes in rice Brown in culture medium with NaCl. *Biol. Plant*, **47**: 67–70.

Oettmeier, W. 1999. Herbicide resistance and supersensitivity in photosystem II. *CMLS. Cell. Mol. Life Sci.*, **55**: 1255–1277.

Ogweno, J.O., Song, X.S., Shi, K., Hu, W.H., Mao, W.H., Zhou, Y.H., Yu, J.Q., and Nogués, S. 2008. Brassinosteroid alleviate heat-induced inhibition of photosynthesis by increasing carboxilation efficiency and enhancing antioxidant systems in *Lycopersicum esculentum*. *J. Plant Growth Regul.*, **27**: 49–57.

Oxborough, K., and Baker, N.R. 1997. Resolving chlorophyll *a* fluorescence images of photosynthetic efficiency into photochemical and non-photochemical components – calculations of qP and Fv′/Fm′ without measuring Fo′. *Photosynthesis Res.*, **54**: 135–142.

Percival, M.P., and Baker, N.R. 1991. Herbicides and photosynthesis. *In* NR Baker, MP Percival (Eds.) Herbicides. Elsevier Sci. Pub. B.V., Amsterdam: pp. 1–26.

Pinto de Carvalho, S.J., Nicolai, M., Rodrigues Ferreira, R., Vargas de Oliveira Figueira, A., and Christoffoleti, P.J. 2009. Herbicides selectivity by differential metabolism: considerations for reducing crop damages. *Sci. Agric. (Piracicaba, Braz.)*, **66**: 136–142.

Piñol, R., and Simón, E. 2009. Effect of 24-epibrassinolide on chlorophyll fluorescence and photosynthetic CO_2 assimilation in *Vicia faba* plants treated with the photosynthesis-inhibiting herbicide terbutryn. *J. Plant Growth Regul.*, **28**: 97–105.

Pshenichnaya, L.A., Khripach, V.A., Volynetz, A.P., Prokhorchik, R.A., Manzhelesova, N.E., and Morozik, G.V. 1997. Brassinosteroids and resistance of barley plants to leaves diseases. *In* Problems of experimental botany, Byelorussian Science, Minsk: pp. 210–217.

van Eerd, L.L., Hoagland, R.E., and Hall, J.C. 2003. Pesticide metabolism in plants and microorganisms. *Weed Sic.* **51**: 472–495.

van Rensen, J.J.S. 1989. Herbicides interacting with photosystem II. *In* AD Dodge (Ed.) Herbicides and plant metabolism. Cambridge University Press, Cambridge: pp. 21–36.

Rohácek, K. 2002. Chlorophyll fluorescence parameters: the definitions, photosynthetic meaning and mutual relationships. *Photosynthetica*, **40**: 13–29.

Sasse, J.M. 2003. Physiological actions of brassinosteroids: an update. *J. Plant Growth Regul.*, **22**: 276–288.

Schreiber, U., and Bilger, W. 1993. Progress in chlorophyll fluorescence research: major developments during the past years in retrospect. *Progress in Botany*, **54**: 151–173.

Shimabukuro, R.H., Walsh, W.C., Lamoureux, G.L., and Stafford, L.E. 1973. Atrazine metabolism in sorghum: chloroform-soluble intermediates in the N-dealkylation and glutathione conjugation pathways. *J. Agric. Food Chem.*, **21**: 1031–1036.

Shukla. A., and Devine, M.D. 2008. Basis of crop selectivity and weed resistance to triazine herbicides. *In* HM LeBaron, JE McFarland, O Burnside (Eds.) The triazine herbicides: 50 years revolutionizing agriculture. Elsevier, Amsterdam: pp. 111–118.

Siminszky, B. 2006. Plant cytochrome P450-mediated herbicide metabolism. *Phytochem. Rev.*, **5**: 445–458.

Simoneaux, B.J., and Gould, T.J. 2008. Plant uptake and metabolism of triazine herbicides. *In* HM LeBaron, JE McFarland, O Burnside (Eds.) The triazine herbicides: 50 years revolutionizing agriculture. Elsevier, Amsterdam: pp. 73–100.

Singh, I., and Shono, M. 2005. Physiological and molecular effects od 24-epibrassinolide, a brassinosteroid on termotolerance of tomato. *Plant Growth Regul.*, **47**: 111–119.

Sivak, M.N., and Walker, D.A. 1985. Chlorophyll *a* fluorescence can it shed light on fundamental questions in photosynthetic carbon dioxide fixation. *Plant Cell Environ.*, **8**: 439–448.

Sommer, A., and Böger, P. 2001. Enzymological studies on recombinant isoforms of glutathione *s*-transferase from corn. *In* JC Hall, RE Hoagland, RM Zablotowicz (Eds.) Pesticide biotransformation in plants and microorganisms: similarities and divergences. ACS Sym. Series 777, Am. Chem. Soc., Washington DC: pp. 253–267.

Strandberg, M.T., and Scott-Fordsmand, J.J. 2002. Field effects of simazine at lower trophic levels – a review. *The Sci. of the Total Environ.*, **296**: 117–137.

Szigeti, Z., and Lehoczki, E. 2003. A review of physiological and biochemical aspects of resistance to atrazine and paraquat in Hungarian weeds. *Pest Manag. Sci.*, **59**: 451–458.

Takematsu, T., and Takeuchi, Y. 1989. Effects of brassinosteroids on growth and yields of crops. *Proc. Japan Acad.*, **65 Ser. B**: 148–152.

Vardhini, B.V., Anuradha, S., and Rao, S.S.R. 2006. Brassinosteroids – new class of plant hormone with potential to improve crop productivity. *Indian J. Plant Physiol.*, **11**: 1–12.

Vert, G., Nemhauser, J.L., Geldner, N., Hong, F., and Chory, J., 2005. Molecular mechanisms of steroid hormone signaling in plants. *Ann. Rev. Cell Dev. Biol.*, **21**: 177–201.

Vidal, D., Martínez-Guijarro, J., and Simón, E. 1995. Chlorophyll fluorescence and photosynthetic O_2 evolution in *Vicia faba* treated with methabenzthiazuron. *Photosynthetica*, **31**: 9–20.

Wakabayashi, K., and Böger, P. 2004. Phytotoxic sites of action for molecular design of modern pesticides (Part 1): the photosynthetic electron transport system. *Weed Biol. Manag.*, **4**: 8–18.

Wang, Z.Y., Wang, Q., Chong, K., Wang, F., Wang, L., Bai, M., and Jia, C. 2006. The brassinosteoird signal transduction pathway. *Cell Res.*, **16**: 427–434.

Wilen, R.W., Sacco, M., Gusta, L.V., and Krishna, P. 1995. Effects of 24-epibrassinolide on freezing and thermotolerance of bromegrass (*Bromus inermis*) cell cultures. *Physiol. Plant.*, **95**: 195–202.

Xia, X.J., Huang, Y.Y., Wang, L., Huang, L.F., Yu, Y.L., Zhou, Y.H., and Yu, J.Q. 2006. Pesticides-induced depression of photosynthesis was alleviated by 24-epibrassinolide pretreatmet in *Cucumis sativus* L. *Pest. Biochem. Physio.*, **86**: 42–48.

Xia, X.J., Wang, Y.J., Zhou, Y.H., Tao, Y., Mao, W.H., Shi, K., Asami, T., Chen, Z., and Yu, J.Q. 2009a. Reactive oxygen species are involved in brassinosteroid-induced stress tolerance in cucumber. *Plant Physiol.*, **150**: 801–814.

Xia, X.J., Zhang, Y., Wu, J.X., Wang, J.T., Zhou, Y.H., Shi, K., Yu, Y.L., and Yu, J.Q. 2009b. Brassinosteroids promote metabolism of pesticides in cucumber. *J. Agric. Food Chem.*, **57**: 8406–8413.

Xiong, J, Hutchison, R.S., Sayre, R.T., and Govindjee 1997. Modification of the photosystem II acceptor side function in a D1mutant (arginine-269-glycine) of *Chlamydomonas reinhardti*. *Biochim. Biophys. Acta*, **1322**: 60–76.

Ye, L., and Sugiura, M. 1992. Domains required for nucleic acid binding activities in chloroplast ribonucleoproteins. *Nucleic Acids Res.*, **20**: 6275–6279.

Yu, S.J. 2004. Induction of detoxification enzymes by triazine herbicides in the fall armyworm, *Spodoptera frugiperda* (JE. Smith). *Pest. Biochem Physiol.*, **80**: 113–122.

Zang, L., Paakkarinen, V., Wijk, K.J. van, and Aro, E.M. 2000. Biogenesis of the Chloroplast-encoded D1 protein: regulation of translation elongation, insertion and assembly into photosystem II. *The Plant Cell*, **12**: 1769–1781.

Zimmermann, K., Heck, M., Frank, J., Kern, J., Vass, I., and Zouni, A. 2006. Herbicide binding and thermal stability of photosystem II isolated from *Thermosynechococcus elongates*. *Biochim. Biophys. Acta*, **1757**: 106–114.

Chapter 12

BRASSINOSTEROIDS: UNDER BIOTIC STRESS

SHAMSUL HAYAT, MOHAMMAD IRFAN AND AQIL AHMAD
Department of Botany, Aligarh Muslim University, Aligarh 202002, India

Abstract: Brassinosteroids are the steroidal plant hormones, key regulators of growth and metabolism in metabolically active tissues. Their interplay ensures either the accelerated growth or differentiation in meristematic tissues or induces the defense responses in biotically challenged tissues, stopping the growth responses in concentration dependent manner. A rationale of highly active growth regulators and classical phytohormones, based on external cues and endogenous developmental program, ascertains the fate of localized tissue for the sake of whole plant, may induce hypersensitive response based PCD, activating LAR and/or SAR. A shift of growth metabolism towards defense strategies is evidenced in several *Arabidopsis* and other BR mutants suggesting their antagonistic role. However, specific interaction of BRs with well known other phytohormones in suppressing the growth responses under biotic stress, have been tried to underline in the present survey.

Key words: Antioxidant activity, biotic-stress, brassinosteroids, hormone interactions

1. INTRODUCTION

Brassinosteroids (BRs) are essential regulators of plant growth and development (Clouse and Sasse, 1998). They are perceived as stress hormones. However, unlike ABA, BRs neither induce dormancy nor senescence. The activity of BRs in metabolically active tissues is to promote and regulate the growth in low concentration/sensitivity dependent manner, controlling the antioxidant system mediated production of ROS. The application of BRs at higher concentrations is reported to mediate induced progression of PCD (internal higher sensitivity of a particular tissue for BRs mediates PCD, under

S. Hayat and A. Ahmad (eds.), *Brassinosteroids: A Class of Plant Hormone*,
DOI 10.1007/978-94-007-0189-2_12, © Springer Science+Business Media B.V. 2011

genetic program, may be induced as per changed external environmental condition). This progression is the result of antagonistic simultaneous effect of BRs interaction with other active species against classical growth hormones which are known for growth promotion rather than defense. The role of former is crucial at the time of biotic challenge where localized cell death becomes necessary to check the further infection. The level of free auxin is recognised to suppress the host defense against pathogens (Chen *et al.*, 2007; Wang *et al.*, 2007) and is also associated with disease development (Ding *et al.*, 2008) that suggests high antioxidant activity based failure of pathogen disposal. A bell shaped progressive curve (initial rise followed by decrease) of antioxidant enzyme activity suggests the click towards PCD (Kuzniak and Sklodowska, 2005) in successful pathogen disposal, followed by symptom development. Such symptoms may be invisible to naked eyes (Figure 1).

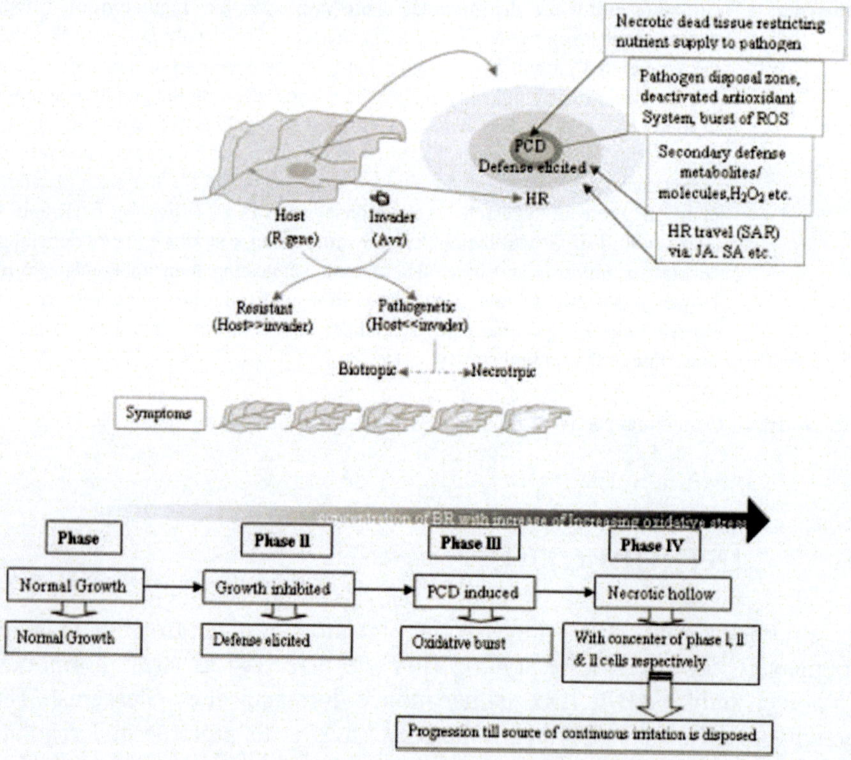

Figure 1. Concentric-layer strategy for checking the progression of infection under plant pathogenic challenge. The development of symptoms depends upon the compatibility of *Avr-R* gene interaction.

Highly active species together with BRs seem to trigger determined step towards PCD. The time of decision for plants at such localized areas, although is very rapid, to assess the recruited protein players under progressively changing regulation of growth-defense regulators in time-lapse study, is presented in this chapter. It is designed keeping in mind the behaviour of antioxidant activity and ROS production regulated by changing cocktail of plant hormones as a strategy of plants to combat biotic challenges.

2. ROLE OF BRs UNDER BIOTIC STRESS

The role of BRs is not only implicated in plant response to environmental stresses but also in alleviating biotic challenges of bacterial, fungal and viral pathogens (Krishna, 2003; Ali *et al.*, 2007; Jager *et al.*, 2008). The application of tissue extracts containing BRs and commercially available BR-analogues have shown to overcome plant biotic challenges (Khripach *et al.*, 1999). The enhanced resistance of tobacco, cucumber, and tomato plants to viral (TMV) and fungal pathogens (*Sphaerotheca fuligenia, Botrytis* etc.) have been observed when these were treated with seed-extract of *Lychnis viscaria,* containing low amount of three PR proteins viz. chitinase, B-1,3 glucanase and peroxidase, the markers of SAR, which supported the activation of plant defense status. Two BR-compounds were also identified from the leaf extract were 24-epi-secasterone and 24-epi-castasterone (Schmidt *et al.*, 1996). However, in *Arabidopsis cpd* mutant of BR biosynthesis, a remarkably decreased expression of PR1, PR2 and PR5 genes has been reported where as in CPD over-expressed transgenic plants these were up-regulated.

24-epibrassinosteroids (24-EBL) protect potato plants against *Phytophthora* mediated late blights. Although the appearance of visible disease symptoms were cultivar dependent, reducing disease incidence between 11% and 34% (Khripach *et al.*, 1997). Studies on barley plants sprayed with 24-EBL, at tillering time considerably reduced the leaf disease incidence, mediated by *Helminthosporium teres* Sacc (Volynets *et al.*, 1997). Recent findings also favoured the protective effects of brassinolide (BL) on rice and tobacco (Nakashita *et al.*, 2003). Application of BL on tobacco plants induced broad range resistance against viral (TMV), bacterial (*Pseudomonas syringae* pv. Tobaci and fungal (*Oidium* species) pathogens, while in rice it induced the resistance against *Maganoprothe grisea* and *Xathomonas oryzae* pv. Orycae (Friebe, 2006). In case of cucumbers also the treatment with BL increased the resistance against fungal peronosporosis. Moreover, the increased activity of peroxidase and polyphenoloxidase enzymes in leaves was correlated with the enhanced resistance of plants with the pathogens (Khripach *et al.*, 1999).

Protective and deprotective both types of roles have been attributed to BRs, depending upon the time and method of application at the point of interaction of host with its pathogen (Korableva *et al.*, 2002). In case of late blight of potato, BRs have been reported to suppress the immune status of plant tissues, inhibiting wound reparations. BRs stimulated the mycelial growth and sporulation intensity of *P. infestans* thereby induced susceptibility of potato tuber tissues. However, on the other hand, post harvest treatment of potato plants or tubers with BRs was shown to prolong the dormancy and subsequent enhancement of resistance to *P. infestans*.

3. PLANT REGULATION STRATEGY AGAINST BIOTIC STRESS

The increased activity of antioxidant system is often used synonymously with the plant protection from the prevailing stress conditions, and so the most molecules up-expressed/induced under such conditions are positively correlated with the protective function. Is it the activation of antioxidant system that always prevents the plant from stress to correspond the greater defense and hence the improved growth and bio-productivity, or this notion is specific under certain conditions? The survival seems through either the activation of antioxidant system or via its suppression. The former condition is well established under several abiotic stress reports. However, latter cases are also reported where certain dispensable/infected localized tissues have been sacrificed under abiotic conditions (Irfan *et al.*, 2010). Here the suppressed activity of antioxidant system ensures the survival, by activating burst of ROS (toxic locally, but required for the activation of SAR, LIR) for systemic resistance, ensuring the escape of distant tissues from stress/infection progression. Kuzniak and Sklodowska (2005) reported that under biotic stress conditions initial activity of antioxidant system increases, followed by the decline, based on gradual shift of redox equilibrium (ratio of ROS and antioxidant activity) towards oxidation (Figure 1). The overproduction of H_2O_2 is also reported in several studies as essential regulator of SAR, and that it mediates the PCD (Neill *et al.*, 2002). The pile of literature suggests that the activity of antioxidant system (therefore the level of production of ROS) is regulated by the growth regulators, working spatially and temporally under developmental and stress conditions. For the former case their production is induced under the control of developmental program (Gapper and Dolan, 2006), while under stress conditions production of ROS is enforced exteriorly by the causing agent (Lamb and Dixon, 1997; Bolwell *et al.*, 2002; Torres *et al.*, 2006).

Different phytohormones are recruited as per their requirement at a particular site of the plant. At all the localized points defense-regulators appear to antagonize growth-regulators. The cases where certain pathogens either carry auxins and/or other classical hormones with them during the time of infection (Spaepen *et al.*, 2007) or they induce them on the infection site within the host support the above statement (Shah, 2009). These two types of regulators are therefore, up-expressed/induced depending upon the equilibrium shift of intrinsic and prevailing extrinsic conditions. The interface of activity of antioxidant system and ROS level in the tissue/cell ascertains their redox status, and the activity of antioxidant system (Foyer, 2005) and expression (Arrigo, 1999), is further regulated by the rationale of different phytohormones and their crosstalk (Overmyer *et al.*, 2003). The BRs have been reported to be synthesized in highly metabolically active or meristematic tissues interacting with other highly active molecules.

Phytohormones are growth regulators, so as the BRs that regulate plant growth under different stress conditions. The lower activity of BRs regulate the level of other growth promoters and facilitates growth enhancement while its higher concentration works as defense-regulators recruiting highly active molecules as NO, ROS etc. at the cost of growth (Heil and Bostock, 2002; Swarbrick *et al.*, 2006). The oxidative state of cells/tissue disposes the invader and elicits subsequent SAR necessary to prevent the further establishment of pathogen. After successful combat, at post-stress state it re-recruits the molecules (ROS, antioxidative enzymes and quenching molecules) at their normal level. The defense proteins and metabolites are metabolized and diverted back towards growth.

4. BRs AND ANTIOXIDANT SYSTEM STATUS UNDER BIOTIC STRESS

Disease causing pathogens inevitably pose serious stress for plants. Therefore, plants evolved a unique self protective system incorporating change in ion flux and membrane potential (Zhang *et al.*, 2005), induction of HR, accumulation of secondary products, physiological alterations. It ultimately affects carbon and nitrogen fixation and growth and productivity. The early response involves the pathogen induced local host cell death (Ross, 1961) followed by build up of induced resistance thus protecting plants from further attack (Kuc, 1982; Mclntyre *et al.*, 1981). Since, ROS (superoxides, H_2O_2 etc.) have been perceived as necessary regulators of biochemical fate of cells, their basal level recognized as essential for the regulation of growth and development (Neill *et al.*, 2002). This basal level, however, is tissue specific. The increased activity of antioxidant system is often correlated with decreased level of notorious level of ROS and therefore tissue survival. The cost of enhanced resistance of *det2* mutants of *Arabidopsis thaliana* lead to various inconsistencies as compared to wild types. The evergreen dwarfs were recognized as 'loss of function' mutants in DET2 gene suggested to be the negative regulator of detoxification of oxygen regulators (Cao *et al.*, 2005) expressing antioxidant enzymes constitutively. The identification of dwarf mutants of *Arabidopsis thaliana* was deduced as noval insertion-mutant of *DET2* gene encoding the protein homologous to 5α-reductases (Chory *et al.*, 1991, Li *et al.*, 1996; Li and Chory, 1997). The metabolic constraints cost either the down-regulation of active growth regulators or their perception leading to dwarf phenotypes. The tissue specific initial production of ROS is regulated by enzymes such as SOD, CAT and POX. Over activity of either of them may disrupt cellular fate and thus normal plant development. As mentioned above sometimes (under biotic stress) localized PCD becomes necessary for the survival of whole plant, inducting defense genes in vicinity and SAR distantly. The 'lower activity' of antioxidant system, however, here protects plants from possible threat of infection. From the induction of antioxidants and synthesis of metabolites, it is clear that BRs induce multiple pathways of protection from oxidative damage in rice (Xia *et al.*, 2009). BRs regulate the activity of SOD, CAT and APX, which protect the leaves from oxidative damage. The expression of peroxidase encoding genes *ATP2* and *ATP24a* has been demonstrated by Goda *et al.* (2002) to be regulated by BRs in *Arabidopsis*.

5. INTERACTION OF BRs WITH OTHER PHYTOHORMONES UNDER BIOTIC STRESS

5.1 The interaction of classical phytohormones with BRs

Auxins appear to be the universal and fundamental growth regulators of all the plant hormones though they themselves are under the influence of other phytohormones, expressed as per external cues at their flanking outskirts. The equilibrium between external boundary perception and internal genotype regulation moulds the shape and size (morph) of an individual. Auxins determine the polarity of the single cell and provide the coherence to the cells of the tissue/organ/plant system to direct the growth to shape up the plant. Auxins mediate the plating of microtubules in particular direction. BRs aid them to elongate the cells in the direction set up by the auxins. Most other responses of the auxins appear to be directed by the plant growth regulators working at sublevel of auxins. However, BRs appears to work at extremely fine level, regulating the redox status of cell by altering the generation of ROS thereby the oxidation status of structural and/or functionality of cellular molecules. BRs and auxin also synergistically promote the production of ethylene in etiolated mung bean hypocotyls (Arteca *et al.*, 1983). The expression of ACC synthase, the rate-limiting step in ethylene biosynthesis, is controlled by multiple regulatory pathways of auxin and BR in mung bean seedlings (Yi *et al.*, 1999). The expression of BR responsive gene *TCH4* was also increased by auxin (Xu *et al.*, 1995). The promotion of lateral root development in *Arabidopsis* is the result of BRs interaction with auxins (Bao *et al.*, 2004).

The crosstalk (Teale *et al.*, 2008; Weiss and Ori, 2007) and coordinated regulation of site specific cocktail of phyohormones co-regulate numerous genes. BRs and ABA both are considered as stress hormones. They have been indicated to co-regulate hundreds of genes controlling many developmental processes. However, BRs appear to antagonize the effect of ABA. ABA induces seed dormancy during embryo maturation and inhibits seed germination while BRs reverts them. The ABA shows antagonistic relationship with GA also during seed dormancy and germination (Finkelstein *et al.*, 2008; Razem *et al.*, 2006). The increased level of ABA is presumed to be the reason for the delay in sprouting of potato tubers under dormancy, while a decline in its concentration with subsequent increase of GA was correlated with the breakdown of dormancy. GA biosynthesis in insensitive mutants was rescued by BR (24-EBL and BL) application. However, different effects of exogenous application of BR and GA on tobacco seed germination depended on a set of conditions such as state of dormancy, internal ABA content, exposure to light and imbibitions (Leubner-Metzger, 2001). The distinct but parallel pathways of the two hormones ABA and GA regulate the growth and embryo development in seeds. BRs seem to promote the seed germination by enhancing the embryo growth thus antagonizing the effect of ABA (Finkelstein *et al.*, 2008; Steber and McCourt, 2001; Leubner-Metzger, 2001). The BR biosynthesis and insensitive (*det2-1 and bril-1*) mutants were shown inhibited more strongly in the presence of ABA (Steber and McCourt, 2001) suggesting that GA (directing downstream or) along with BR is simultaneously needed to substitute ABA induced dormancy. The seed germination and embryo maturation phase of BR deficient and BR perceptional mutants are more sensitive to ABA application (Steber and McCourt, 2001).

The experimental evidence of Zhang *et al.* (2009) suggests that antagonistic effect of ABA on BR signaling is based on phosphorylation state of BES-1 that regulates the BR synthesis. The sensitivity of BR deficient and BR perceptional mutants for ABA during seed germination suggests that they regulate common components but with opposite direction to control seed germination, most probably by regulating BIN-2 kinase activity. The effect of ABA on BR signaling is not through an alteration of BR level. The activity of BIN-2, a negative regulator of BR signaling, can also regulate the activity of *ARF2* (Auxin responsive factor-2) (Vert *et al.*, 2008), further indicating the hormonal crosstalk at signaling level. BES-1 binding to the promoter of *SAUR-AC1* gene activates its expression. The *SAUR* gene (*SAUR-AC1, SAUR16 etc.*) encodes short transcripts that accumulate rapidly after auxins treatment (Park *et al.*, 2007).

Sprouting is accompanied with the decline of ethylene as evident from the increase of endogenous ethylene synthesis prolong the ABA mediated deep dormancy in potato tubers. BRs stimulate the synthesis of ethylene in intact potato tubers and ABA synthesis in cells of tuber apices. In the ethylene sensitive mutant (*ein2*) of *Arabidopsis*, BL were shown not capable to induce resistance. However, it has been discussed earlier discussed above that they induce resistance against a broad range of pathogens. BRs induce ethylene synthesis in tuber tissues by an increase in the level of the terpenoids and phenolics via phenylpropanoid pathway (Bajguz and Hayat, 2009). It appears ethylene plays a crucial interactive role with BR during the induction of disease resistance.

The cross talk regulation of phytohormones is a well established phenomenon. Above discussion suggests that phytohormones are not mere chemical messengers of internal developmental program and external environmental cues, these also co-ordinate and regulate the responses of the whole plant (Figure 4). Unlike animals, plant hormones are not produced in specialized organs to carry out their functions at distant sites and as specifically. They are either produced de novo, de-conjugated and/or if necessary, translocated to the site of action. Their site specific requirement is regulated by the of cocktail-cross talk with already present phytohormones which may however be substituted, up to certain level, to recruit altered assembly of proteins (as cytosolic, nuclear or membrane receptors) as per conditions prevailing. Plant scientists sometimes consider these chemicals as plant growth regulators, the interacting metabolites rather than controlling agents. Therefore, sometimes it is concluded that responses mediated by them are based on tissue sensitivity rather than their amount (Jain *et al.*, 2006).

5.2 BRs interaction with highly active molecules

Most often, different types of stresses follow up the excessive generation of ROS at their cardinal thus altering the redox status to put the architecture of cell (plant) to stress. The consequence of biotic stress is generally characterized by the localized cell(s) death that may or may not be visible from unaided eyes, depending upon the level of compatibility of the system, inducing SAR/ISR as hypersensitive response. In any case, however, a rapid and transient generation of ROS leads to cell(s) death, with transient increase of antioxidant activity followed by its suppression (Kuzniak and Sklodowska, 2005). Post-disposal check of accelerated PCD is important for further loss of tissue. Highly active growth modulators (-regulators) find their role within seconds to minutes to dispose invader, regulating the ROS production. Successful combat metabolizes the excessive ROS and other high active regulators, recruiting elevated concentration of classical hormones to reestablish normal growth and development through the gradual achievement of antioxidant activity. Therefore, it seems that an altered ratio and cross talk of ROS, highly active growth regulators and classical hormones simultaneously contributes the stress response.

The gaseous molecule NO, besides regulating other physiological functions, are known for phytoalexin synthesis, hypersensitive response, defense responses, and apoptosis/programmed cell death (Neill *et al.*, 2003). The excess individual concentrations of BRs and NO have been shown to be toxic, inhibiting enzyme activities, increasing lipid peroxidation and electrolyte leakage. However, the toxicity generated by higher dose of NO was effectively overcome by subsequent BRs application (Hayat *et al.*, 2010). BRs activate the genes BRU1 and TCH4 corresponding to xyloglucan endotransglycosylase (*XETs*) and expansins (Cosgrove, 1997). However, over activity of these enzymes may prove lethal to cells over expressing them, leading to acceleration of PCD.

Tomato mutant study for BR and/or jasmonates (*dpy3/jail-1, curl3*) indicated BRs also antagonize jasmonates (Holton *et al.*, 2007; Campos *et al.*, 2009), that find their crucial role under biotic defense. It appears that BRs action negatively control JA based signaling, upstream to *jail-1.* Also the action of BRs seems more localized at the infection site, away from which JA acts spreading SAR (Figure 2). Their response may appear similar to salicylic acids. However, Nakashita *et al.* (2003) showed BR mediated disease resistance in plant defense is independent of SA mediated response.

6. BRs, SIGNAL PERCEPTION AND ITS TRANSDUCTION

These highly active regulators at their utmost boundaries of plants (root or shoot apex meristems or biotically challenged sites) seem to be recruited more by membrane perceived, external signal. The intrinsically embedded cells within tissues have membrane receptors to perceive the chemical messages produced by boundary cells. The cytosolic/membrane-bound population of receptors is likely the second order regulating factor of the tissue-sensing. The membrane potential of surface cells (basically specialized as epidermis), invariably work in association with membrane proteins (Barbier-Brygoo *et al.*, 1991; Felle *et al.*, 1991; David *et al.*, 2001; Lohse and Hedrich, 1992; Zimmermann *et al.*, 1994), and the signal produced by them elicits the downstream signal cascade (Thomine *et al.*, 1997) to express the systemic and/or localized growth regulators (Figure 3).

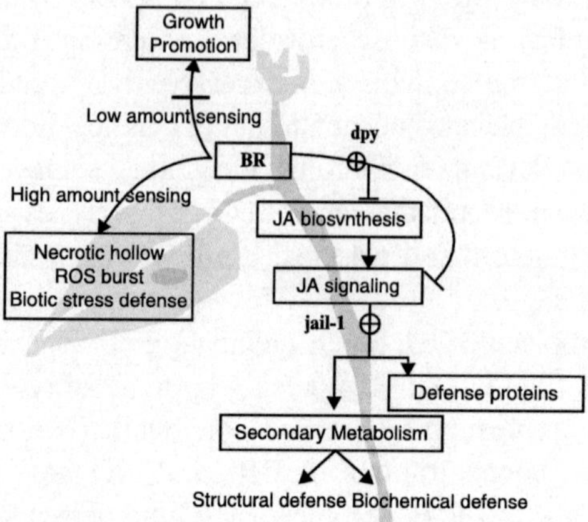

Figure 2. BRs antagonize JA at infection site but away from this site JA signaling elicits SAR at the cost of growth.

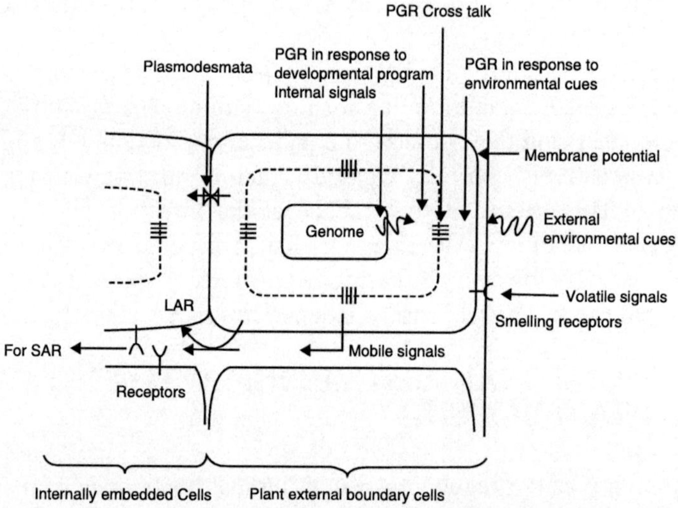

Figure 3. Illustration showing external boundary perception of signals and subsequent production and cross talk of phytohormones to induce systemic resistance.

The activity of membrane bound voltage gated ion channels appears to be the crucial determiner of the cell internal metabolic change to prerequisite the activity of upcoming protein players. Various plasma membrane derived signals include lipid derived secondary signals (Gunther and Scherer, 1996; Scherer, 1996), H_2O_2 (by the activity of NADPH oxidase, Fridovich, 1975), and low activity of antioxidant molecules), and membrane linked kinases.

The response is one of the major determiners of survival of tissue/cell. Steroid regulators are important for the tissues as they elicit the quicker responses than other hormones. The steroids are characterized by their lipophilic property; therefore these can easily pass through the membrane, acting directly at the level of nuclear gene expression (Marcinkowska and Wiedlocha, 2002). They have generally cytosolic or nuclear receptors. The receptors of these steroidal regulators upexpressed in highly active tissues, and so the concentrations/release of their ligands (steroids). The BRs are the new class of phytohormones, characterized by their steroidal (lipophilic) nature, and quick responses at very low concentrations. However, recent findings suggests their 'nongenomic steroid signaling'. Still they are highly active regulators with very low dose requirement. The higher concentration application proves often lethal to seed germination and growth properties of plants. High cellular concentrations of BRs appears to be invariably involved in regulated increase of ROS (by deactivating the antioxidant system activity) in cascadian fashion, thereby, mediating in-vivo PCD (Figure 4).

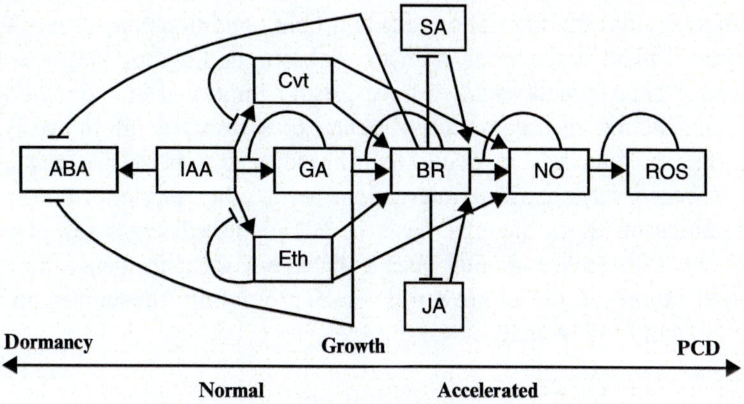

Figure 4. Suggested spatial cross talk of different phytohormones at whole plant level.

BR is a class of phytohormones with its pleotropic interactive responses. The molecular mechanism of action of BRs has not yet been elucidated. However, fragmentary work based on mutant-studies has exposed some part of the mechanism of signal transduction and components of the cascade. Recent advances in molecular understanding led to the identification of mediators participating in BR induced signal transduction cascade (Clouse, 2002; Li, 2003). BRI-1 encoded receptor LRR-RLK (leucine rich repeat receptor like kinases) and BAK1 interaction mediates the BR specific downstream signal cascade. Another peptide hormone systemin in tomato shares the same receptor ortholog tBRI-1. However, partial induction of jasmonic acid (JA) was also suggested to mediate the stress response in tomato in this case. Rice membrane bound MAPK was revealed (Sharma *et al.*, 2001) to be activated in response to exogenous application of brassinolide. MAPKs are known to operate under various abiotic and biotic stress stimuli and in response to JA, ethylene, SA and BRs mediated induced plant responses. The expression of components of phosphatidylinositol pathway have also been implicated by 24-epibrassionolide in cDNA chip based studies (Lin *et al.*, 2004).

7. CONCLUSIONS AND FUTURE PROSPECTS

The yield increasing effects of BRs, synthetic BRs derivatives and BR-containing natural products are considered as protective influences. Often biotic stress together with adverse environmental condition leads to progressive loss of yield and plant growth. Stress priming (low dose) accelerates the defensive potential, preventing the plant to pay off the ecological cost of induced resistance as compared to intensified direct stress. The post-priming increased presence of biochemical signaling components may result in increased and accelerated response to secondary stress. A number of experimental studies under laboratory and field conditions have revealed the activation of plant defense mechanisms to biotic and abiotic stress by BRs. Recent molecular genetic studies have greatly improved our knowledge of signal transduction and regulation of gene expression. The ability of BRs to induce disease tolerance in plants to a broad spectrum of stressful agents seems to result largely from interactions with other phytohormones. The investigations on the molecular basis of BR-mediated stress response and interactions with environmental cues will have a great influence on future field application of these growth-defense promoting substances in crop production and food security.

8. REFERENCES

Ali, B., Hayat, S., and Ahmad, A., 2007. 28-Homobrassinolide ameliorates the saline stress in chickpea (*Cicer arietinum* L.). *Env. Exp. Bot.* **59**: 33–41.

Arrigo, A.P., 1999. Gene expression and the thiol redox state. *Free Radic. Biol. Med.* **27(9–10)**: 936–944.

Arteca, R.N., Tsai, D.S., Schlagnhaufer, C., and Mandava, N.B., 1983. The effect of brassinosteroid on auxin-induced ethylene production by etiolated mung bean segments. *Physiol. Plant.* **59**: 539–544.

Bajguz, A., and Hayat, S., 2009. Effects of brassinosteroids on the plant responses to environmental stresses. *Plant Physiol. Biochem.* **47**: 1–8.

Bao, F., Shen, J., Brady, S.R., Muday, G.K., Asami, T., and Yang, Z., 2004. Brassinosteroids interact with auxin to promote lateral root development in *Arabidopsis*. *Plant Physiol.* **134**: 1624–1631.

Barbier-Brygoo, H., Ephritikhine, G., Kiambt, D., Maurel, C., Palme, K., Schell, J., and Guern, 1991. Perception of the auxin signal at the plasma membrane of tobacco mesophyll protoplasts. *Plant J.* **1**: 83–93.

Bolwell, P.G., Bindschedler, L.V., Blee, K.A., Butt, V.S., Davies, D.R., Gardner, S.L., Gerrish, C., and Minibyeva, F., 2002. The apoplastic oxidative burst in response to biotic stress in plants: a three component system. *J. Exp. Bot.* **53**: 1367–1376.

Cao, S., Xu, Q., Cao, Y., Qian, K., An, K., Zhu, Y., Binzeng, H., Zhao, H., and Kuai, B., 2005. Loss-of-function mutations in *DET2* gene lead to an enhanced resistance of oxidative stress in *Arabidopsis. Physiol. Plant.* **123**: 57–66.

Campos, M.L., Almeida, M.de, Rossi, M.L., Martinelli, A.P. Junior, C.G.L., Figueira, A., Rampelotti-Ferreira, F.T., Vendramim, J.D., Benedito, V.A., and Peres, L.E.P. 2009. Brassinosteroids interact negatively with jasmonates in the formation of anti-herbivory traits in tomato. *J. Exp. Bot.* **60**: 4347–4361.

Chen, Z., Agnew, J.L., Cohen, J.D., He, P., Shan, L., Sheen, J., and Kunkel, B.N., 2007. *Pseudomonas syringae* type III effector AvrRpt2 alters *Arabidopsis thaliana* auxin physiology. *Proc Natl. Acad. Sci. USA* **104**: 10136–20131.

Chory, J., Nagpal, P., and Peto, C.A., 1991. Phenotypic and genetic analysis of *det2*, a new mutant that affects light-regulated seedling development in *Arabidopsis. Plant Cell.* **3**: 445–459.

Clouse, S.D., 2002. Brassinosteroid signal transduction: clarifying the pathway from ligand perception to gene expression. *Mol. Cell* **10**: 973–982.

Clouse, S.D., and Sasse, J.M., 1998. Brassinosteroids: essential regulators of plant growth and development. *Ann. Rev. Plant Physiol. Plant Mol. Biol.* **49**: 427–451.

Cosgrove, D., 1997. Relaxation in a high stress environment: the molecular basis of extensible cell walls and enlargement. *Plant Cell* **9**: 1031–1041.

David, K., Carnero-Diaz, E., Leblanc, N., Monestiez, M., Grosclaude, J., and Perrot- Rechenmann, C., 2001. Conformational dynamics underlie the activity of the auxin-binding protein, *nt-abp1*. *J. Biol. Chem.* **276**: 34517–34523.

Ding, X., Cao, Y., Huang, L., Zhao, J., Xu, C., Li, X., and Wang, S., 2008. Activation of the indole-3-acetic acid-amido synthetase GH3-8 suppresses expansin expression and promotes salicylate- and jasmonate independent basal immunity in rice. *Plant Cell* **20**: 228–240.

Felle, H., Peters, W., and Palme, K., 1991. The electrical response of maize to auxins *Biochim. Biophys. Acta* **1064**: 199–204.

Finkelstein, R., Reeves, W., Ariizumi, T., and Steber, C., 2008. Molecular aspects of seed dormancy. *Ann. Rev. Plant Biol.* **59**: 387–415.

Foyer, C.H., 2005. Redox homeostasis and antioxidant signaling: A metabolic interface between stress perception and physiological responses. *Plant Cell* **17**: 1866–1875.

Fridovich, I., 1975. Superoxide dismutase. *Ann. Rev. Biochem.* **44**: 147–159.

Friebe, A., 2006. Brassinosteroids in induced resistance and induction of tolerances to abiotic stress in plants. Ludwig-Erhard-Allee 2, 531 Bonn, Germany.

Gapper, C., and Dolan, L., 2006. Control of plant development by reactive oxygen species. *Plant Physiol.* **141**: 341–345.

Goda, H., Shimada, Y., Asami, T., Fujioka, S., and Yoshida, S., 2002. Microarray analysis of brassinosteroid-regulated genes in *Arabidopsis. Plant Physiol.* **130**: 1319–1334.

Gunther, F.E., and Scherer, 1996. Phospholipid signaling and lipid-derived second messengers in plants. *Plant Growth Reg.* **18**: 125–133.

Hayat, S., Yadav, S., Ali, B., and Ahmad, A., 2010. Interactive effects of nitric oxide and brassinosteroids on photosynthesis and antioxidant system of *Lycopersicon esculentum Russ. J. Plant Physiol.* **57**: 212–221.

Heil, M., and Bostock, R.M., 2002. Induced systemic resistance (ISR) against pathogens in the context of induced plant defences. *Ann. Bot.* **89**: 503–512.

Holton, N., Cano-Delgado, A., Harrison, K., Montoya, T., Chory, J., and Bishop, G.J., 2007. Tomato BRASSINOSTEROID INSENSITIVE1 is required for systemin-induced root elongation in *Solanum pimpinellifolium* but is not essential for wound signaling. *Plant Cell* **19**: 1709–1717.

Irfan, M., Hayat, S., Hayat, Q., Afroz, S., and Ahmad, A., 2010. Physiological and biochemical changes in plants under waterlogging. *Protoplasma* **241**: 3–17.

Jager, C.E., Symons, G.M., Ross, J.J., and Reid, J.B., 2008. Do brassinosteroids mediate the water stress response? *Physiol. Plant.* **133**: 417–425.

Jain, J.L., Jain, S., and Jain, N., 2006. Fundamentals of biochemistry (6th ed). S. Chand and Company Ltd. N. Delhi: pp. 954–955.

Khripach, V.A., Zhabinskii, V.N., and Groot, A.E., 1999. Brassinosteroids– a new class of plant hormones. Academic Press, SanDiego.

Khripach, V.A., Zhabinskii, V.N., and Malevannaya, N.N., 1997. *Proc. Plant Growth Regul. Soc. Am.* 1997, 101–106.

Korableva, N.P., Platonova, T.A., Dogonadze, M.Z., and Evsunina, A.S., 2002. Brassinolide effect on growth of apical meristems, ethylene production, and abscisic acid content in potato tubers. *Biol. Plant.* **45**: 39–43.

Krishna, P., 2003. Brassinosteroid-mediated stress resistance. *J. Plant Growth Reg.* **22**: 265–275.

Kuc, J., 1982. Induced immunity to plant disease. *BioScience* **32**: 854–860.

Kuzniak, E., and Sklodowska, M., 2005. Fungal pathogen-induced changes in the anti-oxidant systems of leaf peroxisomes from infected tomato plants. *Planta.* **222**: 192–200.

Lamb, C., and Dixon, R.A., 1997. The oxidative burst in plant disease resistance. *Ann. Rev. Plant Physiol. Plant Mol. Biol.* **48**: 251–275.

Leubner-Metzger, G., 2001. Brassinosteroids and gibberellins promote tobacco seed germination by distinct pathways. *Planta* **213**: 758–763.

Li, J., 2003. Brssinosteroids signal through two receptor-like kinases. *Curr. Opin. Plant Biol.* **6(5)**: 494–499.

Li, J.M., and Chory, J., 1997. A putative leucine-rich repeat receptor kinase involved in brassinosteroid signal transduction. *Cell* **90**: 929–938.

Li, J., Nagpal, P., Vitart, V., McMorris, T.C., and Chory, J., 1996. A role for brassinosteroids in light-dependent development of *Arabidopsis. Science* **272**: 398–401.

Lin, W.H., Ye, R., Ma, H., Xu, Z.H., and Xue, H.W., 2004. Genome and expression profile analysis of phsophatidylinositol signaling pathway in higher plants. *Cell Res.*, **14**: 45–52.

Lohse, G., and Hedrich, R., 1992. Characterization of the plasma membrane H+-ATPase from *Vicia faba* cells. *Planta* **188**: 206–214.

Marcinkowska, E., and Wiedlocha, A., 2002. Steroid signal transduction activated at the cell membrane: from plants to animals. *Acta Biochimica Polonica* **49(3)**: 735–745.

Mclntyre, J.L., Dodds, J.A., and Hare, D., 1981. Effects of localized infections of *Nicotiana tabacum* by tobacco mosaic virus on systemic resistance against diverse pathogens and an insect. *Phytopathology* **71**: 297–301.

Nakashita, H., Yasuda, M., Nitta, T., Asami, T., Fujioka, S., Arai, Y., Sekimata, K., Takatsuto, I., and Yoshida, S., 2003. Brassinosteroid functions in a broad range of disease resistance in tobacco and rice. *Plant J.* **33**: 887–898.

Neill, S.J., Desikan, R., Clarke, A., and Hancock, J.T., 2002. Nitric oxide is a novel component of abscisic acid signaling in stomatal guard cells. *Plant Physiol.* **128**: 13–16.

Neill, S.J., Desikan, R., and Hancock, J.T., 2003. Nitric oxide signalling in plants. *New Phytologist* **159**:11–35.

Overmyer, K., Brosche, M., and Kangasjarvi, J., 2003. Reactive oxygen species and hormonal control of cell death. *Trends Plant Sci.* **8(7)**: 1360–1385.

Park, J.E., Park, J.Y., Kim, Y.S., Staswick, P.E., Jeon, J., Yun, J., Kim, S.Y., Kim, J., Lee, Y.H., and Park, C-M., 2007. GH3-mediated auxin homeostasis links growth regulation with stress adaptation response in *Arabidopsis. J. Biol. Chem.* **282**: 10036–10046.

Razem, F.A., Baron, K., and Hill, R.D., 2006. Turning on gibberellin and abscisic acid signaling. *Curr. Opin. Plant Biol.* **9**: 454–459.

Ross, A.F., 1961. Systemic acquired resistance induced by localized virus infections in plants. *Virology* **14**: 340–358.

Scherer, G.F.E., 1996. Phospholipid signalling and lipid-derived second messengers in plants. *Plant Growth Reg.* **18**: 125–133.

Schmidt, J., Böhme, F., and Adam, G.Z., 1996. 24-2pibrassinolide from *Gypsophila perfoliata Z. Naturforsch.* **51c**, 897–899.

Shah, J., 2009. Plants under attack: systemic signals in defence. *Curr. Opin Plant Biol.* **12**: 459–464.

Sharma, A., Matsuoka, M., Tanaka, H., and Komatu, S., 2001. Antisense inhibition of a BRI1 receptor reveals additional protein kinase signaling components downstream to the perception of brassinosteroids in rice. *FEBS Lett.* **507**: 346–350.

Spaepen, S., Vanderleyden, J., and Remans, R., 2007. Indole-3-acetic acid in microbial and microorganism-plant signaling. *FEMS Microbiol. Rev.* **31**: 425–448.

Steber, C.M., and McCourt, P., 2001. A role for brassinosteroids in germination in *Arabidopsis. Plant Physiol.* **125**: 763–769.

Swarbrick, P.J., Schulze-Lefert, P., and Scholes, J.D., 2006. Metabolic consequences of susceptibility and resistance in barley leaves challenged with powdery mildew. *Plant Cell Environ.* **29**: 1061–1076.

Teale, W.D., Ditengou, F.A., Dovzhenko, A.D., Li, X., Molendijk, A.M., Ruperti B., Paponov, B., and Palme, K., 2008. Auxin as a model for the integration of hormonal signal processing and transduction. *Mol. Plant* **1**: 229–237.

Thomine, S., Lelièvre, F., Boufflet, M., Guern, J., and Barbier-Brygoo, H., 1997. Anion channel blockers interfere with auxin responses in dark-grown *Arabidopsis* hypocotyls. *Plant Physiol.* **115**: 533–542.

Torres, M.A., Jones, J.D.G., and Dangl, J.L., 2006. Reactive oxygen species signaling in response to pathogens. *Plant Physiol.* **141**: 373–378.

Vert, G., Walcher, C.L., Chory, J., and Nemhauser, J.L., 2008. Integration of auxin and brassinosteroid pathways by Auxin Response Factor 2. *Proc. Natl. Acad. Sci. USA* **105**: 9829–9834.

Volynets, A.P., Pschenichnaya, L.A., Manzhelesova, N.E., Morozik, G.V., and Khripach, V.A., 1997. The nature of protective action of 24-epibrassinolide on barley plants. *Pl. Grow. Regul. Soc. Am.* **24**: 133–137.

Yi, H.C., Joo1, S., Nam, K.H., Lee, J.S., Kang1, B.G., and Kim, W.T., 1999. Auxin and brassinosteroid differentially regulate the expression of three members of the 1-aminocyclopropane-1-carboxylate synthase gene family in mung bean (*Vigna radiata* L.) *Plant Mol. Biol.* **41**: 443–454.

Wang, D., Pajerowska-Mukhtar, K., Culler, A.H., and Dong, X., 2007. Salicylic acid inhibits pathogen growth in plants through repression of the auxin signaling pathway. *Curr. Biol.* **17**: 1784–1790.

Weiss, D., and Ori, N., 2007. Mechanisms of cross talk between gibberellin and other hormones *Plant Physiol.* **144**: 1240–1246.

Xia, X.J., Zhang, Y., Wu, J.X., Wang, J.T., Zhou, Y.H., Shi, K., Yu, Y.L., and Yu, J.Q., 2009. Brassinosteroids promote metabolism of pesticides in Cucumber. *J. Agric. Food Chem.* **57**: 8406–8413.

Xu, W., Purugganan, M.M., Polisensky, D.H., Antosiewicz, D.M., and Fry, S.C., 1995. *Arabidopsis* TCH4, regulated by hormones and the environment, encodes a xyloglucan endotransglycosylase. *Plant Cell* **7**: 1555–1567.

Zhang, Z., Ramirez, J., Reboutier, D., Brault, M., Trouverie, J., Pennarun, A.-M., Amiar, Z., Biligui, B., Galagovsky, L., and Rona, J.-P., 2005. Brassinosteroids regulate plasma membrane anion channels in addition to proton pumps during expansion of *Arabidopsis thaliana* cells. *Plant Cell Physiol.* **46(9)**: 1494–1504.

Zhang, S., Cai, Z., and Wang, X., 2009. The primary signaling outputs of brassinosteroids are regulated by abscisic acid signaling. *PNAS* **106(11)**: 4543–4548.

Zimmermann, S., Thomine, S., Guern, J., and Barbier-Brygoo, H., 1994. An anion current at the plasma membrane of tobacco protoplasts shows ATP-dependent voltage regulation and is modulated by auxin. *Plant J.* **6**: 707–716.

Chapter 13

THE SIGNIFICANCE OF ETHANOL AS A HORMONE SOLVENT IN EXPERIMENTS ON THE PHYSIOLOGICAL ACTIVITY OF BRASSINOSTEROIDS

ANNA JANECZKO
The Franciszek Górski Institute of Plant Physiology, Polish Academy of Sciences, Niezapominajek 21, 30-239 Krakow, Poland

Abstract: Many compounds acting as regulators in plants are substances insoluble in water. The examples here are steroid compounds of the brassinosteroid group. In biological experiments, where they are exogenously supplied to plants, brassinosteroids are first dissolved in organic solvents such as alcohols or DMSO. These experiments should include appropriate controls, as many research results indicate that alcohols, e.g., methanol or ethanol, are not free of effect on the metabolism in plant cells. In this chapter, examples of experiments on brassinosteroid activity in plants show the significance of using two kinds of controls (absolute ones and ones with a hormone solvent) for the interpretation of results concerning the permeability of cell membranes in oilseed rape leaves during low-temperature stress, changes in protein content in wheat grains, etc. Attention is also paid to the combined effect of exogenous brassinosteroids and ethanol (which is present in trace quantities in aqueous solutions used to treat plants) on brassinosteroid management inside plants. The physiological effects caused by brassinosteroids are thought to depend on many factors (i.e. environmental conditions of plant growth). This chapter shows how the use of different controls may lead to problems in interpretation, and be another factor contributing to the nonreproducibility and/or ambiguity of the results obtained.

Key words: Brassinosteroids, ethanol, hormone solvents, interactions, physiological effects

S. Hayat and A. Ahmad (eds.), *Brassinosteroids: A Class of Plant Hormone*,
DOI 10.1007/978-94-007-0189-2_13,

1. INTRODUCTION

Many regulators acting in plants are insoluble or only sparingly soluble in water. Plants have developed mechanisms that allow the transfer of these regulators among cells, tissues or organs, as this is a basis for their co-ordinated growth and development (Srivastava, 2002). Plant hormones of the brassinosteroid (BR) group can serve as a model example of plant hormones insoluble in water. In biological experiments requiring a selected BR to be applied exogenously, these compounds are first dissolved in a small volume (e.g., 1 cm^3) of organic solvent, e.g., in ethanol of various concentrations (Cerana *et al.*, 1983; Nam and Li, 2004; Janeczko *et al.*, 2010), methanol (Dahse *et al.*, 1990) or in dimethyl sulfoxide – DMSO (Zhang *et al.*, 2009). The thus obtained stock solution is used to obtain the required BR concentration in water or in the culture medium. As shown later in this chapter, these solvents affect the metabolism of plant cells and the use of various types of controls can affect interpretation of the results. Most attention will be directed at the action of ethanol because it is relatively frequently used in experiments with BR.

2. ETHANOL IN PLANTS

2.1 Occurrence of endogenous ethanol in plants

Ethanol is a compound which is naturally produced in plants. The content of ethanol in plants is differentiated and depends on the species, cultivar, type of organ or the stress agent (especially anaerobic stress). Ethanol may be generated by various plant organs; leaves, roots or germinating seeds (Leblová *et al.*, 1969; Kimmerer and Kozlowski, 1982; Chang *et al.*, 1983; Kimmerer and Macdonald, 1987; Obendorf *et al.*, 1990). For instance, in tomato roots, the natural ethanol content ranges from 0.22 to 0.27 $\mu mol\ g^{-1}{}_{FW}$ (Chang *et al.*, 1983). After the roots are placed in an environment composed of 100% of CO_2, the alcohol content increases to 70.52–73.96 $\mu mol\ g^{-1}{}_{FW}$, whereas the exposure of roots to nitrogen leads to an ethanol content ranging from 23.53 to 35.90 $\mu mol\ g^{-1}{}_{FW}$ (Chang *et al.*, 1983). In rice cell suspension cultures ethanol is accumulated both in the cells and in the medium (Igaue and Yagi, 1982). The highest amount of ethanol is accumulated on the 4th day in cells (10 $\mu mole\ g^{-1}{}_{FW}$) and during the stationary growth phase (8th day) (180 mM, ca. 1%) in the medium (Igaue and Yagi, 1982).

2.2 Metabolism of exogenous ethanol in plants

Exogenously applied ethanol is metabolized by plant tissues. In pea cotyledon slices, 2 hours after the exogenous application of ^{14}C ethanol, radioactivity was detected in such components as keto acids, acetaldehyde, acetone, acetic acid, arginine, aspartic acid, glutamic acid, glycine, methionine, serine and malic acid (Cossins and Turner, 1963). The first compounds to appear (the highest radioactivity detected) – only 15 minutes after alcohol application – were acetaldehyde and acetone with ^{14}C, which may be associated with the transformation of ethanol in to acetaldehyde and the interconversion of acetaldehyde to acetone (Cossins and Turner, 1963). Cossins and Beevers (1963) report that radioactivity is also detected in sugars, lipids and in CO_2 released by the plant, and the labelled ethanol is quickly metabolized by carrot disks, pea cotyledons, castor bean endosperm, corn coleoptiles, pea shoots, potato tubers and corn shoots. Experiments carried out by Perata and Alpi (1991) indicated that carrot cells oxidized only small amounts of ethanol to CO_2 and they converted ethanol mainly to acetaldehyde. Transgenic potato tuber tissue, exposed to ethanol solutions, showed a decrease in the level of sugars and a large increase in several organic and amino acids (alanine) (Junker *et al.*, 2003). Therefore, exogenous ethanol can to some extent become a source of carbon in plants.

2.3 Physiological activity of ethanol in plants

The treatment of plants with ethanol results in multi-faceted metabolic effects. The history of the study of the effect of ethanol on plants, which dates back to the beginning of the 20th century, was reviewed by Cossins and Turner (1963). Of particular interest was the observation made by Mer (1958), who found that oat seedlings undergo growth stimulation when subjected to treatment with auxins dissolved in ethanol. The growth, however, did not take place when the plants were treated with potassium and ammonium salts of auxins dissolved in water. In Mer's growth experiments the concentration of ethanol ranged from 0.2% to 0.3%. When applied at concentrations of 0.0002% to 0.2%, ethanol reduced the number of roots in dark-grown cuttings of *Phaseolus aureus* Roxb (Middleton *et al.*, 1978). On the other hand the growth of adventitious roots was slightly stimulated by ethanol (Middleton *et al.*, 1978). They also noted that the use of ethanol at similar concentrations in experiments with auxins involves the induction of changes which may be attributed to auxin. According to Bhattacharya *et al.* (1985) ethanol (at a concentration of about 0.1–0.2%) stimulated rooting in etiolated hypocotyl cuttings *Vigna radiata* L. while at a concentration 0.5%, inhibited it. Bhattacharya *et al.* supposed that ethanol could substitute the requirement for carbohydrates. This was especially so because roots were also formed in the medium containing sucrose, whereas in the medium containing sucrose and ethanol the formation of roots was inhibited. This may prove that the level of carbon source becomes supraoptimal eventually impairing balance (Bhattacharya *et al.*, 1985).

Fruit ripening of the tomato plant was inhibited by tissue concentrations of ethanol that were produced by exposure to exogenous ethanol vapours or its synthesis under anaerobic conditions (Kelly and Saltveit, 1988). A possible explanation for this phenomenon is the inhibition of ethylene action. These authors also note that the action of ethanol, even when used for sterilisation of biological specimens, can sometimes affect experimental results. In another paper on tomato ripening, Saltveit (1989) reports that two mechanisms of ethanol action are possible: at low concentrations (below 0.6 mmol g^{-1} $_{FW}$) ethanol appears to be directly affecting ethylene action, whereas at concentrations above 1.2 mmol g^{-1} $_{FW}$, ethanol appears to have a non-specific toxic effect (decrease in lycopene synthesis or membrane damage). The author discusses the basics of ethanol interaction with cell membranes. Alcohol penetrating the membrane causes structural changes in the lipid system thus increasing the membrane's permeability and denaturation of associated proteins. The author also suggests that as a result of such action of ethanol, the membrane ethylene receptor loses its functionality. It is also known that treatment with ethanol inhibits ethylene synthesis (Podd and Van Staden, 1998). Such phenomena are of practical importance, e.g., in extending the longevity of cut flowers and regulating fruit ripening (Podd and Van Staden, 1998).

Claassens *et al.* (2005) showed that in *in vitro*-cultured growing and maturing potato tubers, ethanol (0.5%) can break dormancy and induce the growth of the apical bud. As an effect of ethanol treatment, the expression of carbohydrate-storage, protein-storage, and cell division-related genes were down-regulated in tuber tissue. Because the effects of ethanol action were blocked by an alcohol dehydrogenase inhibitor, these authors think that this enzyme is an essential player in the metabolic pathways activated by ethanol.

3. THE EFFECT OF METHANOL AND DMSO ON PLANTS

Methanol is one of the major organic compounds in forest air and in the troposphere (Fall and Benson, 1996). Methanol emissions are detected for example from the leaves of *Gossypium hirsutum* L., *Populus deltoides* Bartr. ex. Marsh and *Fagus sylvatica* L. (Hüve *et al.*, 2007). A tissue specific accumulation of methanol is also observed during the *in vitro* culture of immature soybean seeds (Obendorf *et al.*, 1990). The amount of methanol produced by developing soybean cotyledons starts at 37.1 $\mu g\ g^{-1}{}_{FW}$ and decreases over a period of 20 days to 2.95 $\mu g\ g^{-1}{}_{FW}$. Plants emit methanol especially during the early stages of leaf expansion – it is probably produced as a by-product of pectin metabolism during cell wall synthesis (Fall and Benson, 1996). Methanol is an alternative carbon source for quicker and efficient production of the microalgae *Chlorella minutissima* (Kotzabasis *et al.*, 1999). Nonomura and Benson (1992) suggest that plants respond to this alcohol in two or more ways, first utilising photorespiratory and other metabolic pathways available for detoxification and, thereafter, activating a mechanism which improves carbon fixation. The use of this alcohol is reflected in an increase in biomass in algae and in the yield of crop plants, however, this effect depends on such factors as the type of photosynthesis in the plant (C_3–C_4) or growth conditions (good direct sunlight is required), as well as other factors, such as fertilization with nitrogen (Nonomura and Benson, 1992; Kotzabasis *et al.*, 1999). It is interesting that localized methanol treatment of one leaf on a whole plant induces a systemic transmission of effects throughout the plant (Hemming *et al.*, 1995). All the leaves of the plant responded to distal exposure at about the same time. This means that methanol or some component induced by methanol must be transported throughout the plant in a few minutes (Hemming *et al.*, 1995).

DMSO is a widely used commercial solvent derived from trees as a by-product in the production of paper. It was shown that exogenous DMSO can affect plants. When applied to soil at a concentration of 1%, DMSO is toxic for rice cultures (Kumar *et al.*, 1976). Application of DMSO to soil at lower concentrations (0.01% and 0.1%) or as foliar sprays (0.001% and 0.01%), however, slightly increases the dry weight of all plant parts. In this case grain yield is significantly increased by all DMSO treatments. Enzymatic activities, chlorophyll and carotenoid accumulation showed concentration-dependent stimulation or inhibition by DMSO treatments (Kumar *et al.*, 1976). There are also studies concerning the effect of DMSO on the intake and transportation of metals – zinc, manganese, potassium and iron – due to the possible pH change caused by this compound (Estes *et al.*, 1970; Chamel 1972; Kumar *et al.*, 1976). For instance, DMSO significantly increases the uptake of Mn and decreases P uptake by beans when applied to soil at concentrations in excess of 0.01% (Estes *et al.*, 1970).

4. ETHANOL IN EXPERIMENTS WITH BRASSINOSTEROIDS

4.1 Effect of 24-epibrassinolide and ethanol on brassinosteroids content in grains of spring wheat

In the field cultivation of spring wheat, plants were treated with one of the brassinosteroids (24-epibrassinolide – BR_{27}) to improve the yield (Janeczko *et al.*, 2010). Water–ethanol solution of 24-epibrassinolide was applied to the foliage of the plants in the heading stage (BR_{27}: 0.25 mg•dm^{-3}, 0.00625% of ethanol in water) or by pre-sowing seed soaking (BR_{27}: 1 mg•dm^{-3}, 0.025% of ethanol in water). Consequently, two types of controls were used: an absolute control, consisting of completely untreated plants, and two ethanol controls at a concentration of 0.00625% (for plant spraying) and 0.025% (for seed soaking). Three brassinosteroids: brassinolide, castasterone and 24-epicastasterone were found in the harvested grains, using the UPLC-MS technique. The content of castasterone was about 130 pg per gram of grain and did not vary among all the grain samples studied. Significant changes were however noted in the brassinolide and 24-epicastasterone content. In grains of the ethanol control the content of brassinolide increased in both kinds of treatment: an increase was noted as compared to both the absolute control and to the samples treated with brassinosteroid (Figure 1) (Janeczko *et al.*, 2010). In the case of grains harvested from plants obtained from seeds pre-soaked in an ethanol solution, the amount of brassinolide as compared to the absolute control was twice as high (Figure 1). Interestingly, the amount of brassinolide in grains was similar in absolute control samples and in samples from plants pre-treated with 24-epibrassinolide. The amount of

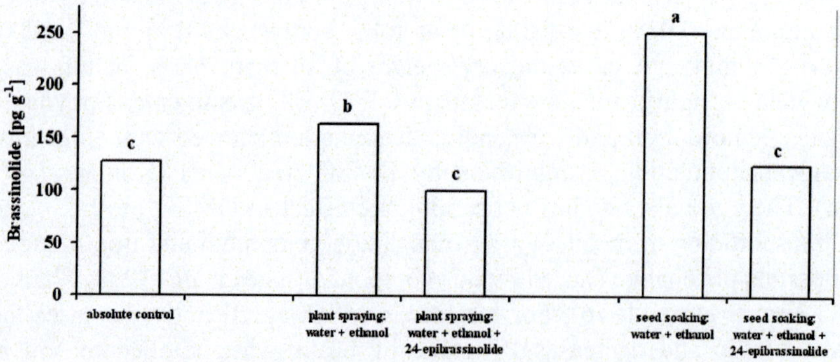

Figure 1. Brassinolide content in grains of spring wheat cv. Torka after treating plants (by spraying) or seeds (by pre-sowing soaking) with various working solutions (based on Janeczko et al., 2010 – modified); absolute control – plants not treated with any solutions. Values marked with the same letters do not significantly differ according to the Duncan test, $P \leq 0.05$.

the third brassinosteroid, 24-epicastasterone, was also changed by the application of solutions containing ethanol and exogenous BR_{27} (Janeczko *et al.*, 2010). The examples presented here show that ethanol can have far-reaching secondary consequences for hormone economy in plants and can even be passed on to the next generation.

4.2 Effect of 24-epibrassinolide and ethanol on qualitative composition of spring wheat grains

In the experiment described above (subsection 4.1.), the same plant specimens were studied for qualitative composition of grains, which involved the content of carbohydrates, lipids, mineral components, proteins and tocopherols (Janeczko *et al.* 2009, 2010). Some effects of ethanol as a hormone solvent were demonstrated as well as its interaction with 24-epibrassinolide in modifying the qualitative composition of grains. Spraying plants with either water + ethanol or water + ethanol + BR_{27} solutions resulted in a similar increase in the starch content in grains, by around 18–20%, as compared to the absolute control (Janeczko *et al.*, 2010). It can be supposed that in this case 24-epibrassinolide had no effect on the content of starch, and the results were solely attributable to ethanol action. It was also established in this experiment that plant treatment with water + ethanol solution decreases the content of soluble proteins in grains. In this case BR_{27} neutralized this effect and increased the content of protein to values found in the absolute control (Janeczko *et al.*, 2010). Figure 2 shows the content of tocopherols in wheat grains harvested from the absolute control and from plants treated with water + ethanol and water + ethanol + 24-epibrassinolide solutions (Janeczko *et al.*, 2009, Janeczko, unpublished data). The content of tocopherols in the absolute control did not differ from their content found in grains, harvested from the spayed plants (Figure 2). In these three specimens, however, a much lower tocopherol content was found than in the grains produced by plants whose seeds had been soaked in water + ethanol and water + ethanol + 24-epibrassinolide solutions. The results obtained for both treatments, pre-sowing seed soaking in ethanol control and in BR_{27}, are similar. In this experiment the method of treatment was probably decisive. Beneficial effects of seed priming are well known (Harris *et al.*, 2001). The pre-sowing soaking of seeds of parental plants increased the content of tocopherols in the next generation of seeds. Comparison of the results obtained with BR_{27}-treated plants (pre-sowing seed soaking with BR_{27}) to the absolute control shows that the hormone significantly increases the content of tocopherols, but when compared to the ethanol control, no action of the hormone was proven.

In the experiments discussed here, no effect of ethanol was found on the content of lipids, total soluble sugars, or selected mineral components, such as magnesium, potassium or sodium (Janeczko *et al.*, 2010).

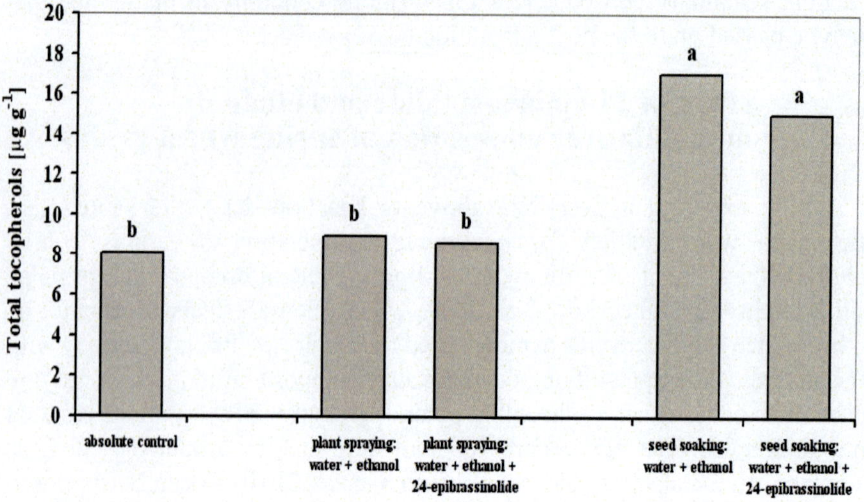

Figure 2. The content of total tocopherols in grains of spring wheat cv. Cytra after application of various working solutions *via* spraying of plants or pre-sowing soaking of seeds (based on Janeczko *et al.*, 2009 – modified; Janeczko, unpublished data); absolute control – plants not treated with any solutions. Values marked with the same letters do not significantly differ according to the Duncan test, $P \leq 0.05$.

4.3 Effect of 24-epibrassinolide and ethanol on permeability of cell membranes

Changes in cell membrane permeability were studied in leaves of oilseed rape after treatment with 24-epibrassinolide (Janeczko *et al.*, 2007). Cell permeability was measured by the conductometric method as a change in ion leakage. An aqueous solution of ethanol (0.467%) containing 24-epibrassinolide (1 µM) was infiltrated into the primary leaves of 21-day old oilseed rape seedlings. Two types of controls were used, plants treated with ethanol and plants treated with water. The ethanol control consisted of leaves infiltrated with an ethanol solution (0.467%) whereas in the water control, leaves were infiltrated only with water. The plants were kept at temperatures of 2°C and 20°C, and after 4 days ion leakage measurements were conducted. An increased membrane permeability was found at 20°C in leaves treated with the water + ethanol + BR_{27} solution as compared to leaves treated with water or water + ethanol solution (Figure 3). Also, the very addition of ethanol increased ion leakage. At a temperature of 2°C the stimulating action of ethanol on ion leakage intensified, but in this case 24-epibrassinolide clearly limited this action (Janeczko *et al.*, 2007).

In the study of thermotolerance of potato, carried out by Confraria *et al.* (2007), 24-epibrassinolide protected *in vitro* cultured plants against heat stress. The effectiveness of protection was however affected by the concentration of ethanol – the solvent for BR_{27} – in the growth medium. The protective effects of 24-epibrassinolide against heat stress (measured by ion leakage) were particularly evident in the presence of 0.1% ethanol, as compared to a lower solvent concentration (Confraria, personal communication).

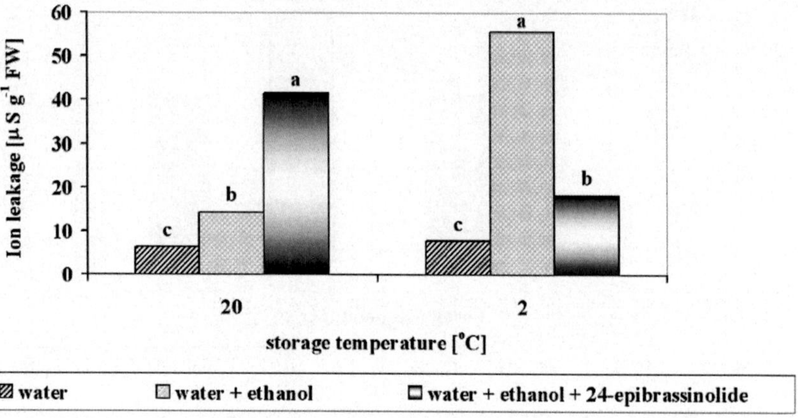

Figure 3. Ion leakage from the primary leaf of oilseed rape seedlings exposed to temperatures of 20°C and 2°C after treatment with various working solutions (based on Janeczko *et al.*, 2007 – modified). Values marked with the same letters do not significantly differ (for each temperature separately) according to the Duncan test, $P \leq 0.05$.

4.4 Effect of 24-epibrassinolide and ethanol on the content of pigments

The effect of 24-epibrassinolide on the content of pigments (chlorophyll *b* and carotenoids) was studied in leaves of oilseed rape kept at 20°C and 2°C (Janeczko *et al.*, 2007). An aqueous solution of ethanol (0.467%) containing 24-epibrassinolide (1μM) was infiltrated into primary leaves of 21-day old oilseed rape seedlings. Leaves of plants of absolute control were not infiltrated. In the case of ethanol control the leaves were injected with an 0.467% aqueous solution of ethanol. Primary leaves were cut off from the oilseed rape seedlings and before the measurement of pigments the leaves were kept in darkness for 7 days at a temperature of 2°C and 3 days at 20°C.

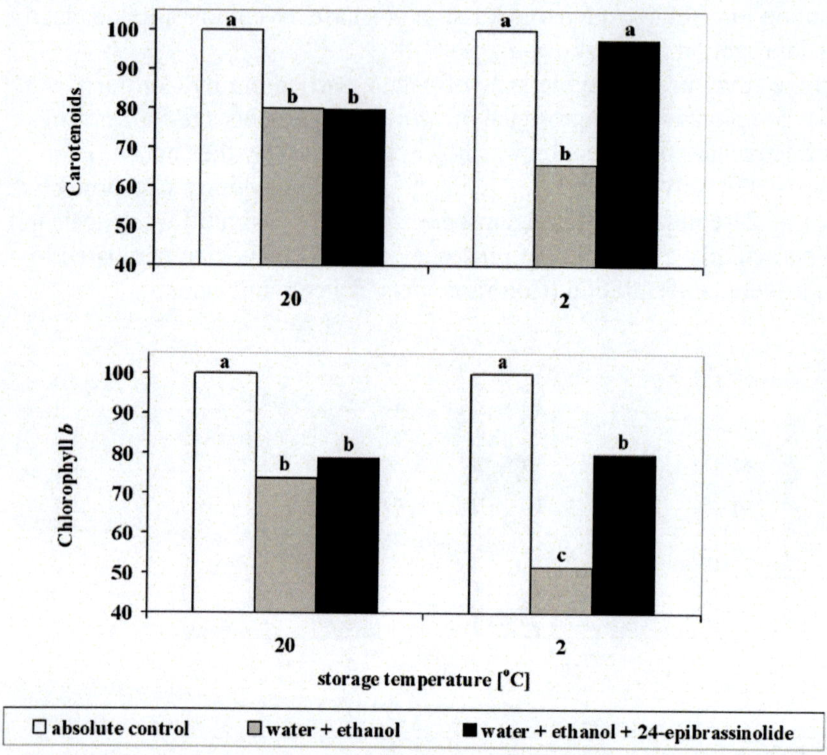

Figure 4. Content of pigments in the primary leaves cut off from the oilseed rape seedlings treated with various working solutions; before the measurement the leaf was kept in darkness for 7 days at a temperature of 2°C and 3 days at 20°C (based on Janeczko *et al*., 2007 – modified); absolute control – plants not treated with any solutions. The results are given as a percentage of the absolute control. Values marked with the same letters (for each temperature separately) do not significantly differ according to the Duncan test, $P \leq 0.05$.

Infiltration with solutions of water + ethanol and water + ethanol + 24-epibrassinolide reduced the contents of chlorophyll *b* and carotenoids at 20°C and 2°C as compared to the untreated (absolute) control (Figure 4), however at 2°C this effect was compensated in the case of carotenoids by 24-epibrassinolide and significantly reduced in the case of chlorophyll *b*.

4.5 Effect of 24-epibrassinolide and ethanol on spring barley heading

Spraying with solutions containing ethanol can also affect plant development. Figure 5 shows the dynamics of barley heading after spraying 18-day old seedlings with solutions containing water + ethanol (0.5%) and water +

ethanol (0.5%) + 24-epibrassinolide (0.5 mg·dm^{-3}) (Pociecha and Janeczko, 2008). Heading of the first plants was observed between the 50th and 55th day of vegetation. On the 54th day, heading was observed in 17% of plants of the absolute control, 10% of the ethanol control and only 2% of the plants sprayed with 24-epibrassinolide. Statistical analysis of the significance of differences in individual pairs of points is given in Table 1. The addition of 24-epibrassinolide caused statistically significant delay in heading. The same tendency was observed for plants sprayed with water–ethanol solution. For instance, on the 64th day of the culture, 74% of plants sprayed with water + ethanol were heading whereas in the absolute control group it was almost 100% plants. Heading of 100% of plants was observed 9 days later in plants sprayed with water + ethanol solution, as compared to the absolute control, and 18 days later in plants sprinkled with water + ethanol + 24-epibrassinolide.

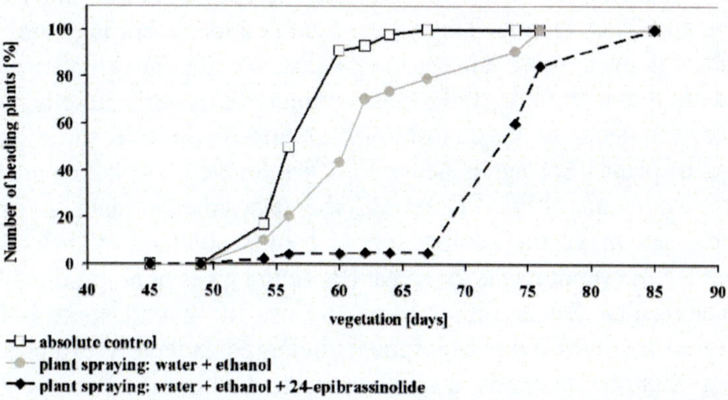

Figure 5. Percentage of heading plants of barley sprayed with various working solutions versus time of vegetation (based on Pociecha and Janeczko, 2008); absolute control – plants not treated with any solutions. The figure reprinted by permission of IPP PAS.

Table 1. Statistical significance of differences in the percentage of heading barley plants for the following pairs: absolute control vs. BR_{27}; ethanol control vs. BR_{27}; absolute control vs. ethanol control was analyzed using the λ^2 test, * $P \leq 0.05$; ** $P \leq 0.01$; *** $P \leq 0.001$; NS - not significant

Comparison of pairs	Vegetation [days]			
	54	64	74	85
Absolute control vs. BR_{27}	*	***	***	NS
Ethanol control vs. BR_{27}	NS	***	***	NS
Absolute control vs. ethanol control	NS	**	*	NS

5. CONCLUSION

Brassinosteroids belong to the group of steroidal regulators that are insoluble in water. In biological experiments, where they are supplied exogenously to plants, brassinosteroids are first dissolved in organic solvents such as alcohols and the thus-obtained stock solution is used to prepare aqueous solutions or nutritional media of required brassinosteroid concentrations. This is associated with the need to use properly treated specimens as controls. The examples presented here show how the use of various controls in experiments designed to study the action of exogenous brassinosteroids can affect the interpretation of the results concerning physiological activity of the compound tested. This results from the fact that the hormone solvent itself affects plant metabolism. Also, it seems that exogenous steroid and ethanol interact in the regulation of some physiological processes. Their action can be additionally affected by ambient conditions (especially temperature), plant genotype and other factors. This issue requires a more thorough study, but what makes it even more interesting is that we already know some selected elements that interact between such compounds as ethylene-ethanol, brassinosteroids-ethylene, or brassinosteroids-auxins-ethylene in the metabolic pathways of plants (Schlagnhaufer *et al.*, 1984; Saltveit, 1989; Lim *et al.*, 2002; De Grauwe *et al.*, 2005). Referring the experimental data to one selected control can make the comparison of results obtained in different laboratories harder and supports the view that BR action gives nonreproducible and ambiguous results. On the other hand, the use of several controls in experiments often involves those experiments being considerably extended. In such cases it might be more advantageous to test whether the solvent, e.g., ethanol, does not affect the experimental results. If an arbitrary control is used without a preliminary test, one should be prepared for some of the effects of the brassinosteroid not being revealed.

6. REFERENCES

Bhattacharya, S., Bhattacharya, N.C., and Bhatnagar V.B. 1985. Effect of ethanol, methanol and acetone on rooting etiolated cuttings of *Vigna radiata* in presence of sucrose and auxin. *Ann. Bot.*, **55**: 143–145.

Cerana, R., Bonetti, A., Marrè, M.T., Romani, G., Lado, P., and Marrè, E. 1983. Effects of a brassinosteroid on growth and electrogenic proton extrusion in Azuki bean epicotyls. *Physiol. Plant.*, **59**: 23–27.

Chamel, A. 1972. Effect of dimethyl sulfoxide on penetration and migration of ^{59}Fe applied to maize leaves. *Physiol. Plant.*, **26**: 170–174.

Chang, L.A., Hammett, L.K., and Pharr, D.M. 1983. Carbon dioxide effects on ethanol production, pyruvate decarboxylase, and alcohol dehydrogenase activities in anaerobic sweet potato roots. *Plant Physiol.*, **71**: 59–62.

Claassens, M.M.J., Verhees, J., van der Plas, L.H.W., van der Krol, A.R., and Vreugdenhil, D. 2005. Ethanol breaks dormancy of the potato tuber apical bud. *J. Exp. Bot.*, **56**: 2515–2525.

Confraria, A., Desikan, R., and Neill, S. 2007. Brassinosteroids protect plants against heat stress. *Comp. Biochem. Physiol. A. Mol. Integr. Physiol.*, **146**(Suppl. 1): S279.

Cossins, E.A., and Beevers, H. 1963. Ethanol metabolism in plant tissues. *Plant Physiol.*, **38**: 375–380.

Cossins, E.A., and Turner, E.R. 1963. The metabolism of ethanol in germinating pea seedlings. *J. Exp. Bot.*, 14: 290–298.

Dahse, I., Sack, H., Bernstein, M., Petzold, U., Müller, E., Vorbrodt, H.M., and Günter, A. 1990. Effects of (22S,23S)–homobrassinolide and related compounds on membrane potential and transport of Egeria leaf cells. *Plant Physiol.*, **93**: 1268–1271.

De Grauwe, L., Vandenbussche, F., Tietz, O., Palme, K., and Van Der Straeten D. 2005. Auxin, ethylene and brassinosteroids: tripartite control of growth in the *Arabidopsis* hypocotyls. *Plant Cell Physiol.*, **46**: 827–836.

Estes, G.O., Mack, H.J., and Willis, D.L. 1970. Influence of dimethyl sulfoxide (DMSO) on nutrient uptake by potatoes and bush beans. *Agron. J.*, **62**: 461–463.

Fall, R., and Benson, A.A. 1996. Leaf methanol – the simplest natural product from plants. *Trends Plant Sci.*, **1**: 296–301.

Harris, D., Pathan, A.K., Gothkar, P., Joshi, A., Chivasa, W., and Nyamudeza, P. 2001. On-farm seed priming: using participatory methods to revive and refine a key technology. *Agricult. Syst.*, **69**: 151–164.

Hemming, D.J.B., Criddle, R.S., and Hansen, L.D. 1995. Effects of methanol on plant respiration. *J. Plant Physiol.*, **146**: 193–198.

Hüve, K., Christ, M.M., Kleist, E., Uerlings, R., Niinemets, Ü., Walter, A., and Wildt, J. 2007. Simultaneous growth and emission measurements demonstrate an interactive control of methanol release by leaf expansion and stomata. *J. Exp. Bot.*, **58**: 1783–1793.

Igaue, I., and Yagi, M. 1982. Alcohol dehydrogenase from cultured rice cells. *Plant Cell Physiol.*, **23**: 213–225.

Janeczko, A., Biesaga-Kościelniak, J., and Dziurka, M. 2009. 24-epibrassinolide modifies seed composition in soybean, oilseed rape and wheat. *Seed Sci. Technol.* **37**: 625–637.

Janeczko, A., Biesaga-Kościelniak, J., Oklešťková, J., Filek, M., Dziurka, M., Szarek-Łukaszewska, G., and Kościelniak, J. 2010. Role of 24-epibrassinolide in wheat production. physiological effects and uptake. *J. Agron. Crop Sci.*, **196**: 311–321.

Janeczko, A., Gullner, G., Skoczowski, A., Dubert, F., and Barna, B. 2007. Effects of brassinosteroid infiltration prior to cold treatment on ion leakage and pigment contents in rape leaves. *Biol. Plant.* **51**: 355–358.

Junker, B.H., Chu, C., Sonnewald, U., Willmitzer, L., and Fernie, A.R. 2003. In plants the *alc* gene expression system responds more rapidly following induction with acetaldehyde than with ethanol. *FEBS Lett.*, **535**: 136–140.

Kelly, M.O., and Saltveit, M.E. Jr. 1988. Effect of endogenously synthesized and exogenously applied ethanol on tomato fruit ripening. *Plant Physiol.*, **88**: 143–147.

Kimmerer, T.W., and Kozlowski, T.T. 1982. Ethylene, ethane, acetaldehyde, and ethanol production by plants under stress. *Plant Physiol.*, **69**: 840–847.

Kimmerer, T.W., and Macdonald, R.C. 1987. Acetaldehyde and ethanol biosynthesis in leaves of plants. *Plant Physiol.*, **84**: 1204–1209.

Kotzabasis, K., Hatziathanasiou, A., Bengoa-Ruigomez, M.V., Kentouri, M., and Divanach, P. 1999. Methanol as alternative carbon source for quicker efficient production of the microalgae Chlorella minutissima: Role of the concentration and frequence of administration. *J. Biotechnol.*, **70**: 357–362.

Kumar, B., Gangwar, M.S., and Rathore, V.S. 1976. Effect of dimethyl sulfoxide (DMSO) on zinc availability (L-value), growth and metabolic activities of rice plants. *Plant Soil*, **45**: 235–246.

Leblová, S., Zimáková, I., Sofrová, D., and Barthová, J. 1969. Occurrence of ethanol in pea plants in the course of growth under normal and anaerobic conditions. *Biol. Plant.*, **11**: 417–423.

Lim, S.H., Chang, S.C., Lee, J.S., Kim S.K., and Kim, S.Y. 2002. Brassinosteroids affect ethylene production in the primary roots of maize (*Zea mays* L.). *J. Plant Biol.*, **45**: 148–153.

Mer, C.L. 1958. Growth-promoting effect of ethanol on oat seedlings. *Nature*, **182**: 1812–1813.

Middleton, W., Jarvis, B.C., and Booth, A. 1978. The effects of ethanol on rooting and carbohydrate metabolism in stem cuttings of *Phaseolus Aureus* Roxb. *New Phytol.*, **81**: 279–285.

Nam, K.H., and Li J. 2004. The *Arabidopsis* transthyretin-like protein is a potential substrate of BRASSINOSTEROID-INSENSITIVE 1. *Plant Cell*, **16**: 2406–2417.

Nonomura, A.M., and Benson, A.A. 1992. The path of carbon in photosynthesis: improved crop yields with methanol. *Proc. Natl. Acad. Sci. USA*, **89**: 9794–9798.

Obendorf, R.L., Koch, J.L., Gorecki, R.J., Amable, R.A., and Aveni, M.T. 1990. Methanol accumulation in maturing seeds. *J. Exp. Bot.*, **41**: 489–495.

Perata, P., and Alpi, A. 1991. Ethanol metabolism in suspension cultured carrot cells. *Physiol. Plant.*, **82**: 103–108.

Pociecha, E., and Janeczko, A. 2008. Antioxidant enzymes activity and development of barley after high temperature stress – impact of 24-epibrassinolide. *Zesz. Probl. Post. Nauk Rol.* **524**: 83–93 (in Polish with English abstract and figures description).

Podd, L.A., and Van Staden, J. 1998. The role of ethanol and acetaldehyde in flower senescence and fruit ripening – a review. *Plant Growth Regul.*, **26**: 183–189.

Saltveit, M.E., Jr. 1989. Effect of alcohols and their interaction with ethylene on the ripening of epidermal pericarp discs of tomato fruit. *Plant Physiol.*, **90**: 167–174.

Schlagnhaufer, C., Arteca, R.N., and Yopp, J.H. 1984. Evidence that brassinosteroid stimulates auxin-induced ethylene synthesis in mung bean hypocotyls between S-adenosylmethionine and 1-aminocyclopropane-1-carboxylic acid. *Physiol. Plant*, **61**: 555–558.

Srivastava, L.M. 2002. Plant growth and development. Hormones and the environment Oxford: Academic Press.

Zhang, S., Cai, Z., and Wang X. 2009. The primary signaling outputs of brassinosteroids are regulated by abscisic acid signalling. *Proc. Nat. Acad. Sci. USA*, **106**: 4543–4548.

Chapter 14

IMMUNOASSAYS OF BRASSINOSTEROIDS

V.A. KHRIPACH, V.N. ZHABINSKII AND R.P. LITVINOVSKAYA
Institute of Bioorganic Chemistry, Academy of Sciences of Belarus, Kuprevich str., 5/2, 220141 Minsk, Belarus

Abstract: Identification of brassinosteroids (BS) in plants is rather time consuming and laborious process involving many operations. The highest measured level of BS in plant tissues is about 10^{-5}%, and the expenditure rates in agricultural use vary in a range of 5–50 mg/ha. Such concentrations create a serious problem for measuring BS in plant sources that is of crucial importance for investigation of their biosynthesis and metabolism and for their efficient practical application. The known chromatographic methods of BS analysis suffer from the relatively long analysis time, laborious and expensive sample cleanup prior to analysis and the need of rather sophisticated equipment. As a result, they are not suitable for routine analysis of samples. A good alternative to chromatographic methods is the immunochemical approach because of simplicity of equipment and measuring procedure. There is no doubt that namely this type of assay is the most realistic for wide-scale practical applications. The present chapter summarizes the latest developments in this field with an emphasis on contributions made in authors' laboratory.

Key words: Brassinosteroids, plant hormones, immunoanalysis, ELISA, haptens, immunogen, labeling

1. INTRODUCTION

Brassinosteroids (BS) have been already known for more than 30 years as plant growth hormones showing essentiality or beneficial functions for growth and development of plants. BS have profound physiological effects at extremely low concentrations. The highest measured level of BS in plant

S. Hayat and A. Ahmad (eds.), *Brassinosteroids: A Class of Plant Hormone*,
DOI 10.1007/978-94-007-0189-2_14,

tissues is about 10^{-5}%, and the expenditure rates in agricultural use vary in a range of 5–50 mg/ha. In this respect, analytical aspects of BS are of crucial importance in achieving reliable results both in scientific research and in practical application.

In the early period of BS studies the most fruitful analytical approaches were those based on gas chromatography – mass spectrometry and high-performance liquid chromatography of BS derivatives. The main drawback of all such apparatus approaches is the relatively long analysis time, laborious and expensive sample cleanup prior to analysis and the need of rather sophisticated equipment. As a result, these methods can not be applied for online monitoring and a large number routing analysis of BS in plant samples.

A good alternative to the apparatus approaches seems to be immunochemical analysis. Being low molecular weight compounds, BS cannot generally stimulate an immunogenic response of their own. They can be analyzed if coupled to larger carrier proteins (e.g., bovine serum albumin) to give a synthetic antigen capable of inducing antibody formation.

The first attempt to produce antibodies to BS was described by Horgen *et al.* (1984). Monoclonal antibodies directed against a synthetic brassinosteroid non-covalently bound to a carrier protein were produced in CAF_1 mice. However, the resulting antibodies exhibited very high (up to 48%) cross-reactivity to abundant plant sterols and were essentially useless for analytical purposes.

It is evident that the design of corresponding hapten for conjugation with a protein needs a greater level of effort during the development of BS immunoassay. The obtained hapten should preserve the most significant structural and stereochemical features of the parent molecule. The linker required for coupling to a carrier protein should be of a sufficient length and with a position of attachment allowing appropriate presentation of all important recognition elements.

There are two main strategies that can be implemented in the preparation of haptens: to use the native BS functionality for coupling with a protein or to introduce an additional functional group for that purpose. Evidently, the first approach can be realized much more easily. However, it is known that antibody's specificity is maximal for a part of hapten furthest located from the functional group to which the protein is linked (Law, 2005). With haptens prepared by attaching the carrier protein at BS functional group, the latter will be sterically hindered thus preventing its specific recognition. That is the main disadvantage of such an approach, which nevertheless is the most widely used because of its simplicity.

2. IMMUNOGEN PREPARATION

2.1 O-(Carboxymethyl)oxime haptens linked in the C-6 position

The first assay of that kind was developed by Yokota *et al.* (1990b). Antiserum against castasterone was prepared by immunizing a rabbit with the corresponding conjugate **1** of bovine serum albumin and carboxymethyloxime of castasterone (Figure 1).

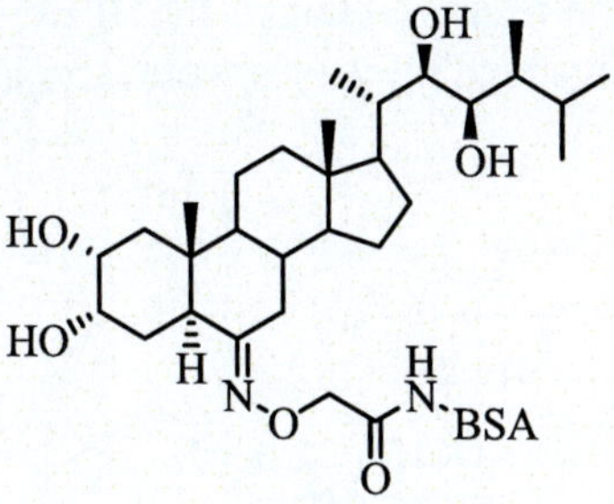

1

*Figure 1. Structure of conjugate **1**.*

The antiserum was shown to recognize a number of natural BS with varying specificities. Maximal cross-reactivity was observed for BS having a 6-keto group or a lactone function in cycle B. The antibodies were able to distinguish between natural 22R,23R- and unnatural 22S,23S-isomers. The affinity to 24S-BS was the highest, but compounds differing from castasterone by an alkyl substituent at C-24 still exhibited considerable cross-reactivity. It was concluded also that antibodies were not sensitive to the stereochemistry of the hydroxy groups in the ring A. Surprisingly, a very low cross-reactivity was found for 6-deoxoderivatives. The developed radioimmunoassay was used to measure endogenous BS in seeds and stems of *Phaseolus vulgaris* L. (Yokota *et al.*, 1990b). Comparison with bioassay method of BS analysis showed that the RIA gave similar results in monitoring BS fractions and required much smaller amount of material. The antibodies were also used for localization studies in developing and mature pollen of rye-grass *Lolium perenne* L. (Taylor *et al.*, 1993). They showed that BS were increasingly accumulated through the developmental sequence of pollen in starch granules indicating that these were storage organelles for BS.

Further investigations were directed toward broadening the utility of this methodology to BS with other substitution pattern at C-24. Two groups reported results of their studies on immunoassay of (24R)-BS (Khripach *et al.*, 2005a–d, 2006, 2007; Swaczynová *et al.*, 2006). Synthesis of the immunogenic conjugate **5** was performed starting from 24-epicastasterone **2** *via* the corresponding 6-*O*-(carboxymethyl)oxime **3** and activated ester **4** (Figure 2). Bovine serum albumin (BSA) was used as a protein antigen for antibodies production. As expected, the obtained antibodies were highly specific to 24-epicastasterone and 24-epibrassinolide (Khripach *et al.*, 2007; Swaczynová *et al.*, 2007). The applicability of the elaborated HPLC-ELISA technique was illustrated through the results of BS detection in young tissue extracts (Swaczynová *et al.*, 2007). The level of 24-epibrassinolide in *A. thaliana* seedlings (9.5 pmol/g of fresh weight) was found to be in a good correlation with the results obtained by HPLC-MS (7.56 pmol/g). For plants *D. carota* the corresponding values were 2.56 and 1.37 pmol/g, correspondingly.

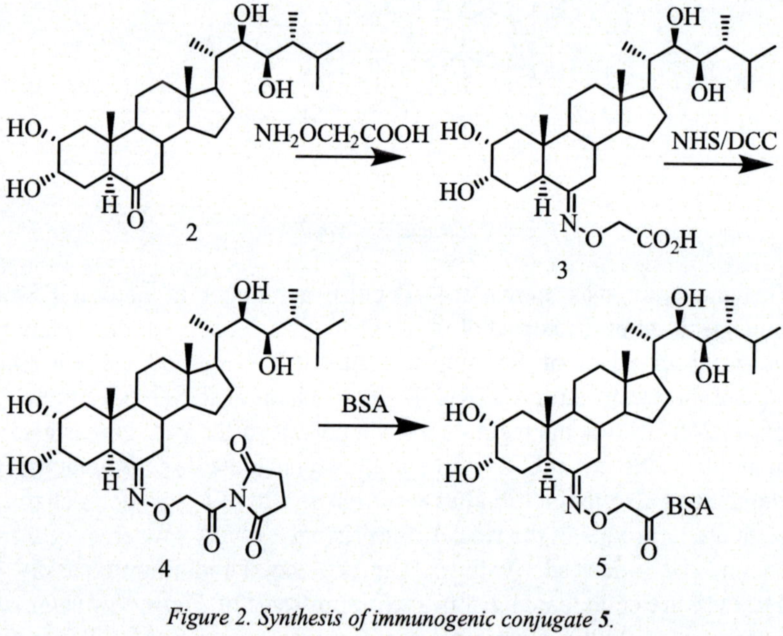

Figure 2. Synthesis of immunogenic conjugate 5.

Essentially no cross-reactivity was observed with non-BS compounds and only marginal for (22S,23S)-isomers. Antibodies could distinguish readily between (24R)-BS and those having other alkyl substituents at C-24. That prompted to develop similar immunoenzymatic assays for (24S)-methyl- and (24S)-ethyl-BS.

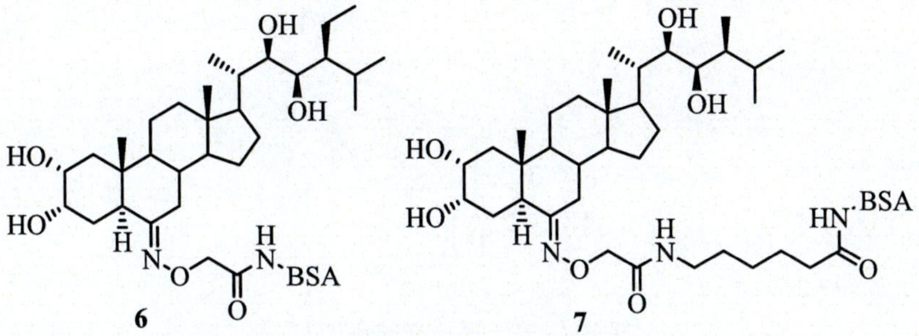

*Figure 3. Structures of immunogenic conjugates **6** and 7.*

Immunogenic conjugate **6** (Figure 3) for induction of antibodies against (24S)-ethyl-BS was prepared from 28-homocastasterone (Khripach *et al.*, 2008a,c–f) in a manner similar to that of the (24R)-methyl-BS (Khripach *et al.*, 2007). In the case of (24S)-methyl-BS linker was elongated with one aminocaproic unit to afford conjugate **7** (Khripach *et al.*, 2009).

Although 6-*O*-(carboxymethyl)oxime based immunoassays proved to be very practical, search for a new design of BS haptens was a necessity. The main disadvantage of antibodies produced from the BS-BSA conjugates having a linker at C-6 position is serious limitations in the structural differentiation between major groups of natural BS (6-oxo-7-oxa-, 6-oxo- and 6-deoxo-).

2.2 Haptens linked through hydroxy groups

BS contain up to 4 hydroxy groups which potentially could be used for the attachment with a protein. The first attempt to realize such an idea was undertaken by Schlagnhaufer *et al.* (1988, 1991). However, the obtained results were not completely satisfactory. The immunogenic conjugate was prepared by coupling goat or mouse albumin to a carboxylated BS derivative. The latter was obtained by refluxing 22S,23S-24-epibrassinolide with succinic anhydride in pyridine for 12 h. Although no information was reported on the structure of the carboxylated BS, one can expect formation of 2-succinylated compound **8** (Figure 4) based on relative reactivity of BS hydroxy groups and authors' statement that this was a monosuccinate derivative. The resulting antiserum from mice exhibited relatively low affinity for natural BS, what in addition to limited data on cross-reactivity made difficult a direct comparison with other studies.

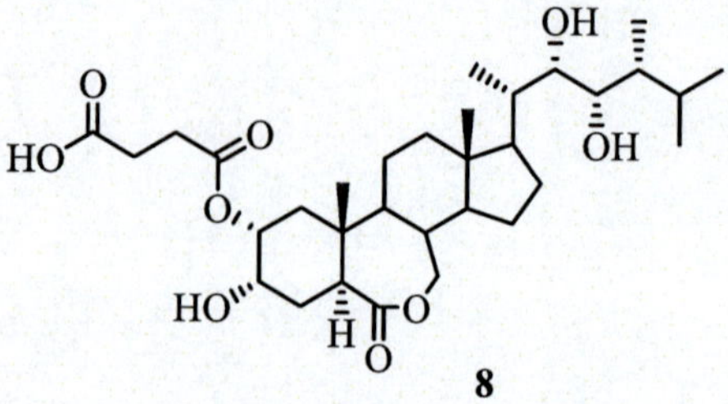

8

Figure 4. Structure of compound 8.

All the above immunoassays have the same drawbacks associated with a poor recognition of functional group(s) adjacent to a linker. The solution of the problem was seen in the application of haptens containing a full set of BS inherent functional groups. This meant that an additional functional group should be introduced into BS molecule for coupling with a protein. Among many possibilities for such a functionalization, introduction of an additional hydroxy group in terminal part of the side chain was supposed to be a good choice. Coupling the protein through spacer at C-26 should allow both the cyclic part and the side chain to be exposed for recognition by the antibodies to maximize their specificity in respect to the cycle B. With such an approach, specific antibodies could be raised to all major cyclic structures for their selective determination.

Steroidal ester **10** accessible from stigmasterol **9** was selected as a key intermediate for the preparation of 26-functionalized BS analogue (Khripach *et al.*, 2008b). Hydride reduction of the ester **10** followed by acetylation gave compound **11** (Figure 5), subsequent acidic treatment of which resulted in regeneration of cyclic part characteristic of sterols. Further functionalization of the molecule was done in a manner similar to that applied for the preparation of natural BS (Khripach *et al.*, 1999). Tosylation of the alcohol **12** followed by *i*-steroidal rearrangement of the formed tosylate and Jones oxidation of the intermediate alcohol led to a cycloketone **13**. Its transformation into Δ^2-6-ketone **14** was achieved by heating in dimethylacetamide with pyridinium bromide followed by removal of acetyl protecting group.

Next compound **15** should be transformed into the corresponding lactone that could be done by its Baeyer-Villiger oxidation. However, this reaction proceeds with formation of both regioisomeric lactones, which can be more easily separated when all hydroxy groups are protected as acetates. Therefore **15** was first protected as pentaacetate and then subjected to Baeyer-Villiger oxidation to give pentaacetoxylactone **16**. Its saponification followed by relactonization led to the lactone **17**. The hemisuccinate group was introduced by treatment of the diacetonide **18** with succinic anhydride in pyridine in the presence of 4-dimethylaminopyridine. Removal of isopropylidene protection by acidic hydrolysis afforded the derivative **20** containing a succinic moiety. The reaction of the hapten **20** with *N*-hydroxysuccinimide led to activated *N*-succinimide ester **21**. Its interaction with BSA in aqueous dioxane solution led after lyophilization to conjugate **22**.

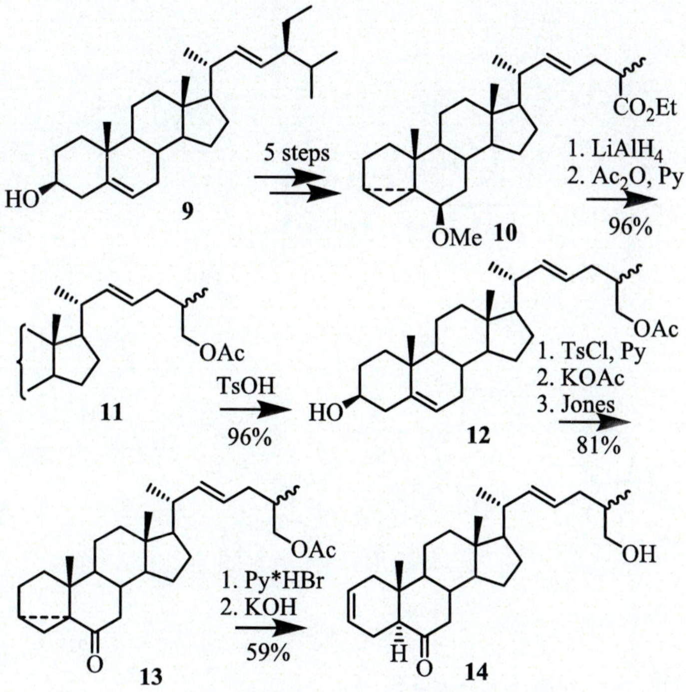

Figure 5. Synthesis of compound 14.

Introduction of both diol groups characteristic of BS was done by Sharpless dihydroxylation of the dienone **14** proceeding with formation of pentol **15** as the main product (Figure 6).

Another BS hapten containing an ethyl group at C-24 was prepared in a similar manner (Figure 7). The construction of the required carbon skeleton of the desired compound was achieved by coupling the aldehyde **23** with lithium butyne followed by Claisen rearrangement of the allylic alcohol **25** (Litvinovskaya *et al.*, 2009).

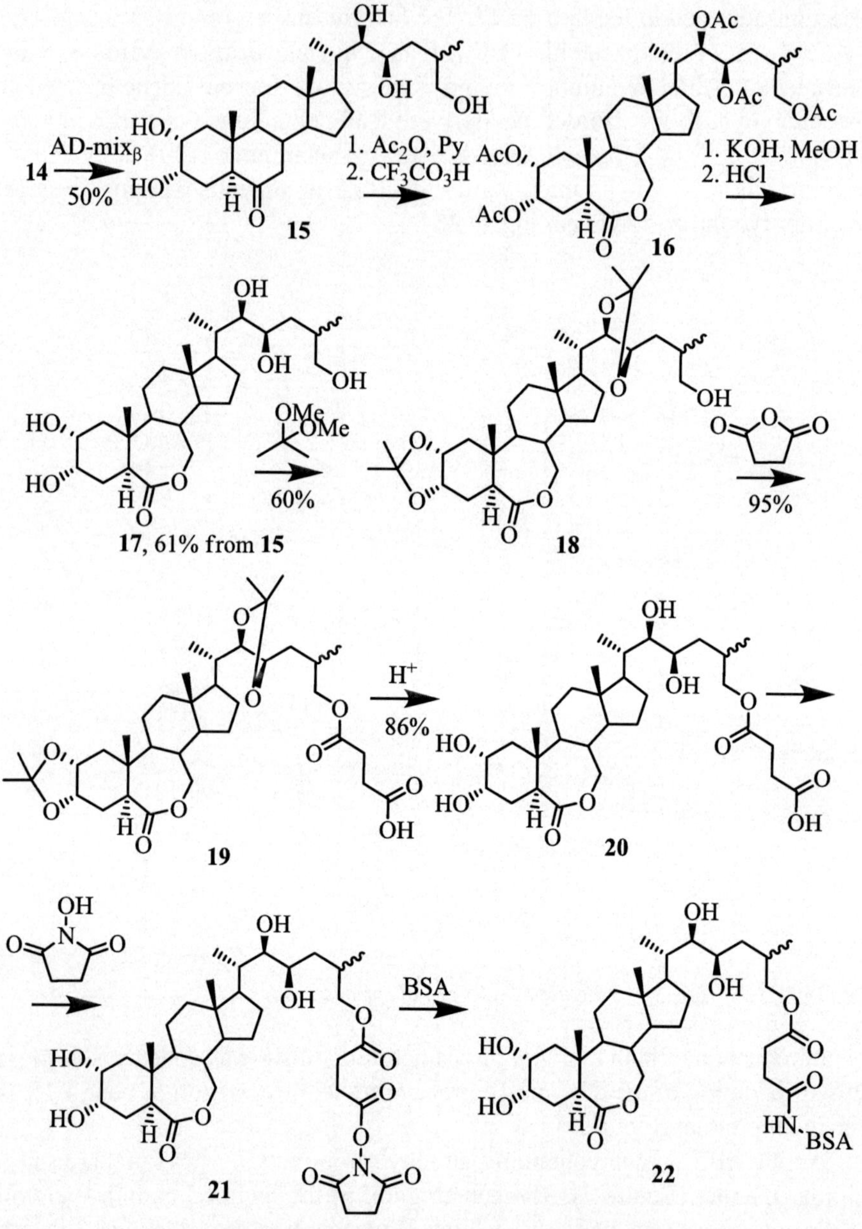

Figure 6. Synthesis of immunogenic conjugate 22.

Transformation of the cyclic part according to the aforementioned procedure gave the dienone **27**. Its Sharples dihydroxylation proceeded with the introduction of the expected 2α,3α- and 22R,23R-diol groups. However, under the reaction conditions ester reacted with the adjacent hydroxy group to give lactone **28**. Its treatment with base followed by a careful acidification afforded ring-opened hydroxy acid **29**. Activation of the carboxylic acid with N-hydroxysuccinimide (NHS) produced ester **30**, which is a versatile hapten for BS immunoassay.

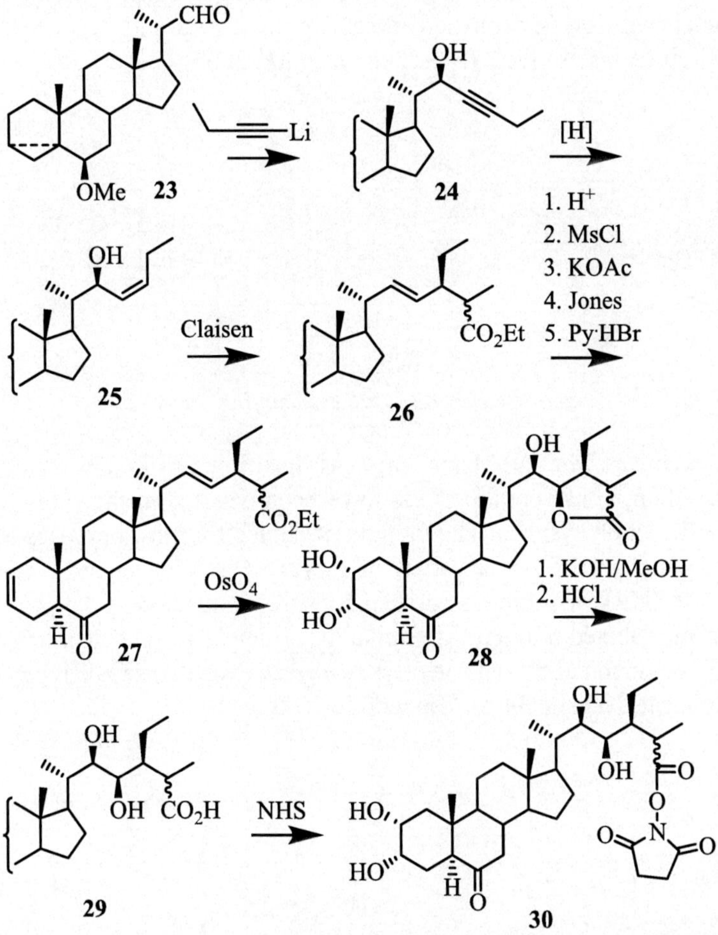

*Figure 7. Synthesis of N-hydroxysuccinimide ester **30**.*

3. ANTIGEN LABELING

Isotopically labeled antigen is an ideal compound for the use as an internal standard, because of its identical immunological properties to the analyte. Synthesis of a number of radioisotope labeled BS has been published to date, and all of them can be potentially used for immunoanalysis (Kolbe *et al.*, 1992, 1998; Seo *et al.*, 1989; Yokota *et al.*, 1990a). In practice, the most simple procedures of BS labeling were employed. Radiolabeled brassinolide **31** was prepared by a catalytic reduction of dolicholide over platinum using tritium gas (Yokota *et al.*, 1990b) (Figure 8). An advantage of such an approach is that the tritium atoms were introduced into stable and non-exchangeable position. Labeling of BS 6-ketones can be easily achieved by a base-catalyzed isotope exchange at α-positions to a carbonyl group as shown for [^{3}H]-epicastasterone **32** (Swaczynová *et al.*, 2007).

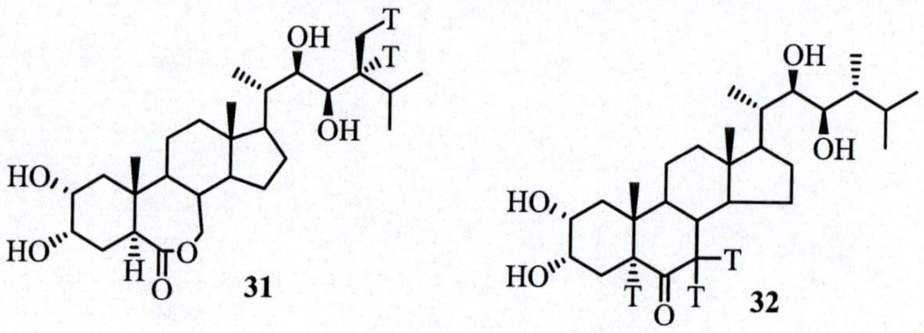

*Figure 8. Structures of tritiated antigens **31** and **32**.*

To eliminate the problems such as health hazard and environment contamination, considerable efforts have been devoted to the study of non-isotopic detection systems. The method of choice for most purposes is enzymatic labeling. Out of a wide variety of enzymatic labeling agents, horseradish peroxidase (HRP) is commonly used for BS immunoassay. Preparation of the enzyme labeled antigens is similar in principle to that of immunogens. Thus, condensation of the activated esters **4** and **21** with horseradish peroxidase gave conjugates **33** and **34** correspondingly (Figure 9).

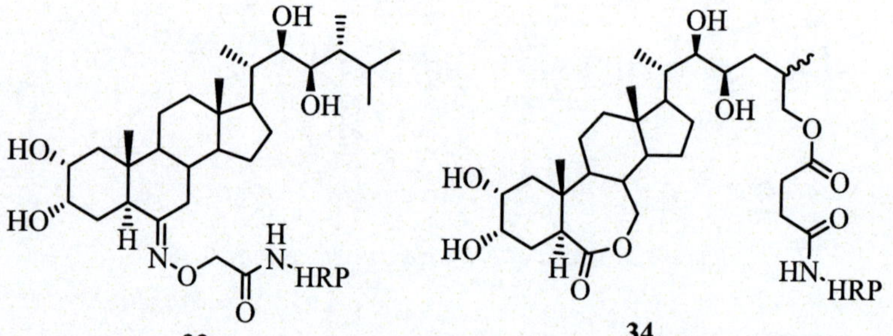

*Figure 9. Structures of enzyme labeled antigens **33** and **34**.*

Although ELISA is quite reliable diagnostic tool, in some cases it is not the ideal method for BS measuring. Under the conditions when the enzyme conjugate may be damaged (e.g. by enzymatic hydrolysis with plant proteases), the analytical results will be incorrect. The solution of the problem is seen in using a non-enzyme label. One of the best alternatives seems to be fluorescence labeling because of its high sensitivity, ease of operation, and in situ applicability.

The dansylhydrazide **35** was obtained by the reaction of 24-epicastasterone **2** with dansylhydrazine in a mixture of dimethylformamide and acetic acid (Borisevich *et al.*, 2008). Its fluorescence energetic and time characteristics were shown to be suitable for developing a method for immunofluorescent analysis of the corresponding BS (Figure 10).

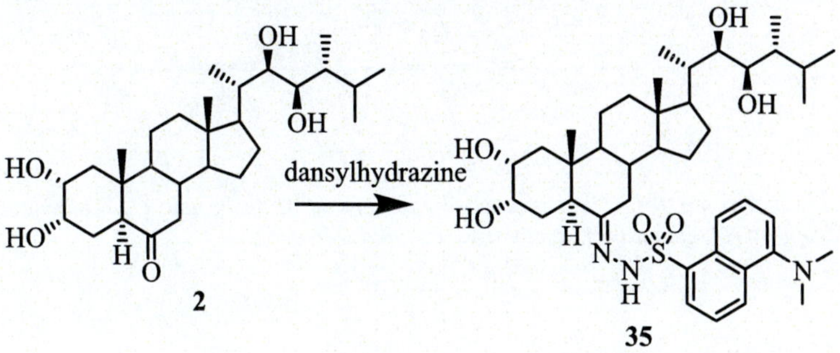

Figure 10. Synthesis of dansylhydrazide 35.

Its absorption properties are shown at Figure 11. The absorption spectra of **35** in most solvents are very similar and in fact are sum of the corresponding spectra of dansylhydrazine and 24-epicastasterone **2**. This means that no specific interaction between the label and brassinosteroid occurs. Compound **35** exhibits intense fluorescence (Figure 12). The fluorescence spectra shift to longer wavelength as the dielectric constant of the solvents increases. The fluorescence spectrum of the aqueous solution of the conjugate **35** consists of two bands with λ_{max} = 539 and 490 nm. This may be the result of the existence of the hydrazide **35** in two isomeric forms. The quantum yield γ = 0.4 and fluorescence lifetime τ = 13.1 ns were measured for the conjugate in ethanol. The τ values were similar in other solvents: 12.92 (acetonitrile), 16.23 (diethylether), and 16.22 ns (tetrahydrofuran). The fluorescence decay was exponential. The fluorescence excitation spectrum, which was measured within the first absorption band, coincided with the absorption spectrum.

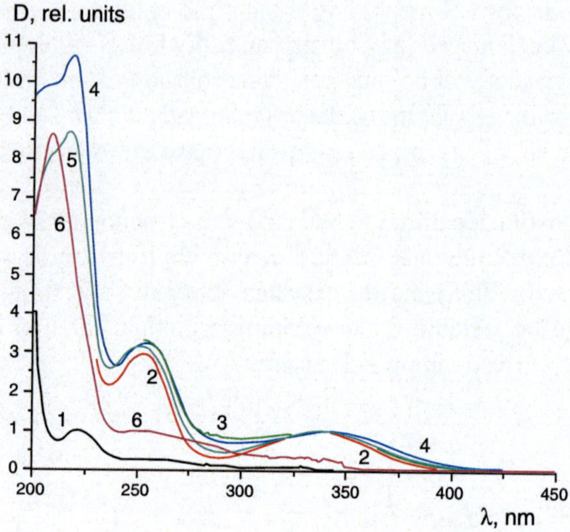

Figure 11. Electronic absorption spectra of solutions of **35** *in hexane (1), diethylether (2), tetrahydrofuran (3), ethanol (4), acetonitrile (5), and water (6).*

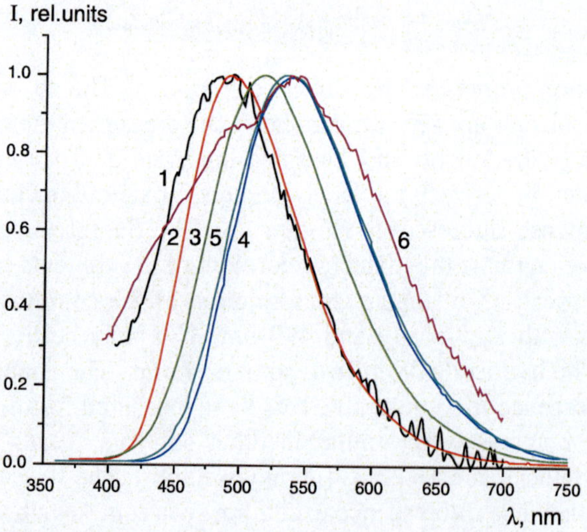

Figure 12. Electronic fluorescence spectra of solutions of **35** *in hexane (1), diethylether (2), tetrahydrofuran (3), ethanol (4), acetonitrile (5), and water (6).*

An interesting conjugate **38** (Figure 13) containing four steroidal residues was prepared by the reaction of 24-epibrassinolide **36** with porphyrin boronate **37** in chloroform (Raichyonok *et al.*, 2009). It should be noted that its formation in this reaction was somewhat surprising. 24-Epibrassinolide **36** has two vicinal diol functions, but only those of the side chain (containing sterically more hindered 22 and 23 hydroxy groups) reacted with **37**. The outcome of this reaction can be explained by higher stability of 22,23-boronate cyclic ethers in comparison with 2,3-boronate cyclic ethers.

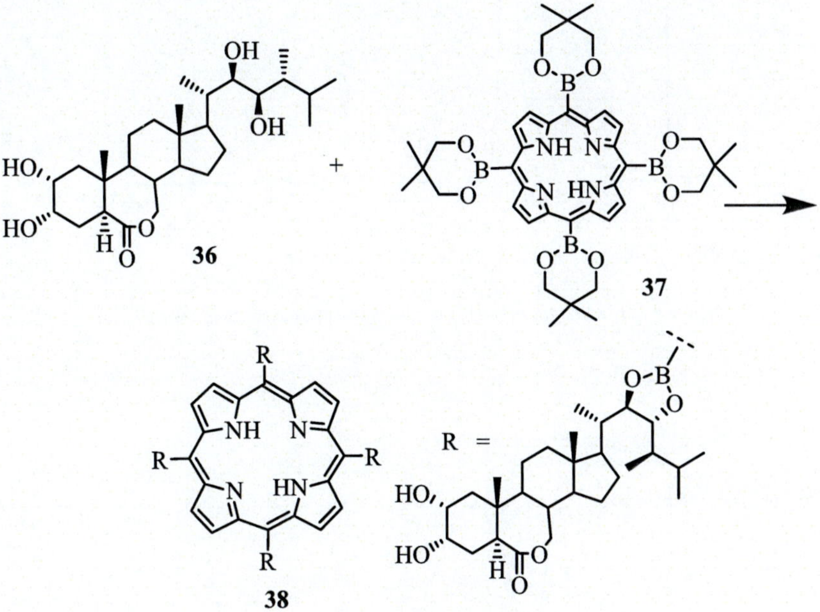

36

37

38

*Figure 13. Synthesis of porphyrin conjugate **38**.*

The absorption properties of conjugate **38** and starting porphyrin **37** are almost identical (Figure 14), indicating the absence of macrocycle deformation in **38** because of steric interference. The fluorescence spectrum of **38** is shown at Figure 15. The quantum yield $\gamma = 0.048$ and fluorescence lifetime $\tau = 10.6$ ns were determined for **38** in tetrahydrofuran. It should be mentioned that fluorescence properties of **38** underwent certain changes after keeping the solution for 2–3 days at room temperature. This can be explained by formation of diprotonated form of porphyrin.

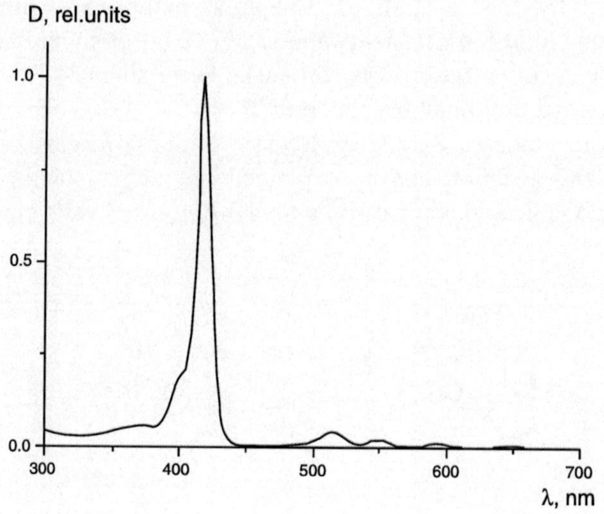

*Figure 14. Electronic absorption spectrum of solutions of **38** in THF.*

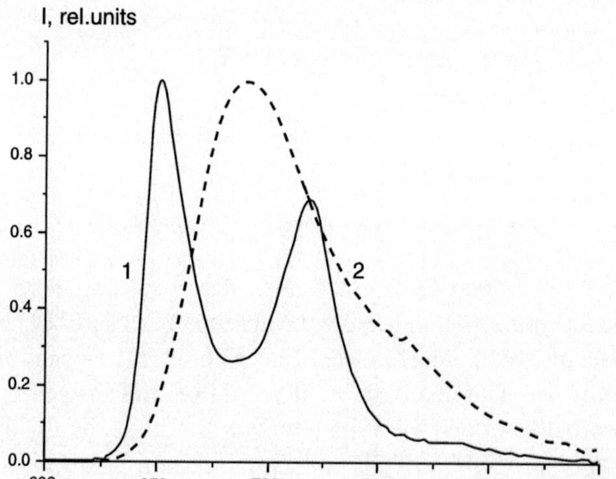

*Figure 15. Fluorescence spectrum of solutions of **38** in THF (1 – free base form, 2 – diprotonated form).*

4. BINDING PROPERTIES OF ANTIBODIES

Specificity is an important issue in immunoassay analysis of BS which are present in plants as a group of structurally related compounds together with other steroids occurring at much higher level. Table 1 summarizes

Table 1. Relative cross reactivities of different immunoassays toward BS and other steroids, %

Compound	Antigen used for antibodies' production					
	1[a]	7[b]	5[c]	5[d]	6[e]	22[f]
Lactones						
Brassinolide	105	100	0.45	1.3	9	180
24-Epibrassinolide		8	95	24.3	1.48	100
28-Homobrassinolide	31	14	0.37	1	100	100
28-Norbrassinolide					0.01	100
25-Methylbrassinolide	22					
Dolicholide	1					
6-Ketones						
Castasterone	100	100	3		13.8	8.0
24-Epicastasterone	13	5	100	100	2.5	2.9
28-Homocastasterone	36	18	15	0.11	100	1.9
28-Norcastasterone	27				2.4	4.3
Typhasterol	29					
Teasterone	45					
3-Epicastasterone	129					
2-Epicastasterone	36					
22S,23S-BS						
22S,23S-24-Epibrassinolide			6.4	0.2	0.6	100
22S,23S-28-Homobrassinolide	<0.001			0.3		
22S,23S-Castasterone	<0.001					
22S,23S-24-Epicastasterone			7.2	0.4	0.15	2.6
22S,23S-28-Homocastasterone	<0.001		0.04	0.1	1.0	
6-Deoxo-BS						
6-Deoxocastasterone	0.3	12.5			2.2	0.85
6-Deoxo-28-homocastasterone					5.7	<0.01
6-Deoxo-24-epicastasterone		2	100		0.27	<0.01
6-Deoxo-24-norcastasterone					0.3	0.5
Non-brassinosteroid compounds						
Campesterol		0.01	<0.01		<0.01	<0.01
Stigmasterol		0.04	<0.01	<0.01	<0.01	<0.01
Ergosterol		0.09	<0.01	<0.01		
Cholesterol		0.03	<0.01	0.05	<0.01	<0.01
Pregnenolone		0.09	<0.01		<0.01	<0.01
Androstenolone		0.01	<0.01			<0.01
Ecdysterone			0.3			<0.01

[a]Relative to castasterone (Yokota *et al.*, 1990b).
[b]Relative to castasterone (Khripach et al., 2009).
[c]Relative to 24-epicastasterone (Khripach *et al.*, 2007).
[d]Relative to 24-epicastasterone (Swaczynová *et al.*, 2007).
[e]Relative to 28-homocastasterone (Khripach *et al.*, 2008a).
[f]Relative to 28-norbrassinolide (Khripach *et al.*, 2008b).

cross-reactivities of BS antibodies developed by various investigators against a range of steroids.

It should be noted that for all antibodies no or only marginal cross-reactivity was observed with non-brassinosteroid compounds like sterols or ecdysteroids. In most cases, 6-*O*-(carboxymethyl)oxime derived antibodies showed quite different specificity for BS differing by an alkyl substituent at C-24. Thus, in immunoassay to 24R-methyl-BS developed by Swaczynová *et al.* (2006, 2007) only a small cross-reactivity (<1.3%) was observed for natural phytohormones. Somewhat lower but acceptable specificity (up to 15% for 28-homocastasterone) was obtained for similar immunoassay by Khripach *et al.* (2006, 2007). An appreciable cross-reactivity (up to 45%) with natural BS displayed antibodies for 24S-methyl BS (Yokota *et al.*, 1990b). Better results for similar immunoassay were obtained by Khripach *et al.* (2009) (up to 18% for 28-homocastasterone). In nearly all experiments no or a very small cross-reactivity was observed for unnatural 22S,23S-BS (with the exception of 22S,23S-24-epibrassinolide (Khripach *et al.*, 2008b)). Very good results were achieved for immunoassay based on hapten linked through 26-hydroxy group (Khripach *et al.*, 2008b). This assay proved to be highly specific for compounds containing a lactone moiety in the B-ring. The antibodies showed a strong cross-reactivity for all studied lactones (up to 180%). At the same time, very small (if at all) cross-reaction was showed for 6-deoxo-BS. The obtained antibodies cross-reacted weakly with BS of 6-keto series with potencies ranging from 1.9% for 28-homocastasterone to 8% for castasterone.

5. ACKNOWLEDGEMENT

The authors are grateful to the Belorussian Republican Foundation for Fundamental Research (Grants X07K-045 and X08P-064) and ISTC (Project 1332) for financial support.

6. REFERENCES

Borisevich, N.A., Raichenock, T.F., Khripach, V.A., Zhabinskii, V.N., and Ivanova, G.V., 2008. Solution electronic spectra of brassinosteroid and a synthesized conjugate of a steroid and a fluorescent label. *J. Appl. Spectr.*, **75**: 75–79.

Horgen, P.A., Nakagawa, C.H., and Irvin, R.T., 1984. Production of monoclonal-antibodies to a steroid plant-growth regulator. *Can. J. Biochem. Cell Biol.*, **62**: 715–721.

Khripach, V.A., Zhabinskii, V.N., and de Groot, A. (1999) *Brassinosteroids. A new class of plant hormones*. Academic Press. San Diego.

Khripach, V.A., Sviridov, O.V., Litvinovskaya, R.P., Pryadko, A.G., Drach, S.V., Matveentsev, V.D., and Demchenko, N.I., 2005a. Conjugate of 24-epicastasterone with bovine serum albumine as immunogen. Pat. 9770 (BY).

Khripach, V.A., Sviridov, O.V., Litvinovskaya, R.P., Pryadko, A.G., Drach, S.V., Matveentsev, V.D., and Novik, T.V., 2005b. Conjugate of 24-epicastasterone with horse radish peroxidase as a labeled antigen for immunoenzymatic analysis of 24R-methylbrassinosteroids. Pat. 9771 (BY).

Khripach, V.A., Sviridov, O.V., Litvinovskaya, R.P., Pryadko, A.G., Drach, S.V., Zhabinskii, V.N., Zavadskaya, M.I., 2005c. {[(22R,23R,24R)-2α,3α,22,23-Tetrahydroxy-24-methyl-5α-cholest-6-ylidenoamino]-oxy}acetic acid as intermediate in the synthesis of conjugate of 24-epicastasterone with bovine serum albumine. Pat. 10120 (BY).

Khripach, V.A., Sviridov, O.V., Litvinovskaya, R.P., Pryadko, A.G., Drach, S.V., Mikhailopulo, K.I., Matveentsev, V.D., Zhabinskii, V.N., Zavadskaya, M.I., Novik, T.V. 2005d. A method for quantitative determination of 24R-methylbrassinosteroids (24-epicastasterone and 24-epibrassinolide) and a composition for its realization. Pat. 10512 (BY).

Khripach, V.A., Sviridov, O.V., Litvinovskaya, R.P., Pryadko, A.G., Drach, S.V., and Zhabinskii, V.N., 2006. Analysis of brassinosteroids. *Polish J. Chem*, **80**: 651–654.

Khripach, V.A., Sviridov, O.V., Pryadko, A.G., Litvinovskaya, R.P., Drach, S.V., Matveentsev, V.D., Novik, T.V., Mikhailopulo, K.I., Zhabinskii, V.N., Zavadskaya, M.I., Averkova, M.A., Drachenova, O.A., and Chaschina, N.M., 2007. Immunoenzymatic analysis of (24*R*)-brassinosteroids. *Russ. J. Bioorg. Chem.*, **33**: 347–353.

Khripach, V.A., Litvinovskaya, R.P., Raiman, M.E., Drach, S.V., Zhabinskii, V.N., Sviridov, O.V., Pryadko, A.G., and Novik, T.V., 2008a. Synthesis and immunochemical determination of 28-homobrassinosteroids. *Vestsi of the National Academy of Sciences of Belarus, ser. khim. sci.*, 47–58.

Khripach, V., Zhabinskii, V., Antonchick, A., Litvinovskaya, R., Drach, S., Sviridov, O., Pryadko, A., Novik, T., Matveentsev, V., and Schneider, B., 2008b. A new type of modified brassinosteroids for enzyme-linked immunosorbent assay. *Nat. Prod. Commun.*, **3**: 735–748.

Khripach, V.A., Litvinovskaya, R.P., Raiman, M.E., Drach, S.V., Zhabinskii, V.N., Sviridov, O.V., Pryadko, A.G., and Novik, T.V., 2008c. {[(22R,23R,24S)-2α,3α,22,23-Tetrahydroxy-24-ethyl-5α-cholest-6-ylidenoamino]-oxy}acetic acid as hapten in the synthesis of conjugate of 28-homocastasterone with bovine serum albumine. Pat. 11831 (BY).

Khripach, V.A., Litvinovskaya, R.P., Raiman, M.E., Drach, S.V., Zhabinskii, V.N., Sviridov, O.V., Pryadko, A.G., and Novik, T.V., 2008d. Conjugate of 28-homocastasterone with bovine serum albumine as immunogen. Pat. 11832 (BY).

Khripach, V.A., Litvinovskaya, R.P., Raiman, M.E., Drach, S.V., Zhabinskii, V.N., Sviridov, O.V., Pryadko, A.G., and Novik, T.V., 2008e. A method for quantitative determination of 28-homobrassinosteroids and a composition for its realization. Pat. 11833 (BY).

Khripach, V.A., Litvinovskaya, R.P., Raiman, M.E., Drach, S.V., Zhabinskii, V.N., Sviridov, O.V., Pryadko, A.G., and Novik, T.V., 2008f. Conjugate of 28-homocastasterone with horse radish peroxidase as a labeled antigen for immunoenzymatic analysis of 28-homobrassinosteroids. Pat. 11835 (BY).

Khripach, V.A., Litvinovskaya, R.P., Drach, S.V., Averkova, M.A., Zhabinskii, V.N., Sviridov, O.V., Pryadko, A.G., Novik, T.V., and Matveentsev, V.D., 2009. Immunoenzymatic analysis of (24S)-methylbrassinosteroids. *Doklady of the National Academy of Sciences of Belarus*, **53**: 74–77.

Kolbe, A., Marquardt, V., and Adam, G., 1992. synthesis of tritium labeled 24-epibrassinolide. *J. Label. Compd. Radiopharm.*, **31**: 801–805.

Kolbe, A., Schneider, B., Voigt, B., and Adam, G., 1998. Labelling of biogenetic brassinosteroid precursors. *J. Label. Compd. Radiopharm.*, **41**: 131–137.

Law, B. (ed.), 2005. *Immunoassay. A practical guide*. Taylor & Francis. U.K.

Litvinovskaya, R.P., Raiman, M.E., and Khripach, V.A., 2009. Synthesis of 28-homobrassinosteroids modified in the 26-position. *Chem. Nat. Comp.*, **45**: 647–652.

Raichyonok, T.F., Khripach, V.A., Zhabinskii, V.N., Konstantinova, O.V., Drasar, P.B., and Monti, D. 2009. Synthesis and spectral-luminescence properties of the conjugate of 24-epibrassinolide with porphyrin. *J. Appl. Spectr.*, **76**: 542–546.

Schlagnhaufer, C.D., Arteca, R.N., and Philips, A.T., 1988. Monoclonal antibodies against brassinosteroid. A new plant growth regulating compound. *Plant Physiol.*, **86**: 679.

Schlagnhaufer, C.D., Arteca, R.N., and Phillips, A.T., 1991. Induction of anti-brassinosteroid antibodies. *J. Plant Physiol.*, **138**: 404–410.

Seo, S., Nagasaki, T., Katsuyama, Y., Matsubara, F., Sakata, T., Yoshioka, M., and Makisumi, Y., 1989. Synthesis of (22R,23R)- and (22S,23S)-[4-^{14}C]-24-Epibrassinolide. *J. Label. Compd. Radiopharm.*, **27**: 1383–1393.

Swaczynová, J., Šíša, M., Hniličková, J., Kohout, L., and Strnad, M., 2006. Synthesis, biological, immunological and anticancer properties of a new brassinosteroid ligand. *Polish J. Chem.*, **80**: 629–635.

Swaczynová, J., Novák, O., Hauserová, E., Fuksová, K., Šíša, M., Kohout, L., and Strnad, M., 2007. New techniques for the estimation of naturally occurring brassinosteroids. *J. Plant Growth Regul.*, **26**: 1–14.

Taylor, P.E., Spuck, K., Smith, P.M., Sasse, J.M., Yokota, T., Griffiths, P.G., and Cameron, D.W., 1993. Detection of brassinosteroids in pollen of *Lolium perenne* L. by immunocytochemistry. *Planta*, **189**: 91–100.

Yokota, T., Ogino, Y., Takahashi, N., Saimoto, H., Fujioka, S., and Sakurai, A., 1990a. brassinolide is biosynthesized from castasterone in catharanthus-roseus crown gall cells. *Agric. Biol. Chem.*, **54**: 1107–1108.

Yokota, T., Watanabe, S., Ogino, Y., Yamaguchi, I., and Takahashi, N., 1990b. Radioimmunoassay for brassinosteroids and its use for comparative-analysis of brassinosteroids in stems and seeds of *Phaseolus vulgaris*. *J. Plant Growth Regul.*, **9**: 151–159.

Chapter 15

TRANSCRIPTOMICS AND PROTEOMICS STUDY IN REGULATION OF BRASSINOSTEROIDS

H.F. JUAN
Department of Life Science, Institute of Molecular and Cellular Biology, and Graduate Institute of Biomedical Electronics and Bioinformatics, National Taiwan University, No. 1, Sec. 4, Roosevelt Rd., Taipei 106, Taiwan

Abstract: The transcriptomics (microarray) technique is a powerful, high-throughput method for accurately determining changes in global gene expressions in functional genomics. It can measure gene expressions of tens of thousands of discrete sequences in a single array. This technique has been used for novel genes discoveries, gene function determinations and pathway dissections. Genes with a similar expression pattern often function in the same biological processes. Genes are the blueprints but proteins are the functional entities of the cell, regulating which genes are activated and when, relaying signals within and between cells, and driving metabolic processes. Proteomics technique is the other powerful high-throughput method to describe the study of the complete set of proteins (proteome) that is expressed at a given time in a cell, tissue, organ or organism. Proteomics characterizes cellular proteins and as well as its abundance, state of modification, protein complexes and interactions. Therefore, the holistic study of biological transformations will enable more rapid advances in elucidating biochemical pathways. The plant steroid hormones brassinosteroids (BRs) play an important role in a wide range of developmental and physiological processes. In this chapter, how to use transcriptomics and proteomics techniques to study the regulation of brassinosteroids in plants will be introduced.

Key words: Transcriptomics, proteomics, brassinosteroids, clustering, gene network, regulation

S. Hayat and A. Ahmad (eds.), *Brassinosteroids: A Class of Plant Hormone*,
DOI 10.1007/978-94-007-0189-2_15, © Springer Science+Business Media B.V. 2011

1. INTRODUCTION

Brassinosteroids (BRs) are growth-promoting natural products found at low levels in pollen, seeds, and young vegetative tissues throughout the plant kingdom (Clouse and Sasse, 1998). Combination with HPLC, GC-MS techniques and recent achievements in genetic research, such as mutants with deficient BR biosynthesis and signaling pathways from *Arabidopsis*, pea, tomato and rice, the biosynthetic pathway (including early and late C6-oxidation) and signaling components (BRI1, BAK 1, BZR1 and BES1) have been identified (Schumacher and Chory, 2000; Bishop and Koncz, 2002; Wang and He, 2004; He *et al.*, 2005; Tang *et al.*, 2008b; Tang *et al.*, 2009). In addition to the growth-promoting activity, the correlations of BR to seeds germination, flowering, senescence and stress tolerance are also reported (Krishna, 2003). These studies confirmed predictions of plant physiologists that BRs are essential for plant growth and must be considered along with auxin, cytokinins, gibberellins, abscisic acid, and ethylene in any model of plant development (Rao *et al.*, 2002). Thus, it is consensus among scientists to regard BR as the sixth class of phytohormone. There are 40 free BRs and 4 BRs conjugates have been characterized from 44 plant species, including 37 angiosperms (9 monocots and 28 dicots), 5 gymnosperms, 1 alga, and 1 pteridophyte (Fujioka and Yokota, 2003). In the physiological studies, brassinolide (BL), 24-epibrassinolide, (EBS) and 28-homobrassinolide are the three biologically active BRs that have been widely used (Rao *et al.*, 2002).

The mechanisms of BR-induced physiological responses and the downstream components, including gene transcription, protein expression and intermediate metabolites have been widely discussed (Schluter *et al.*, 2002; Yang and Komatsu, 2004). The two major physiological responses with regard to molecular regulations of BR include rice lamina inclination and bean second internode elongation (Rao *et al.*, 2002). The mechanisms of BR-regulated growth and development were discovered in the cDNA microarray and proteomic analysis of rice lamina joints (Komatsu *et al.*, 2003; Yang and Komatsu, 2004; Hirano *et al.*, 2008). The mechanisms of internode elongation were only illustrated in transcriptome while absent in proteome (Clouse *et al.*, 1992; Zurek and Clouse, 1994). In the reason of the post-transcriptional modification and alternative splicing, the levels of mRNA do not necessarily predict the levels of corresponding proteins in physiological responses. A comprehensive analysis of proteins regulated by BR in the highly elongated stem is necessary (Zurek *et al.*, 1994; Goda *et al.*, 2002). Additionally, BRs like the other nonpeptide small organic compounds have been thought to mediate plant cell-cell interactions (Motose *et al.*, 2009). In this chapter, I will introduce transcriptomics and proteomics as well as bioinformatics approach such as clustering and gene network construction to study the regulation of brassinosteroids (Figure 1).

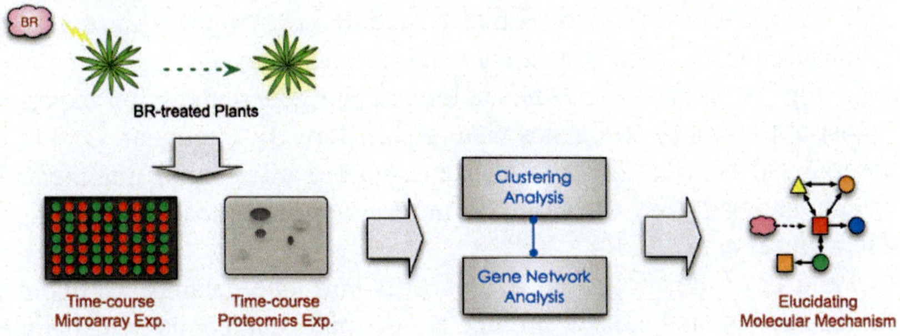

Figure 1. A diagram to study the regulation of brassinosteroid (BR) in plant using combinations of multiple approaches that include transcriptomics, proteomics, clustering and gene network.

2. TRANSCRIPTOMICS STUDY IN BR REGULATION

The transcriptomics (microarray) technique is a powerful, high-through put method for accurately determining changes in global gene expressions in functional genomics (Yue *et al.*, 2001; Ideker *et al.*, 2001). It can measure gene expressions of tens of thousands of discrete sequences in a single array (Hughes *et al.*, 2000). This technique has been used for novel genes discoveries (Seo *et al.*, 2000), gene function determinations, and pathway dissections (Naour *et al.*, 2001; Hughes *et al.*, 2000). Genes with a similar expression pattern often function in the same biological processes. Genes are the blueprints but proteins are the functional entities of the cell, regulating which genes are activated and when, relaying signals within and between cells, and driving metabolic processes.

Many reports provided comprehensive view of the physiological functions of BRs using microarray analysis (Goda *et al.*, 2002; Clouse, 2002; Mockaitis and Estelle, 2004; He *et al.*, 2005; Osakabe *et al.*, 2005; Nemhauser *et al.*, 2004; Nakamura *et al.*, 2006; Nemhauser *et al.*, 2006; Wu *et al.*, 2008; Kim *et al.*, 2010). One public available microarray data were produced by AtGenExpress Consortium which is a multinational effort designed to uncover the transcriptome of the multicellular model organism *Arabidopsis thaliana*. The project is coordinated by Detlef Weigel, Thomas Altmann and Lutz Nover with funding from the German Arabidopsis Functional Genomics Network (AFGN), and includes contributions from Germany, supported by DFG, as well as substantial contributions by RIKEN (Japan), NSF (USA; via funding of TAIR and the 2010 program), BBSRC (UK; via funding of the GARNET initiative), and the Max-Planck-Society. Nemhauser *et al.* (2006) used the public microarray data to study that different hormones appear to regulate distinct members of protein families and conclude that there is not a core transcriptional growth-regulated module in young *Arabidopsis* seedlings. Nemhauser *et al.* (2004) examines interactions between the auxin and BR signaling pathways in both physiological and gene-expression experiments. Transcript levels of 638 genes were affected by BR treatment. Of 342 genes upregulated by BR, 82 were also upregulated after auxin treatment. Their results showed that the two hormones appeared to act synergistically (Nemhauser *et al.*, 2004).

Peng *et al.* developed an *Arabidopsis* hormone database (http://ahd.cbi.pku.edu.cn/) which aims to provide a systematic and comprehensive view of genes participating in plant hormonal regulation, as well as morphological phenotypes controlled by plant hormones (Peng *et al.*, 2009). This database collects 1026 *Arabidopsis* genes related to the actions of eight plant hormones including auxin, gibberellins, cytokinin, abscisic acid, ethylene, jasmonic acid, salicyclic acid and BR (Peng *et al.*, 2009). Users can get access to information about these hormone-related genes, including sequences, functional category, mutant information, phenotypic description, microarray data and linked publications.

Genes controlling hormone levels have been used to increase grain yields in wheat and rice (Wu *et al.*, 2008). Wu *et al.* used microarray and photosynthesis analysis to reveal evidence of enhanced CO2 assimilation, enlarged glucose pools in the flag leaves, and increased assimilation of glucose to starch in the seed. Their results suggested that BRs stimulate the flow of assimilate and regulate grain filling in rice (Wu *et al.*, 2008).

Transcriptomics data not only provide comprehensive view to study the regulation of BRs but also let us understand the molecular mechanisms by BRs. Additionally, the transcriptome results will be useful in the characterization of new mutants and new growth-regulating compounds that are associated with BR function, furthermore, they could be used to further increase grain yield in crop plants.

3. PROTEOMICS STUDY IN BR REGULATION

Proteomics technique is the other powerful high-throughput method to describe the study of the complete set of proteins (proteome) that is expressed at a given time in a cell, tissue, organ or organism. Proteomics characterizes cellular proteins and as well as its abundance, state of modification, protein complexes and interactions (Naour *et al.*, 2001; Gygi and Aebersold, 2000).

The global changes in cellular protein expression can be visualized by two-dimensional gel electrophoresis and identified by mass spectrometry analysis (Sinchaikul *et al.*, 2001). Therefore, the holistic study of biological transformations will enable more rapid advances in elucidating biochemical pathways (Juan *et al.*, 2002).

A comprehensive analysis of total protein samples using 2-D DIGE identified a large number of proteins that responded to a relative long time of BR treatment (Deng *et al.*, 2007). While these late BR-responsive proteins are likely to mediate downstream cellular responses, the study failed to detect any of the known BR-signaling components, which are phosphorylated or dephosphorylated within minutes of BR treatment. Using proteomics study, mung bean epicityls whose growth was initially suppressed by chilling partly recovered their ability to elongate after treatment with 24-epibrassinolide, a kind of BRs, the proteins involved in methionine assimilation, ATP synthesis, cell wall construction and the stress response are up-regulated (Huang *et al.*, 2006).

Proteomics studies of plasma membrane proteins in *Arabidopsis* lead to the identification of three homologous BR-signaling kinases (BSK1, BSK2, and BSK3) (Tang *et al.*, 2008a). The BSKs are phosphorylated by BRI1 *in vitro* and interact with BRI1 *in vivo* (Tang *et al.*, 2008a). BRI1 is a receptor-like kinase (RLK) that functions as the major receptor for steroid hormones BRs (Johnson and Ingram, 2005). BRs bind the extracellular domain of BRI1 to activate its kinase activity, initiating a signal transduction cascade that regulates nuclear gene expression and a wide range of developmental and physiological processes (Vert *et al.*, 2005; Tang *et al.*, 2008b). Recently, proteomics studies identified new components that bridge the last gap in the genetically defined BR-signaling pathway, establishing the first complete pathway from an RLK to transcription factors in plants (Tang *et al.*, 2009).

4. CLUSTERING FOR ELUCIDATING REGULATION OF BRs

Clustering, a natural basis for organizing gene expression data, is to group together genes with similar patterns of expression (Eisen *et al.*, 1998; Hirano *et al.*, 2008; Huang *et al.*, 2008). Because of the large number of genes and the complexity of biological networks, clustering is a useful exploratory technique for the analysis of gene expression data. This clustering information helps researchers to better understand biological processes (Luo *et al.*, 2005), and to investigate biological mechanisms (Sadlier *et al.*, 2004). The ultimate purpose of the gene expression experiments is to produce biological knowledge. Before producing biological knowledge, we have to interpret the microarray data. The interpretation of these microarray data can be facilitated by well-presented functional annotations or classifications, which provide an overview of the dominate functions for differentially expressed or clustered genes.

Many softwares can be used for clustering analysis such as GeneSpring and Cluster 3.0. Cluster 3.0 is an open source clustering software; however, GeneSpring is commercial one. Here I would like to introduce more about Cluster 3.0. For hierarchical clustering analysis, you can combine two softwares, one is Cluster 3.0 data clustering software developed at Tokyo University (de Hoon *et al.*, 2004), and the other one is Java Treeview visualization software (Saldanha, 2004) to display the clustering result. For proteomics data analysis, at first, the intensities of protein expression in time instances measured by PDQuest software need to be normalized and subjected to 1-D hierarchical clustering, and the dendrograms will be generated based on the pair-wise calculation of uncentered correlation and average-linked clustering (Juan *et al.*, 2006). The clustering result from Cluster 3.0 will be imported into Java Treeview and displayed in a graded color scheme. Transition of color for each protein from light to dark indicates a gradual decrease in expression over time, and from dark to light indicates an up-regulation of protein expression.

Using microarray combined clustering analysis, Hirano *et al.* analyzed the global expression profiles of genes related to seven hormones including BRs (Hirano *et al.*, 2008). They showed the representative gene expression profiles of the BR biosynthesis, deactivation and signaling pathways in the microspore/pollen (MS/POL) and tapetum (TAP) of rice. Most of the BR signaling genes were preferentially expressed during the early stages of MS/POL development and constitutively in TAP, except for *OsBIN2/OsGSK1*, which demonstrated an expression pattern with preferential expression at later stages of MS/POL development (Hirano *et al.*, 2008).

5. GENE NETWORK FOR REGULATION OF BRs

Inferring gene network architecture from time-course data generated from high-throughput experimental technologies, such as transcriptomics or proteomics, can help us understand the system behavior of living organisms. Networks of interacting proteins and gene regulations can provide researchers rudimentary understanding in cellular mechanisms; therefore, it is possible to understand the regulation of BRs.

5.1 Inferring gene network

An interactive tool, GeneNetwork (Wu *et al.*, 2004) provides four reverse-engineering models and three data interpolation approaches to infer relationships between genes from time-course expression data. GeneNetwork enables a user to readily reconstruct genetic networks based on microarray data without being intimate knowledge of the mathematical models. A simple graphical user interface enables rapid, intuitive mapping and analysis of the reconstructed network, allowing biologists to explore gene relationships at the system level.

5.2 Significant functional networks and biological pathway annotations

The differentially expressed genes can be annotated by Ingenuity Pathway Analysis (IPA; Ingenuity Systems, Redwood City, CA, USA). In brief, the identified genes are mapped onto available functional networks and specific biological pathways, and then ranked by score. The score is based on a P-value calculation; for example, if the score is 3, then the corresponding P-value is 10^{-3}, meaning there is a 1/1000 chance that the focus genes are in a network due to random chance. The significance of the association between the data set and the canonical pathway can be measured in two ways, ratio and P-value. Ratio is displayed as the number of genes from the data set that map to the pathway divided by the total number of genes that map to the canonical pathway. Fischer's exact test is used to calculate a P-value determining the probability that the association between the genes in the dataset and the canonical pathway is explained by chance alone. The statistical analysis of biofunctions and disturbed pathways will be sorted on the order of matching significance by using IPA web tool.

6. CONCLUSIONS

BRs are plant hormones that are essential for a wide range of developmental processes in plants. The transcriptomics and proteomics techniques are powerful and high-throughput methods for accurately determining changes in global gene and protein expressions related to the effects of BRs, respectively. Clustering and gene network analysis can help to easily analyze more than tens of thousands gene and protein expression profiles. Combining transcriptomics, proteomics and bioinformatics approaches opens a novel way of elucidating hormonal targets for growth of plants. This may not only elucidate the molecular mechanism of BR regulation but also be further applied to plant breeding.

7. ACKNOWLEDGMENTS

The author is grateful to National Science Council of Taiwan and National Taiwan University for financial support.

8. REFERENCES

Bishop, G.J., and Koncz, C. 2002. Brassinosteroids and plant steroid hormone signaling. *Plant Cell*, S97–S110.

Clouse, S.D. 2002. Brassinosteroid signal transduction: clarifying the pathway from ligand perception to gene expression. *Mol. Cell*, **10**: 973–982.

Clouse, S.D., and Sasse, J.M. 1998. Brassinosteroids: essential regulators of plant growth and development. *Ann. Rev. Plant Physiol. Plant Mol. Biol.*, **49**: 427–451.

Clouse, S.D., Zurek, D.M., McMorris, T.C., and Baker M.E. 1992. Effect of brassinolide on gene expression in elongating soybean epicotyls. *Plant Physiol.*, **100**: 1377–1383.

Deng, Z., Zhang, X., Tang, W., Oses-Prieto, J.A., Suzuki, N., Gendron, J.M., Chen, H., Guan, S., Chalkley, R.J., Peterman, T.K., Burlingame, A.L., and Wang, Z.Y. 2007. A proteomics study of brassinosteroid response in *Arabidopsis*. *Mol. Cell. Proteomics*, **6**: 2058–2071.

Eisen, M.B., Spellman, P.T., Brown, P.O., and Botstein, D. 1998. Cluster analysis and display of genome-wide expression patterns. *Proc. Natl. Acad. Sci. USA*, **95**: 14863–14868.

Fujioka, S., and Yokota, T. 2003. Biosynthesis and metabolism of brassinosteroids. *Ann. Rev. Plant Biol.*, **54**: 137–164.

Goda, H., Shimada, Y., Asami, T., Fujioka, S., and Yoshida, S. 2002. Microarray analysis of brassinosteroid-regulated genes in *Arabidopsis*. *Plant Physiol.*, **130**: 1319–1334.

Gygi, S.P., and Aebersold, R. 2000. Mass spectrometry and proteomics. *Curr. Opin. Chem. Biol.*, **4**: 489–494.

He, J.X., Gendron, J.M., Sun, Y., Gampala, S.S., Gendron, N., Sun, C.Q., and Wang, Z.Y. 2005. BZR1 is a transcriptional repressor with dual roles in brassinosteroid homeostasis and growth responses. *Science*, **307**: 1634–1638.

Hirano, K., Aya, K., Hobo, T., Sakakibara, H., Kojima, M., Shim, R.A., Hasegawa, Y., Ueguchi-Tanaka, M., and Matsuoka, M. 2008. Comprehensive transcriptome analysis of phytohormone biosynthesis and signaling genes in microspore/pollen and tapetum of rice. *Plant Cell Physiol.*, **49**: 1429–1450.

de Hoon, M.J., Imoto, S., Nolan, J., and Miyano, S. 2004. Open source clustering software. *Bioinformatics*, **20**: 1453–1454.

Huang, B., Chu, C.H., Chen, S.L., Juan, H.F., and Chen, Y.M. 2006. A proteomics study of the mung bean epicotyl regulated by brassinosteroids under conditions of chilling stress. *Cell Mol. Biol. Lett.*, **11**: 264–278.

Huang, D., Wu, W., Abrams, S.R., and Cutler, A.J. 2008. The relationship of drought-related gene expression in *Arabidopsis thaliana* to hormonal and environmental factors. *J. Exp. Bot.*, **59**: 2991–3007.

Hughes, T.R., Marton, M.J., Jones, A.R., Roberts, C.J., Stoughton, R., Armour, C.D., Bennett, H.A., Coffey, E., Dai, H., He, Y.D., Kidd, M.J., King, A.M., Meyer, M.R., Slade, D., Lum, P.Y., Stepaniants, S.B., Shoemaker, D.D., Gachotte, D., Chakraburtty, K., Simon, J., Bard, M., and Friend, S.H. 2000. Functional discovery via a compendium of expression profiles. *Cell*, **102**: 109–126.

Ideker, T., Thorsson, V., Ranish, J.A., Christmas, R., Buhler, J., Eng, J.K., Bumgarner, R., Goodlett, D.R., Aebersold, R., and Hood, L. 2001. Integrated genomic and proteomic analyses of a systematically perturbed metabolic network. *Science*, **292**: 929–934.

Johnson, K.L., and Ingram, G.C. 2005. Sending the right signals: regulating receptor kinase activity. *Curr. Opin. Plant Biol.* **8**: 648–656.

Juan, H.F., Lin, J.Y.C., Chang, W.H., Wu, C.Y., Pan, T.L. Tseng, M.J., Khoo, K.H., and Chen, S.T. 2002. Biomic study of human myeloid leukemia cells differentiation to macrophages using DNA array, proteomic, and bioinformatic analytical methods. *Electrophoresis*, **23**: 2490–2504.

Juan, H.F., Wang, I.H., Huang, T.C., Li, J.J., Chen, S.T., and Huang, H.C. 2006. Proteomics analysis of a novel compound: cyclic RGD in breast carcinoma cell line MCF-7. *Proteomics*, **6**: 2991–3000.

Kim, S.Y., Kim, B.H., Lim, C.J., Lim, C.O., and Nam, K.H. 2010. Constitutive activation of stress-inducible genes in a brassinosteroid-insensitive 1 (bri1) mutant results in higher tolerance to cold. *Physiol. Plant*, **138**: 191–204.

Komatsu, S., Konishi, H., Shen, S., and Yang, G. 2003. Rice proteomics. *Mol. Cell Proteomics*, **2**: 2–10.

Krishna, P. 2003. Brassinosteroid-mediated stress response. *J. Plant Growth Regul.*, **22**: 289–297.

Luo, X., Ding, L., Xu, J., Williams, R.S., and Chegini, N. 2005. Leiomyoma and myometrial gene expression profiles and their responses to gonadotropin-releasing hormone analog therapy. *Endocrinology*, **146**: 1074–1096.

Mockaitis, K., and Estelle, M. 2004. Integrating transcriptional controls for plant cell expansion. *Genome Biol.*, **5**: 245.

Motose, H., Iwamoto, K., Endo, S., Demura, T., Sakagami, Y., Matsubayashi, Y., Moore, K. L., and Fukuda, H. 2009. Involvement of phytosulfokine in the attenuation of stress response during the transdifferentiation of zinnia mesophyll cells into tracheary elements. *Plant Physiol.*, **150**: 437–447.

Nakamura, A., Nakajima, N., Goda, H, Shimada, Y., Hayashi, K., Nozaki, H., Asami, T., Yoshida, S., and Fujioka, S. 2006. *Arabidopsis* Aux/IAA genes are involved in brassinosteroid-mediated growth responses in a manner dependent on organ type. *Plant J.*, **45**: 193–205.

Naour, F.L., Hohenkirk, L., Grolleau, A., Misek, D.E., Lescure, P., Geiger, J.D., Hanash, S., and Beretta, L. 2001. Profiling changes in gene expression during differentiation and maturation of monocyte-derived dendritic cells using both oligonucleotide microarrays and proteomics. *J Bio Chem*, **276**: 17920–17931.

Nemhauser, J.L., Hong, F., and Chory, J. 2006. Different plant hormones regulate similar processes through largely nonoverlapping transcriptional responses. *Cell*, **126**: 467–475.

Nemhauser, J.L., Mockler, T.C., and Chory, J. 2004. Interdependency of brassinosteroid and auxin signaling in *Arabidopsis*. *PLoS Biol.*, **2**: E258.

Osakabe, Y., Maruyama, K., Seki, M., Satou, M., Shinozaki, K., and Yamaguchi-Shinozaki, K. 2005. Leucine-rich repeat receptor-like kinase1 is a key membrane-bound regulator of abscisic acid early signaling in *Arabidopsis*. *Plant Cell*, **17**: 1105–1119.

Peng, Z.Y., Zhou, X., Li, L., Yu, X., Li, H., Jiang, Z., Cao, G., Bai, M., Wang, X., Jiang, C., Lu, H., Hou, X., Qu, L., Wang, Z., Zuo, J., Fu, X., Su, Z., Li, S., and Guo, H. 2009. Arabidopsis Hormone Database: a comprehensive genetic and phenotypic information database for plant hormone research in *Arabidopsis*. *Nucleic Acids Res.*, **37**: D975–D982.

Rao, S.S.R., Vardhini, B.V., Sujatha, E., and Anuradha, S. 2002. Brassinosteroids- a new class of phytohormones. *Curr. Sci.,* **82**: 1239–1245.

Sadlier, D.M., Connolly, S.B., Kieran, N.E., Roxburgh, S., Brazil, D.P., Kairaitis, L., Wang, Y., Harris, D.C., Doran, P., and Brady, H.R. 2004. Sequential extracellular matrix-focused and baited-global cluster analysis of serial transcriptomic profiles identifies candidate modulators of renal tubulointerstitial fibrosis in murine adriamycin-induced nephropathy. *J. Biol. Chem.*, **279**: 29670–29680.

Saldanha, A.J. 2004. Java Treeview – extensible visualization of microarray data. *Bioinformatics*, **20**: 3246–3248.

Schluter, U., Kopke, D., Altmann, T., and Mussig, C. 2002. Analysis of carbohydrate metabolism of CPD antisense plants and the brassinosteroid-deficient cbb1 mutant. *Plant Cell Environ.*, **25**: 783–791.

Schumacher, K., and Chory, J. 2000. Brassinosteroid signal transduction: still casting the actors. *Curr. Opin. Plant Biol.*, **3**: 79–84.

Seo, J., Kim, M., and Kim, J. 2000. Identification of novel genes differentially expressed in PMA-induced HL-60 cells using cDNA microarrays. *Mol. Cells*, **10**: 733–739.

Sinchaikul, S., Sookkheo, B., Topanuruk, S., Juan, H.F., Phutrakul, S., and Chen, S.T. 2001. Bioinformatics, functional genomics, and proteomics study of *Bacillus sp. J. Chromatogr. B*, **771**: 261–287.

Tang, W., Deng, Z., and Wang, Z.Y. 2009. Proteomics shed light on the brassinosteroid signaling mechanisms. *Curr. Opin. Plant Biol.*, **13**: 1–7.

Tang, W., Kim, T.W., Oses-Prieto, J.A., Sun, Y., Deng, Z., Zhu, S., Wang, R., Burlingame, A.L., and Wang, Z.Y. 2008a. BSKs mediate signal transduction from the receptor kinase BRI1 in *Arabidopsis*. *Science*, **321**: 557–560.

Tang, W., Deng, Z., Oses-Prieto, J.A., Suzuki, N., Zhu, S., Zhang, X., Burlingame, A.L., and Wang, Z.Y. 2008b. Proteomics studies of brassinosteroid signal transduction using prefractionation and two-dimensional DIGE. *Mol. Cell. Proteomics*, **7**: 728–738.

Vert, G., Nemhauser, J.L., Geldner, N., Hong, F., and Chory, J. 2005. Molecular mechanisms of steroid hormone signaling in plants. *Ann. Rev. Cell Dev. Biol.*, **21**: 177–201.

Wang, Z.Y., and He, J.X. 2004. Brassinosteroid signal transduction – choices of signals and receptors. *Trend Plant Sci.*, **9**: 91–96.

Wu, C.C., Huang, H.C., Juan, H.F., and Chen, S.T. 2004. GeneNetwork: an interactive tool for reconstruction of genetic networks using microarray data. *Bioinformatics*, **20**: 3691–3693.

Wu, C.Y., Trieu, A., Radhakrishnan, P., Kwok, S.F., Harris, S., Zhang, K., Wang, J., Wan, J., Zhai, H., Takatsuto, S., Matsumoto, S., Fujioka, S., Feldmann, K.A., and Pennell, R.I. 2008. Brassinosteroids regulate grain filling in rice. *Plant Cell*, **20**: 2130–2145.

Yang, G., and Komatsu, S. 2004. Microarray and proteomic analysis of brassinosteroid- and gibberellin- regulated gene and protein expression in rice. *Geno. Prot. Bioinfo.*, **2**: 77–83.

Yue, H., Eastman, P.S., Wang, B.B., Minor, J., Doctolero, M.H., Nuttall, R.L., Stack, R., Becker, J.W., Montgomery, J.R., Vainer, M., and Johnston, R. 2001. An evaluation of the performance of cDNA microarrays for detecting changes in global mRNA expression. *Nucleic Acids Res.*, **29**: e41.

Zurek, D.M., and Clouse, S.D. 1994. Molecular cloning and characterization of a brassinosteroid-regulated gene from epicotyl elongating soybean (*Glycine max* L.) epicotyls. *Plant Physiol.*, **104**: 161–170.

Zurek, D.M., Rayle, D.L., McMorris, T.C., and Clouse, S.D. 1994. Investigation of gene expression, growth kinetics, and wall extensibility during brassinosteroid-regulated stem elongation. *Plant Physiol.*, **104**: 505–513.

Chapter 16

BRASSINOSTEROIDS FOR PHYTOREMEDIATION APPLICATION

M. BARBAFIERI AND E. TASSI
National Research Council, Institute of Ecosystem Study, Pisa Section, Italy

Abstract: Phytoremediation is a plant-based family of technologies for remediating and/or containing contamination. It has good public acceptance and is economical, compared to traditional and engineering technologies for soil treatment. Phytohormones have the specific ability to increase and support plant physiology, and are well-known and applied in horticulture, floriculture, fruit farming and other agricultural fields. The main processes of phytoremediation involve plant physiology within the plant and/or its immediate surroundings (rhizosphere); thus it can take advantage of any "assistants" that improve the efficiency of the physiological mechanisms that can make phytoremediation process more efficient. We can call "assisted phytoremediation by plant growth regulators" that phytoremediation aided by phyto-hormone treatment. Such treatment is harmless to the environment, practical and economically viable. Only recently a very few studies have revealed the possibility of applying phytohormones for phytoremediation purposes. Phytohormones in this context should increase plant resistance to stress, increase plant biomass production, increase plant metal uptake, and increase organic degradation. Brassinosteroids could enter into this class of phytohormone for "assisted phytoremediation by plant growth regulators". This may open a new research field, intriguing experts in both phytoremediation and phytohormones.

Key words: Phytohormone, environmental stress, plant growth regulators, plant biomass, assisted phytoremediation, environmental remediation

S. Hayat and A. Ahmad (eds.), *Brassinosteroids: A Class of Plant Hormone*,
DOI 10.1007/978-94-007-0189-2_16, © Springer Science+Business Media B.V. 2011

1. INTRODUCTION

Environmental contamination of media by various substances has spurred the scientific world to seek new techniques for rehabilitating contaminated areas, and restoring them for social use in clean and/or profitable ways. Very often such techniques are very difficult to use, requiring complicated engineering and mechanical tools; moreover, costs can be prohibitive. Other techniques can be considered "soft" when they are not destructive and are less expensive, such as biological techniques. In particular, techniques for resolving metal contamination are very destructive for the matrix (metals can be extracted from soil particles by means of strong chemical and/or mechanical treatments (soil flushing, soil washing, electroremediation, etc.).

So when phytoremediation, an environmentally friendly and economically feasible technique was conceived, it attracted enormous scientific and technical interest. Phytoremediation technologies offer a broad range of applications, while other remediation technologies are specific for certain classes of contaminants or types of media. Such technologies can be used for all kinds of contaminants (organic and inorganic) and media (soil, sediment, water, wastes, etc.). Plants are crucial to this technology. In many studies plants have displayed their potential and efficiency in treating contaminants, giving us back clean media. In the case of organics, results can be achieved in a short time and with complete degradation, while with inorganics it can take long time; in this case an appropriate plan for dismantling residue should be planned when plants are very rich in toxic metals.

Phytoremediation is basically a modified "agronomic" practice. It is not used for food production but for "cleaning" and/or rendering contaminated media less harmful. Consequently, different plants and plant abilities are needed. Moreover, different agronomic inputs to increase and maximize phytoremediation efficiency must be set up and applied.

Thus, some research is focused on improving phytoremediation applicability and efficiency. However, weak points of the technology are the survival of plants in unfavorable environmental conditions (contaminant toxicity, low fertility, etc.) and the often lengthy time it takes to reduce contaminants to the level of natural content or to the limit established by legislation; this is due to the plant growth times and to "low" ability to treat the contaminants (usually referred to as metal uptake).

We speak of "assisted phytoremediation" when additives are used to "push" plants to increase their cleaning performance. For metals, chemical additives are usually applied to the soil to make them easily available for plant uptake, or plants with rapid and high biomass production are adopted.

Only recently a few studies have addressed the use of phytohormones applied as additives to plants for phytoremediation purposes.

Phytohormones are already available in patented agrochemical products and are used in horticulture, floriculture, fruit farming and agriculture. Phytohormones can interact with plant physiology and stimulate the necessary processes to produce better yields and cultivation quality. Phytohormones can be easily applied and are cost effective and environmentally friendly.

The most appropriate phytohormones for phytoremediation still need to be identified, as well as the correct application methods and timing, in order to reap the maximum benefit for phytoremediation needs.

Therefore, brassinosteroids are interesting compounds due to their important characteristics in enhancing crop yields and increasing stress tolerance (Divi and Krishna 2009).

This chapter aims to: (1) illustrate various phytoremediation mechanisms and (2) discuss the potential of phytohormones (specifically brassinosteroids) to implement plants' abilities to increase the efficiency of this technology. Moreover, these studies could implement our knowledge of the contaminant metabolism process in plants.

2. PHYTOREMEDIATION TECHNOLOGIES: PROCESSES AND PHYSIOLOGICAL MECHANISMS

Phytoremediation is a versatile treatment technique that uses selected or specialized plants and their associated microorganisms to extract, sequester and/or detoxify nearly all pollutants from water, sediment, soil and air. The remediation process carried out by these selected or specialized plants can be associated with a sustainable system of 'pumping and treating', which uses solar energy and the root system, where roots are the primary sites for absorbing pollutants. Phytoremediation as an *in situ* solar-driven technique is considered an innovative low-cost remediation procedure, able to reduce the exposure of humans, wild life and environment to contaminants. It is also well-accepted by the surrounding communities and regulatory agencies (Cunningham and Berti, 1993; McCutcheon and Schnoor, 2003).

The two major classes of pollutants in the environment are organic and inorganic contaminants. Organic pollutants are mostly man-made and are released via spills (fuel, solvents), military activity (explosives, chemical weapons), agriculture (pesticides, herbicides), industry (chemical, petro-chemical), wood treatment, etc. (Chiou, 2002). Inorganic pollutants such as heavy metals, metalloids and radionuclides are generally toxic even in very low concentrations. The primary sources of this pollution are the mining and smelting of metalliferous ores, metallurgical industries, burning of fossil fuels, sewage sludge treatment, warfare and military training, waste disposal sites, municipal waste, agricultural fertilizers and electronic industries (Nriagu, 1979). In addition to sites contaminated by human activity, natural mineral deposits containing large quantities of heavy metals are present in many regions of the globe (Nriagu, 1979; Plumlee and Longsdon, 1999), contributing to the diffusion of contaminants.

Since xenobiotics can be degraded by plants in the root zone or taken up and followed by degradation, sequestration or volatilization, they have been successfully phytoremediated (Paterson *et al.*, 1990; Schnoor *et al.*, 1995). On the contrary, inorganics cannot be degraded but they can be stabilized or sequestered in harvestable plant tissues. Depending on the class of pollutant, different phytoremediation technologies are defined: phytodegradation, phytovolatilization and phytostimulation or rhizodegradation for organic contaminants while phytoextraction, phytostabilization, rhizofiltration and phytovolatilization are suitable for the inorganic contaminants.

The phytotechnologies described in Sections 2.1 and 2.2 are suitable for the different classes of pollutants; they make use of special plant properties, and typically, different plant species are used for each. Generally, favourable plants for phytoremediation need to have fast-growing, high biomass and tolerance to pollution.

2.1 Phytoremediation technologies for organic contaminants

Phytoremediation of organic contaminants is focused on three classes of compounds: chlorinated solvents (Shang *et al.*, 2003), explosives (Subramanian and Shanks, 2003; McCutcheon *et al*, 2003) and petroleum hydrocarbons (Hutchinson *et al.*, 2003). Moreover, plants were addressed also to treat more recalcitrant pollutants such as polynuclear aromatic hydrocarbons (PAHs) and polychlorinated biphenyls (PCBs) (Harms *et al.*, 2003).

Actually, the concept of using plants to remediate soils contaminated with organic pollutants was based on observations that plants accelerate the disappearance of organic chemicals compared to non-vegetated bulk soils (Alkorta and Garbisu, 2001; Tassi *et al.*, 2004a). Plants with an extensive, dense root system will be advantageous for uptake and/or degradation of organics from soil. The phytoremediation of organics is an attractive technology, since in most cases pollutants can be removed from the site without the need for plant harvesting and disposal.

The various phytoremediation technologies described below are suitable for different classes of organic contaminants, and are not mutually exclusive but can occur simultaneously for the same pollutant.

2.1.1 Rhizodegradation

Rhizodegradation (or Phytostimulation) is the result of the symbiosis between plants and microorganisms from the rhizosphere (bacteria and mycorrhizal fungi). Plants secrete organic compounds (sugar, organic acids, proteins, etc) and enzymes through the root system.

The growth and survival of microorganisms in the rhizosphere that are capable of degrading organic pollutants are supported by some of the root exudates that serve as sources of carbon and nitrogen. In fact, it has been observed that the density of rhizospheric microorganisms is from two to four orders of magnitude greater than the populations in the surrounding bulk soil, although the chemical composition of root exudates and rates of exudation differ considerably among plant species (Salt *et al.*, 1998). For example, some plant species exude phenols capable of supporting polychlorbiphenyls (PCB)-degrading bacteria (Fletcher and Hedge, 1995; Leigh *et al.*, 2006). Microorganisms from the rhizosphere may also increase the production of humic substances (Weber and Huang, 2003) or accelerate the remediation process by volatilization of organic pollutants such as polycyclic aromatic hydrocarbons (PAHs).

In addition to exudation compounds, degradative enzymes can be released by plants (ex.: nitroreductases, laccases, nitrilases, halogenases) and are also responsible for the pollutant degradation processes found, for example, in ammunition wastes, triaminotoluene, 4-chloro-benzonitrile and hexa-chloroethane (Schnoor *et al.*, 1995; Alkorta and Garbisu, 2001).

2.1.2 Phytodegradation

Phytodegradation (or phytotransformation) is a process by which plants can directly absorb organic compounds from soil solution or from water without the participation of microorganisms present in the rhizosphere. For this, the contaminant, in a bioavailable form, is in direct contact with the roots and can be degraded or transformed inside the cells into simpler molecules that can be useful for plant growth. The breakdown of complex organic compounds into simpler molecules is catalyzed by different enzymes (dehalogenases, oxygenases and nitroreductases) that act as a detoxification mechanism, producing the degradation by-products. The enzymatic degradation of organics can occur in both root and shoot tissue. Degradation within plant tissues is generally attributed to the plant, but may in some cases involve endophytic microorganisms (Barac *et al.*, 2004).

The fate of the degradation by-products may be one of the following:

- lignification – a process by which the plant stores the degradation products via differentiation of xylem and where lignin is deposited within the cell wall
- mineralization or complete degradation of organic compounds to convert them into water and carbon dioxide.

Since the organic compound has to be in a direct contact with the roots, the primary factors governing the uptake of xenobiotics are the physico-chemical characteristics of the compound. Two important chemical properties of a pollutant that affect its movement in soil are hydrophobicity and volatility.

Hydrophobicity is usually expressed as the octanol-water partition coefficient (log K_{ow}), where a high log K_{ow} value corresponds to high hydrophobicity (Briggs *et al.*, 1982; Ryan *et al.*, 1988). Thus, compounds with lipophilicity close to that to the respective plant root can be transferred from soil or water to the plant spontaneously by a diffusion-driven process. It was stated that organic compounds with log K_{ow} ranging from 1 and 3 (moderately hydrophobic) are mostly taken up by plants (Briggs *et al.*, 1982, Trapp, 2000). Compounds exhibiting a log $K_{ow}<1$ will have high water solubility, will be able to migrate in the soil solution to an extent, and will only sparsely penetrate the root epidermis. Those compounds with log $K_{ow}>3$ are extremely hydrophobic (ex.: PCBs, PAHs) and are increasingly retained in the soil's organic matter or on the root epidermis and the mucilage surrounding the root; they do not dissolve in the soil solution or water.

Pollutant volatility also measures the pollutant's ability to move in water. It is expressed as Henry's law constant (H) and measures the partition to air relative to water (Bromilow and Chamberlain, 1995). Pollutants with $H>10^{-4}$ tend to move in the air spaces between soil particles and those with $H<10^{-6}$ move predominantly in water. The class of pollutants with Henry's constant between 10^{-4} and 10^{-6} can move in both water and air and diffuse passively through plants, but water-mobile organic contaminants are mainly phyto-degraded.

Other properties such as polarity, acidity constant (pK_a), concentration, etc. could be important to the pollutant's bioavailability. Non-polar molecules with molecular weight < 500 will be absorbed onto the root surfaces, whereas polar molecules will enter the root and be translocated (Tropp *et al.*, 1986).

2.1.3 Phytovolatilization

This process refers to the uptake and transpiration of contaminants, and is usually associated with the phytodegradation of organic compounds and volatile organic compounds. Volatile pollutants, mobile in both air and water

(Henry's constant ranging from 10^{-6} to 10^{-4}), can be removed from soil via the transpiration stream to the atmosphere. For instance, chlorinated solvents and fuel additives such as trichloroethene (TCE) (Ma and Burken, 2003) and methyl tet-butyl ether (MBTE) (Xingmao *et al.*, 2004) respectively, can be volatilized by plants. Although volatilization is a passive process it can be maximized using phreatophyte species with high transpiration rate. Moreover, the transpiration rate can be promoted through sufficient irrigation preventing the stomatal closure or by the application of exogenous plant growth regulators that also act at the stomata level (Tassi *et al.*, 2008), even if specific studies on field application are not available on this topic.

Phytovolatilization can take place at the rhizosphere level with the transformation of contaminants by the action of associated microorganisms or can start after its absorption into the plant cells and translocation by transpiration stream to the stems, trunks and leaves. The poplar tree is the most common species for phytovolatilization of volatile organic compounds. Phytovolatilization is also associated with the absorption of inorganics such as selenium, mercury and arsenic.

2.2 Mechanisms associated with the phytoremediation of organic contaminants

The removal of organic xenobiotics by phytodegradation and phytovolatilization depends on the pollutants' metabolism and the plant's metabolic networks. Once organic pollutants have reached the plant rhizosphere, they must migrate into the root where the lipophilicity can limit uptake into the plant or lead to accumulation in the partially suberised root cortex. The pollutant will be detoxified and metabolized after it overcomes the barrier in the root endodermis. The barrier (the suberised Casparian Strip) limits its entrance and the transportation by transpiration stream in the xylem into the roots and shoots.

Studies of the metabolism of the organic compounds concerned pesticide and herbicide detoxification mechanisms in crop species and focused on their potential contamination. Studies on phytoremediation have expanded the research beyond crops to look at trees, wetland plants and wild flora. These studies suggested the existence of enzyme systems forming a cascade metabolism of xenobiotics analogous to the human hepatic metabolism (Sandremann, 1994). For this reason this network of reactions was called 'green liver' with three distinct phases: (I) activation of the xenobiotic, (II) conjugation and (III) sequestration.

Activation (phase I) is the initial transformation of organic compounds in plants which comprises many different reactions catalyzed by specific enzymes. Oxidations comprise the majority of transformations; they can increase solubility and assist the conjugation of xenobiotics. One important reaction is hydroxylation (addition of OH^-), in most cases catalyzed by the enzymes p450 mono-oxygenases (localized in the membrane fractions of plant cells) or by the peroxidases (localized in the apoplast and in the cytosol). Other enzymes, such us peroxygenases and carboxylesterases are also active in the initial transformation processes. Reduction in the transformation phase was also detected, where the reduction of nitroaromatic compounds and explosives is of special interest for phytoremediation (Burken *et al.*, 2000, Schröder and Collins, 2002, McCutcheon *et al.*, 2003). The more highly halogenated contaminants (e.g., tetrachlorethylene) are also phytoreduced, whereas in trichloroethylene and less chlorinated compounds, phyto-oxidation is the main reaction in this phase. Poplar trees were used in several studies of phyto-transformation metabolites of chlorinated solvents (Newman *et al.*, 1999; Schwitzguebel and Vanek, 2003).

Conjugation (phase II) is the next step, where molecules such as sugars, amino acids or glutathione may be transferred to the activated xenobiotic. The activated functional groups (OH^-, NH_2^-, SH^-, and $COOH^-$) serve as a trigger to the transfer of the aforementioned molecules by the action of enzymes. One important group of enzymes is represented by the Glutathione S-transferases (GSTs) that are involved in the conjugation of glutathione inactivating toxic chemicals in plants, mammalians and other animals. GSTs are also important in the management of plant hormones (e.g., auxin, brassinosteriod) or secondary plant metabolites, in response to oxidative stress, in lipid peroxidation and in defence against pathogens. Other enzymes also involved in the conjugation phase such as glucosyltransferases, malonyl-transferases, etc. were identified in crops (Schröder and Collins, 2002).

The third phase is the sequestration or compartmentation of conjugated molecules. In this process the resulting molecule could either be stored in the vacuole or the apoplast, or bound covalently to cell walls. For example, the lignification of triazoles in phytoremediation studies using *Helianthus annuus* has been observed (Castro *et al.*, 2001).

2.3 Phytoremediation technologies for inorganic contaminants

Among tolerant plants, three general classes of vegetation were defined and recognized as growing on contaminated and metalliferous soils: *metal excluders* – plants that avoid the translocation to upper parts over a broad range of metal concentrations in the soil, but can have large amounts in their roots; *metal indicators* – plants that accumulate metals in their above-ground tissues where the metal levels in the plant reflect metal levels in the soil; *metal accumulators* – usually refers to hyperaccumulators that concentrate metals in the above-ground tissues to levels higher than 1% of Zn, Mn; 0.1% of Ni, Co, Cu, Cr, Pb, Al; 0.01% of Cd, Se or 0.001% of Hg on a dry weight basis irrespective of the metal in the soil (Baker and Brooks, 1989).

Several technologies are available to recover sites contaminated by inorganics where (hyper)accumulator or excluder plants can be used.

2.3.1 Phytoextraction or phytoaccumulation

This process refers to the translocation of metal contaminants from soil into the above-ground tissues by the root system. After plants have grown for a certain period, they are harvested and may be incinerated to recycle the metals. This procedure, repeated several times, brings soil contaminant levels down to permissible limits. The time required for remediation depends on the type and extent of metal contamination, duration of the growing season, the biomass, and the plants' capacity for metal accumulation. Two different strategies can be used: *continuous phytoextraction* – using natural metal hyperaccumulator plants which absorb, translocate and accumulate an enormous amount of metals during their entire life period without visible toxicity symptoms; *assisted phytoextraction* – the accumulation process is induced in tolerant plants by the increased contaminant bioavailability in soil. Synthetic amendments such as chelates (e.g., EDTA, EDDS, NTA), organic acids (e.g., citric acid), or ion competitors (e.g., phosphate – Tassi *et al.*, 2004b) added to the soil enhance metal bioavailability, although the soil microbial community is usually neglected and there is a potential risk of leaching of metals to groundwater (Dickinson *et al.*, 2009).

Generally, phytoextraction is applicable to sites containing low-to-moderate levels of metal contamination.

Effective phytoextraction requires both plant genetic ability and optimal soil and crop management practices (Di Gregorio *et al.*, 2006; Tassi *et al.*, 2008; Pedron *et al.*, 2009). *Thlaspi caerulescens* (Cd and Zn hyperaccumulator) and *Brassica juncea* (heavy metal accumulator) are examples of species that well represent the two phytoextraction strategies described above. Metals such as Ni, Zn, Cu, and As are the best candidates for removal by phytoextraction although Cd, Pb, etc. have been extensively studied. Genetic engineering studies have been performed to manipulate plant accumulation with the overexpression or knockdown of membrane transporter proteins (Rogers *et al.*, 2000).

The accumulation of hazardous plant biomass must be disposed of in order to minimize environmental risk. The waste volume can be reduced by thermal, microbial, physical or chemical means such as composting, compactation or thermo-chemical conversion processes (combustion, gasification and pyrolysis). Recycling the biomass from phytoextraction for fuel and other uses cuts down on the need for landfills and provides the contaminated site with an economical value. Added value to the phytoextraction process could be obtained by combining the biomass produced as an energy source, resulting in an ore after incinerating the residual biomass. This would be possible in the case of phytomining, a particular example of phytoextraction.

Phytomining involves the exploitation of sub-economic ore bodies using hyperaccumulating plants. For instance, the species *Alyssum bertolonii*, *Berkheya coddii* have greater potential in extracting Ni because of their high biomass and a Ni concentration of 1% in the dry matter (Robinson *et al.*, 2003). Other metals such as gold, thallium, and cobalt have been exploited from tailings or other residues of low commercial value (LaCoste *et al.*, 2001; Keeling *et al.*, 2003).

2.3.2 *Phytostabilization*

This process concerns the immobilization of contaminants in certain plant species through absorption and accumulation by roots, adsorption onto roots or precipitation within the root zone and physical stabilization of soil. It reduces the mobility of contaminants to groundwater or air and can re-establish a vegetative cover preventing erosion, leaching or runoff. It is possible to promote phytostabilization with direct seeding of the selected plants (grasses and legumes, metal-tolerant and salt-tolerant ecotypes, etc) or transplanting adapted native species. Frequently, the stabilization also requires the use of amendments such as lime, organic matter and fertilizers.

The biological availability of metals can be reduced by altering the physicochemical properties of the complex formed between metal and soil. For instance, the addition of humified organic matter (compost) or lime immobilizes heavy metals by increasing the soil pH and improving soil conditions for revegetation (Brown *et al.*, 2003; Clemente *et al.*, 2003). Plants with potential for phytostabilization were found to have a high bioconcentration factor (metal concentration ration of plant roots to soil) and low translocation factor (metal concentration factor of plant shoots to roots) (Yoon *et al.*, 2006).

A combination of shrubs or trees and grasses can be used in a vegetative cap for the long-term stabilization and containment of tailings, where fast-transpiring trees such as poplar maintain the upward flow to prevent downward leaching and grasses prevent wind erosion and lateral runoff with their dense root system. Shrubs or trees provide a high nutrient environment for grasses while reducing moisture stress and improving soil physical characteristics in arid and semi-arid climates (Mendez and Maier, 2008).

2.3.3 *Rhizofiltration or phytofiltration*

This term is basically used to define the extraction of inorganic contaminants (metals and radionuclides) from groundwater, surface water and waste water. Contaminants as a result of biotic or abiotic processes are absorbed, precipitated and concentrated in the roots or in the rhizosphere and removed by harvesting the plants grown in aerated water (Dushenkov and Kapulnik, 2000). For the treatment of contaminated water (e.g., industrial–process water) plants were hydroponically cultivated until a large root system developed, then were transplanted into the metal-polluted waters where the contaminant concentration decreased, mainly due to absorption by the roots (the fastest and often most prevalent mechanism). In addition to surface absorption, slower mechanisms such as intracellular uptake, deposit in vacuoles, translocation to shoots or precipitation of metals by plant exudates may be present in the rhizofiltration process. As the roots become saturated, they are harvested and disposed off safely (Gardea-Torresdey *et al.*, 2004; Prasad, 2004).

Rhizofiltration often make uses of terrestrial plants instead of aquatic plants due to the much larger fibrous root systems covered by root hairs, providing extremely large surface area. Terrestrial plants commonly used to remove metals from water are *Brassica juncea* and *Helianthus annuus* (Dushenkov and Kapulnik, 2000), although several popular aquatic plants such as water hyacinth (*Eichhornia crassipes*), cattail (*Typha* sp.), and duckweed (*Lemna* sp) have been tested for rhyzofiltration in constructed wetlands. Poplar (*Polulus* sp.) and willow (*Salix* sp.) trees can be used on the edges of wetlands (Mkandawire and Dudel, 2005). These plants can take up heavy metals such as Pb, Cu, Cd, Fe, As, and Hg. Moreover, *H. annuus* has been shown to be efficient in the extraction of radionuclides (U, ^{137}Cs, ^{90}Sr) from contaminated waters (Dushenkov *et al.*, 1997; Gardea-Torresdey *et al.*, 2004).

2.3.4 *Phytohydraulic or hydraulic control*

Phytohydraulic exploits the ability of plants to evapotranspire enormous amounts of sources of water or groundwater. Trees with deep root system and exceptional transpiration streams can be used as a hydraulic barrier to create upward water flow in the root zone. Roots can reach the water-table and prevent the contamination from leaching down, or prevent a contaminated groundwater plume from spreading horizontally. Thus, phreatophyte trees, which are deep-rooted, high-transpiring, water-loving trees and can survive in conditions of temporary saturation (such as poplars, willows and cottonwoods), have been widely studied to function as hydraulic control in groundwater where contaminants were absorbed into roots, accumulated, degraded or volatilized from this aquatic environment (EPA, 2009).

2.4 Mechanisms associated with the phytoremediation of inorganic contaminants

The uptake of inorganic soil constituents such as salts, metals, and radionuclides into plants depends on the redox state, chemical speciation in the soil, sediment or groundwater, and the plant species. Generally, it is agreed that the uptake of ions by plant roots consists of two phases, adsorption and accumulation.

The adsorption is predominantly related to cations in soil solution and is a non-metabolic process, whereas accumulation is an active process with specific carrier transporters. More specifically, the availability to plants of metal contaminants is governed by a variety of reactions that include complexation with organic and inorganic ligands, ion exchange and adsorption, precipitation and dissolution of solids, and acid-base equilibria. Thus, the extraction of metals by plants is usually limited by its availability in the soil.

The rates of redistribution of metals among the solid and liquid phases and their binding intensity are affected by the metal species, loading levels, aging and soil properties.

The more bioavailable fractions for plant uptake are generally those with higher water solubility, i.e., metal fractions following the order: exchangable > carbonate specifically adsorbed > Fe-Mn oxide > organic sulphide > residual (Barbafieri *et al.*, 1996; Li and Thornton, 2001). These fractions can be estimated by various extraction procedures (generally sequential extraction schemes are used to isolate specific fractions). Soft extractants such as non-buffered salt solutions, diluted acids and complexant agents better correlate with the fraction for plant uptake (Lindsay and Norwell, 1978). Neutral salts mainly dissolve the cation exchangable fraction. Diluted acids partially dissolve trace elements associated with different fractions such as exchangable, carbonates, iron and manganese oxides and organic matter. Complexing agents dissolve not only exchangable element fractions but also the element fraction forming organic matter complexes and the element fraction fixed on soil hydroxides (Rauret, 1998; Barbafieri, 2000)

As a general rule, readily bioavailable inorganics for plant uptake include As, Cd, Cu, Ni, Se, and Zn. Moderately bioavailable metals are Co, Fe, and Mn, whereas Cr, Pb, and U are not very bioavailable (EPA, 2009).

Inorganic contaminants may be localized differently in roots and shoots, depending on the type of plant (sensitive or tolerant, i.e., metal excluder, indicator or accumulator species). Recent studies have identified potential cellular/molecular mechanisms involved in the resistance and tolerance of plants to high concentrations of inorganic contaminants in the environment. Generally, the plants avoid metal accumulation in the cytosol to prevent toxicity symptoms. Plants present various mechanisms and most of them are shared by several common metal-tolerant plants. However, it seems that specific mechanisms are employed for specific metals in particular species (Hall, 2002; Clemens, 2006; Hanikenne *et al.*, 2008). Nickel hyper-accumulation, for example, was found in some specialized flora as the *Alyssum murale* that has colonized Ni-rich serpentine soils. It was also shown that more than one mechanism may be involved in reducing the toxicity of a particular metal (Salt *et al.*, 2000; Hartley-Whitaker *et al.*, 2001). Contrary to organic contaminants, which tend to be taken up and move into plant tissues by diffusion, the inorganics (usually ions) are taken up by biological processes via membrane transporter proteins. The transporter proteins have unique properties (transport rate, substrate affinity, substrate specificity, etc.) that are regulated by the level of specific metabolites or regulatory proteins. The abundance of each transporter also depends on the tissue type and on environmental conditions (Pilon-Smits, 2005).

The different mechanisms for the detoxification of metals and thus for the increased tolerance to metal stress may be subdivided into mechanisms at the extracellular level and those at the cellular level.

Extracellular strategies include roles for mycorrhizas, the cell wall and extracellular exudates. They can also involve the plasma membrane, which reduces the uptake or stimulates the efflux out of the cytosol.

Mycorrhizas and particularly ectomycorrhizas can be effective in reducing the effects of metal toxicity on host plants (Jentschke and Godbold, 2000), but elucidation of their mechanisms is difficult, due to the high specificity of metals and fungal species. These mechanisms consist of various exclusion processes (reduced access to the apoplast due to the hydrophobicity of fungal sheath, chelation by fungal exudates and adsorption onto the external mycelium) that restrict either the metal's movement to the host roots or the absorption of metals.

Metals in the soil solution are in direct contact with the root cell wall but its adsorption has a limited effect on metal activity at the surface of the plasma membrane. However, root exudates can play a role in the tolerance to metals. For example, Salt *et al.* (2000) showed that root exudates of non-hyperaccumulating plants (histidine and citrate) promote the accumulation of Ni-chelating compounds in roots, and state that exudates could have a role in a Ni-detoxification strategy. Ma *et al.* (2001) showed that buckwheat plants secrete oxalic acid from the roots in response to Al stress and a nontoxic Al-oxalate compound was found in its leaves. Phytosiderophores (biosynthesized from nicotianamine) are mucigenic acids released by roots and together with the proper nicotianamine, the organic acids (citrate, malate, histidine) and the thiol-rich peptides (glutathione – GSH and phytochelatines – PCs) have the ability to chelate metals at the root level. Chelated metals in roots may be stored in vacuoles or exported to the shoot via the xylem.

Heavy metals could affect the plasma membrane and its damage would produce increased leakage of solutes from cells. The cell membrane of metal-tolerant plants may play an important role in preventing or reducing entry into the cell or through efflux mechanisms. One example of reduced uptake as an adapted tolerance mechanism was found in *Holcus lanatus* and arsenic toxicity. Phosphate and arsenate appear to be taken up by the same system (Tassi *et al.*, 2004b). The arsenate-tolerant genotype showed a much lower uptake for both anions than the non-tolerant genotype. Moreover, it was suggested that arsenate tolerance in *H. lanatus* is due to both this adaptive suppression transport system and constitutive phytochelatin production, since arsenate can still accumulate to high levels in tolerant plants (Meharg and Macnair, 1992; Hartley-Whitaker *et al.*, 2001).

The efflux of metal ions is another strategy for controlling extracellular metal levels at the plasma membrane. Membrane transporters are responsible for the uptake, efflux, translocation and sequestration of essential and non-essential mineral nutrients. Although there is little direct evidence of plasma membrane efflux transporters in metal tolerance in plants, recent research has revealed that several classes of metal transporters could play a key role in tolerance. These include the Zn-regulated transporter (ZRT), Fe-regulated transporter (IRT)-like proteins (ZIP), natural resistance-associated macrophage proteins (Nramp) and cation diffusion facilitator (CDF) (Kim *et al.*, 2004; Lee *et al.*, 2007).

Root-to-shoot translocation requires an intermediate step of membrane transport. Membrane transporter proteins are needed to promote the movement of inorganics from the root endodermis to the root xylem. It is an area under study and is still unclear via which transporter proteins most metals are exported to the root xylem. The bulk flow in the xylem from root to shoot is driven by the transpiration stream from the shoot, where negative pressure pulls up water, solutes and the bioavailable inorganic contaminants which have access through root membrane. Chelators such as organic acids (histidine, malate, citrate), nicotianamine and thiol-rich peptides can bind metals during xylem transport. Another membrane transport step is required to import metals into the leaf cells from the leaf xylem. Again, specific membrane transporter proteins are needed. Once inside the leaf symplast, the metal may be kept away from metabolic processes by compartmentation at the tissue level (epidermis or trichomes) or at the cellular level (vacuoles or cell wall). At the cellular level, other potential mechanisms exist for the repair of stress-damaged proteins (heat shock proteins or metallotioneins), for the chelation of metals in cytosol by organic acids, amino acids or peptides and for transport away from metabolic processes with compartmentation in the vacuole. At the tissue level, metals were sequestered by chelators to form conjugates and aid in the detoxification process. Chelators such as the tripeptide

GSH (glu-cys-gly) and the PCs can bind metals, and by the action of active transporters molecules, the metal-chelate complex can be transported to the vacuole where it can be further complexed, for example by sulphide on PC-Cd complexes (Hall, 2002; Pilon-Smits, 2005; Memon and Schroder, 2009).

The induction of metal-chelating proteins such as metallothioneins and phytochelatins increased the level of cell tolerance to an excess of metal ions by the sequestration of metals and accumulation in plant cells. Moreover, plants with high transporter activities from cytosol to vacuole can be more efficient at storing toxic inorganic contaminants. Thus, the overexpression or knockdown of membrane transporter proteins or the alteration of plant chelator production by genetic engineering allows us to manipulate plant accumulation (Memon and Schroder, 2009). Selected or engineered plants that have high levels of transporters involved in the uptake of metals from the xylem into the leaf symplast are recommended for phytoextraction, while selected or engineered plants with higher production of chelators or conjugates that confer an enhanced sequestration and tolerance could be useful for phytostabilization.

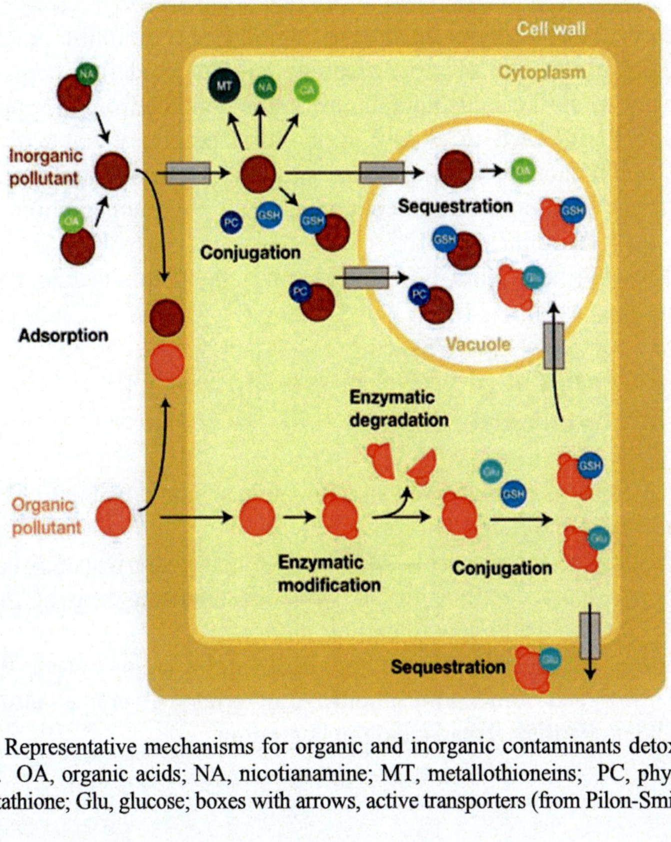

Figure 1. Representative mechanisms for organic and inorganic contaminants detoxification in plant cell. OA, organic acids; NA, nicotianamine; MT, metallothioneins; PC, phytochelatins; GSH, glutathione; Glu, glucose; boxes with arrows, active transporters (from Pilon-Smits, 2005).

An overview of the main mechanisms regarding organic transformation and inorganic detoxification are schematically represented in Figure 1.

3. BRASSINOSTEROIDS AND THEIR POSSIBLE ROLE IN PHYTOREMEDIATION MECHANISMS

Phytohormones regulate and control virtually every aspect of plant growth and development. Phytohormone research is a crucially important area of plant science; phytohormones are one of the key systems that integrate metabolic and developmental events in the entire plant and the response of plants to external factors. Thus, they influence the yield and quality of crops.

As agriculture becomes more mechanized and science increases the possibilities for using various inputs to enhance production, the role of plant growth regulators (PGRs) becomes vital. PGRs in agriculture and horticulture provide agriculture professionals and researchers with the information needed to effectively tap these versatile resources to enhance crop production and quality.

Phytoremediation can be considered a modified agronomic practice, not used for food production but for "cleaning" and/or rendering contaminated media less harmful. It can also obtain profitable benefits from the application of the same PGRs used in classic agricultural practices, but with special PGR and application protocols. Thus, phytoremediation that is aided by phyto-hormone treatment can be called "assisted phytoremediation by plant growth factor (phyto-hormone)".

The following highlights, how PGRs can support this technology based on the plant system.

3.1 Potential of phytohormone application in phytoremediation

Phytoremediation is considered an innovative technology, but must expand its applicability in the field to become a useful one. Gaps still exist regarding the field testing of the concepts. Although much basic research has been performed, we must continue to increase our understanding of the plant mechanisms involved and how to exploit them.

Phytohormones could benefit phytoremediation in various ways. We have listed and discussed some points below, but others of equal validity and interest will surely arise. This is just the beginning.

Different classes of contaminants (organics and inorganics) involve different plant mechanisms (see previous section). The following lists the aims of implementing efficient phytoremediation technologies:

For treatment of organic contaminants:

- increase survival and plant development in stressful conditions; contaminated sites are often quite poor in fertility regarding soil structure and nutrient content
- increase tolerance to toxic effects of different organic compounds and level of concentration present in contaminated media
- increase the rate of contaminant degradation

For treatment of inorganic contaminants:

- increase survival and plant development in stressful conditions
- increase tolerance to toxic effects of various inorganic contaminants and the level of concentration present in contaminated media
- increase metal uptake and accumulation in the aerial part of plants for favorable plant harvesting

In the case of phytostabilization/revegetation (no contaminant treatment, only ecological site restoration):

- increase survival and plant development in stress conditions
- increase tolerance to toxic effects of different inorganic contaminants and level of concentration present in the contaminated media;
- increase root development for media stabilization

To our knowledge, in the case of phytoremediation for organics, no studies have been published regarding the possible use of exogenous phytohormone application to directly implement the technologies' performance and hasten the complete degradation of organic contaminants.

For inorganics, few studies have been conducted to verify the possible use of phytohormones for phytoremediation purposes. Most of them focus on increasing stress resistance to metal toxicity and its uptake for phytoextraction as an alternative strategy for increasing the efficiency of the "assisted phyto-extraction" (Ouzounidou and Ilias, 2005; Tassi *et al.*, 2008).

Some studies have examined the combined effects of PGRs and heavy metals, most performed in hydroponics (Ouzounidou and Ilias, 2005) or in spiked soil (Sayed, 1997; Khan and Chaudhry, 2006). Only a few publications (Fuentes *et al.*, 2000; Liphadzi *et al.*, 2006) have focused on the application of phytohormones in crop plants grown in highly contaminated soils to improve phytoextraction. Fuentes *et al.*, (2000) treated corn (*Zea mays*) by spraying either indolebutyric acid (IBA) or naphthylacetic acid (NAA), which resulted in a 41.2% increase in Pb uptake using IBA and a 127.4% increase in Pb and a 59.5% increase in Zn uptake using NAA. Nevertheless, these results led to a high mortality (up to 45% of the treated plants), a huge PGR concentration (NAA sprayed at 1000 mg kg^{-1}) and a decrease in overall biomass. Liphadzi *et al.* (2006) used IAA to increase the root growth of sunflower plants and observed an increase in Cd and Pb accumulation in leaves with and without EDTA treatment in soil.

Cytokinins (CKs) could prove to be interesting phytohormones for phytoremediation. They are plant growth regulators that play a major role in cell division and cell differentiation. CKs can stimulate shoot initiation, bud formation, the growth of lateral buds, leaf expansion, and chlorophyll synthesis. CKs can also delay leaf senescence, enhance resistance to salinity, low temperatures and drought and induce stomatal opening in some species (Letham *et al.*, 1978; Pospisilova *et al.*, 2000; Werner *et al.*, 2001). Transpiration in plants is driven by a combination of abiotic (climate, soil water availability, ground water depth, etc.) and biotic (leaf area, stomatal functions, root amount and distribution, hormone synthesis, etc.) conditions (Letham *et al.*, 1978). The rate of transpiration is also directly related to whether the stomata are open or closed, and it accounts for the movement of water from roots to shoots by subsequent water loss as vapor through the stomata. An increase in transpiration rate has been observed from 20 to 40% when stomata are wide open (Vose *et al.*, 2003). The chemical regulation of stomatal behavior is one strategy for improving water and contaminant uptake as the water absorbed at the roots by osmosis carries any dissolved mineral nutrients and/or soluble contaminants through the xylem. Thus, transpiration is considered a key process for the success of phytoremediation in soil and groundwater since the vegetation must transpire enough water to control or take up contaminants (Rock, 2003). The application of exogenous CKs can increase the transpiration rate and consequently the absorption of contaminants present in the soil solution. An increased transpiration rate in the excised leaves of *Brassica*, *Helianthus*, *Anthephora*, *Avena*, *Hordeum*, *Triticum* and *Vigna* was found after the addition of exogenous CKs (kinetin, N6-benzyladenine and zeatin) (Pharmavati *et al.*, 1998; Pospisilova, 2003). Kinetin also markedly increases the stomatal aperture in *Tradescantia* and *Paphiopedilum tonsum* leaves (Irving *et al.*, 1992).

Results reported in Tassi *et al.* (2008) indicate that in treated sunflower plants the exogenous application of a mixture of CKs had a positive effect on the aerial biomass and phytoextraction efficiency, and increased the transpiration rate. CK treatment showed higher Pb phytoextraction efficiency in leaves and shoots than the other two PGRs (about 120% and 50%, respectively) when compared to the control plants. Also for Zn, again the treatment showed a high increase, which was about 100% in leaves and 20% in shoots, compared to the control plants. When the EDTA treatment is added to increase metal bioavailability in the soil, CK treatment increased the transpiration rate by about 50% but the treatment did not produce a meaningful increase in the overall aerial biomass, but induced a higher increase in the biomass of leaves (about 30%), which are the aerial tissues with the greatest Pb accumulation.

The most positive result in phytoextraction efficiency was achieved when both CKs and EDTA were used. This result supports the hypothesis that phytoextraction can be improved by increasing both the dry mass and (to a lesser extent) the metal accumulation in the upper parts of the plants. The regulation of stomatal opening (Dodd, 2003) is also an important effect of CKs. This in turn can increase the transpiration rate and consequently the flux (via xylem sap) of water-soluble soil components or contaminants to the upper parts of plants. Results show that the application increased the transpiration rate of *H. annuus*.

The increased transpiration rate and aerial biomass of plants treated with PGRs indicate that water and micronutrients are efficiently absorbed from the soil and translocated to the parts of the plant that are above ground, but were only able to remove a limited fraction of soluble metals. This suggests that phytohormones can increase the metal translocation of an already absorbed metal fraction, as demonstrated in combined treatments with PGRs and EDTA.

Increasing plant transpiration alone does not lead directly to an increase in metal uptake. Many indubitably confirm the fact that metal absorbing processes are strictly regulated by complex metal homeostasis in the plant (Clemens, 2001; Clemens *et al*., 2002; Verbruggen *et al*., 2009) as previously described in Section 2.4 (Mechanisms associated with the phytoremediation of inorganic contaminants).

The combined effects of EDTA and the growth-promoting 3-indole acetic acid (IAA) on Pb uptake in *Medicago sativa* plants have been studied in hydroponics. Increased Pb content in leaves of plants exposed to Pb and EDTA/IAA is about 2800% compared to those treated with Pb alone and about 600% compared to those treated with Pb and EDTA (Lopez *et al*., 2005).

3.2 Brassinosteroids for phytoremediation: potential application

Brassinosteroids (BRs), first identified 20 years ago, play an important role in regulating plant growth and development. They have also been implicated in plant response to environmental stresses and in plant defense against bacterial, fungal, and viral pathogens.

BRs are a group of plant steroidal hormones that are structurally similar to animal and insect steroids. BRs are ubiquitous in the plant kingdom and regulate a wide range of physiological responses, including cell elongation, photo-morphogenesis, xylem differentiation, seed germination (Sasse, 2002) and stress response (Krishna, 2003). The growth-promoting properties of BRs have been known since the early 1970s.

This section focuses on the potential of brassinosteroids to improve phytoremediation technology by crop yield, stress tolerance and increased efficiency of the technology.

Table 1, extracted from the work of Khripach *et al.* (2000) summarizes the most important physiological effects of BRs at both the cellular level and the whole plant level, thus increasing crop growth and quality. This highlights the potential of BRs to improve plants' capacity to remediate contaminated sites.

Table 1. Some physiological effects of brassinosteroids in plants (Khripach *et al.*, 2000)

Cell level	**Whole plant level**
Stimulation of elongation and fission	Growth promotion
Effect on hormonal balance	Increased success of fertilization
Effect on enzyme activity; H.-pump activation	Shortened period of vegetative growth
Activation of protein and nucleic acid synthesis	Increased size and quantity of fruit
Effect on the protein spectrum and on amino acid composition of proteins	Effect on the content of nutritive components and improved fruit quality
Effect on fatty acid composition and on the properties of membranes	Increased resistance to unfavorable environmental factors, stress and disease
Enhancement of photosynthetic capacity and translocation of products	Increase crop yield

3.3 Brassinosteroids increase germination

In the beginning, the practical application of BRs was to promote the germination of seeds treated with BRs before sowing.

It is well known that germination has been promoted by treatment of seeds with BRs before sowing.

The effects on seeds were expressed as stimulation of germination, increasing germination percentage, promotion of rooting and root growth, increased plant growth of the seedling and increased tillers and vegetative buds. Experiments on *Arabidospis* by Steber and McCourt (2001) have shown BR signal is required to reverse ABA-induced dormancy and stimulate germination in *Arabidopsis*.

These characteristics are of great interest in phytoremediation application since germination capability is often hampered by toxic effects of contamination and by difficulties derived from unfavorable soil and weather conditions.

3.4 Brassinosteroids increase plant yield

Due to the substantial effect of BRs on plant growth and development, the economic potential of BRs in agriculture was recognized as early as the 1980s. The synthesis of BR analogs confirmed structure–activity relationships and provided a method for the preparation of large enough quantities of active BRs for greenhouse and field evaluations (Takatsuto *et al.*, 1996).

Extensive testing of a synthetic BR, 24-epibrassinolide (EBR), in China, Japan and Russia showed that while exogenous BR has the ability to substantially increase yields in a variety of plant species, the results can be variable depending on the mode of application, growth stage of application and environmental conditions (Khripach *et al.*, 2000; Ikekawa and Zao, 1991).

Such BRs have a positive impact on both phytostabilization and phytoextraction technologies. Contaminated sites are often very low in fertility properties and contaminants are often present at toxic levels, so plants suffer in germination and in different growth phases, which negatively impacts plant growth. In particular, in the case of phytoextraction biomass is one of the critical factors for its efficiency, along with metal accumulation in plant tissue.

Applicability studies such as those conducted by McGrath and Zhao (2003) have shown that the bioconcentration factor (ratio of metal concentration in plant shoot to that in the soil) and biomass production in tons per year are the two key factors for determining phytoextraction efficiency. Figure 2 explains how these factors are correlated in estimating the number of growing cycles to reduce the metal concentration in contaminated soil. Having plants with a high bioconcentration factor and high biomass production optimize the phytoextraction efficiency, reducing the number of crops needed to decontaminate the soil.

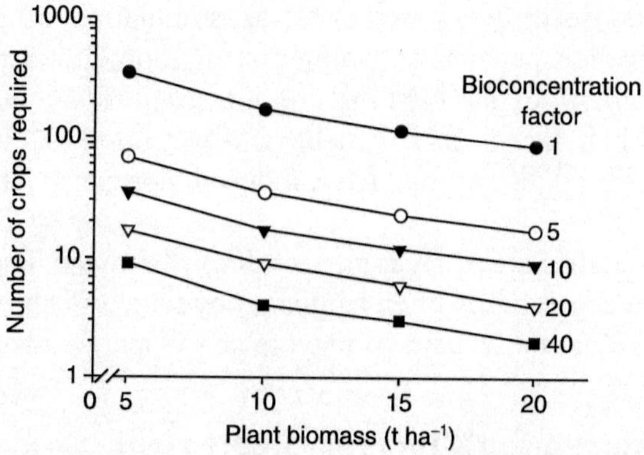

Figure 2. Model calculations of the number of crops (harvests) required to halve metal concentrations in the top soil, assuming that the metal taken up by plants is from the top 20 cm of soil. The results show that both metal hyperaccumulation and a good biomass yield are essential for efficient phytoextraction (from McGrath and Zhao, 2003).

3.5 Brassinosteroids increase environmental stress tolerance

In addition to their role in plant development, BRs protect plants from a variety of environmental stresses, including high and low temperatures, drought, salinity and pathogen attack (Krishna, 2003; Divi and Krishna, 2009a,b, Muhammad and Muhammad, 2007). Most studies demonstrating the anti-stress effects of BRs have treated plants with exogenous BRs. Although exogenous BR treatment is neither a conventional greenhouse nor field agricultural method, it is instrumental in the laboratory for studying molecular changes induced by BRs that might be contributing to increased stress tolerance in treated vs untreated plants.

The BR-induced molecular changes that are related to stress tolerance include a higher level of expression of stress-responsive genes (Kagale *et al.*, 2007a,b), maintenance of protein synthesis (Dhaubhadel *et al.*, 2002; Ozdemir *et al.*, 2004) induction of other hormone responses, increased activity of antioxidant enzymes, greater accumulation of osmoprotectants (Divi and Krishna, 2009a,b) and higher photosynthetic efficiency (Krishna, 2003). Results of a global gene expression study of untreated and BR-treated *Arabidopsis* under stress and no-stress conditions also indicate that BR affects a myriad of cellular processes either directly or through crosstalk with other plant hormone pathways, resulting in better performance of plants under stress conditions (Krishna and co-workers, unpublished data). Clearly, altering endogenous BRs activity through transgenic technology offers great promise of generating crops with increased stress tolerance.

Further, BRs also influence the electrical properties of membranes and ion transport by altering membrane permeability and structure, and the stability and activity of membrane enzymes. The reduction of toxicity by BRs was associated with lesser uptake of ions and enhanced levels of soluble proteins and nucleic acids with the increasing activity of ATPase, an enzyme responsible for acid secretion and changes in membrane level (Bajguz, 2000a). The proton pump generates an H+ electrochemical gradient and provides a driving force for the rapid ion fluxes required for the uptake of various nutrients such as K^+, Cl^-, NO_3^-, amino acids and sucrose across the plasma membrane (Sze *et al.*, 1999). Regulation of H^+-ATPase activity (Kasamo, 2003) not only allows nutrient uptake in plant cells but also controls water fluxes (Sondergaard *et al.*, 2004). However, little is known about the role of ion transport systems during BR-induced cell expansion. Other studies have reported proton secretion induced by BL when applied to azuki bean epicotyls or apical root segments of maize (Zhang *et al.*, 2005). This proton secretion was accompanied by an early hyperpolarization of the plasma membrane, indicating that proton pumps could be targets of BRs (Zhang *et al.*, 2005). The high biological activity of BRs suggests an important role in the regulation of physiological processes in plants and also in anti-stress activity.

3.6 Brassinosteroids and inorganic contaminant action

Another recently discovered aspect of the influence of BRs is their ability to regulate the uptake of ions into the plant cell. BRs can be used to reduce the accumulation of heavy metals and radioactive elements when plants are grown in areas that are polluted by these contaminants (Khripach *et al.*, 2000).

Sharma and Bhardwaj (2007) reported that EBR has been observed to improve the growth of *B. juncea* plants compared to metal treated and non-treated plants. Similarly, BR stimulated leaf elongation of wheat and mustard plants (Braun and Wild, 1984). In addition, Clouse *et al.* (1992) reported that BR (0.1 M) induced a measurable increase in the length of soybean epicotyls. Franck-Duchenne *et al.* (1998) found that EBR at a concentration of 0.1 M promoted stem induction and elongation, and produced large leaves in two cultivars of sweet peppers. EBR has also been found to reduce the Cu ion uptake and accumulation in *B. juncea* L plants. Other studies on the accumulation of heavy metals under the influence of BRs also showed reduction in metal toxicity in barley, tomato and radish plants (Volynets *et al.*, 1997). Bajguz (2000b) revealed that 24-epibrassinolide blocked heavy metal accumulation in *Chlorella vulgaris* cells.

It has been observed that BRs in combination with Pb caused stimulation of phytochelatins synthesis in *C. vulgaris*. BRs stimulated growth and photosynthetic parameters after blocking the bioabsorption of lead in *C. vulgaris* (Bajguz, 2002).

The protective effect of epibrassinolide on winter rape plants under Cd stress was investigated by Janeczko *et al.* (2005). Bilkisu *et al.* (2003) reported that BL during Al-related stress stimulated growth in *Phaseolus aureus*.

It was shown that changes in the ion/metal content were influenced by EBR and dependent on the stage of plant development when the seeds were treated. The content of 137Cs in the plants at the flowering stage was even higher than in the untreated control. Fully matured plants showed some decrease in 137Cs content, especially in the vegetative organs, probably as a result of vegetative dilution (Khripach *et al.*, 1999). The mechanism involved in reducing the toxicity may be the chelation of the metal ion by a ligand. Such ligands include organic acids, amino acids, peptides or polypeptides (Bajguz, 2002).

Reduction of metal stress symptoms were detected in studies performed on Cd treatment by Hayat *et al.* (2007, 2010). Authors showed the ameliorative effect of hormonal treatment on production of antioxidant enzymes. Sprayed plants have been shown to reduce heavy metal (cadmium, copper, lead, and zinc) accumulation. This has been tested in different plants, for example Indian mustard, tomato, radish, and barley (Fariduddin et al. 2009). This effect could be improper in improve phytoextraction technology but can be useful for the revegetation/stabilization of contaminated soil where interested process is the avoidance of metal transfer to plants limiting the metal transfer to food chain and environment. In other cases were agricultural practices cannot be stopped for needy countries treatment with BRs could reduce the metal transfer to human and reduce dramatic toxic effects on animal and human as reported in case of seleniferous soil in northwestern India (Dhillon and Dhillon, 1997).

3.7 Brassinosteroids and organic contaminant action

Scientists worldwide have been seeking new ways of minimizing pesticide residues that remain in food crops after harvest. Only recently researchers from the Zhejiang University in China (Xia *et al.*, 2006, 2009a,b) reported that application of BRs to crops can help plants to eliminate residues of certain pesticides. Xia and colleagues treated cucumber plants with EBR, then treated the plants with various pesticides, including chloropyrifos, a broad-spectrum commercial insecticide. EBR significantly reduced their toxicity and residues in the plants. In this paper the author stated that BRs may be promising, environmentally friendly natural substances suitable for extensive application to reduce the risk of human and environmental exposure to pesticides. At present, there is no direct evidence for harmful effects of EBRs on human health.

However, little is known about the involvement of phytohormones and in particular EBRs in plant responses to organic pollutants such as pesticides. The exogenous application of EBRs reduced the phytotoxic effect of herbicides, fungicides, and insecticides on cucumber leaves; moreover BRs can reduce the damage caused by simazine, butachlor, or pretilachlo in rice (Bajguz and Hayat, 2009). How EBRs reduce the toxicity of various pesticides, however, is unclear.

On the basis of their effects on the toxicity of pesticide, BRs can be considered safeners, which are known to induce the activity of numerous plant P450s and enhance glutathione conjugation involved in the biodegradation of herbicides (Hatzios and Burgos, 2004). Indeed, Hatzios and Burgos (2004) have previously suggested that plant hormones are involved in the induction of plant defense and detoxification genes in response to pesticides. Furthermore, transcriptional analyses of BR-deficient or BR-treated *Arabidopsis* and cucumber plants have shown that BR-regulated genes include those pesticide detoxification genes encoding P450 monooxygenase, glutathione S-transferase, and UDP-glycosyltransferase (Mussig *et al.*, 2002; Xia *et al.*, 2009b).

The recent studies of Xia *et al.* (2009a,b) clearly showed that application of 24-epibrassinolide (EBR) accelerated the degradation of various fungicides and insecticides and that this effect was associated with the enhanced expression of detoxification genes.

These results strongly suggest that BRs enhance plant tolerance to pesticides by modulating the metabolic process of these pesticides.

The same route could be involved in other organic compounds (such as PAH, PCB etc.) and may have similar results in stimulating their degradation. This in turn could bring benefits to the phytoremediation technologies focused on degradation of organic compounds

3.8 Applying Brassinosteroids in the field

From Khripach *et al.* (2000) it is reported that in Russia and Belarus, EBR is the active ingredient of the plant growth promoter Epin2, which has been officially registered since 1992 and is recommended for treatment of agricultural plants such as tomato, potato, cucumber, pepper and barley. Moreover a preparation with the same active ingredient was extensively investigated in Japan in the middle of the 1980s (Nippon Kayaku Co., 1988). It was found to be efficient for increasing the yield of many crops. Another chemical,

Tianfengsu, has been developed in China as the result of a joint research program between Japanese and Chinese scientists (Ikekawa and Zhao, 1991). This preparation is mainly a mixture of natural 24-epibrassinolide and its unnatural 22S, 23S isomer; Tianfengsu is widely used in China to increase yield in crops such as rice, maize, wheat, cotton, tobacco, vegetables and fruit.

The potential role of BRs in countries with highly developed agriculture such as the USA, Japan and Europe is worth mentioning. Information regarding any official registration of EBR in Japan is lacking, probably because of the variable responses obtained in field trials. American researchers have also reported erratic responses to EBR in field evaluations, the main reason for the little interest in developing BRs for agricultural use in the US. The relative effects of BRs may be low when the conditions under which plants are growing are generally favorable. This could explain the consistently good results obtained in China, compared to those in Japan (Ikekawa and Zhao, 1991). However, it has become clear that not only crop yields, but also the quality of crops and their resistance to disease, could be positively influenced by the application of BRs. This allows a more environmentally-friendly type of agriculture, which may become an issue in the near future.

The growth-promoting activity of BRs usually takes place only after treatment of plants in the appropriate phase of development and within a certain concentration range, which is different for each plant species and type of BR. For large-scale field application, two modes of BRs application are possible: seed soaking and foliar spray. Both methods have been investigated extensively, but results with the latter method were found to be highly dependent on the phase of plant development. Generally, better results can be obtained when young rather than old plants are treated. The formulation of the spraying solution is very important, and additives are necessary to facilitate the spread of active substances, to prevent early drying and to ensure penetration of BRs via cell walls. The working solution should be prepared by dilution with water shortly before application, which can be done with normal agricultural sprayers. The ability to combine BR treatment of plants with other pesticides, for example with fungicides for potato treatment, allows BRs to be used with existing technologies for plant protection, with no significant additional expense.

A new variant of BR application has been developed recently, based on the use of combinations of BRs with mineral fertilizers. Field trials showed increased yield and improved quality of crops of barley, oat, potato, winter rye and wheat. In this approach, EBR was recommended as an addition (10^{-4}–10^{-6}%) to nitrogen-phosphorus-potassium fertilizers.

The development of a marketable growth regulator generally takes longer than the development of other plant treatment agents; determination of the

Table 2. Effects of brassinosteroids on plants for phytoremediation to be tested and verified in selected species and on contaminated sites

Inorganics		Organics
Phytostabilization	**Phytoextraction**	
Increase stress tolerance	Increase stress tolerance	Increase stress tolerance
Increase germination and growth development	Increase germination and growth development	Increase germination and growth development
Reduce metal uptake	Increase metal uptake and accumulation	Increase contaminant degradation
Increase root apparatus	Increase biomass production Increase root apparatus	Increase root apparatus

dose rate and the best time for application is often critical and requires extensive outdoor experimentation under varying conditions.

4. CONCLUSION AND FUTURE DIRECTIONS

The studies reviewed here have enhanced our knowledge of the effects of phytohormones (and more specifically of BRs) in plants that demonstrate their potential applicability to increase effectiveness in phytoremediation technologies (still largely unexplored but with intriguing factors to verify and interesting chances of success).

From the current review (summarized in Tables 1 and 2) it is clear that the ability of BRs to enhance yield and increase stress tolerance in plants, are of great interest in improving application of phytoremediation technologies, and that genetic manipulation of BRs activity could be a practical strategy for generating “high-phytoremediating” transgenic plants.

Since BRs control several important agronomic traits such as time of flowering, number of branches, plant height, vegetative and seed yield and stress tolerance, it might not be possible to obtain optimal BR effects for all of the traits controlled by BR. However, even if a subset of these traits could be impacted by BRs, either through a single nucleotide change (mutation) or introduction of a transgene, the accomplishment would be significant. It should be noted that the currently available transgenic plants with altered BRs activity according to the literature have been tested in the field under only a limited set of environmental conditions. Results of rigorous testing in these directions will be the ultimate test of the potential of BRs in crop improvement for remediation techniques.

Phytoremediation uses natural or genetically modified plants (GMPs) to extract a wide range of heavy metals and organic pollutants from the soil. Initial experiments with transgenic plants have shown that they are indeed efficient in drawing metals from heavily contaminated soils. However, despite this and other advantages, the progress and application of this technology to tackle widespread environmental problems is being hampered by ideology-driven, restrictive legislation over the use and release of GMPs in Europe.

Nonetheless, BR application remains viable via the simple EBR spray solution, an easy and inexpensive agronomic practice.

In conclusion, the application of BRs and more generally of phytohormones as a concept for improving phytoremediation technologies shows great promise. As with any other emerging technology, it is a prerequisite that the technology be implemented as part of a well-designed and well-thought-out environmental engineering concept.

Their crucial role in plant development promises not only exciting results but also exciting opportunities to apply new techniques. However, the potential uses of agrochemicals for phytoremediation still remain unexplored. It requires specific tests focused on verifying effectiveness, and represents an open door that could lead to interesting positive results for increasing the phytoremediation technologies efficiency and applicability.

This chapter raises the possibility that the effects of BRs could provide positive interactions with plants for a more efficient application of phytoremediation technologies. BRs may be promising, environmentally friendly, natural substances suitable for widespread application; and thus the risk of human and environmental exposure to contaminated sites can be reduced through phytoremediation technologies.

5. REFERENCES

Alkorta, I., and Garbisu, C. 2001. Phytoremediation of organic contaminants in soils. *Biores. Technol.* **79**: 273–276.

Bajguz A. 2000a. Effects of brassinosteroids on nucleic acids and protein in cultured cells in *Chlorella vulgaris*. *Plant Physiol. Biochem.* **38**: 209–215.

Bajguz A. 2000b. Blockage of heavy metal accumulation in *Chlorella vulgaris* cells by 24epibrassinolide. *Plant Physiol. Biochem.* **38**: 797–801.

Bajguz A. 2002. Brassinosteroids and lead as stimulators of phytochelatins synthesis in *Chlorella vulgaris*. *J. Plant Physiol.* **159**: 321–324.

Bajguz, A., and Hayat, S. 2009. Effects of brassinosteroids on the plant responses to environmental stresses. *Plant Physiol. Biochem.* **47**: 1–8.

Baker, A.J.M., and Brooks, R.R. 1989. Terrestrial higher plants which hyper accumulate metallic elements–review of their distribution, ecology and phytochemistry. *Biorecovery*, **1**: 81–126.

Barac, T., Taghavi, T., Borremans, B., Provoost, A., Oeyen, L., *et al.* 2004. Engineered endophytic bacteria improve phytodegradation of water-soluble, volatile, organic pollutants. *Nat. Biotechnol.* **22**: 583–588.

Barbafieri, M. 2000. The importance of nickel phytoavailable chemical species characterization in soil for phytoremediation applicability. *Int. J. Phytorem.* **2**: 105–115.

Barbafieri, M., Lubrano, L., and Petruzzelli, G. 1996. Characterization of pollution at heavy metal contaminated sites: a proposal. *Annali di Chimica* **86**: 585–594.

Bilkisu, A.A., Xiao-Gang, G., Qing-Lei, G., and Yong-Hua, Y. 2003. Brassinolide amelioration of aluminium toxicity in mungbean seedling growth. *J. Plant Nutr.* **26**: 1725–1734.

Braun, P., and Wild, A. 1984. The influence of brassinosteroid on growth and parameters of photosynthesis of wheat and mustard plants. *J. Plant Physiol.* **116**: 189–196.

Briggs, G.G., Bromilow R.H., and Evans A.A. 1982. Relationship between lipophilicity and root uptake and translocation of non-ionized chemicals by barley. *Pestic. Sci.* **13**: 495–504.

Bromilow, R.H., and Chamberlain, K. 1995. Principles governing uptake and transport of chemicals. In: Plant contamination: modelling and simulation of organic chemical processes (Ed. S. Trapp and J.C. McFarlane). Boca Raton, Lewis, pp. 37–68.

Brown, S.L., Henry, C.L., Chaney, R., Compton H., and DeVolder, P.M. 2003. Using municipal biosolids in combination with other residuals to restore metal contaminated mining areas. *Plant Soil* **249**: 203–215.

Burken, J.G., Shanks, J.V., and Thompson, P.L. 2000. Phytoremediation and plant metabolism of explosives and nitroaromatic compounds. In: Biodegradation of nitroaromatic compounds and explosives (Ed. J.C. Spain, J.B. Hughes, and H.J. Knackmuss). Boca-Raton, FL, CRC Press, pp. 239–275.

Castro, S., Davis, L.C., and Erikson, L.E. 2001. Plant enhanced remediation of glycol-based aircraft deicing fluids. Pract. Periodical Hazard., Toxic, Radioact. *Waste Manag.* **5**: 141–152.

Chiou, C.T. 2002. Contaminant partition and bioconcentration. In: Partition and adsorption of organic contaminants in environmental systems (Ed. C.T. Chiou). New Jersey, Wiley, pp. 86–105.

Clemens, S. 2001. Molecular mechanisms of plant metal tolerance and homeostasis. *Planta* **212**: 475–486.

Clemens, S. 2006. Toxic metal accumulation, response to exposure and mechanism of tolerance in plants. *Biochimie* **88**: 1707–1719.

Clemens S., Plamgren M.G., Kramer, U. 2002. A long way ahead; understanding and engineering plant metal accumulation. *Trends Plant Sci.* **7**: 309–315.

Clemente R., Walker, J.D., Roig, A., and Bernal, P.M. 2003. Heavy metal bioavailability in a soil affected by mineral sulphides contamination following the mine spillage at Aznalcollar (Spain). *Biodegradation* **14**: 199–205.

Clouse, S.D., Zurek, D.M., McMorris, T.C., and Baker, M.E. 1992. Effect of brassinolide on gene expression in elongating soybean epicotyls. *Plant Physiol.* **100**: 1377–1383.

Cunningham, S.D., and Berti, W.R. 1993. Remediation of contaminated soils with green plants: an overview. *In Vitro Cell Dev. Biol.* **29**: 207–212.

Dhaubhadel, S., Browning, K.S., Gallie, D.R., and Krishna, P. 2002. Brassinosteroid functions to protect the translational machinery and heat-shock protein synthesis following thermal stress. *Plant J.* **29**:681–691.

Dhillon, K.S., and Dhillon, S.K. 1997. Distribution of seleniferous soils in north-west India and associated toxicity problems in the soil–plant–animal–human continuum. *Land Contam. Reclam.* **5**: 313–322.

Dickinson, N.M., Baker, A.J.M., Doronila, A., Laidlaw, S., and Reeves, R.D. 2009. Phytoremediation of inorganics: realism or synergies. *Int. J. Phytorem.* **11**: 97–114.

Di Gregorio, S., Barbafieri, M., Lampis, S., Sanangelantoni, A.M., Tassi, E., and Vallini, G. 2006. Combined application of Triton X-100 and *Sinorrhizobium sp.* Pb002 inoculum for the improvement of lead phytoextraction by *Brassica juncea* in EDTA amended soil. *Chemosphere* **63**: 293–299.

Divi, U.K., and Krishna, P. 2009a. Brassinosteroids confer stress tolerance. In: Plant stress biology (Ed. H. Hirt). Weinheim, Wiley-VCH Verlag GmbH & Co. KGaA, pp. 119–135.

Divi, U.K., and Krishna, P. 2009b. Brassinosteroids: a biotechnological target for enhancing crop yield and stress tolerance. *New Biotechnology* **26**: 131–136.

Dodd, I.C. 2003. Hormonal interactions and stomata responses. *J. Plant. Growth Reg.* **22**: 32–46.

Dushenkov, S., and Kapulnik, Y. 2000. Phytofiltration of metals. In: Phytoremediation of toxic metals – using plants to clean-up the environment (Ed. I. Raskin and B.D. Ensley). New York, Wiley, pp. 89–106.

Dushenkov, S., Vasudev, D., Kapulnik, Y., Gleba, D., Fleisher, D., Ting, K.C., *et al.* 1997. Removal of uranium from water using terrestrial plants. *Environ. Sci. Technol.* **31**: 3468–3474.

E.P.A. 2009. Technical/Regulatory Guidance – Phytotechnology Technical and Regulatory Guidance and Decision Trees, revised. http://www.itrcweb.org/Documents/PHYTO-3.pdf Assessed in February 2010.

Fariduddin, Q., Yusuf, M., Hayat, S., and Ahmad, A. 2009. Effect of 28-homobrassinolide on antioxidant capacity and photosynthesis in *Brassica juncea* plants exposed to different levels of copper. *Environ. Exp. Bot.* **66**: 418–424.

Fletcher, J.S., and Hedge, R.S. 1995. Release of phenols by perennial plant roots and their potential importance in bioremediation. *Chemosphere* **31**: 3000–3016.

Franck-Duchenne M., Wang Y., Tahar B.S., and Beachy N. 1998. *In vitro* stem elongation of sweet pepper in media containing 24-epibrassinolide. *Plant Cell Tissue and Organ Cult* **53**: 79–84.

Fuentes, H.D., Khoo, C.S., Pe, T., Muir, S., and Khan, A.G. 2000. Phytoremediation of a contaminated mine site using plant growth regulators to increase heavy metal uptake. In: Proceedings of the 5th International Conference on Clean Technologies for the Mining Industry, **1**: 749–756.

Gardea-Torresdey, J.L., de la Rosa, G., and Peralta-Videa, J.R. 2004. Use of phytofiltration technologies in the removal of heavy metals: a review. *Pure and Appl. Chem.* **76**: 801–813.

Hall, J.L. 2002. Cellular mechanisms for heavy metal detoxification and tolerance. *J. Exp. Bot.* **53**: 1–11.

Hanikenne, M., Talke, I.N., Haydon, M.J., Lanz, C., Nolte, A., Motte, P., Kroymann, J., Weigel, D., and Kramer U. 2008. Evolution of metal hyperaccumulation required cis- regulatory changes and triplication of HMA4. *Nature* **453**: 391–395.

Harms, H., Boken, M., Kolb, M., and Bock, C. 2003. Transformation of organic contaminants by different plant systems. In: Phytoremediation: transformation and control of contaminants (Ed. S.C. McCutcheon and J.L. Schnoor). New York, Wiley, pp. 285–316.

Hartley-Whitaker, J., Ainsworth, G., Vooijs, R., Ten Bookum, W., Schat, H., and Meharg, A.A. 2001. Phytochelatins are involved in differential arsenate tolerance in *Holcus lanatus*. *Plant Physiol.* **126**: 299–306.

Hatzios, K.K., and Burgos, N. 2004. Metabolism-based herbicide resistance: regulation by safeners. *Weed Sci.* **52**: 454–467.

Hayat, S., Ali, B., Hasan, S.A., and Ahmad, A. 2007. Brassinosteroid enhanced the level of antioxidants under cadmium stress in *Brassica juncea. Environ. Exp. Bot.* **60**: 33–41.

Hayat, S., Hasan, S.A., Hayat Q., and Ahmad, A. 2010. Brassinosteroids protect *Lycopersicon esculentum* from cadmium toxicity applied as shotgun approach. *Protoplasma* **239**: 3–14.

Hutchinson, S.L., Schwab, A.P., and Banks, M.K. 2003. Biodegradation of petroleum hydrocarbons in the rhizosphere. In: Phytoremediation: transformation and control of contaminants (Ed. S.C. McCutcheon and J.L. Schnoor). New York, Wiley, pp. 355–386.

Ikekawa, N., and Zao, Y.J. 1991. Application of 24-epibrassinolide in agriculture. In: Brassinosteroids: chemistry, bioactivity, and applications (Ed. H.G. Cutler, T. Tokota, and G. Sadam) ACS Symposium Series, Vol. 474, pp. 280–291, American Chemical Society.

Irving, H.R., Gehring, C.A., and Parish, R.W. 1992. Changes in cytosolic pH and calcium of guard cells precede stomatal movements. *Plant Biology, Proc. Natl. Acad. Sci. USA* **89**, 1790–1794.

Janeczko, A., Koscielniak, J., Pilipowicz, M., Szarek-Lukaszewska, G., and Skoczowski, A. 2005. Protection of winter rape photosystem 2 by 24-epibrassinolide under cadmium stress. *Photosynthetica* **43**: 293–298.

Jentschke, G., and Godbold, D.L. 2000. Metal toxicity and ectomycorrhizas. *Physiol. Plant.* **109**, 107–116.

Kagale, S., Divi, U.K., Krochko, J.E., Keller, W.A., and Krishna, P. 2007a. Brassinosteroids confer tolerance in *Arabidospsis thaliana* and *Brassica napus* to a range of abiotic stress. *Planta* **225**: 353–364.

Kagale, S., Divi, U.K., and Krochko, J.E. 2007b. Brassinosteroid confers tolerance in *Arabidopsis thaliana* and *Brassica napus* to a range of abiotic stresses. *Planta* **225**: 353–364.

Kasamo, K. 2003. Regulation of plasma membrane H^+-ATPase activity by the membrane environment. *J. Plant Res.* **16**: 517–523.

Khan, A.S., and Chaudhry, N.Y. 2006. GA3 improves flower yield in some cucurbits treated with lead and mercury. *Afr. J. Biotechnol.* **5**(2): 149–153.

Khripach, V.A., Zhabinskii, V.N., and de Groot, A.E. 1999. Brassinosteroids: A new class of plant hormones. Academic, San Diego, p. 456.

Khripach, V.A., Zhabinskii V.N., and de Groot, A.E. 2000. Twenty years of brassinosteroids: steroidal plant hormones warrant better crops for the XXI century. *Ann. Bot.* **86**: 441–447. 447.

Keeling, S.M., Stewart, R.B., Anderson, C.W.N., and Robinson, B.H. 2003. Nickel and cobalt phytoextraction by the hyperaccumulator *Berkheya coddii*: implications for polymetallic phytomining and phytoremediation. *Int. J. Phytorem.* **5**: 235–244.

Kim, D., Lahner, B., Persans, M.W., Baek, D., Yun, D.J., and Salt, D.E. 2004. The plant CDF family member TgMTP1 from Ni/Zn hyperaccumulator *Thlaspi goesingense* acts to enhance efflux of Zn at the plasma membrane when expressed in *Saccharomyces cerevisiae*. *Plant J.*, **39**: 237–251.

Krishna, P. 2003. Brassinosteroid mediated stress response. *J Plant Growth Regul.* **22**: 289–297.

LaCoste, C., Robinson B.H., and Brooks, R.R. 2001. Thallium uptake by vegetables: Its significance for human health, phytoremediation and phytomining. *J. Plant Nutr.* **24**: 1205–1216.

Lee, S., Kim, Y-Y., and An, G. 2007. Rice P1B-type heavy metal ATPase, OsHMA9, is a metal efflux protein. *Plant Physiol.* **145**: 831–842.

Leigh, M.B., Prouzova, P., Mackova, M., Macek, T., Nagel, D., and Fletcher, J.S. 2006. Polychlorinated biphenyl (PCB)-degrading bacteria associated with trees in a PCB-contaminated site. *Appl. Environ. Microbiol.*, **72**: 2331–2342.

Letham, D.S., Goodwin, P.B., and Higgins, T.J.V. 1978. Phytohormones and related compounds: a comprehensive treatise. In: Phytohormones and the development of higher plants (Ed. D.S. Letham, P.B. Goodwin, and T.J.V. Higgins). Vol 2. Amsterdam, Elsevier, North-Holland Biomedical Press.

Li, X.D., and Thornton, I. 2001. Chemical partitioning of trace and major elements in soil contaminated by mining and smelting activities. *Appl. Geochem.*, **16**: 1693–1706.

Lindsay, W., and Norwell, W.A. 1978. Development of DTPA soil test for Zn, Fe, Mn, Cu. *Soil Sci. Soc. Am. J.* **42**: 421–428.

Liphadzi, M.S., Kirkham, M.B., and Paulsen, G.M. 2006. Auxin-enhanced root growth for phytoremediation of sewage-sludge amended soil. *Environ. Technol.* **27**: 695–704.

Lopez, M.L., Peralta-Videa, J.R., Benitez, T., and Gardea-Torresdey, J.L. 2005. Enhancement of lead uptake by alfalfa (*Medicago sativa*) using EDTA and a plant growth promoter. *Chemosphere* **61**: 525–598.

Ma, J.F., Ryan P.R., and Delhaize, E. 2001. Aluminium tolerance in plants and the complexing role of organic acids. *Trends Plant Sci.* **6**: 273–278.

Ma, X., and Burken, J.G. 2003. TCE diffusion to the atmosphere in phytoremediation applications. *Environ. Sci. Technol.* **37**: 2534–2539.

McCutcheon, S.C., Medina, V.F., and Larson, S.L. 2003. Proof of phytoremediation for explosives in water and soil. In: Phytoremediation: transformation and control of contaminants (Ed. S.C. McCutcheon and J.L. Schnoor). New York, Wiley, pp. 429–480.

McCutcheon, S.C. and Schnoor, J.L. (Eds). 2003. Phytoremediation: Transformation and Control of Contaminants, *Wiley Interscience*, **987** pp.

McGrath, P.S., and Zhao, F.J. 2003. Phytoextraction of metals and metalloids from contaminated soils. *Curr. Opin. Biotechnol.* **14**: 277–282.

Meharg, A.A., and Macnair, M.R. 1992. Genetic correlation between arsenate tolerance and the rate of influx of arsenate and phosphate in *Holcus lanatus*. *Heredity.* **69**: 336–341.

Memon, A.D., and Schroder, P. 2009. Implication of metal accumulation mechanisms to phytoremediation. *Environ. Sci. Pollut. Res.* **16**: 162–175.

Mendez, M.O., and Maier, R.M. 2008. Phytostabilization of mining tailings in arid and semiarid environments–an emerging remediation technology. *Environ. Health Persp.* **116**: 278–283.

Mkandawire, M., and Dudel, E.G. 2005. Accumulation of arsenic in *Lemna gibba* L. (duckweed) in mailing waters of two abandoned uranium mining sites in Saxony, Germany. *Sci. Tot. Environ.* **336**: 81–89.

Muhammad S., and Muhammad A.L. 2007. Influence of exogenous application of brassinosteroid on growth and mineral nutrients of wheat (*Triticum aestivum* L.) under saline conditions. *Pak. J. Bot.* **39**: 513–522.

Mussig, C., Fisher, S., and Altman, T. 2002. Brassinosteroids-regulated gene expression. *Plant Physiol.* **130**: 1319–1334.

Newman, L.A., Wang, X.P., Muiznieks, I.A., Ekuan, G., Ruszaj, M., Cortellucci, R., *et al.* 1999. Remediation of trichloroethylene in an artificial aquifer with tree: a controlled field study. *Environ. Sci. Technol.* **33**: 2257–2265.

Nippon Kayaku Co., Ltd. 1988. JRDC-694 (epi-brassinolide). Technical information.

Nriagu, J.O. 1979. Global inventory of natural and anthropogenic emissions of trace metals to the atmosphere. *Nature* **279**: 409–411.

Ouzounidou, G., and Ilias, I. 2005. Hormone-induced protection of sunflower photosynthetic apparatus against copper toxicity. *Biol. Plant.* **49**: 223–228.

Ozdemir, F., Bor, M., Demiral, T., and Turkan, I. 2004. Effects of 24-epibrassinolide on seed germination, seedling growth, lipid peroxidation, proline content and antioxidative system of rice (*Oryza sativa* L.) under salinity stress. *Plant Growth Regul.* **42**: 203–211.

Paterson, S., Mackay, D., Tam, D., and Shiu, W.Y. 1990. Uptake of organic chemicals by plants: a review processes, correlations and models. *Chemosphere* **21**: 297–331.

Pedron, F., Petruzzelli, G., Barbafieri, M., and Tassi, E. 2009. Strategies to use phytoextraction in very acidic soil contaminated by heavy metals. *Chemosphere* **75**: 808–814.

Pharmavati, M., Billington, T., and Gehring, C.A. 1998. Stomatal guard cell responses to kinetin and natriuretic peptides are cGMP dependent. *Cell. Mol. Life Sci.* **54**: 272–276.

Pilon-Smits, E. 2005. Phytoremediation. *Ann. Rev. Plant Biol.* **56**: 15–39.

Plumlee, G.S., and Longsdon, M.J. 1999. The environmental geochemistry of mineral deposits. Society of Economic Geologists, Inc., MI, USA.

Pospisilova, J. 2003. Participation of phytohormones in the stomatal regulation of gas exchange during water stress. *Biol. Plant.* **46**, 491–506.

Pospisilova, J., Synkova, H., and Rulcova, J. 2000. Cytokinins and water stress. *Biol. Plant.* **43**, 321–328.

Prasad, M.N.V. 2004. Phytoremediation of metals and radionuclides in the environment: the case for natural hyperaccumulators, metal transporters, soil-amending chelators and transgenic plants. In: Heavy metal stress in plants–from biomolecules to ecosystems (Ed. M.N.V. Prasad). Berlin, Heidelberg, Springer-Verlag, pp. 345–391.

Rauret, G. 1998. Extraction procedures for the determination of heavy metals in contaminated soil and sediment. *Talanta* **46**: 449–455.

Robinson, B.H., Fernández, J.E., Madejón, P., Marañón, T., Murillo, J.M., Green, S.R., and Clothier, B.E. 2003. Phytoextraction: an assessment of biogeochemical and economic viability. *Plant and Soil* **249**: 117–125.

Rock, S.A. 2003. Vegetative covers for waste containment. In: advances in biochemical engineering/biotechnology (Ed. T. Scheper). Vol. 78. Berlin, Heidelberg, Spring-Verlag, pp. 157–170.

Rogers, E.E., Eide, D.J., and Guerinot, M.L. 2000. Altered selectivity in an *Arabidopsis* metal transporter. *Proc. Natl. Acad. Sci. USA* **97**: 12356–12360.

Ryan, J.A., Bell, R.M., Davinson, J.M., and O'Conor, G.A. 1988. Plant uptake of non-ionic organic chemicals from soils. *Chemosphere* **17**: 2299–2323.

Salt, D.E., Kato, N., Kramer, U., Smith, R.D., and Raskin, I. 2000. The role of root exudates in nickel hyperaccumulation and tolerance and non-accumulator species of *Thlaspi*. In: Phytoremediation of contaminated soil and water (Ed. N. Terry and G. Banuelos). CRC Press LLC, pp. 189–200.

Salt, D.E., Smith, R.D., and Raskin, I. 1998. Phytoremediation. *Ann. Rev. Plant Physiol. Plant Molec. Biol.* **49**: 643–668.

Sandremann, H. 1994. High plant metabolism of xenobiotics: the 'green liver' concept. *Pharmacogentics* **4**: 225–241

Sasse, J. 2002. Physiological actions of brassinosteroids: an update. J. *Plant Growth Regul.* **22**: 276–288.

Sayed, S.A. 1997. Effect of cadmium and kinetin on transpiration rate, stomatal opening and leaf relative water content in safflower plants. *J. Islamic Acad. Sci.* **10**: 73–80.

Schnoor, J.L., Licht, L.A., McCutcheon, S.C., Wolfe, N.L., and Carreira, L.H. 1995. Phytoremediation of organic and nutrient contaminants. *Environ. Sci. Technol.* **29**: 318–323.

Schröder, P., and Collins, C.J. 2002. Conjugating enzymes involved in xenobiotic metabolism of organic xenobiotics in plants. *Int. J. Phytorem.* **4**: 1–15.

Schwitzguebel, J.P., and Vanek, T. 2003. Some fundamental advances for xenobiotic chemicals. In: Phytoremediation: transformation and control of contaminants (Ed. S.C. McCutcheon and J.L. Schnoor). New York, Wiley, pp. 123–157.

Shang, T.Q., Newman, L.A., and Gordon, M.P. 2003. Fate of tricholorethylene in terrestrial plants. In: Phytoremediation: transformation and control of contaminants (Ed. S.C. McCutcheon and J.L. Schnoor). New York, Wiley, pp. 529–560.

Sharma, P., and Bhardwaj, R. 2007. Effects of 24-epibrassinolide on growth and metal uptake in *Brassica juncea* L. under copper metal stress. *Acta Physiol Plant* **29**: 259–263.

Sondergaard, T.E., Schulz, A., and Palmgren, M.G. 2004. Energization of transport processes in plants. Roles of the plasma membrane H^+-ATPase. *Plant Physiol* **136**: 2475–2482.

Steber, C.M., and McCourt, P.A. 2001. A role for brassinosteroids in germination in *Arabidopsis. Plant Physiol.* **125**: 763–769.

Subramanian, M., and Shanks, J.V. 2003. Role of plants in the transformation of explosives. In Phytoremediation: transformation and control of contaminants (Ed. S.C. McCutcheon and J.L. Schnoor). New York, Wiley, pp. 389–407.

Sze, H., Li, X., and Palmgren, M.G. 1999. Energization of plant cell membranes by H^+-pumping ATPases: regulation and biosynthesis. *Plant Cell* **11**: 677–689.

Takatsuto, S., Kamuro, Y., and Watanabe, T. 1996. Synthesis and plant growth promoting effects of brassinosteroid compound TS303. *Proc. Plant Growth Regul. Soc. Amer.* **23**: 15–20.

Tassi, E., Barbafieri, M., Cervelli, S., Petruzzelli, G., Pedron, F., and Szymura, I. 2004a. Phytoremediation test in PAH contaminated soil. *Agrochimica*, **XLVIII**: 73–76.

Tassi, E., Pedron, F., Barbafieri M., and Petruzzelli, G. 2004b. Phosphate-assisted phytoremediation in As-contaminated soil. *Eng. Life Sci.* **4**: 341–346.

Tassi, E., Pouget J., Petruzzelli G., and Barbafieri, M. 2008. The effects of exogenous plant growth regulators in the phytoextraction of heavy metals. *Chemosphere*, **71**: 66–73.

Trapp, S. 2000. Modelling uptake into roots and subsequent translocation of neutral and ionisable organic compounds. *Pest. Manag. Sci.* **56**: 767–778.

Tropp, E., Scheunert, I., Attar, A., and Korte, F. 1986. Factors affecting the uptake of 14C-labeled organic chemicals by plants from soil. *Ecotoxicol. Environ. Safe*, **11**: 219–228.

Verbruggen, N., Hermans, C., and Schat, H. 2009. Molecular mechanism of metal hyperaccumulation in plants. *New Phytologist* **181**: 759–776.

Volynets, A.P., Pschenichanye, L.A., and Khripach, V.A. 1997. The nature of protective action of 24-epibrassinolide on barley plants. *Plant Growth Regul. Soc. Am.* **24**: 133–137.

Vose, J.M., Harvey, G.J., Elliott, K.J., and Clinton, B.D. 2003. Measuring and modelling tree and stand level transpiration. In: Phytoremediation: transformation and control of contaminants (Ed. S.C. McCutcheon and J.L. Schnoor). New York, Wiley.

Weber, W.J., and Huang, Q. 2003. Inclusion of persistent organic pollutants in humification processes: direct chemical incorporation of phenathrene via oxidative coupling. *Environ. Sci. Technol.* **37**: 4221–4227.

Werner, T., Motyka, V., Strnad, M., and Schmülling, T. 2001. Regulation of plant growth by cytokinin. *Plant Biol.* **98**: 10487–10492.

Xia, X.J., Huang, Y.Y., Wang, L., Huang, L.F., Yu, Y.L., Zhou, Y.H., and Yu, J.Q. 2006. Pesticides induced depression of photosynthesis was alleviated by 24-epibrassinolide pretreatment in *Cucumis sativus* L. *Pest Biochem. Physiol.* **86**: 42–48.

Xia, X.J., Zhang, Y., Wu, X.J., Wang, T.J., Zhou, Y.H., Shi, W., Yu, Y.L., and Yu, J.Q. 2009a. Brassinosteroids promote metabolism of pesticides in cucumber. *J. Agric. Food Chem.* **57**: 8406–8413.

Xia, X.J., Wang, Y.J., Zhou, Y.H., Tao, Y., Mao, W.H., Shi, K., Asami, T., Chen, Z., and Yu, J.Q. 2009b. Reactive oxygen species are involved in brassinosteroids-induced stress tolerance in *Cucumis sativus*. *Plant Physiol.*, **150**: 801–814.

Xingmao, M., Richter, A.R., Albers, S., and Burken, J.G. 2004. Phytoremediation of MBTE with hybrid poplar trees. *Int. J. Phytorem.* **6**:157–167.

Yoon, J., Cao, X., Zhou, Q., and Ma, L.Q. 2006. Accumulation of Pb, Cu and Zn in native plants growing on a contaminated Florida site. *Sci. Tot. Environ.* **368**: 456–464.

Zhang, Z., Ramirez, J., Reboutier, D., Brault, M., Trouverie, J., Pennarun, A.M., Amiar, Z., Biligui, B., Galagovsky, L., and Rona, J.P. 2005. Brassinosteroids regulate plasma membrane anion channels in addition to proton-pumps during expansion of *Arabidopsis thaliana* cells. *Plant Cell Physiol* **46**: 1494–1504.

Chapter 17

PROSPECTS OF BRASSINOSTEROIDS IN MEDICINAL APPLICATIONS

R. BHARDWAJ, N. ARORA, P. UPPAL, I. SHARMA AND M.K. KANWAR
Department of Botanical & Environmental Sciences, Guru Nanak Dev University, Amritsar 143005, India

Abstract: Steroids are an imperative group of hormones which play a key role in the transmission of signals that mediate growth and physiological responses in most pluricellular organisms. Brassinosteroids (BRs), a class of plant-specific steroid hormones, control many of the developmental and physiological processes like their animal counterparts, including regulation of gene expression, cell division and expansion, differentiation, programmed cell death, and homeostasis. Recent studies have indicated that these hormones have antiviral, antifungal, antiproliferative, antibacterial, neuroprotective and immunomodulatory properties in animal system. BRs analogues have been reported to have antiviral activity against herpes simplex virus type 1 (HSV-1), arenaviruses as well as against replication of vesicular stomatitis virus (VSV) in Vero cells. Also, antiherpetic activities both in a human conjunctive cell line (IOBA-NHC) and murine herpetic stromal keratitis (HSK) experimental models have been reported. In human cells, anticancer structure-activity relationship of natural BRs revealed their high cytotoxic activity. Since, BRs and their analogues are reported to inhibit cell growth in cancer cell lines, they may be considered as promising phytohormones for potential anticancer drugs. The use of pollens in folk medicine also indicates scope of steroids of plant pollens in medicines. An attempt has been made in this paper to document the information available on the prospects of BRs in therapeutics.

Key words: Antiviral, anticancer, brassinosteroids, medicines

S. Hayat and A. Ahmad (eds.), *Brassinosteroids: A Class of Plant Hormone*,
DOI 10.1007/978-94-007-0189-2_17,

1. INTRODUCTION

Steroids are vital for both plants and animals because they act as hormones. A number of sterols and steroids are produced in plants (Geuns, 1978). Brassinosteroids (BRs) are a group of naturally occurring polyhydroxy steroidal hormones with significant growth promoting activity (Clouse, 1996; Rao *et al.*, 2002). Grove *et al.* (1979) isolated Brassinolide (BL) from 40 kg of bee-collected rape (*Brassica napus* L.) pollen with growth promoting activities. In 1982, another steroidal substance namely, castasterone was isolated from insect galls of chestnut (*Castanea crenata*) (Yokota *et al.*, 1982; Mandava, 1988). Since there are a number of BRs that have been reported in plants and it was assumed that they are ubiquitous occurrence in plant kingdom and are characterized from 44 plant species including angiosperms, gymnosperms, pteridophyte and an alga (Clouse and Sasse, 1998; Fujioka *et al.*, 1998; Yokota *et al.*, 1998; Rao *et al.*, 2002). Till date, there are 70 BRs that have been isolated from 27 families of higher plants and three families of lower plants (Bajguz and Tretyn, 2003). BRs have a common 5α cholestane skeleton, but their functional variations are due to the different orientations of functionalities on the basic skeleton (Fujioka and Sakurai, 1997). Classification of these steroids is done on the basis of alkyl-substitution pattern of the side chain (Yokota *et al.*, 1997).

BRs promote cell elongation, cell division, differentiation, disease resistance, stress tolerance and senescence in plants (Kauschmann *et al.*, 1996) (Clouse 2002; Mussig *et al.*, 2002). In plants BRs also confer resistance against biotic and abiotic stresses (Khripach *et al.*, 2000; Ikekawa and Zhao, 1991). They provide resistance against wide spectrum of environmental stresses produced by low and high temperature (Dhaubhadel *et al.*, 1999), drought (Li and Van Staden, 1998), salt (Sasse *et al.*, 1995), infection and pesticides (Sasse, 1999) and heavy metals (Bajguz, 2000; Janeczko *et al.*, 2005; Bhardwaj *et al.*, 2007). The focus of attention during the past few decades was on the physiological properties of brassinosteroids (Krishna, 2003). BRs have multiple involvement in the regulation of plant physiological activities such as cell expansion, cell division, alteration of membrane properties, vegetative growth, reproductive biology, senescence, seed germination and stress management (Ozdemir *et al.*, 2004; Khripach *et al.*, 2000; Bhardwaj *et al.*, 2006). The plethora of stresses against which the brassinosteroids act are thermal, drought, heavy metals, infection, pesticides, salt and even viruses (Dhaubhadel *et al.*, 1999, 2002; Janeczko *et al.*, 2009; Wachsman *et al.*, 2000, 2002, 2004a, 2004b; Krishna, 2003; Haubrick and Assmann, 2006).

According to Khripach *et al.* (2000) the metabolism of BRs in mammals has not yet been investigated. It may be speculated, however, that a normal catabolism of the steroidal skeleton will take place. Being normal constituents of practically all plants, BRs have been, and are, consumed by mammals, and so additional harmful effects are not likely from their use in agriculture. This assumption is an important prerequisite for considering BRs as ecologically safe, non-toxic chemicals for agriculture. However, conformation of their safety can be obtained from toxicological studies. A study on animal system reveals their role as antiviral, antifungal, antiproliferative, antibacterial, neuroprotective and immunomodulatory properties in animal systems. BRs analogues have been reported to have antiviral activity against herpes simplex virus type 1 (HSV-1), arenaviruses as well as against replication of vesicular stomatitis virus (VSV) in Vero cells. Antiherpetic activities both in a human conjunctive cell line (IOBA-NHC) and murine herpetic stromal keratitis (HSK) experimental models have been reported. In human cells, anticancer structure-activity relationship of natural BRs revealed their high cytotoxic activity. Since, both BRs and analogues are reported to inhibit cell growth in cancer cell lines thus BRs may be considered as promising phytohormones for potential anticancer/antiviral/antibacterial drugs. In agriculture several abiotic stresses have been reported to occur simultaneously, rather than a particular stress condition, which are most lethal to crops but the co-occurrence of different stresses is rarely addressed by molecular biologists (Mittler, 2006).

1.1 Brassinosteroids and abiotic stress

BRs have the ability to protect the plants from various environmental stresses such as drought, extreme temperatures, heavy metals, herbicidal injuries and salinity (Sasse, 1999; Bajguz and Hayat, 2008). 24-Epibrassinolide (EBL) treatments increased drought tolerance in *Arabidopsis thaliana* and *Brassica napus* seedlings by changing the expression of drought responsive genes (Kagale *et al.*, 2007). In our earlier studies, we have reported that EBL and 28-Homobrassinolide (HBL) treatments (presowing) improved the shoot emergence and plant biomass production in *Brassica juncea* seedlings and plants under heavy metal stress (Cu, Zn, Mn, Co and Ni). EBL and HBL have also been found to reduce the heavy metal uptake and accumulation in *B. juncea* seedlings and plants (Sharma and Bhardwaj, 2007a,b; Sharma *et al.*, 2007, 2008; Bhardwaj *et al.*, 2007, 2008). HBL protected chickpea (*Cicer arietinum*) from Cd toxicity by enhancing the activities of antioxidant enzymes (Superoxide dismutase, Catalase and Guaiacol peroxidase), proline content, nitrate reductase and carbonic anhydrase and also by increasing the plant growth, leghemoglobin content, nodule number, nitrogen and carbohydrate contents in the nodules and leaf chlorophyll content, which were decreased proportionately with the increasing concentrations of Cd (Hasan *et al.*, 2008). BRs have been reported to be effective in reducing damage caused by pesticides (simazine, butachlor, or pretilachor) in rice (Sasse, 2003).

The phytotoxic effect on cucumber leaves of nine pesticides including three herbicides (paraquat, fluazifop-p-butyl and haloxyfop), three fungicides (flusilazole, cuproxat and cyazofamid) and three insecticides (imidacloprid, chlorpyrifos and abamectin) has been examined (Xia *et al.*, 2006). BRs have been reported to overcome the salt stress by regulating the activities of key antioxidative enzymes and levels of malondialdehyde (MDA) and proline. Our earlier studies highlight that HBL ameliorated the salinity stress in *Zea mays* seedlings and plants by increasing the activities of antioxidative enzymes and by reducing the level of lipid peroxidation (Arora *et al.*, 2008). Jager *et al.* (2008) studied the endogenous level of BR and Absicic acid (ABA) in wild type (WT), BR-deficient mutant *(lkb)* and BR-perceptive mutant (*lka*) pea plants exposed to water stress.

1.2 Brassinosteroids and biotic stress

BRs stirred inner potentials of plants that are supportive not only in better survival in stress conditions, but also in receding biotic stress caused by pathogens such as viruses, fungi and bacteria (Nakashita *et al.*, 2003; Wachsman *et al.*, 2004a; Swaczynová *et al.*, 2006; Romanutti *et al.*, 2007; Ohri *et al.*, 2008). The potential of BRs to enhance plant resistance against fungal pathogen infection has been documented in several studies (Khripach *et al.*, 2000). Vasyukova *et al.* (1994) reported that the treatment of potato plants with BRs, essentially reduced the incidence of *Phytophthora* infection. BRs and their synthetic derivatives were reported to be good inhibitors of herpes simplex virus type 1(HSV-1) and arena virus replication in cell culture (Wachsman *et al.*, 2004a, 2004b). Further, Michelini *et al.* (2004) reported the *in vitro* and *in vivo* antiherpetic activity of three new synthetic BRs analogues {(22S,23S)-3β-bromo-5α,22,23-trihydroxystigmastan-6-one, (22S,23S)-5α-fluoro-3β-22, 23-trihydroxystigmastan-6-one, (22S,23S)-3β-5α, 22,23-trihydroxystigmastan-6-one. Effects of BRs on insect development, particularly on molting, were reviewed by Zullo and Adam (2002). The treatment of brassinosteroids to root knot nematodes (*M. incognita*) also stimulated their antioxidative defence system (Ohri *et al.*, 2007). Further Ohri *et al.* (2008) studied the influence of EBL on the development of *Meloidogyne incognita* and observed that egg hatching and juvenile emergence in root knot nematode was stimulated by BRs treatments.

2. BIOLOGICAL ACTIVITIES OF BRASSINOSTEROIDS IN VARIOUS TEST SYSTEMS

The potential implications of BRs as a future medicine is acclaimed not only on their antiviral, antibacterial, antifungal, ecdysteroidal, antigenotoxic properties, but also attributed to their anticancerous and antiproliferative activities. As a result, it may become possible to employ these phytohormones in traditional medicines for treating cancer, fungal, bacterial and viral infections etc. (Nakashita *et al.*, 2003; Wachsman *et al.*, 2004a; Swaczynová *et al.*, 2006; Malíková *et al.*, 2008; Romanutti *et al.*, 2007; Ohri *et al.*, 2008;). Although these possible medicinal properties of BRs have been reported recently, but the detailed and systematic investigations in these directions may lead to their use in therapeutics in the forthcoming years.

2.1 Ecdysteroidal activity of brassinosteroids

Ecdysteroids exert widespread effects on insect growth and development. These include roles in morphogenesis, proliferation, programmed cell death, cuticle synthesis, oogenesis and developmental timing (Robertson, 1936; Riddiford, 1993). It is intriguing that some aspects of these pathways share features in common with the wide range of developmental and physiological responses to BRs in plants, which also include promotion of cell division, expansion, and programmed cell death and modulation of reproductive development (Tummel and Chory, 2002). For example, both BL and ecdysteroids are required for cell shape, changes associated with maturation—although they exert this effect in different ways. As described above, BL induces the expression of a range of cell wall-associated proteins that are implicated in cell expansion, pro viding molecular basis for understanding the role of BRs in directing cell elon gation and plant growth. Similarly, ecdysteroids trigger the morphogenesis of adult structures during metamorphosis through coordinated changes in cell shape manifested at the level of the actin cytoskeleton (Von Kalm *et al.*, 1995). It is interesting to speculate that these two responses reflect the basic architectural differences that define plant and animal cells. Thus, the presence of a rigid cel wall in plants demands changes at the level of cell wall-associated proteins to control changes in overall cell shape. Similarly, the integrity of an insect cell is defined by an internal cytoskeleton, which is the target for ecdysteroid triggered changes in cell shape.

Another similarity in steroid responses between plants and insects is programmed cell death. Ecdysteroids trigger the massive death of larval tissues during the early stages of metamorphosis, ridding the animal of these obsolete tissues to make way for their adult counterparts (Robertson, 1936).

This response has been extensively studied in *Drosophila* and shown to occur by autophagy with hallmark features of apoptosis, including DNA fragmentation and caspase activation (Jiang *et al.*, 1997; Jochova *et al.*, 1997; Lee and Baehrecke, 2001). There is evidence that BRs induce programmed cell death during xylogenesis. The specialized xylem vessels that conduct water through plants are made up of individual dead cells called tracheary elements (Roberts and McCann 2000). The BR-deficient *Arabidopsis* mutants *cpd* and *dwf7* have abnormal xylem, implicating the hormone in xylogenesis, although these phenotypes have not been examined in detail (Szekeres *et al.*, 1996; Choe *et al.*, 1999). In addition, Clouse and Zurek (1991) observed that exogenously supplied BL promotes both tracheary element differentiation and cell division in cultured tuber explants of Jerusalem artichoke. Using a zinnia system (*Zinnia elegans* L. cv Canary Bird) in which single mesophyll cells can differentiate directly into tracheary elements, it was observed that exogenously supplied uniconazole (an inhibitor of both gibberellin and BR biosynthesis) prevents uncommitted cells from differentiating into tracheary elements and that BL but not gibberellin overcomes this inhibition (Iwasaki and Shibaoka, 1991). Moreover, BRs appear to act specifically during the final stage of xylogenesis, which involves secondary wall formation and cell death. During this time, the levels of BRs rise dramatically (Yamamoto *et al.*, 2001). The role of BRs in apoptosis and programmed cell death make them potential candidates in the management of lifecycles and different developmental stages of insects and nematodes which spoils the agricultural crops, thereby, helping in sustainable agriculture.

2.2 Antifungal activity of brassinosteroids

Exogenous application of BRs stimulated inner potentials of plants that is helpful not only in better survival in stressful conditions and quality improvement, but also in diminishing disease damage. The potential of BRs to enhance plant resistance against fungal pathogen infection was documented in several studies (Khripach *et al.*, 2000). Vasyukova *et al.* (1994) carried out investigations on the interaction between *Phytophthora infestans* and potato tubers. The treatment of potato plants with BRs, essentially reduced the incidence of *Phytophthora* infection. The increase in resistance in BRs treated potato tubers was associated with enhancement of ABA and ethylene levels and the presence of phenolic and terpenoid substances. BR-induced disease resistance was also noted in barley and cucumber plants. Spraying barley plants at tillering phase with EBL decreased an extent of leaf disease induced by *Helminthosporium teres* Sacc. and increased grain yield even at a dose of 5 mg ha^{-1} (Pshenichnaya *et al.*, 1997; Volynets *et al.*, 1997a,b). The protective effects of EBL in cucumber against fungi were studied by Churikova and Vladimirova (1997). The increased activities of peroxidase and polyphenoloxidase enzymes, which are involved in the metabolism of polyphenols, was suggested as a factor contributing to BR induced disease resistance in cucumber plants. The antifungal potential of BRs indicates their prospects for antifungal formulations to plants against several fungal diseases, but the study needs to be carried for other fungal infections in plants.

2.3 Antigenotoxic activity of brassinosteroids

EBL found to show the antigenotoxic properties and this was evaluated through *Allium cepa* chromosomal aberration bioassay. Howell *et al.* (2007) has studied the effect of EBL on the mitotic index and growth of onion (*Allium cepa*) root tips. Low doses of EBL (0.005 ppm) nearly doubled the mean root length and the number of mitosis over that of controls. Intermediate doses of EBL (0.05 ppm) also produced mean root lengths and number of mitosis that were significantly greater than those of the controls. But the highest dose of EBL (0.5 ppm) produced mean root lengths and number of mitoses that were less than control values. Sondhi *et al.* (2008) isolated and characterized the EBL from leaves of *Aegle marmelos* Correa. (Rutaceae) which was further evaluated for the antigenotoxicity against maleic hydrazide (MH) induced genotoxicity in *Allium cepa* chromosomal aberration assay. It was observed that the percentage of chromosomal aberrations induced by maleic hydrazide (0.01%) declined significantly with 24-epibrassinolide treatment. EBL (10^{-7} M) proved to be the most effective concentration with 91.8% inhibition.

The acute toxicity data obtained at the Sanitary-Hygienic Institute of Belarus for 24-epibrassinolide are: LD 50 (orally and dermally) in rats (male/female) is more than 2000 mg per litre. Dermal toxicity in rats (male/female) is more than 2000 mg per litre. The formulation, Epin (0.025% solution of 24-epibrassinolide), in mice and rats (orally and dermally) has an LD of more than 5000 mg per litre. Repeated experiments confirmed the value of LD 50 for 24-epibrassinolide orally in mice and showed a value for Epin which was higher than 15,000 mg per litre (white rats, orally or intranasally). In concentrations of 0.2%, 24-epibrassinolide did not irritate mucous membranes of rabbits' eyes; this compound, or a solution of Epin, did not irritate the skin. The Ames test for mutagenic activity carried out at the Scientific Research Center of Toxicologic and Hygienic Regulation of Biopreparations of Russia, with or without metabolic activation, was negative (*Salmonella typhimurium* TA 1534, TA 1537, TA 1950, TA 98, TA 100). In micro-nuclear or chromosome abberation tests (mice CBAB1/6) neither 24-epibrassinolide nor Epin caused spontaneous mutations. Complex biological testing on *Tetrahymena pyriformis* carried out at the Sanitary-Hygienic Institute of Belarus has confirmed the genetic safety of 24-epibrassinolide and the absence of mutagenic activity over seven generations. In acute, subacute, and chronic experiments, 24-epibrassino-lide showed low toxicity and very little cumulative effect. In prolonged experiments, 24-epibrassinolide showed no toxicity but a pronounced adaptogenic effect (increasing adaptive ability of the population).

2.4 Antiviral and antibacterial activities of brassinosteroids

One of the important aspects of the protective action of BRs in plants is related to their ability to stimulate resistance to viruses (Rodkin *et al.*, 1997; Bobrick *et al.*, 1998). It has been reported that BR treatment reduced virus infection in the starting plant material, various stages of plant development, and the first and second tuber generations of potato. The plants obtained from BR treated sowing material increased the crop yield by 56% and significantly reduced virus infection. The tobacco plants when given treatment of BRs against tobacco mosaic virus (TMV), the bacterial pathogens *Pseudomonas syringae*, and the fungal pathogen *Odium* species, showed lowered infection and better growth. Similarly in rice, the infection caused by *Magnaporthe grisea* and *Xanthomonas oryzae* which caused rice blast and bacterial blight respectively, was significantly reduced by BR treatments (Nakashita *et al.*, 2003). Potato cuttings cultured in a medium containing brassinolide, 24-epibrassinolide or 28-homobrassinolide were more resistant to viral infection through all stages of development (Khripach *et al.*, 2000).

BRs and their synthetic derivatives are reported to be good inhibitors of herpes simplex virus Type 1 (HSV-1) and arena virus replication in cell culture. The arena virus was susceptible to the compounds throughout its replicative cycle, and the HSV-1 was likely affected at a late step in multiplication (Wachsman *et al.*, 2000). Twenty-seven BRs derivatives/ analogues when tested against measles virus were found to possess antiviral activity. The selectivity index (SI) values of BRs were higher than those obtained with the reference drug ribavirin (Wachsman *et al.*, 2002). Furthermore, BRs and their synthetic derivatives were reported to be good inhibitors of herpes simplex virus type 1(HSV-1) and arena virus replication in cell culture. Wachsman *et al.* (2004b) described synthetic methods to obtain BRs analogues and report the scope of antiviral activity of these compounds against RNA and DNA viruses. Some of the compounds showed selectivity indexes (SI) 10- to 18-fold higher than ribavirin, a broad spectrum antiviral compound, when tested against Junin virus (JV) (*Arenaviridae*); a good antiviral activity against measles virus (MV) (*Paramixoviridae*), with SI values also higher than ribavirin used as reference drug, and a similar or lower activity

Table 1. Cytotoxic and antiviral activities of brassinosteroid derivatives against measles virus (source: Wachsman *et al.*, 2004b)

S. No.	**Brassinosteroids derivatives (IUPAC name)**	**CC50* (mM)**	**EC50** (mM)**
1	(22R,23R)-2α,3a, 22,23-Tetrahydroxy-5α-stigmastan-6-one	62	42
2	(22S,23S)-2α,3α, 22,23-Tetrahydroxy-5α-stigmastan-6-one	259	65
3	(22E)-2α,3α-Dihydroxystigmast-22-en-6-one	158	40
4	(22R,23R)-2α,3α,22,23-Tetrahydroxy-β-Homo-7-oxa-stigmastan-6-one	427	8
5	(22R,23R)-2α,3α,5α,22,23-Pentahydroxystigmastan-6-one	262	131
6	(22S,23S)-2α,3α,5α,22,23-Pentahydroxystigmastan-6-one	364	58
7	(22R,23R)-3β-Acetoxy-22,23-dihydroxy-5α-stigmastan-6-one	238	40
8	(22S,23S)-3β-Acetoxy-22,23-dihydroxy-5α-stigmastan-6-one	226	30
9	(22R,23R)-3β-Acetoxy-5α,22,23-trihydroxystigmastan-6-one	230	21
10	(22S,23S)-3β-Acetoxy-5α,22,23-trihydroxistigmastan-6-one	461	23
11	(22R,23R)-3β-Bromo-22,23-dihydroxy-5α-stigmastan-6-one	114	76
12	(22S,23S)-3β-Bromo-22,23-dihydroxy-5α-stigmastan-6-one	152	38
13	(22R,23R)-3β-Bromo-5α,22,23-trihydroxystigmastan-6-one	36	13
14	(22S,23S)-3β-Bromo-5α,22,23-trihydroxystigmastan-6-one	185	4
15	(22E)-3β-Bromo-5α-hydroxystigmast-22-en-6-one	139	5
16	(22R,23R)-3β, 5α,22,23-Tetrahydroxystigmastan-6-one	819	30
17	(22S,23S)-3β, 5α,22,23-Tetrahydroxystigmastan-6-one	1044	26
18	(22E)-3β,5α-Dihydroxystigmast-22-en-6-one	901	585
19	(22R,23R)-3β-Fluoro-22,23-dihydroxystigmastan-6-one	215	7
20	(22S,23S)-3β-Fluoro-22,23-dihydroxystigmastan-6-one	43	1
21	(22S,23S)-3α-Fluoro-22,23-dihydroxystigmastan-6-one	301	38
22	(22R,23R)-3β-Fluoro-5α,22,23-trihydroxystigmastan-6-one	42	16
23	(22S,23S)-3β-Fluoro-5α,22,23-trihydroxystigmastan-6-one	250	6
24	(22R,23R)-5α-Fluoro-3β,22,23-trihydroxystigmastan-6-one	100	75
25	(22S,23S)-5α-Fluoro-3β,22,23-trihydroxystigmastan-6-one	160	3
26	(22E)-3β-Fluoro-5α-chlorostigmast-22-en-6-one	858	32
27	(22E)-5α-Fluorostigmast-2,22-dien-6-one	935	701
28	(22E)-3β-Hydroxystigmast-5,22-diene (stigmasterol)	479	117
29	Ribavirin {1-β-D-ribofuranosyl-1-2-4 triazole-3-carboxamide}	840	52

*Compound concentration required to reduce cell viability by 50% after 24 h incubation at 37°C using stationary-phase cells. ** Compound concentration required to inhibit virus yield (plaque forming units) by 50%. [IUPAC, International Union of Pure and Applied Chemistry.] (Source: Wachsman *et al.*, 2004b) against herpes simplex type 1 and 2 (HSV-1 and HSV-2) (*Herpesviridae*) when compared to foscarnet or acyclovir, respectively. Structure activity relationship studies (SAR) were analyzed, in order to detect which stereo-chemistry, type and position of functional groups were needed to develop a selective class of virus inhibitors. A variety of chemical structures are currently known as inhibitors of pathogenic virus replication. Nucleoside synthetic analogues are the most successful ones, in clinical use, against infections caused by most perilous viruses disseminated among human population.

Several other molecules have been designed to interrupt specifically human immunodeficiency virus (HIV) replication by inhibiting viral proteases or, in the case of influenza virus, to mimic the substrate for sialidase enzymatic activity. Unfortunately, viruses respond to antiviral treatment with a rapid selection of drug resistant mutant particles, compelling virologists to search for new active compounds. Animal viruses tested for their susceptibility to BRs analogues comprised two RNA monocistronic viral families, *Paramyxoviridae* and *Arenaviridae,* and one DNA virus family *Herpesviridae*, all of them are important human pathogens. Preliminary studies by Wachsman *et al*. (2004b) with brassinolide have shown significant *in vitro* antiviral activity against various RNA and DNA viruses (Tables 1 and 2).

Table 2. Antiviral Activity of Natural Brassinolide against several RNA and DNA Viruses (Source: Wachsman *et al*., 2004b)

S.No.	**Viruses**	**Brassinolide (% inhibition)***
1.	Poliovirus type I	96
2.	VSV Indiana strain	100
3.	HSV-1 F strain (tk+)	96
4.	Tacaribe TRLV 11573 strain	55
5.	Junin IV 4454 strain	74

*Vero cells (African green monkey kidney cells) were grown as monolayers in minimum essential medium (MEM) supplemented with 5% inactivated calf serum and 50 g/ml gentamycin. Maintenance medium (MM) consisted of MEM + 2% calf inactivated serum. The monolayers were infected at a moi (multiplicity of infection) of 1 with the different viruses. After 1 h adsorption at 37°C, the cultures were covered with MM or with MM containing brassinolide at a concentration of 1 μM. After 24 h of incubation at 37°C, supernatants were harvested and titrated by a plaque assay. The inhibition of virus replication was calculated by comparison with virus titres obtained in cultures without the presence of the brassinolide. (Source: Wachsman *et al*., 2004b)

Michelini *et al*. (2004) have reported the *in vitro* and *in vivo* antiherpetic activity of three new synthetic BRs analogues. Chemical synthesis of three new synthetic BRs analogues like (22S,23S)-3β-bromo-5α,22,23-trihydroxy-stigmastan-6-one, (22S, 23S)- 5α-fluoro-3β-22,23-trihydroxystigmastan-6-one, (22S, 23S)-3β-5α,22,23-trihydroxystigmastan-6-one and their antiherpetic activity both in human conjunctive cell lines (IOBA-NHC) and in the murine herpetic stromal keratitis (HSK) experimental model were tested. All compounds prevented HSV-1multiplication in NHC cells in a dose dependent manner when added after infection with no cytotoxicity. Significantly, *in vitro* studies showed that EBL is capable of reducing or even arresting the growth of the HIV in cultured infected cells. Khripach *et al.* (2005) reported that EBL may be used in the prevention and cure of HIV infection and related conditions (AIDS related complex), both symptomatic and asymptomatic, or when exposure to HIV virus was suspected. Further, Talarico *et al.* (2006) investigated that the replication of herpes simplex virus (HSV) Type 1 in Vero cells is inhibited in the presence of (22S,23S)-3β-bromo-5α,22,23-trihydroxystigmastan-6-one (6b), a synthetic brassinosteroid derivative. Since a late step of virus multiplication is hindered by 6b, a study was performed on drug-drug combination with acyclovir (ACV) and foscarnet (FOS). It was determined that 6b would act synergistically with low concentrations of ACV and moderate concentrations of FOS against HSV. The best drug combination tested in this study resulted in an increase of 29.3 and 47.2% in antiviral activity for ACV (0.036 μM) and FOS (37.5 μM) in the presence of 14.8 and 6.9 μM of 6b, respectively.

Vesicular stomatitis virus (VSV), a member of the Rhabdoviridae family, is an enveloped single-strand RNA virus that causes an economically important disease in cattle, horses and swine. Romanutti *et al.* (2007) reported the antiviral effects of a synthetic BR ((22S, 23S)-3β-bromo-5α,22,23-trihydroxystigmastan-6-one) against the replication of vesicular stomatitis virus (VSV) *in vero* cells. Synthetic BR affected the late event of the virus growth cycle and inhibited virus protein synthesis and viral mature particle formation. Time-related experiments showed that 6b mainly affects a late event of the virus growth cycle. Virus adsorption, internalization and early RNA synthesis are not the target of the inhibitory action. Results obtained indicate that the antiviral compound adversely affects virus protein synthesis and viral mature particle formation. Synergistic *in vitro* Interactions between (22S,23S)-3β-Bromo-5α,22,23-Trihydroxystigmastan-6-one and Acyclovir or Foscarnet against Herpes simplex Virus Type 1 were studied.

A number of biologically active steroids bearing unusual side chains, isolated from marine sponges, have been studied for their antiviral activity. Orthoesterols A, B and C have been reported to be active against feline leukemia virus (FeLV), mouse influenza virus and mouse coronavirus. Weinbersterols A and B also isolated from a sponge exhibited *in vitro* activity against FeLV with IC50 (inhibitory concentration) values of 40 and

52 µg/ml, respectively and also showed activity against HIV with IC50 value of 10 µg/ml. The effect of brassinolide on the replication of several viruses, initially tested at a concentration of 1 µM (Table 1), showed that this compound displays a broad spectrum of antiviral activity, with higher inhibition values than that reported for other natural steroidal molecules. Some synthetic BRs were tested against herpes simplex virus type 1(HSV-1) by Michelini *et al.* (2008) which induced an ocular chronic immunoinflammatory syndrome named herpetic stromal keratitis that might lead to vision impairment and blindness in mice.

2.5 Antiproliferative and anticancer activities of brassinosteroid

Franěk *et al.* (2003) observed that EBL at subnanomolar concentrations modulated growth and production characteristics of a mouse hybridoma. A mouse hybridoma was cultured either in standard serum-free medium, or in medium diluted to 30%, in which the cells underwent nutritional stress. Steady-state parameters of semicontinuous cultures conducted at EBL concentrations from 10^{-16} to 10^{-9} mol l^{-1} were evaluated. Typical effects of the EBL found both in standard and in diluted media were increased in the value of mitochondrial membrane potential, drop of intracellular antibody level, increase in the fraction of the cells in the G_0/G_1 phase, and decrease in the fraction of the cells in the S phase. Viable cell density was significantly higher as compared to control at EBL concentrations ranging from 10^{-13} and 10^{-12} mol l^{-1}. So EBL might induce perturbations in the cell division mechanism, in mitochondria performance, and in secreted protein synthesis in a mammalian cell line.

Swaczynová *et al.* (2006) studied the anticancer properties of BRs. Natural types of BRs affected the viability, proliferation, differentiation, apoptosis and expression of some cell cycle related proteins in cancer cell lines. Cytotoxic activity of BRs were tested *in vitro* by Calcein AM assay. IC_{50} values were estimated for human breast adenocarcinoma cell lines (MCF-7–estrogen-sensitive, MDA-MB-468–estrogen-insensitive), human acute lymphoblastic leukemia cell line (CEM) and human myeloma cell line (RPMI 8226). TUNEL, DNA ladder assay and immunoblotting techniques were used for the analysis of changes of cell viability, proliferation, differentiation and apoptosis. 28-Homocastasterone inhibited the viability of cancer cell lines and significantly reduced or induced the expression of *p21*, *p27*, *p53*, cyclins, proteins of Bcl-2 family and ER-alpha. The antiproliferative properties could be used for the development of new brassinosteroid-derived generation of anticancer drugs.

There have been few reports on the effects of BRs on cell division of mammalian cells. Up to now the inhibitory effects of BRs on mammalian cell division are unknown. To determine basic anticancer structure–activity relationships of natural BRs on human cells, several normal and cancer cell lines have been used. Several of the tested BRs were found to have high cytotoxic activity. Hence, in this regard Malìková *et al.* (2007) tested the effects of the most promising and readily available BR analogues with interesting anticancer properties, 28-homocastasterone and 24-epibrassinolide, on the viability, proliferation, and cycling of hormone-sensitive/insensitive (MCF-7/MDA-MB-468) breast and (LNCaP/DU-145) prostate cancer cell lines to determine whether the discovered cytotoxic activity of BRs could be, at least partially, related to brassinosteroid-nuclear receptor interactions. Both BRs inhibited cell growth in a dose dependent manner in the cancer cell lines. Flow cytometry analysis showed that BR treatment arrested, MDA-MB-468, LNCaP and MCF-7 cells in G1 phase of the cell cycle and induced apoptosis in MDA-MB-468, LNCaP, and slightly in the DU-145 cells. These results proved that natural BRs, at micromolar concentrations, can inhibit the growth, at micromolar concentrations, of several human cancer cell lines without affecting the growth of normal cells. Therefore, these plant hormones are promising leads for potential anticancer drugs. Compounds 6b [(22S,23S)-3β-bromo5α,22,23-trihydroxystigmastan-6-one], 1d [(22R,23R)-2α,3α,22,23-tetrahydroxy–Homo-7-oxa-stigmastan-6-one], 8a [(22R,23R)-3β-fluoro-22,23dihydroxystigmastan-6-one], 9b [(22S,23S)-3β-fluoro-5α,22, 23-trihydroxystigmastan-6-one] and 10b [(22S,23S)-5α-fluor-3 β,22,23-trihydroxystigmastan-6-one], with selectivity indexes (SI) of 40, 57,31, 37 and 53, are the derivatives with good antiviral activity against MV. These SI values are higher than those obtained with ribavirin (used as reference drug). A comparative analysis of 50% cytotoxic concentration (CC50) values, using confluent non-growing cells, gives an indication of structure–activity relationship. According to their degree of cytotoxicity the compounds were divided in three groups: low, intermediate and high cytotoxicity. By observing the chemical structures of compounds belonging to the first group we can see that less cytotoxic activities are related to the presence of a 3β-hydroxy group on C-3 (ring A) and a double bond between C-22 and C- 23 (side chain). The replacement of a 5α-hydroxy group by a 5α-fluoro group enhances cytotoxicity. Halogenated brassinosteroid derivatives in C-3 position are more cytotoxic than those with an acetoxy group in the same position. For compounds 1d, 6b, 10b and ribavirin, cytotoxicity measurements were also done with replicating cells; CC50 values were low, but they still competed favourably with ribavirin against MV.

2.6 Brassinosteroids and cotton leafworm *Spodoptera littoralis* G.

The two important plant hormones 24-epibrassinolide and 24-epicastasterone showed 50% competition for binding at IC50 of 1–3.6 μM with [^{3}H] ponasterone A using cultured imaginal wing discs from last-instar larvae of the cotton leafworm, *Spodoptera littoralis* (Boisduval). But, the culture of imaginal wing discs in different concentrations of brassinosteroids, even up to 100 μM, demonstrated no induction of evagination. In contrast, 20E and the non-steroidal agonist RH-5992 competed respectively about 23- and 42-times more effectively with labeled ponasterone A, and their ability (IC50) to induce disc evagination *in vitro* was 158 and 87 nM, respectively. Injections of 10 μg of brassinosteroids in latest moulted last-instar larvae did not cause mortality above controls. Higher mortalities were found when brassinosteroids were injected late in the last instar.

2.7 Brassinosteroids and insects

Schmidt *et al.* (2000) reported that after feeding 24-epi-castasterone to the cockroach *Periplanata Americana,* an organospecific epimerization of the brassinosteroid to 2, 24-di-epi-castasterone could be detected in female insects. The metabolite being observed only in the ovaries and not in the testes of the insect was identified by GC/MS in comparison with a synthesized authentic sample. Contrary, 24-epi-brassinolide is not metabolized in sexual organs of *Periplanta Americana*. Effects of BRs on insect development, particularly on molting, were reviewed by Zullo and Adam (2002). EBL or 24-epicastasterone did not affect the evagination of imaginal wing discs, nor was there any effect on intact last instar larvae of the cotton leaf worm, *Spodoptera littoralis*, after oral feeding (Smagghe *et al*., 2002). Similarly, treatment of root knot nematodes (*Meloidogyne incognita*) with BL revealed much higher percentage of hatching in treated egg masses as compared to control (Ohri *et al*., 2002). Ohri *et al*. (2004, 2005) further revealed enhanced juvenile emergence of *M. incognita* by brassinosteroids treatments. The treatment of brassinosteroids to root knot nematodes (*M. incognita*) also stimulated their antioxidative defence system (Ohri *et al*., 2007). Further Ohri *et al*. (2008) studied the influence of EBL on the development of *Meloidogyne incognita*. EBL enhanced the percentage of hatching in treated egg masses as compared to control. EBL treated juveniles induced more gall numbers and larger size of galls in roots of tomato plants.

3. NEUROACTIVITIES

Rupprecht and Holsboer (1999) reported that steroids influence neuronal function by binding to intracellular receptors that can act as transcription factors and regulate gene expression. In addition, some so-called 'neuroactive steroids' are potent modulators of an array of ligand-gated ion channels and of distinct G-protein coupled receptors via nongenomic mechanisms, and they can influence sleep and memory. They described how these neuroactive steroids modulate neurotransmitter receptors and address the neuro-psycho-pharmacological potential that arose from the intracellular crosstalk between genomic and nongenomic steroid effects. Neuroactive steroids could also have a role in the response to stress and the treatment of psychiatric disorders, such as depression, and, as they affect a broad spectrum of behavioral functions through their unique molecular properties, they could constitute a yet unexploited class of drugs.

4. CONCLUSIONS

Steroid hormones are essential for normal growth and development of plants and animals. Brassinosteroids (BRs) an important group of plant steroid hormones being mainly concentrated in plant pollens are reported to be involved in biostimulation in folk medicine. They also form the basis for the production of some anti-inflammatory and metabolism stimulative medicines, which are especially recommended for children and elderly people with chronic infections (Mashkovskii, 1997). Further, they have bright prospects as potential drugs against viruses like measles, herpes and arena viruses; bacteria, fungi and other pathogens. Although it is too early to predict their clinical utility, but there are evidences that BRs are emerging group of compounds as anti-cancer and antiviral drugs. Further illustrative studies to elucidate eventual anti-inflammatory, anti-cancerous and medicinal effects of BRs in animal models would confirm their use as therapeutics in traditional medicines.

5. REFERENCES

Arora, N., Bhardwaj, R., Sharma, P., Arora, H.K., and Arora, P. 2008. Amelioration of Zn toxicity by 28-homobrassinolide in *Zea mays* L. *Canadian J. Pure and Applied Sci.* **2**(3): 503–509.

Bajguz, A. 2000. Effect of brassinosteroids on nucleic acids and protein content in cultured cells of Chlorella vulgaris. *Plant Physiol. Biochem.* **38**: 209–215.

Bajguz, A., and Hayat, S. 2008. Effects of brassinosteroids on the plant responses to environmental stresses. *Plant Physiol.Biochem.* doi:10.1016/j.plaphy.2008.10.002.

Bajguz, A., and Tretyn, A. 2003. The chemical characteristic and distribution of brassinosteroids in plants. *Phytochemistry*. **62**: 1027–1046.

Bhardwaj, R., Arora, H.K., Nagar, P.K., and Thukral, A.K. 2006. Brassinosteroids-A novel group of plant hormones. In: Plant Molecular Physiology-Current scenario and future Projections (Ed. P.C. Trivedi). Aaviskar Publisher, Jaipur, pp. 58–84.

Bhardwaj, R., Arora, N., Sharma, P., and Arora, H.K. 2007. The effects of 28-homobrassinolide on seedling growth, lipid peroxidation and antioxidative enzyme activities under nickel stress in seedlings of *Zea mays* L. *Asian Journal of Plant Sciences*.**6**(5): 765–772.

Bhardwaj, R., Sharma, P., Arora, H.K., and Arora, N. 2008. 28-Homobrassinolide regulated Mn-uptake and growth of *Brassica juncea* L. *Can. J. Pure Appl. Sci.* **2**(1): 149–154.

Bobrick, A.O., Khripach, V.A., Zhabinskii, V.N, Zavadskaya M.I., and Litvinovskaya, R.P. 1998. A method of production of sanitated seed potato. Pat. Appl. BY. **19**: 981–189.

Choe, S., Noguchi, T., Fujioka, S., Takatsuto, S., Tissier, C.P., Gregory, B.D., Ross, A.S., Tanaka, A., Yoshida, S., and Tax, F.E. 1999. The *Arabidopsis dwf7/ste1* mutant is defective in the _7 sterol C-5 desaturation step leading to brassinosteroid biosynthesis. *Plant Cell*. **11**: 207–221.

Churikova, V.V., and Vladimirova, I.N. 1997. Effect of epin on activity of enzymes of oxidative metabolism of cucumber in peronosporous epiphytotia conditions. In: 4th Conference (Abstract) on Plant Growth and Development Regulators. Moscow, pp. 78.

Clouse, S.D. 1996. Molecular genetic studies confirm the role of brassinosteroids in plant growth and development. *Plant J*. **10**: 1–8.

Clouse, S.D. 2002. Brassinosteroid signal transduction: clarifying the patyway from ligand perception to gene expression. *Molecular Cell*. **10**(5): 973–982.

Clouse, S.D., and Sasse, J.M. 1998. Brassinosteroids: essential regulators of plant growth and development. *Ann. Rev. Plant Physiol. Plant Mol. Biol*. **49**: 427–451.

Clouse, S.D., and Zurek, D. 1991. Molecular analysis of brassinolide action in plant growth and development. *In* HG Cutler, T Yokota, G Adam, eds, Brassinosteroids: Chemistry, Bioactivity and Applications. ACS Symposium Series 474. American Chemical Society, Washington, DC, pp 122–140.

Dhaubhadel, S., Browning, K.S., Gallie, D.R., and Krishna, P. 2002. Brassinosteroid functions to protect the translational machinery and heat shock protein synthesis following thermal stress. *Plant J*. **29**(6): 681–691.

Dhaubhadel, S., Chaudhary, S., Dobinson, K.F., and Krishna, P. 1999. Treatment with 24-epibrassinolide, a brassinosteroid, increases the basic thermotolerance of *Brassica napus* and tomato seedlings. *Plant Mol. Biol*. **40**: 333–342.

Franěk, F., Eckschlager, T., and Kohout, L. 2003. 24-Epibrassinolide at subnanomolar concentrations modulates growth and production characteristics of a mouse hybridoma. *Collect. Czech. Chem. Commun*. **68**: 2190–2200.

Fujioka, S., and Sakurai, A. 1997. Brassinosteroids. *Nat. Prod. Rep*. **14**: 1–10.

Fujioka, S., Noguchi, T., Yokota, T., Takatsuto, S., and Yoshida, S. 1998. Brassinosteroids in *Arabidopsis thaliana. Phytochemistry*. **48**: 595–599.

Geuns, J.M.C. 1978. Steroid hormones and plant growth and development. *Phytochem*. **17**: 1–14.

Grove, M.D, Spencer, G.F, Rohwedder, W.K, Mandava, N., Worley, J.F., Warthen, J.D., Steens, G., Flippen-Anderson, J.L., and Cook, J.C. 1979. Brassinolide, a plant growth-promoting steroid isolated from *Brassica napus* pollen. *Nature*. **281**: 216–217.

Hasan, S.A., Hayat, S., Ali, B., and Ahmad, A. 2008. 28-Homobrassinolide protects chickpea (Cicer arietinum) from cadmium toxicity by stimulating antioxidants. *Environ. Pollut*. **151**: 60–66.

Haubrick, L.L., and Assmann, S.M. 2006. Brassinosteroids and plant function: some clues, more puzzles. *Plant Cell Environ*. **29**: 446–457.

Howell, W.M., Keller, G.E., Kirkpatrick, J.D., Jenkins, R.L., Hunsinger, R.N., and McLaughlin, E.W. 2007. Effects of the plant steroidal hormone, 24-epibrassinolide, on the mitotic index and growth of onion (*Allium cepa*) root tips. *Genet. Mol. Res.* **6**: 50–58.

Ikekawa, N., and Zhao, Y.J. 1991. Application of 24-epibrassinolide in agriculture. In: Brassinosteroids. Chemistry, Bioactivity, and Applications (Ed. H.G., Cutler, T. Yokota, and G. Adam). ACS Symposium Series, *American Chemical Society*, Washington, pp. 280–291.

Iwasaki, T., and Shibaoka, H. 1991. Brassinosteroids act as regulators of tracheary-element differentiation in isolated *Zinnia* mesophyll cells. *Plant Cell Physiol.* **32**: 1007–1014.

Jager, C.E., Symons, G.M., Ross, J.J., and Reid, J.B. 2008. Do brassinosteroids mediate the water stress response? *Physiol. Plant.* **133**(2): 417–425.

Janeczko, A., Hura, K., and Skoczowski, A. 2009. Temperature-dependent impact of 24-epibrassinolide on the fatty acid composition and sugar content in winter oilseed rape callus. *Acta Physiol Plant.* **31**(1): 71–79.

Janeczko, A., Koscielniak, J., Pilipowicz, M., Szarek-Lukaszewsa, G., and Skoczowspi, A. 2005. Protection of winter rape photosystem II by 24-epibrassinolide under cadmium stress. *Photosynthetica.* **43**: 293–298.

Jiang, C., Baehrecke, E.H., and Thummel, C.S. 1997. Steroid regulated programmed cell death during *Drosophila* metamorphosis. *Development* **124**: 4673–4683.

Jochova, J., Zakeri, Z., and Lockshin, R.A. 1997. Rearrangement of the tubulin and actin cytoskeleton during programmed cell death in *Drosophila* salivary glands. *Cell Death Differ.* **4:** 140–149.

Kagale, S., Divi, U.K., Krochko, J.E., Keller, W.A., and Krishna, P. 2007. Brassinosteroid confers tolerance in *Arabidopsis thaliana* and *Brassica napus* to a range of abiotic stresses. *Planta.* **225**: 353–364.

Kauschmann, A., Adam, G., Jessop, A., Konecz, C., Szekeres, M., Voigt, B., Willmitzer, L., and Altmann, T. 1996. Genetic evidence for an essential role of a brassinosteroids in plant development. *Proc. Plant Growth Regul. Soc. Am.* **23**: 13.

Khripach, V., Zhabinskii, V.N., and DeGroot, A. 2000. Twenty years of brassinosteroids: steroidal plant hormones warrant better crops for the XXI century. *Ann. Bot.* **86**: 441–447.

Khripach, V.A., Zhabinskii, V.N., Konstantinova, O.V., Khripach, N.B., Antonchik, A.V., Antonchik, A.P., and Schneider, B. 2005. Preparation of (25R)- and (25S)-26-functionalized steroids as tools for biosynthetic studies of cholic acids. Steroids, **70**(8): 551–562.

Krishna, P. 2003. Brassinosteroids- mediated stress responses. *J.Plant Grow.Regul.* **22**: 289–297.

Lee, C.Y., and Baehrecke, E.H. 2001. Steroid regulation of autophagic programmed cell death during development. *Development* **128**: 1443–1455.

Li, L., and Van Staden, J. 1998. Effects of plant growth regulators on the antioxidant system in callus of two maize cultivars subjected to water stress. *Plant Growth Regul.* **24**: 55–66.

Malíková, J., Swaczynová, J., Kolář, Z., and Strnad, M. 2008. Anticancer and antiproliferative activity of natural brassinosteroids. *Phytochemistry.* **69**: 418–426.

Mandava, N.B. 1988. Plant growth-promoting brassinosteroids. *Annu. Rev. Pl. Physiol. Pl. Mol. Biol.* **39**: 23–52.

Mashkovskii, M.D. 1997. Medicinal Substances. Topsing, kharkov.

Michelini, F.M., Berra, A., and Alché, L.E. 2008. The in vitro immunomodulatory activity of a synthetic brassinosteroid analogue would account for the improvement of herpetic stromal keratitis in mice. *J. Steroid Biochem. Mol. Biol.* **108**: 164–170.

Michelini, F.M., Ramirez, J.A., Berra, A., Galagovsky, L.R., and Alché, L.E. 2004 In vitro and *in-vivo* antiherpetic activity of three new synthetic brassinosteroids analogues. *Steroids.* **69**: 713–720.

Mittler, R. 2006. Abiotic stress, the field environment and stress combination. *Trends Plant Sci.* **11**(1): 15–19.

Mussig, C., Fischer, S., and Altmann, T. 2002. Brassinosteroid-regulated gene expression. *Plant Physiol.* **129**: 1241–1251.

Nakashita, H., Yasuda, M., Nitta, T., Asami, T., Fujioka, S., Arai, Y., Sekimata, K., Takatsuto, S., Yamaguchi, I., and Yoshida, S. 2003. Brassinosteroids function in a broad range of disease resistance in tobacco and rice. *Plant J.* **33**(5): 887–898.

Ohri, P., Bhardwaj, R., and Khurma, U.R. 2002. Role of brassinolide on juvenile emergence of *Meloidogyne incognita. Int. J. Nematol.* **12**: 175–178.

Ohri, P., Kaur, S., Bhardwaj, R., and Khurma, U.R. 2004. Role of brsssinosteroids on juvenile emergence of *Meloidogyne incognita. Int. J. Nematol.* **34**(2): 165–168.

Ohri, P., Sohal, S.K., Bhardwaj, R., and Khurma, U.R. 2008. Studies on on the root-knot nematode, Meloidogyne incognita (Kofoid & White) Chitwood under the influence of 24-epibrassinolide. *Ann. Plant Protec. Sci.* **16**(1): 198–202.

Ohri, P., Sohal, S.K., Bhardwaj, R., and Khurma, U.R. 2005. Effects of 28-homobrassinolide and 24-epibrassinolide on the development of *Meloidogyne incognita. Int. J. Nematol.* **15**(1): 43–47.

Ohri, P., Sohal, S.K., Bhardwaj, R., and Khurma, U.R. 2007. Morphogenetic and biochemical responses of the root-knot nematode, *Meloidogyne incognita* (Kofoid & White) Chitwood to isolated brassinosteroids. *Ann. Pl. Protec. Sci.* **15**(1): 226–231.

Ozdemir, F., Bor, M., Demiral, T., and Turkan, I. 2004. Effects of 24-epibrassinolide on seed germination, seedling growth, lipid peroxidation, proline content and antioxidative system of rice (*Oryza sativa* L.) under salinity stress. *Plant Growth Regul.* **42**: 203–211.

Pshenichnaya, L.A., Khripach, V.A., Volynets, A.P., Prokhorchik, R.A., Manzhelesova, N.E., and Morozik, G.V. 1997. Brassinosteroids and resistance of barley plants to leaf diseases. In: Problems of Experimental Botany, Belorussian Science, Minsk. pp. 210–217.

Rao, S.S.R., Vardhini, B.V., Sujatha, E., and Anuradha, S. 2002. Brassinosteriods – a new class of phytohormones. *Curr. Sci.*, **82**: 1239–1245.

Riddiford, L.M. 1993. Hormones and *Drosophila* development. In: The development of drosophila melanogaster (Ed. M. Bate and A. Martinez-Arias). Cold Spring Harbor Laboratory Press, Cold Spring Harbor, NY, pp. 899–939.

Roberts, K., and McCann, M.C. 2000. Xylogenesis: The birth of a corpse. *Curr. Opin. Plant. Biol.* **3**: 517–522.

Robertson, C.W. 1936. The metamorphosis of *Drosophila melanogaster*, including an accurately timed account of the principal morphological changes. *J. Morphol.* **59**: 351–399.

Rodkin, A.I., Konovalova, G.I., and Bobrick, A.O. 1997. Efficiency of application of biologically active substances in primary breeding of potato. In: 4th Conference (Abstract) on Plant Growth and Development Regulators, Moscow. pp. 317–318.

Romanutti, C., Castilla, V., Coto, C.E., and Wachsman, M.B. 2007. Antiviral effect of a synthetic brassinosteroid on the replication of vesicular stomatitis virus in vero cells. Inter. *J. Antimicrob. Agents.* **29**: 311–316.

Rupprecht, R., and Holsboer, F. 1999. Neuroactive steroids: mechanisms of action and neuropsychopharmacological perspectives. *Trends Neurosci.* **22**: 410–416

Sasse, J.M. 1999. Physiological actions of brassinosteroids. In: Brassinosteroids: steroidal plant hormones (Ed. A. Sakurai, T. Yokota, and S.D. Clouse). Springer-Verlag, Tokyo, pp. 137–161.

Sasse, J.M. 2003. Physiological actions of brassinosteroids: an update. *J. Plant Growth Regul.* **22**: 276–288.

Sasse, J.M., Smith, R., and Hudson, I. 1995. Effect of 24-epibrassinolide on germination of seeds of *Eucalyptus camaldulensis* in saline conditions. *Proc. Plant Growth Regul. Soc. Amer.* **22**: 136–141.

Schmidt, J., Richter, K., Voigt, B., and Adam, G., 2000. Metabolic transformation of the brassinisteroid 24-epi-castasterone by the cockroach *Periplaneta americana. Z. Naturforsch* **55**(c): 233–239.

Sharma, P., and Bhardwaj, R. 2007a. Effect of 24-Epibrassinolide on seed germination, seedling growth and heavy metals uptake in *Brassica juncea* L. *Gen. Appl. Plant Physiology*. **32**: 1–2.

Sharma, P., and Bhardwaj, R. 2007b. Effects of 24-Epibrassinolide on growth and metal uptake in *Brassica juncea* L. under copper metal stress. *Acta Physiol. Plant*. **29**: 259–263.

Sharma, P., Bhardwaj, R., Arora, H.K., and Arora, N. 2008. 28- homobrassinolide-regulated effects of nickel on metal uptake, protein content and antioxidative defence system in *Brassica juncea* L. *Biol. Plant*. **52**(4): 767–770.

Sharma, P., Bhardwaj, R., Arora, N., and Arora, H.K. 2007. Effect of 28- homobrassinolide on growth, Zn metal uptake and antioxidative enzyme activities in *Brassica juncea* L. seedlings. *Brazil. J. Pl. Physiol.* **19**(3): 203–210.

Smagghe, G., Decombel, L., Carton, B., Voight, B., Adam, G., and Tirry, L. 2002. Action of brassinosteroids in the cotton leafworm *Spodoptera littoralis*. *Insect Biochem. Mol. Biol.* **32**: 199–204.

Sondhi, N., Bhardwaj, R., Kaur, S., Kumar, N., and Singh, B. 2008. Isolation of 24-epibrassinolide from leaves of *Aegle marmelos* and evaluation of its antigenotoxicity employing *Allium cepa* chromosomal aberration assay. *Plant Growth Regul.* **54**: 217–224.

Swaczynová, J., Šíša, M., Hniličková, J., Kohout, L., and Strnad, M. 2006. Synthesis, biological, immumological and anticancer properties of a new brassinosteroid ligand. *Polish J. Chem.* **80**: 629–635.

Szekeres, M., Nameth, K., Kalmar, Z.K., Mathur, J., Kauscrman, A., Altmann, T., Redei, G.P., Nagy, F., Scrall, J., and Koncz, C. 1996. Brassinosteroids rescue the deficiency of CYP 90, a cytochrome P450, controlling cell elongation and de-etiolating in *Arabidopsis*. *Cell*. **85**: 171–182.

Talarico, L.B., Castilla, V., Ramirez, J.A., Galagovsky, L.R., and Wachsman, M.B. 2006. Synergistic *in vitro* interactions between (22S,23S)-3-Bromo-5,22,23-trihydroxystigmastan-6-one and acyclovir or foscarnet against herpes simplex virus type 1. *Chemotherapy*. **52**: 38–42.

Thummel, C.S., and Chory, J. 2002. Steroid signaling in plants and insects-common themes, different pathways. *Genes Dev.* **16**: 3113–3129.

Vasyukova, N.I., Chalenko, G.I., Kaneva, I.M., Khripach, V.A., and Ozeretskovskaya, O.L. 1994. Brassinosteroids and potato late blight. Prikl. Biokhim. *Micribiol*. **30**: 464–470.

Volynets, A.P., Pschenichnaya, L.A., Manzhelesova, N.E., Morozik, G.V., and Khripach, V.A. 1997a. Method of protection of barley plants from leave disease. Pat. BY. 970,680

Volynets, A.P., Pschenichnaya, L.A., Manzhelesova, N.E., Morozik, G.V., and Khripach, V.A. 1997b. The nature of protective action of 24-epibrassinolide on barley plants. *Proc. Plant Growth Regul. Soc. Am.* **24**: 133–137.

Von Kalm, L., Fristrom, D., and Fristrom, J. 1995. The making of a fly leg: a model for epithelial morphogenesis. *Bioessays*. **17**: 693–702

Wachsman, M.B., Castilla, V., Talarico, L.B., Ramirez, J.A., Galagovsky, L.R., and Coto, C.E. 2004a. Antiherpetic mode of action of (22S,23S)-3-bromo-5,22,23-trihydroxystigmastan-6-one in vitro. *Inter. J. Antimicrob. Agents*. **23**: 524–526.

Wachsman, M.B., Lopez, E.M., Ramirez, J.A., Galagovsky, L.R., and Coto, C.E.. 2000. Antiviral effect of brassinosteroid against herpes virus and arena viruses. *Anti. Chem. Chemotherapy*. **11**: 71–77.

Wachsman, M.B., Ramirez, J.A., Galagovsky, L.R., and Coto, C.E. 2002. Antiviral activity of brassinosteroid derivatives against measles virus in cell cultures. *Anti. Chem. Chemotherapy*. **13**: 61–66.

Wachsman, M.B., Ramirez, J.A., Talarico, L.B., Galagovsky, L.R., and Coto, C.E. 2004b. Antiviral activity of natural and synthetic brassinosteroids. *Curr. Med. Chem.-Anti-Infective Agents*. **3**: 163–179.

Xia, X.J., Huang, Y.Y., Wang, L., Huang, L.F., Yu, Y.L., Zhou, Y.H., and Yu, J.Q. 2006. Pesticides-induced depression of photosynthesis was alleviated by 24-epibrassinolide pretreatment in *Cucumis sativus L. Pestic. Biochem. Physiol.* **86**(1): 42–48.

Yamamoto, R., Fujioka, S., Demura, T., Takatsuto, S., Yoshida, S., and Fukuda, H. 2001. Brassinosteroid levels increase drastically prior to morphogenesis of tracheary elements. *Plant Physiol.* **125**: 556–563.

Yokota, T., Arima, M., and Takahashi, N. 1982. Castasterone, a new phytosterol with plant-hormone potency from chestnut insect gall. Tetrahed. *Lett.* **23**: 1275–1278.

Yokota, T., Normura, T., and Nakayama, M. 1997. Identification of brassiaosteraids that appear to be derived from campesterol and cholesterol in tomato shoots. *Pl. Cell Physiol.* **38**: 1291–1294.

Yokota, T., Higuchi, K., Takahashi, N., Kamuro, Y., Watanabe, T., and Takatsuto, S. 1998. Identification of brassinosteroids with epimerization substituents and /or the 23-oxo group in pollen and anthers of Japanese cedar. *Biosci. Biotech. Biochem.* **62**: 526–531.

Zullo, M.A.T., and Adam, G. 2002. Brassinosteroids phytohormones–structure bioactivity and applications. *Braz. J. Plant Physiol.* **14**: 143–181.

Index

S. Hayat and A. Ahmad (eds.), *Brassinosteroids: A Class of Plant Hormone*,
DOI 10.1007/978-94-007-0189-2,